Pollution Prevention and Sustainability

Pollution Prevention and Sustainability

Edited by Anne Offit

SYRAWOOD
PUBLISHING HOUSE
New York

Published by Syrawood Publishing House,
750 Third Avenue, 9th Floor,
New York, NY 10017, USA
www.syrawoodpublishinghouse.com

Pollution Prevention and Sustainability
Edited by Anne Offit

International Standard Book Number: 978-1-68286-560-6 (Hardback)

Cataloging-in-Publication Data

Pollution prevention and sustainability / edited by Anne Offit.
p. cm.
Includes bibliographical references and index.
ISBN 978-1-68286-560-6
1. Pollution prevention. 2. Sustainability. I. Offit, Anne.
TD191.5 .P65 2018
628.5--dc23

Printed in China.

TABLE OF CONTENTS

PREFACE

This book has been a concerted effort by a group of academicians, researchers and scientists, who have contributed their research works for the realization of the book. This book has materialized in the wake of emerging advancements and innovations in this field. Therefore, the need of the hour was to compile all the required researches and disseminate the knowledge to a broad spectrum of people comprising of students, researchers and specialists of the field.

The pollution of natural resources by either natural or human activities is known as environmental pollution. Some of the popularly known forms of environmental pollution include air pollution, noise pollution, water pollution, soil contamination, etc. Environmental pollution results in ecological imbalance and also adversely affects human health. The methods adopted to monitor as well as mitigate pollution are referred to as the techniques for sustainability. This book delves into the diverse areas of environmental pollution and brings forth innovative techniques for minimizing its adverse impacts. It will serve as a resource guide for environmentalists, conservationists, ecologists, environmental engineers, researchers and students associated with this field at various levels.

At the end of the preface, I would like to thank the authors for their brilliant chapters and the publisher for guiding us all-through the making of the book till its final stage. Also, I would like to thank my family for providing the support and encouragement throughout my academic career and research projects.

Editor

1

Beneficial reuse: A field trial to remediate and a bench-scale test to revegetate coal seam gas dam sediments from Queensland

Lee Fergusson

Principal Consultant, Prana World Consulting, Gold Coast, Australia

Email address:
lee@pranaworldgroup.com

Abstract: Coal seam gas (CSG) has become an increasingly common method of extracting methane from coal in Australia, with more than 20,000 wells expected to enter production by 2020. However, large quantities of "produced water" also come to the surface with gas, and these several thousand litres of water per day per well have to be managed sustainably. Furthermore, up to five percent of produced water is composed of suspended or dissolved solids, most typically present in the form of salty brines and a range of other elements, sometimes including benzene and other hydrocarbons like phenols. CSG solids therefore have a high pH and total alkalinity, and hence have elevated electrical conductivity. As a consequence, the settled solids from CSG extraction have no proven beneficial reuse value, and successful revegetation of dams and untreated sediments is limited to salt-tolerant grass species but is often impossible using any species under any condition. The purpose of this study is to investigate the remediation of CSG dam sediments from Queensland for the purposes of determining their potential beneficial reuse as "clean, usable soil" in revegetation projects. Experiment #1, a field trial conducted in the Bowen Basin, examined the impact of various additives to two different types of CSG dam sediments. Experiment #1 found that both types of sediment could be remediated, examples of which include decreases in pH from 10.0 to 7.4, chloride from 19,900mg/kg to 1,770mg/kg, cation-exchange capacity (CEC) from 23meq/100g to 4.0meq/100g, and sodium adsorption ratio (SAR) from 931meq/100g to 44meq/100g, and increases in total phosphorus from 27mg/kg to 855mg/kg and total nitrogen from 950mg/kg to 3,520mg/kg. These findings confirm that contaminated CSG sediments have beneficial reuse potential in dam decommissioning and revegetation projects. Experiment #2, a bench-scale test utilizing samples of treated sediments from Experiment #1, examined the revegetation potential of these remediated CSG sediments. Experiment #2 showed that both types of CSG dam sediment could be effectively revegetated using non-salt-tolerant grass species, while untreated sediments were not suitable for revegetation. However, the design and scale of this work need to be expanded, and variables such as sediment pH, CEC and SAR should be monitored and controlled more carefully before fully reliable conclusions can be made about the revegetation potential of treated CSG dam sediments.

Keywords: Beneficial Reuse, Coal Seam Gas, CSG, Sediments, Remediation, Revegetation

1. Introduction

At the turn of the last century, a significant shortfall in natural gas supplies was forecast for Australia, with the predicted shortage to be made up by gas piped to the mainland across Torres Strait from Papua New Guinea (PNG) in the north and from the gas fields of Bass Strait in the south [1]. Greater industrial and regulatory focus, particularly in the states of Queensland and New South Wales, was therefore placed on the development of indigenous, unconventional gas resources on the Australian mainland, with coal seam gas (CSG) extraction finding particular favor. As a consequence, the need for importing gas from PNG and Bass Strait has been obviated in the last ten years, and no shortfalls are now predicted [2].

In fact, the primary sedimentary coal basins in eastern Australia have now proven and probable reserves of natural gas derived from CSG totaling 28,000 pentajoules (PJ), representing more than 60% of the known gas reserves [1]. To put this figure into context, total Australian natural gas reserves were 5PJ in 1996, with CSG production increasing from 1PJ per annum in 1996 to 167PJ per annum by 2008 to a further 264PJ per annum by 2013 [3].

The "natural gas" referred to in these figures is comprised of two basic types of CSG: "dry gas", which is composed of >95% methane (CH_4); and "wet gas", which is composed of <95% methane. Both dry and wet gas contain varying amounts of ethane (C_2H_6), propane (C_3H_8), and butane (C_4H_{10}), along with other secondary gases, including carbon dioxide (CO_2), hydrogen sulfide (H_2S), hydrogen (H), nitrogen (N), helium (He), and argon (Ar). Perhaps more importantly for the Australian economy, the rapid expansion of CSG production has resulted in not only the obviation of gas imports but a net gain in gas production levels, generating an entirely new economic opportunity for the export of liquefied natural gas (LNG) from Australia to markets in Asia, with multi-billion dollar gas export terminals and rail infrastructure being built throughout Queensland.

These growth trends are not dissimilar to the growth trends in the natural gas economy of the U.S. currently being driven by shale oil gas from which much of the Australian experience has been derived.

The presence and potential build-up of methane in underground coal mines has been well documented in Australia for over 100 years; however, methane was typically extracted and flared prior to coal being mined safely rather than being used as a fuel source. It was not until the late 1930s that CSG was first recognized as a resource in its own right, being commercially produced at an abandoned colliery in New South Wales during the Second World War [4].

Given that gas extracted by "unconventional" methods like CSG typically resides as much shallower depths than "conventional" natural gas reserves (i.e., generally <1,000m below the surface, in layers or seams ranging in thicknesses from a few millimetres to tens of metres), its commercial potential has been well understood since the mid-twentieth century, but it has taken technological developments in geology, engineering, and chemistry 60 years to fully realize this potential.

In Queensland, 50 significant gas fields have been identified, including the smaller Styx Basin, the Laura Basin and the Cooper Basin, with the Bowen Basin (including the Bandanna, Baralaba, and Mornabah Coal Measures), the Surat Basin (including the Walloon Coal Measure), and the Galilee Basin (predicted to be one of the largest coal basins in the world, including the Betts Creek and Aramac Coal Measures) among the most productive or potentially productive fields in the State [4, 5]. For example, declared in June, 2014 by the Queensland Government, the so-called Galilee Basin State Development Area has been created to support the development of coal and gas reserves of the Galilee Basin, with an estimated $28 billion investment in infrastructure expected over the next ten years, including a new airport, rail corridors and export hub at Abbot Point.

Both CSG and conventional gas originate as decomposing organic matter in coal as a result of thermogenic and biogenic processes. However, where conventional gas migrates through permeable rock and eventually lodges in stratigraphic layers (usually at depths >1,000m), CSG adsorbs into the coal matrix and is held in place by fluid pressure, thereby making it less economic to extract when compared to conventional gas. This so-called "unminable gas" can be effectively liberated from coal using either CSG extraction processes, including hydraulic rock fracturing (or "fracking", where a variety of polymer-based foaming gels, liquids and gases are used as fracturing fluids), or by underground coal gasification (UCG).

Whereas UCG utilizes combustion of gas at high temperature to produce synthetic gas or "syngas" (i.e., a combination of mostly carbon monoxide [CO], hydrogen [H], and carbon dioxide [CO_2]), CSG liberates methane through the depressurization of coal seams by either removing (when gas resides in aquifers) or injecting (when the seam is dry) fresh or treated water. As a consequence, CSG processes generate large quantities of so-called "produced water" in addition to gas, and this water, often contaminated with phenols, heavy metals (including lead, mercury and arsenic [6]), benzene, and other organic and inorganic contaminants, including ions like sodium and chloride [7, 8], must be managed sustainably.

For example, an average CSG well in Queensland generates between 3,000 litres and several hundred thousand litres of produced water per day, depending on the type of seam formation [9]. With about 4,000 CSG wells in operation throughout the State, a total of 16.9GL of produced water was generated in 2012-2013 [3].

However, given that the number of CSG wells in production is projected to reach 20,000-40,000 by 2020 [8], the volume of wastewater generated the industry may reach more than 100GL per year within six years.

Figure 1. *Photographic examples of an operating CSG dam (A) and a dam after evaporation of produced water (B) in the Bowen Basin in central Queensland, 2013.*

CSG produced water is stored in open evaporation dams, such as those shown in Figures 1 and 2. These dams, sometimes called "turkey's nests" because they are strung together to form large overflowing evaporative pans, can be up to 100ha in size and may contain several gigalitres of produced water, most of it in a moderate to acute saline state when measured by total dissolved solids (TDS). The TDS of CSG produced water is normally between a few hundred parts per million and 10,000 mg/L, composed mainly of sodium (Na), chloride (Cl), and potassium (K). To put these levels of salinity into context, standard drinking water contains <500 mg/L TDS and global seawater averages are 35,000mg/L TDS, with the majority of salinity originating from sodium (Na), chloride (Cl), magnesium (Mg), calcium (Ca), potassium (K), and sulfate (SO_4).

Moreover, about 3-5% of CSG produced water by volume is actually a solid, typically present as suspended, colloidal particles or as dissolved solids. Most of the colloidal solids are composed of indigenous alkanic hydrocarbons or clay particles, and most of the dissolved solids are so-called "salty brines", which do not form part of the gas fraction and must be partitioned after the gas-water mix is brought to the surface. As a consequence, high levels of sodium, chloride, potassium and other ions accumulate in the evaporation dams as a dense brine, because these "salts" are concentrated in dams when water evaporates.

The standard design of CSG dams means that produced water either evaporates, is recycled as stock drinking water, irrigation water, or cooling water after treatment with reverse osmosis (RO), or is disposed as "trade waste" to a wastewater or sewage treatment plant [10]. Supplementing potable water with treated produced water has been considered, but the concept has not gained regulatory or public approval due to the fact that low weight molecules and uncharged organic chemicals may be rejected by RO membranes, thereby potentially entering the water supply chain [6].

Figure 2. Photographic examples of CSG evaporation dam after dewatering (A) and close up of CSG sediment (B) in the Bowen Basin, 2013.

Moreover, given transpoevaporation rates in Queensland heavily favor water rather than evaporation, particularly in the summer months from December through March, the result for most CSG sites is a net accumulation of settled solids, as shown in Figures 1B and 2 after dewatering. With about 500 million tonnes of solids generated in Queensland during 2012 and a further projected 2.5 billion tonnes accumulating by 2020, the question of how to sustainably manage CSG sediments is a growing concern to regulators, industry and local farming communities.

As a consequence, operators are required to develop environmental plans which manage the high volumes of produced water and waste sediments that accumulate as a result of CSG production. There are, however, no formal compliance standards for contaminated CSG liquid or solid wastes in Queensland, with each operator negotiating a "sustainable" management plan on a site-by-site basis coupled with self-reporting of results and compliance [11].

Despite the fact that only a few years ago construction of new evaporation dams was being phased out and decommissioning the encouraged option [12], today ever larger dams are being constructed to handle the exponential increase in water and sediment volumes. Current best practice means that when a dam becomes no longer viable, the most preferred practice is to remove residual wastewater, cover the sediment with clean fill, contour the dam into a stable landform, revegetate with salt-tolerant grass species, and leave the untreated sediment *in-situ*.

Attempts at deep-well injection and offshore submarine disposal of CSG sediments have been considered, but these methods are in their infancy and have not gained widespread public acceptance [6].

Moreover, beneficial reuse of CSG sediments as "clean fill" for levy, dam wall and road construction, and other applications in agriculture (such as forestry using salt-tolerant species like western wattle [*Acacia saligna*] and black oak [*Casaurina cristata pauper*] or biofuel production using salt-tolerant safflowers [*Carthamus tinctorius*] and sunflowers [*Helianthus annuus*] as sources of polyunsaturated fatty acids) have been contemplated, but most potential reuse solutions have been found wanting either due to the highly saline properties of sediments or low tolerance levels for innovation from local farming communities and the wider society due to fears of downstream environmental contamination [8].

Therefore, in order to investigate whether treated CSG dam sediments are amendable to beneficial reuse, this study asked the following research questions: Experiment #1—can two different types of CSG dam sediments from Queensland be remediated to Australian standards of "clean, usable soil" using two rates of direct addition of chemicals and natural fertilizers; and Experiment #2—are remediated dam sediments responsive to revegetation using non-salt-tolerant grass species?

2. Method

2.1. Experiment #1

A site with 70+ operating CSG wells and four evaporation dams in the Bowen Basin was chosen for the field trial, which was conducted during December 2013. At this site, 7.0m^3 samples of sediment were excavated from two evaporation dams. Based on previous records, the sediments from these two dams would represent the low and high ends of the salinity range. Sediment A was excavated using a backhoe from a decommissioned, but untreated, dam in which 1,900m^3 sediments had been atmospherically dried and stockpiled over a three-month period; this dam was located at the centre of the site. As shown in Figure 3A, Sediment A was a fine-grained, silty grey clay of medium density, with obvious white patches of salt on its surface; the sediment contained a significant portion of organic matter and 2-10mm mixed gravels, but did not emanate any odour.

Sediment B was excavated from under the water column of a second evaporation dam 15km to the south of the first; the second dam contained an estimated 6,300m^3 of sediment and was still in production at the time of the field trial (the dam is shown in Figure 1A). Sediment B was atmospherically dewatered for three days prior to the trial at an average daytime temperature of 25°C. As shown in Figure 6A,

Sediment B was a fine-grained, silty grey-brown clay of medium density; the sediment contained little organic matter and did not emanate any odour. The two CSG dam sediment samples were transported to a central 70m x 70m designated treatment area (DTA) in the centre of the site where the field trial was conducted.

As shown in Tables 1 and 2, untreated Sediment A had a lower moisture content than Sediment B (27.4% versus 49.3%), consistent with the sources of sediment on this site, a higher electrical conductivity (EC, 9.39mS/cm versus 2.06mS/cm) due to higher total concentrations of Na (16,100mg/kg versus 4,090mg/kg) and Cl (19,900mg/kg versus 5,400mg/kg), and a higher total alkalinity (6,290mg/kg versus 1,320mg/kg). CEC and SAR are particularly relevant in this study because both relate to soil fertility. Cation-exchange capacity (CEC) measures the total number of cations available for exchange with water as a result of negatively charged particles in the sediment holding positively charged cations by electrical attraction.

The four most important cations in this study are calcium (Ca^+), magnesium (Mg^+), potassium (K^+), and sodium (Na^+); these four cations are held to the sediment by negatively charged particles present within the sediment matrix. A CEC of 24meq/100g for Sediment A and 25meq/100g for Sediment B is normal for clay-like material with a pH ±10.0. Similarly, SAR measures sodicity in sediments which can be extracted by water and is therefore also an important factor in determining soil structure and fertility (SAR equals the total concentration of Na in the sediment divided by the square root of Ca added to Mg divided by two). As soils with a SAR >6.0meq/100g are classified as "sodic" in Australia [13], both Sediment A at 931meq/100g and Sediment B at 189meq/100g would be classified as extremely sodic.

From the data presented in Tables 1 and 2, it is therefore reasonable to conclude that Sediments A and B are both sodic (due to the presence of high concentrations of Na^+) *and* saline (due to the presence of high concentrations of Na^+ and Cl^-). Of note also is the presence of heavy metals in the sediments (particularly arsenic, chromium, copper, lead, nickel and zinc), although metals were universally non-leachable and non-bioavailable; the presence of heavy metals in sediments were therefore predicted to have no adverse impact on revegetation.

Leachability of metals was ascertained by the Toxic Characteristic Leaching Procedure (TCLP) analyzed by inductively coupled plasma atomic emission spectrometry (ICP-AES). As no evidence was found for the presence of phenols, benzene, toluene, ethylbenzene, xylene, polycyclic aromatic hydrocarbons, alkanic hydrocarbons or organochlorine pesticides in either Sediments A or B, these data have not been included in Tables 1 and 2.

Each sample was divided into two equal parts, with each 3.5 m^3 labelled: Sediment A1; Sediment A2; Sediment B1; and Sediment B2. The weight of each biopile of excavated sediment was estimated to be 5.25t (bulk density = 1.5), and measured approximately 2.0m L x 2.0m W x 1.0m H when placed as piles in the DTA. For the purposes of consistency, all additive measurements in this study are given on a dry wt/wt basis.

Treatment of Sediment A1 using a backhoe followed this procedure:

Add 150kg of modified alumina refinery residue (MARR) (3% wt/wt) to stabilize pH, add micronutrients such as sulfate (SO_4), and consolidate physical properties of sediment through the addition of silica (Si) and iron (Fe);

Mix thoroughly until visually homogenous;

Add approximately 100L of water as required to aid mixing;

Add 15kg of ferrous sulfate heptahydrate ($FeSO_4$) (0.3% wt/wt) to reduce pH;

Add 25kg of nitrogen-phosphorus-potassium fertilizer (0.5% wt/wt) to provide macronutrients;

Mix thoroughly until visually homogenous;

Add approximately 100L water as required to aid mixing;

Add 80kg of locally produced oaten farm hay as a carbon source and mix thoroughly;

Collect four x 500g samples from the front left, front right, back left, and back right of pile and commingle to form one x 2.0kg sample; and

Send 500g commingled sample to an independent laboratory in Brisbane, and analyze sample for analytes listed in Table 1.

Treatment of Sediment A2 followed the same procedure, except:

Add 1,000kg of MARR (20% wt/wt);

Add 300kg of $FeSO_4$ (6% wt/wt); and

Add 100kg of NPK fertilizer (2% wt/wt).

Treatments of Sediment B1 followed the same procedure as Sediment A1, and Sediment B2 followed the same procedure as Sediment A2.

Alumina refinery residue, sometimes referred to as bauxite residue or Bayer Process residue, has been used in the mining industry to neutralize acid and sequester heavy metals in tailings [14, 15]; its modification for reuse in environmental remediation and industrial waste treatment projects as MARR has been outlined elsewhere [16, 17].

In this study, MARR had a pH of 9.5 (i.e., roughly the same as the untreated sediments in the field trial), and was high in carbonate, hydroxide and hydroxycarbonate minerals. MARR also contained a cocktail of metals and minerals, including hematite (Fe_2O_3), beohmite (AlOOH), gibbsite ($Al[OH]_3$) and sodalite ($Na_4Al_3Si_3O_{12}Cl$), anatase (TiO_2), aragonite ($CaCO_3$), brucite ($Mg[OH]_2$), diaspore ($Al_2O_3.H_2O$), ferrihydrite ($Fe_5O_7[OH].4H_2O$), gypsum ($CaSO_4.2H_2O$), hydrocalumite ($Ca_2Al[OH]_7.3H_2O$), hydrotalcite ($Mg_6Al_2CO_3[OH]_{16}.4H_2O$), and p-aluminohydrocalcite ($CaAl_2[CO_3]_2[OH]_4.3H_2O$). Of significance in these formulae was the presence of oxyhydroxide compounds as well as negatively charged iron-, aluminium-, magnesium- and titanium-based molecules, which not only adsorb positively charged cations but also lead to precipitation and isomorphic substitution reactions; these reactions are responsible for long-term sequestration phenomena of positively charged cationic and negatively charged anionic species. These and other relevant phenomena identified with MARR at metaliferous mine sites around the world have been discussed elsewhere [18], and applications utilizing these and related reagents in the treatment of coal waste and in industrial site remediation have been examined [19, 20, 21]. To this author's knowledge, MARR has never been applied to CSG produced water or sediments.

Figure 3. Sediment A (A) and Sediment A1 (B).

The $FeSO_4$ applied in step D was a commercial-grade, >98% pure, grey-white, soluble solid with a pH of 3.5; the NPK fertilizer applied in step E was a commercially available, light brown, soluble fertilizer containing N = 23.8%, P = 3.7%, K = 13%, and sulfur (S) = 4.1% with a pH of 4.0; and the oaten hay (a combination of *Avena byzantine* [red oats] and *Avena sativa* [white oats]) applied in step H was sourced locally from an adjoining farm and had a pH of ± 5.0.

2.2. Experiment #2

At the completion of Experiment #1, 300g samples of Sediment A, Sediment A1, Sediment A2, Sediment B, Sediment B1, and Sediment B2 were collected; all six samples were placed in separate plastic trays measuring 200mm x 200mm; large pieces of oaten hay, clods of clay and gravel were removed to allow for improved sample consolidation. Each tray had approximately 15 x 4.0mm holes drilled through the base of the tray prior to sediment addition to allow for efficient drainage. This resulted in approximately 50mm of sediment in each tray, photographs of which are shown in Figures 4 and 5.

A standard, non-salt-tolerant grass species (i.e., Munns over-sow lawn seed) was sprinkled onto the surface of all six samples, with approximately 150 seeds added per tray. The seeds were then covered with 0.5cm of clean soil (to replicate what would occur as part of the site's sustainable rehabilitation program) to aid in germination. Trays were lightly watered each day for ten days and left in the open sunlight; temperatures during the bench-scale test averaged 25-30°C throughout the experimental period. Photographs of each tray were taken on day one (see Figures 3 and 4) and after seeds had germinated on day ten in order to document establishment height and density rates.

Figure 4. Sediment B (A); Sediment B1 (B); and Sediment B2 (C) on day one before seeding and topsoil.

Figure 5. Sediment B (A); Sediment B1 (B); and Sediment B2 (C) on day one before seeding and topsoil.

3. Results

3.1. Experiment #1

There are no guidelines for what constitutes "clean soil" in Queensland. Therefore, two other Australian standards were used in this study to determine the analytical baseline from which an empirical measure of "clean, usable soil" could be applied. First, the Victorian Government's Environment Protection Authority soil hazard categorization for "clean fill" [22]; this standard is used throughout Australia as a criterion for determining soil which is not contaminated.

Second, the National Environment Protection assessment of site contamination, the so-called NEPM guidelines; this study applied health investigation Level 'F' of the guidelines, which relates to determining "clean industrial soil" [23]. While there is no limit for total phosphorus (TP) under NEPM Level 'F', there is a 2,000mg/kg TP limit under its Ecological Investigation Level for urban environments, which has therefore been applied in this study. Sediments A1, A2, B1 and B2 were compared to these two standards to determine whether they met the criterion of "clean, usable soil".

Table 1. Physical and chemical properties of untreated Sediment A, treated Sediments A1 and A2, compared to Victorian EPA soil criteria for "clean fill" and NEPM criteria for "clean industrial soil".

Parameter	Untreated Sediment A	Treated Sediment A1	Treated Sediment A2	Victorian EPA Clean Fill	NEPM Clean Soil
Moisture content (%)	28	14	30	NL	NL
pH	10.0	8.5	7.4	NL	NL
EC (mS/cm)	9.3	5.9	8.2	NL	NL
Calcium (mg/kg)	10	200	1,740	NL	NL
Magnesium (mg/kg)	<10†	90	530	NL	NL
Potassium (mg/kg)	340	380	1,710	NL	NL
Sodium (mg/kg)	16,100	7,380	9,970	NL	NL

Parameter	Untreated Sediment A	Treated Sediment A1	Treated Sediment A2	Victorian EPA Clean Fill	NEPM Clean Soil
Chloride (mg/kg)	19,900	9,450	1,770	NL	NL
Phosphorus (mg/kg)	27	554	855	2,000	2,000
Nitrogen (mg/kg)	656	1,440	2,550	NL	NL
Cation-exchange capacity (meq/100g)	24	11	4.0	NL	NL
Sodium adsorption ratio (meq/100g)	931	68	44	NL	NL
Total alkalinity (mg/kg)	6,290	1,800	391	NL	NL
Arsenic (mg/kg)	<5.0†	5.0	9.0	20	500
Cadmium (mg/kg)	<1.0†	<1.0†	<1.0†	3.0	100
Chromium (mg/kg)	11	32	110	NL	NL
Copper (mg/kg)	10	10	10	100	5,000
Lead (mg/kg)	10	14	20	300	1,500
Mercury (mg/kg)	<1.0†	<1.0†	<1.0†	1.0	75
Nickel (mg/kg)	6.0	10	14	60	3,000
Zinc (mg/kg)	21.0	19.0	20	200	35,000
Leachable Arsenic (mg/L)	<0.1†	<0.1†	<0.1†	<0.7	NL
Leachable Cadmium (mg/L)	<0.05†	<0.05†	<0.05†	<0.2	NL
Leachable Chromium (mg/L)	<0.1†	<0.1†	<0.1†	<5.0	NL
Leachable Copper (mg/L)	<0.1†	<0.1†	<0.1†	<200	NL
Leachable Lead (mg/L)	<0.1†	<0.1†	<0.1†	<1.0	NL
Leachable Mercury (mg/L)	<0.001†	<0.001†	<0.001†	<0.1	NL
Leachable Nickel (mg/L)	<0.1†	<0.1†	<0.1†	<2.0	NL
Leachable Zinc (mg/L)	<0.1†	<0.1†	<0.1†	<300	NL

† Below level of detection; NL = no limit.

Figure 3B shows a close-up photograph of Sediment A1, which was visually identical to Sediment A2. Clearly evident in this image is the red-coloured MARR (30-40% of which is composed of Fe_2O_3) and oaten hay; some evidence for a mottled salty precipitate is also visible in this photograph. Table 1 presents the analytical data for Sediments A, A1 and A2, and compares these data to the Victorian EPA guidelines for "clean fill" and the NEPM guidelines for "clean industrial soil". From these data it can be seen that the pH of Sediment A was reduced from 10.0 to 8.5 in Sediment A1 and to a near-neutral 7.4 in Sediment A2; EC was reduced from 9.3mS/cm to 5.9mS/cm in Sediment A1 but remained largely unchanged in Sediment A2.

CEC was reduced from 24meq/100g to 11meq/100g (a 50% reduction) in Sediment A1 and to 4.0meq/100g (an 84% reduction) in Sediment A2, and SAR was reduced from 931meq/100g to 68meq/100g (a 93% reduction) in Sediment A1 and to 44meq/100g (a 95% reduction) in Sediment A2. Total heavy metal concentrations remained largely unchanged in both treated sediments as would be expected (although Cr increased for reasons which are not obvious, although cross-contamination from MARR and NPK are possible), and leachable metals were below detection limits both before and after treatment.

Total alkalinity decreased from 6,290mg/kg to 1,800mg/kg (an 84% reduction) in Sediment A1 and 391mg/kg (a 94% reduction) in Sediment A2, Ca, Mg and K increased in both treated sediments, which would be expected given MARR contains significant amounts of both Ca and Mg and NPK contains 13% K, but Na decreased from 16,100mg/kg to 7,380mg/kg (a 55% reduction) in Sediment A1 and 9,970mg/kg (a 39% reduction) in Sediment A2, and Cl decreased from 19,900mg/kg to 9,450mg/kg (a 53% reduction) in Sediment A1 and 1,770mg/kg (a 91% reduction) in Sediment A2. Total phosphorus (TP) and total nitrogen (TN) both increased significantly due to the addition of NPK.

Figure 6B shows a close-up photograph of Sediment B1, which was visually identical to Sediment B2. Clearly evident in this image is the red-coloured MARR and oaten hay; some evidence of a mottled salty precipitate is also visible in this photograph. Table 2 presents the analytical data for Sediments B, B1 and B2, and compares these data to the Victorian EPA guidelines for "clean fill" and the NEPM guidelines for "clean industrial soil".

Figure 6. *Sediment B (A) and Sediment B1 (B).*

Table 2. *Physical and chemical properties of untreated Sediment B, treated Sediments B1 and B2, compared to Victorian EPA soil criteria for "clean fill" and NEPM criteria for "clean industrial soil".*

Parameter	Untreated Sediment B	Treated Sediment B1	Treated Sediment B2	Victorian EPA Clean Fill	NEPM Clean Soil
Moisture content (%)	49	16	22	NL	NL
pH	9.8	8.4	7.4	NL	NL
EC (mS/cm)	2.0	3.5	9.5	NL	NL
Calcium (mg/kg)	<10	120	2,670	NL	NL
Magnesium (mg/kg)	<10†	50	670	NL	NL
Potassium (mg/kg)	80	300	1,750	NL	NL
Sodium (mg/kg)	4,090	4,600	9,900	NL	NL
Chloride (mg/kg)	5,400	5,440	8,880	NL	NL
Phosphorus (mg/kg)	270	318	1,030	2,000	2,000
Nitrogen (mg/kg)	950	1,950	3,520	NL	NL
Cation-exchange capacity (meq/100g)	25	34	27	NL	NL
Sodium adsorption ratio (meq/100g)	189	78	34	NL	NL
Total alkalinity (mg/kg)	1,320	1,850	340	NL	NL
Arsenic (mg/kg)	7.0	10	9.0	20	500
Cadmium (mg/kg)	<1.0†	<1.0†	<1.0†	3.0	100
Chromium (mg/kg)	20	35	93	NL	NL
Copper (mg/kg)	17	25	16	100	5,000
Lead (mg/kg)	11	17	20	300	1,500
Mercury (mg/kg)	<1.0†	<1.0†	<1.0†	1.0	75
Nickel (mg/kg)	12	21	17	60	3,000
Zinc (mg/kg)	24	39	33	200	35,000
Leachable Arsenic (mg/L)	<0.1†	<0.1†	<0.1†	<0.7	NL
Leachable Cadmium (mg/L)	<0.05†	<0.05†	<0.05†	<0.2	NL
Leachable Chromium (mg/L)	<0.1†	<0.1†	<0.1†	<5.0	NL
Leachable Copper (mg/L)	<0.1†	<0.1†	<0.1†	<200	NL
Leachable Lead (mg/L)	<0.1†	<0.1†	<0.1†	<1.0	NL
Leachable Mercury (mg/L)	<0.001†	<0.001†	<0.001†	<0.1	NL
Leachable Nickel (mg/L)	<0.1†	0.2	<0.1†	<2.0	NL
Leachable Zinc (mg/L)	<0.1†	0.1	<0.2	<300	NL

† Below level of detection; NL = no limit.

From these data it can be seen the pH of Sediment B was reduced from 9.8 to 8.4 in Sediment B1 and to a near-neutral 7.4 in Sediment B2; however, EC increased from 2.0mS/cm to 3.5mS/cm in Sediment B1 and 9.5mS/cm in Sediment B2, presumably as a result of MARR addition, although it is not clear why this would occur in one sediment and not the other.

CEC remained largely unchanged in both sediments, but SAR was reduced from 189meq/100g to 78meq/100g (a 59% reduction) in Sediment B1 and to 34meq/100g (an 83% reduction) in Sediment B2. As observed in Sediments A1 and A2, total heavy metals remained largely unchanged in both treated sediments as would be expected (although Cr and Pb increased for reasons which are not immediately obvious, although cross-contamination from MARR and NPK are possible), and leachable metals were below detection limits both before and after treatment for most metals.

Total alkalinity increased by 40% from 1,320mg/kg to 1,850mg/kg in Sediment B1, but decreased to 340mg/kg (a 75% reduction) in Sediment B2, Ca, Mg and K increased in both treated sediments, which would be expected given MARR contains significant amounts of both Ca and Mg and NPK contains 13% K, but Na remained unchanged in Sediment B1 but increased by 60% from 4,090mg/kg to 9,900mg/kg in Sediment B2, and Cl similarly remained unchanged in Sediment B1 and increased by 40% from 5,400mg/kg to 8,880mg/kg in Sediment B2. TP and TN both increased due to the addition of NPK.

3.2. Experiment #2

Most or all 150 lawn seeds appear to have germinated in Sediments A1, A2, B1 and B2, but not in untreated Sediments A and B, with the average height of seedlings at 30mm in Sediments A and B, 60mm in Sediments A1 and B1, and 90mm in Sediments A2 and B2.

Figure 7. *Sediment A (A); Sediment A1 (B); and Sediment A2 (C) on day ten.*

Figure 8. *Sediment B (A); Sediment B1 (B); and Sediment B2 (C) on day ten.*

Figures 7 and 8 show the results of the bench-scale test to revegetate Sediments A, A1, A2, B, B1 and B2, in which it is clear that seeds had germinated and established in all six sediment trays by day ten. In order of greatest height and density of seedlings, these can visually be rated for Sediment A: 1) untreated Sediment A; 2) treated Sediment A1; and 3) treated Sediment A2, with the same observation for Sediment B. It is of interest to note that seedlings in both untreated sediments in Figures 7A and 8A only established in about 25% of the tray, despite being spread evenly over the surface of the sediment, whereas seedlings in 7B, 7C, 8B and 8C have established across the entire surface of the tray.

4. Conclusions

The CSG dam sediments used in this study were both saline and sodic, rendering them unsuitable for beneficial reuse, particularly in revegetation projects and agriculture. However, it is reasonable to conclude that the remediation of CSG dam sediments from Queensland described in this field trial using the direct addition of MARR, $FeSO_4$, NPK fertilizer, and oaten hay was effective in converting contaminated CSG sediments into potentially reusable soil media when comparing the results to Victorian EPA guidelines for clean fill and NEPM guidelines for clean industrial soil. In fact, concentrations of contaminants in Sediments A1, A2, B1 and B2 were lower than the recommended limits for all parameters, thereby meeting the "clean, usable soil" criterion set forth in this study.

However, it should be stated that both sets of guidelines do not include a comprehensive list of all relevant soil fertility and structure factors, particularly EC, CEC, SAR, and total sodium and chloride. For example, electrical conductivity is a key measure of soil salinity not included in either the Victorian EPA or NEPM guidelines. The recommended limit of EC for "salt-sensitive" and "moderately salt-sensitive" crops and pasture species (such as potatoes and clover) is 0-2.0mS/cm and for "moderately salt-tolerant species" (such as sorghum and lucerne) is 2.0-4.0mS/cm [24]. Only Sediment B1 would be suitable for beneficial reuse with moderately tolerant species but not for salt-sensitive species. Similarly, Sediment A1 would be suitable for "salt-tolerant species" (such as sunflower and ryegrass), which can survive in soil with an EC in the 4.0-8.0mS/cm range, but Sediments A2 and B2 would only be suitable for reuse with "very salt-tolerant" species (such as barley and wheat grass) in the EC 8.0-16mS/cm range. Any soil with an EC >16mS/cm is too saline for crops to be productive, but some pasture species, including salt bush and inland salt-grass, can survive [24, 25].

Similarly, further investigation is needed to shed light on the function of CEC and its relation to revegetation in CSG dam sediments. As noted above, the CEC for both Sediments A and B was normal at 24-25meq/100g before treatment and would be suitable when considering the beneficial reuse of sediments in revegetation. However, while treatments B1 and B2 had no significant effect on CEC, as a result of remediation, the CEC was significantly reduced by 55% from 24meq/100g to 11meq/100g in Sediment A1 and by 84% to 4.0meq/100g in Sediment A2. These levels of CEC may not be advantageous for revegetation over the long term.

A reduction to pH can also reduce the CEC of soils and sediments, however the reductions in pH observed in this study cannot explain the reduction to CEC in Sediment A because the pH of Sediments B1 and B2 were reduced by the same orders of magnitude as Sediments A1 and A2 without the same commensurate changes to CEC. Increases in organic matter, which occurred in all four treatments as a result of hay addition, should have also increased CEC due to increases in pH [13], but the phenomenon was not observed in this study for Sediments A1 and A2. If the oaten hay used in this field trial had been reduced in size by shredding prior addition it may have liberated carbon and affected pH (and hence CEC) of remediated sediments.

Moreover, while the reductions to CEC in Sediments A1 and A2 apparently did not hinder revegetation as shown in Experiment #2, the interactions between exchangeable Ca (which was 15meq/100g in Sediment A but 5.0meq/100g in Sediment A1 and 1.5meq/100g in Sediment A2) and exchangeable Na (which was 7.0meq/100g in Sediment A but 3.4meq/100g in Sediment A1 and 1.2meq/100g in Sediment A2) should be the subject of further inquiry, because soil fertility (specifically, spontaneous and mechanical soil dispersion) is related to an increase in exchangeable Mg and K at the expense of Ca in the presence of Na in clay-like soil media [13].

Likewise, while the elevated levels of SAR apparently did not hinder revegetation of remediated sediments in this study as shown in Experiment #2, long-term growth rates and crop yields have not been investigated, and it is possible these would be compromised by the presence of high concentrations of Na in the remediated sediments.

From this data is can be concluded that a 3-4% wt/wt addition rate of chemical reagents and additives used in this field trial resulted in more viable revegetation when measuring density and height of plant growth compared to untreated sediment. It can also be concluded that a 28% wt/wt addition rate resulted in the establishment of even more viable revegetation compared to the 3-4% addition rate, suggesting that with higher addition rates, and hence availability of more macro- and micro-nutrients, CSG dam sediments become more amendable to revegetation. To verify this statement, larger scale and longer term studies should be conducted, and the use of more salt-tolerant grass species, such as katambora rhodes (*Chloris gayana*), gayndah buffel (*Cenchrus ciliaris*), bisset creeping blue (*Bothriochloa insculpta*), and bambatsi panic (*Panicum coloratum*), which thrive in pastures around the evaporation dams at the field trial site, should be tested in future.

However, it is reasonable to conclude that the remediation of two different types of CSG dam sediments from Queensland using the direct addition and mixing of a combination of MARR, $FeSO_4$, NPK fertilizer, and oaten hay was effective in converting salt-contaminated sediments into a potentially beneficial reusable soil medium, but further investigation into the suitability of treated CSG sediments as "clean, usable soil" is required before sound, scientific conclusions can be drawn.

References

[1] Day, R.W. (2009). Coal seam gas booms in eastern Australia. Feature Paper, *Preview*, June 2009.

[2] Baker, G. and Slater, S. (2008). *Coal seam gas: An increasingly significant source of natural gas in eastern Australia*, PESA EABS III Symposium, Sydney, September 14-17, 2008, 378-391.

[3] Department of Natural Resources and Mines (2014). *Queensland's coal seam gas overview*. Department of Natural Resources and Mines, Report No. B/Brochures/CSG/CC13-PET001, Queensland Government, Brisbane, January 2014.

[4] Cook, P.J. (2013). *Life cycle of coal seam gas projects: Technologies and potential impacts*. Report for the New South Wales Office of the Chief Scientist and Engineer, Sydney, June 2013.

[5] Sydney Catchment Authority (2012). *Literature review: Coal seam gas impacts on water resources*. Sydney Catchment Authority, New South Wales Government, Penrith, New South Wales, December 2012.

[6] Khan, S. and Kordek, G. (2013). *Coal seam gas: Produced water and solids*. Prepared for the New South Wales Office of the Chief Scientist and Engineer, School of Civil Environmental Engineering, The University of New South Wales, Sydney, June 2013.

[7] Cubby, B. (2010). Origin stops coal seam gas drilling after chemicals found in water. *The Sydney Morning Herald*, October 21, 2010.

[8] Lloyd-Smith, M. and Senjen, R. (2011). *Hydraulic fracturing in coal seam gas mining: The risks to our health, communities, environment and climate*. National Toxics Network, Bangalow, NSW, July 2011.

[9] Commonwealth Scientific and Industrial Research Organisation (2012). *Coal seam gas: Produced water and site management*. Commonwealth Scientific and Industrial Research Organisation (CSIRO), Clayton South, Victoria.

[10] Egan G (2012). CSG waste water trucked to Ipswich. *The Queensland Times*, June 13, 2012.

[11] Department of Environment and Heritage Protection (2012). *CSG/LNG compliance plan 2012-2013*. Energy Resources and Enforcement Branch of the Environmental Services and Regulation Division, Department of Environment and Heritage Protection, State Government of Queensland, Brisbane, November 2012.

[12] Glynn, P. (2009). Treatment options for water produced by CSG extraction, *Gas Today*, November 2009.

[13] Murray R.S., and Grant, C.D. (2007). *The impact of irrigation on soil structure*. School of Earth & Environmental Sciences, The University of Adelaide, National Program for Sustainable Irrigation (Land & Water Australia), July 2007.

[14] Fergusson, L. (2009). *Commercialisation of environmental technologies derived from alumina refinery residues: A ten-year case history of Virotec*. Commonwealth Scientific and Industrial Research Organisation (CSIRO), Project ATF-06-3 "Management of Bauxite Residues", Department of Resources, Energy and Tourism (DRET), Commonwealth Government of Australia, Asia-Pacific Partnership on Clean Development and Climate, Canberra.

[15] Fergusson, L. (2012). *ViroMine technology: A solution to the world's mining megawaste*, Prana World Publishing, Gold Coast, Australia, 170pp.

[16] Fergusson, L. (2007). The conversion and sustainable use of alumina refinery residues: Global solution examples, in *Light Metals 2007*, edited by T.J. Galloway, The Minerals, Metals & Materials Society, 2007, pp. 105-112.

[17] Taylor, K., Mullett, M., Adamson, H., Wehrli, J. and Fergusson, L. (2011). Application of nanofiltration technology to improve sea water neutralization of Bayer process residue, in *Light Metals 2011*, edited by Stephen J. Lindsay, John Wiley & Sons, Inc., Hoboken, New York, 2011, 79-87.

[18] Fergusson, L. (2010). *Virotec: A ten-year story of success in environmental remediation*, Prana World Publishing, Gold Coast, Australia, 189pp.

[19] Clark, M.W., McConchie, D., Berry, J., Caldicott, W., Davies-McConchie, F., and Castro, J. (2004). Bauxsol technology to treat acid and metals: Applications in the coal industry. In J. Skousen and T. Hilton (Eds.), proceedings of the *Joint Conference of the American Society of Mining and Reclamation* and the *25th West Virginia Surface Mine Drainage Task Force*, Morgantown, West Virginia, April 18-24, 2004, 292-313.

[20] Fergusson, L. (2013). An industrial legacy now gone, *Water Management and Environment*, *24*(1), 40.

[21] McConchie, D., Clark, M., Maddocks, G., Davies-McConchie, F., Pope, S., and Caldicott, W. (2003). The use of Bauxsol technology in mine site management and remediation. In proceedings of the *CIM Mining Industry Conference*, Montreal, May 2003, Disk Record s33a1141, 20pp.

[22] Environment Protection Authority (2009). *Industrial waste resource guidelines: Soil hazard categorisation and management*. Environment Protection Authority, Victorian Government, Melbourne, 2009.

[23] National Environment Protection Council (2013). *National environment protection (assessment of site contamination) amendment measure 2013*, National Environment Protection Council, Canberra.

[24] Department of Natural Resources Queensland (1997). *Salinity management handbook*. Department of Natural Resources, Brisbane, Queensland, 1997.

[25] Wagner, R. (1987). *Dryland salinity in the south-east region*. Soil Conservation Service of NSW, Sydney, NSW.

2

Administrative and Legislative Scrutiny of Existing Computer Waste Management Practices in Pakistan

Yawar Abbas[1], Fiza Sarwar[1,3], Muhammad Idrees[1], Ishtiaque Hussain[3], Syed Umairullah Jamil[1], Attarad Ali[2,*]

[1]Department of Earth and Environmental Sciences, Bahria University, Islamabad, Pakistan
[2]Department of Biotechnology, Quaid-i-Azam University, Islamabad, Pakistan
[3]Department of Environmental Sciences, Quaid-i-Azam University, Islamabad, Pakistan

Email address:
attarad.ali@kiu.edu.pk (Attarad Ali)

Abstract: Electronic-waste (E-waste) management is one of the global emerging issues. The current study has been carried out to evaluate the existing computer waste management (CWM) in the twin cities of Rawalpindi and Islamabad, Pakistan. Study focuses on managerial gap analysis and determination of potential occupational impacts from current CWM practices. System observation and waste quantification are the tools used in field, while interviews and questionnaire based surveys were also conducted. The results reveal that approximately 624.15 tons of computer waste (CW) is generated in the twin cities per year. Approximately 825kg of monitors and 1125kg of CPUs wastes are generated in Twin cities per day. Associated authorities like CDA, Ministry of Climate Change and EPA were interviewed for legal compliance. There is the lack of rules implementation on E-waste management. However, legislations made by concerned authorities exist in the country but are not practically enforced. Further concerns observed during the study include lack of using basic PPEs and employing child labor in their adolescence. Implementation of existing laws along with basic occupational health and safety rules in the system may play a better role to attenuate the load of waste from the very important cities of Pakistan.

Keywords: Computer Waste Management (CWM), Health and Safety, Generation, Legislation, Administration, Hazards, Electronic-Waste

1. Introduction

Globalization has caused people to be more connected than ever before. It has opened up innumerable economic opportunities for both post-industrialized and developing countries. It has also fashioned a commodification of electronic waste. The term electronic waste (EW) is a generic term encompassing various forms of electrical and electronic equipment that are superannuated. EW includes discarded electronic or electric waste. Among EW, computer waste is utmost significance for being linked to socioeconomic, health, safety and environmental issues. Production of EW is growing rapidly due to the increased utilization of computers in domestic and offices. Every year, thousands of expired computers, mobile phones, television sets and radio equipment are discarded, most of which either end up in open dump site or unauthorized recycling (Saoji, 2012). In Europe it was estimated that e-waste generation was 6 million tons in 1988. In 2005 it was estimated that 8.3 and 9.1 million tons of e-waste generated. In 2020 it will increase to 12.3 million tons annually. The annual growth rate of e-waste is about 2.5-2.7% (Sawhney, Henzler, Melnitzky, & Lung, 2008) (Bashir et al., 2005).

Electronic Equipment Waste generated from non-usable or old electronics contain heavy elements that have adverse effects on the environment and human health (Asiime, E. N). Computer waste is most hazardous than other solid wastes because it contains thousands of components made up of deadly chemical and material like poly vinyl chloride (PVC, Bromated Flame, Lead, Mercury, Cadmium, Chromium, Copper). Long term exposures to these chemicals have potential for disruption of nervous system, kidneys and bones (Saoji, 2012). Many adults get exposure Lead from the environment, so this leads to nervous system and blood impairments. The informal practices are due to lack of awareness (Pirzada & Pirzada, 2010). In the United States, it was estimated that over 315 million computers would become obsolete between 1997 and 2004 and that an additional 185 million computers will become obsolete between 2004 and 2007 (Leung, Cai, & Wong, 2006) (EPA, 1997). According to a report of Confederation of Indian Industries, the total waste generated by obsolete or broken down electronic and electrical equipment in India has been estimated to be 1,46,000 tons per year (Joseph, 2007). The dismantling practice may be risk to the health and safety of the workers. In developing countries all the dismantling is done with the bare hand and only with the help of hammer and screw drivers (Mittal, Goel, & Rani, 2012). Different countries around the globe have made legislation on the e-waste. Japan deals e-waste in two ways, one is law for the promotion of effective utilization of the resources (LPUR), which focuses on the recycling of the goods and reduction of the waste. The other is the recycling of home Appliances (Chung & Murakami-Suzuki, 2008). US Congress has failed to pass Federal Legislation specially targeting either the National management or the export of e-waste. As result only two Regulations addresses on e-waste and its export, The Resource Conservation and Recovery Act 1976 (RCRA) and Environmental Protection Agency (EPA's) (Skinner, Dinter, Lloyd, & Strothmann, 2010).

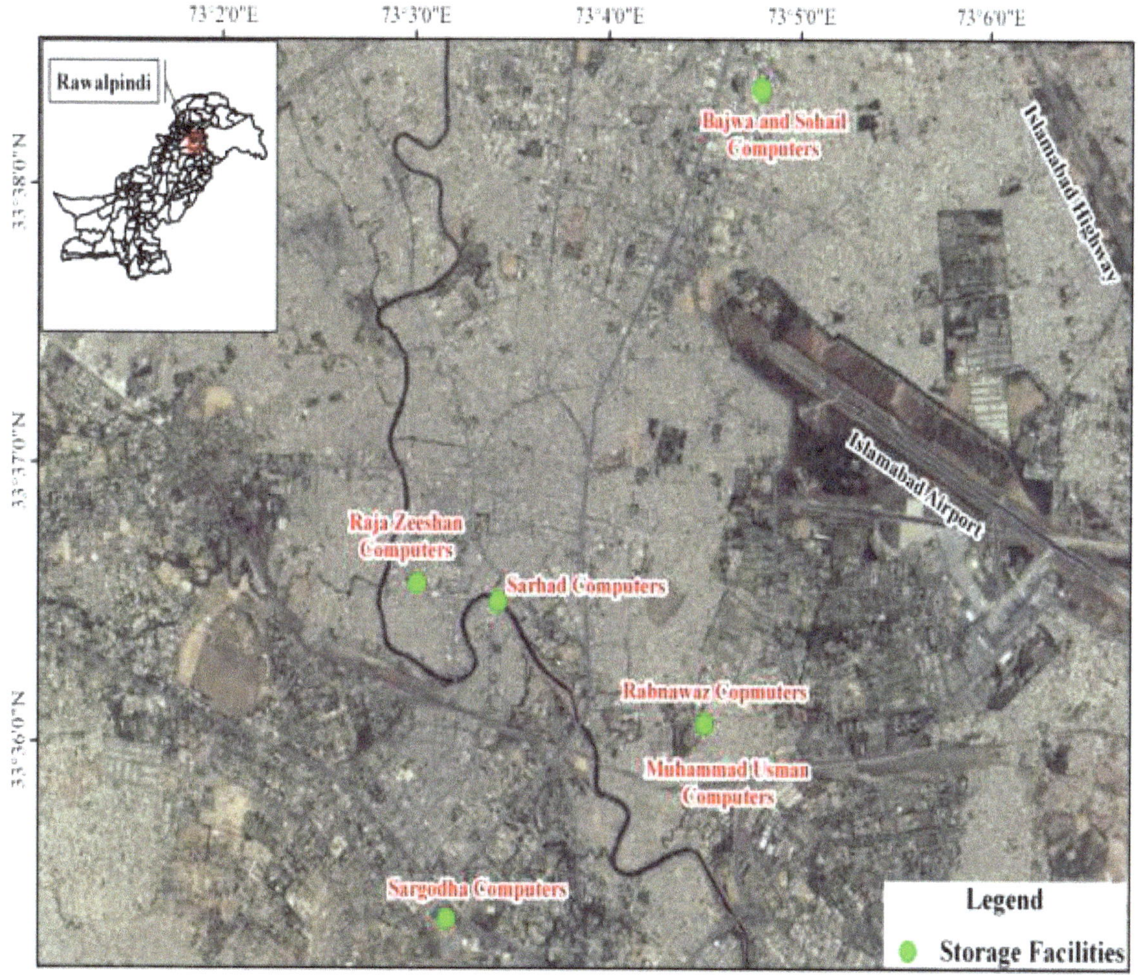

Figure 1. *Site Map and Locations of Storage Facilities.*

Pakistan, although has no specific legislation on e-waste import or disposal but legislations on management of hazardous substances including, Section 11, section 13 and section 14 of Pakistan Environment Protection Act, 1997, Hazardous Substance Rule 2003, National Biosafety Guidelines 2005, Import Policy Order 2003 (Bashir, et al., 2005) (Aggarwal, 2006) (EPA, 1997). Moreover, legal binding on trans-boundary movement of hazardous waste are also applicable to import of refused e-goods from developed world (Cronin et al., 2003). Pakistan is a signatory to the Basel Convention on trans-boundary movement of hazardous waste. Pakistan has huge market for the refurbished and second hand products because second hand equipment's is affordable for majority. Most of second equipment's brought to the Pakistan. Many times waste send to Pakistan as donation for the Government Schools. Pakistan is the third largest importer of e-waste or third waste dump fill site (Umair, Björklund, & Petersen, 2013).

The computers and other obsolete electronic goods are however imported under the pretext of 'second-hand equipment'. Barely a small percentage of the items imported are usable. In practice, after removing the working machines and usable parts the bulk of the consignment is sent to the recycling industry (Umair, Björklund, & Petersen, 2015). This is how fundamentals of Basal Convention are ignored (Pirzada & Pirzada, 2010). The mismanaged e-waste results in circulation of health hazards into the environmental matrices rendering the life more helplessly vulnerable (Godwin, 1993).

2. Materials and Methods

Study Area

The current research was designed and carried out for the assessment of computer waste management (CWM), administrative structure, occupational safety and legislative gaps analysis of management practices for the Computer Waste in Rawalpindi and Islamabad. Research focused on the generation of CW in the storage facilities. The area of twin cities is famous for the computer business. The area was chosen because here, the computer industry flourishes day by day with addition of new technology (see figure 1).

Data was collected through study of existing CWMS employing 'walk through survey' as a tool. Moreover, quantification of computer waste was done through onsite weighing. The extant information is collected through weighing of the computer waste in the storage areas and closely observing the management practice of CW for the period of seven months (January 2014-July 2014). For the ease of interpretation, codes are assigned to the CW handling facilities (see table 3).

Table 1. *Parameters for data analysis.*

s/no	Parameters
1	Legislative Framework
2	CW/EW Management System
3	Responsible Authority
4	Role of Government Bodies
5	Public Private Partnership
6	Occupational Hazards in CWM

Table 2. *Role and Responsibilities of Government Organizations.*

s/no	Organization	Parameters for investigation of responsibilities and current works	Justification
1	CDA (Capital Development Authority)	CWM	CDA is working in Islamabad for waste management
2	MoCC (Ministry of Climate Change)	Legislation and management Practices	MoCC is working for the safeguard of environment and formulation of laws
3	Pak-EPA (Pakistan Environment protection Agency)	Legislation and management Practices	Pak-EPA is regulatory body for implementation of laws and management practices
4	Provincial EPA	Waste management	To acquire data on the CWM

Table 3. *Generation of CW in Rawalpindi-Islamabad.*

s/n	Name of Facility	Location	Type of Waste	Generation of CW per day	Generation of CW weekly	Generation of waste Monthly	Annual Generation	Code of facility
1	Bajwa Computers	6th Road Rawalpindi	Monitor Waste CPU Waste Total Waste	150-200 kg 230 kg 390 kg	2730 kg	10920 kg (10.920 tons)	142350 kg (142.350 tons)	a
2	Sohail Computer	6th Road Rawalpindi	Monitor Waste CPU Waste Total Waste	50-100 kg 100-150 kg 230 kg	1610 kg	6440 kg (6.440 tons)	83950 kg (83.950 tons)	b
3	Sarghoda Computer	Sadar, Rawalpindi	Monitor Waste CPU Waste Total Waste	50-100 kg 150-200 kg 240 kg	1680 kg	6720 kg (6.720 tons)	87600 kg (87.600 tons)	c
4	Rabnawaz Computer	Tipu Road, Rawalpindi	Monitor Waste CPU Waste Total Waste	50-100 kg 150-200 kg 250 kg	1750 kg	7000 kg (7 tons)	91250 kg (91.250 tons)	d
5	Usman Computer	Tipu Road, Rawalpindi	Monitor Waste CPU Waste Total Waste	50-100 kg 100-150 kg 200 kg	1400 kg	5600 kg (5.6 tons)	73000 kg (73 tons)	e
6	Raja Zeeshan Computer	Raja Bazar, Rawalpindi	Monitor Waste CPU Waste Total Waste	50-100 kg 50-100 kg 150 kg	1050 kg	4200 kg (4.2 tons)	54750 kg (54.750 tons)	f
7	Sarhad Computer	Gawalmandi Rawalpindi	Monitor Waste CPU waste Total Waste	100-150 kg 100-150 kg 250 kg	1750 kg	7500 kg (7.500 tons)	91250 kg (91.250 tons) 6624.15 tons)	g

For compliance assessment questionnaire based surveys and interviews were carried out. The random sampling includes workforce (30 respondents), storage facility owners (07 respondents) and competent authorities (06) separately. The C.W generation and disposal practices (the Administrative framework) are correlated with the existing legal framework. The gap analysis is done on the basis of six selected parameters tabulated in table 1, however table 2 presents the competent authorities surveyed to evaluate the management system responsibilities and legal compliance.

3. Results and Discussions

3.1. Computer Waste Quantification

The best possible mean of computer waste quantification, i.e. 'on-site weighing' was chosen for the purpose. In Rawalpindi-Islamabad there are seven storage facilities classified as a, b, c, d, e, f and g in table 3. On average, the facility (a) receives 150-200 kg of monitor waste on daily basis. On the other side facility (g) receives 100-150 kg waste per day. Facilities (b, c, d, e and f) receive 50-100 kg of monitor waste per day.

According to the facility classification shown in table 3, the facility (a) receives 200-250 kg of CPU waste per day. The amount of CPU fluctuates on daily basis. Facility (b, e and g) receive 100-150 CPU waste per day. Facility (c and d) receive 150-200 kg on daily basis while, facility (f) receives 50-100 kg of CPU waste. Most of the CPU waste is being sold in the markets due to advancement of technologies day by day and people prefer to buy new technologies.

Approximately 624.15 tons of CW generates in the Rawalpindi-Islamabad annually. 57% of storage owners receive 51-100 tons, 28.6% 1-50 tons of CW annually and 14.3% of storage facilities receive 101-150 tons of CW annually. This variation among facilities depends not only on size of storage facilities but also the areas from where the computer waste is brought to them.

According to system observation, on average, 49% of storage facilities send 1-50 kg of CW for recycling per day. About 28.6% of facilities send 50-100 kg of CW for recycling and further processing. 14.3% of storage facilities send 150-200 kg. 14.3% of facilities send more than 200 kg of waste for recycling per day.

Through system examination it is found 57.1% of storage facilities send 1-50 tons of CW for recycling annually. While 14.3% of facilities sent 50-100 tons of CW. 14.3% of facilities send 100-150 tons. Moreover 14.3% sent 150-200 tons of CW for recycling annually. The fluctuations in quantities are always expected depending on the advancement in technology and resulting increase or decrease in lots of rejected goods received from developed world (Umair, et al., 2015).

3.2. Legislative and Administrative Evaluation

The core elements of waste management system i.e. quantification, segregation, management and disposal are found to be virtually absent in Rawalpindi and Islamabad. The private sector plays key role for the management of the computer waste. No government agency is involving in the CWM. There are no criteria for waste segregation for reuse and recycling in the system. The storage and dismantling facilities exist but there is no recycling facility for the CW (Umair, et al., 2015). The whole computer waste management system (CWM) is operated by private and no public participation has been recorded.

The system analysis revealed that the CWM process is operating by storage owners without license. Moreover, the informal and unauthorized owners of the facilities do not have license of handling, treatment and transportation of hazardous waste. A proper landfill for the hazardous waste is missing in the Rawalpindi-Islamabad. In Hazardous waste Rules 2003, it is mandatory to get license for handling and treatment of hazardous material. The informal handling and treatment of the hazardous waste may lead towards the worse condition for the health of work force and environment (Cronin, et al., 2003).

Pakistan has signed Basal Convention on Hazardous waste. The e-waste contains hazardous substances and majority of the e-waste being imported from the developed countries (Godwin, 1993). It is mandatory for every country that has signed convention to recycle their waste in their jurisdiction (Umair, et al., 2015). The work forces in the facilities are prone to the health and safety risks. Majority of CW being generated is in the form of second hand computer which is donated by developed countries. This leads to burden of computer waste in the country.

Capital development authority (CDA) and Rawalpindi development authority (RDA) that are responsible for solid waste management in the twin cities, have not devised some management for e-waste so far. Although the environmental and occupational hazards of computer waste handling and disposal are known, the waste is being dealt with as municipal waste, i.e. non-hazardous solid waste. Unfortunately the government agencies did not ever arrange any awareness campaign for the workers.

As far as the role of Pakistan Environment Protection Agency (EPA) is concerned, Sindh-EPA and Gilgit-Baltistan EPA has formulated legislation on computer waste management (CWM) but Federal administration and Punjab EPA have not taken such an initiative yet. This is probably due to lack of data and improper system. No doubt the hidden unlawful handling of hazardous materials and lack of awareness of local dealers and traders on associated environmental and health issues are the guilty parts of the picture.

3.3. Occupational Health and Safety

The workforce belonging to almost all age groups is involved in CWM process. Although 9.09% belong to the age group of 10-15years, 27.3% fall in age group 16-20 years. 9.09% workers are 21-25 years, while 27.3% are 26-30 years and 27.3% fall in age group of more than 30 years, working in CWM. Teenagers are engaged in the dismantling of CW without protective measures. Teenagers are working in this industry due to lack of job opportunities and education. Section 3 of the Employment of Children Act (ECA), 1991 bans employment of children under-14 in occupations connected with hazardous labor works that seems being widely not complied with in this industry (Fasih, 2007). Majority of the work force involved in the CWM are having primary level. A very small percentage of work forces are only secondary school qualified.

The Chapter 3 of Factories Act, 1934, talks about various safety arrangements at work place including provision of personal protective equipment and safe working conditions. Through system surveillance, it is determined that the personnel engaged in computer waste handling, transporting, dismantling and recycling are unaware of health hazards associated with their work (Umair, et al., 2015). There is no concept of personal protective equipment (PPEs). None of them found to be using PPEs or taking any precautionary measure during dismantling specially. This non-compliance is not an issue of one industrial unit but a matter of overall system ignorance on existing legal framework.

4. Recommendations

- Government agencies may pay keen attention to solve this problem with the collaboration of non-government organization (NGO).
- The proper training may be conducted for those who are directly involved in this profession in order to enhance the technical expertise of the work force.
- Government may develop a monitoring system to minimize Second hand computers donated by the developed nation to the Government Schools and institutions.
- Basal convention may implement formally to stop illegal import of hazardous waste in the country.
- Compliance of Child labor Act may ensure in this profession, most of dismantling of the computer waste is done by the teenagers.
- The Government of Pakistan may raise the education level among the children.
- Responsible authorities related to environment may ensure hazardous substance rules.
- Section 13 and section 14 if PEPA may enact where there is computer waste handling and dismantling
- CWM may be centralized and implementation of the polluter pays principle and there may a monitoring system for the better management.
- Ministry of Commerce may be directly involved for the management of the electronic waste.
- OHS policies may be implemented in this occupation for the minimal injuries.
- Use of alternate material and equipment having fewer hazards may be beneficial for the workers who are associated with CWM.

5. Conclusion

The overall CW-management is under the unofficial and informal organizations. Computer waste management (CWM) in Rawalpindi-Islamabad is a difficult task, if the Government Agencies do not adopt a pragmatic approach. The computer waste contains hazardous material and may pose a risk to the human health and environment. There exists an urgent need for the detail assessment of existing and future scenarios including generation, characteristics, existing disposal practices and environmental impacts. Institutions need to develop mechanism for collection, transportation, storage, segregation recycling, reuse and treatment of computer waste. Institutions which are engaged in the environmental services do not have any accurate data on generation of computer waste in Islamabad and disposal of computer waste. In short, instead of focusing on non-existence of a direct law on the subject, the need of the hour is to bridge the gaps between existence of indirect legislation and its proper compliance. This includes the issues of import, marketing of computer waste and occupational safety as well. The concerned administrative and legislative competent authorities need to focus on the issue before it gets complicated. In this regard, current case-study may be used as an initial authentic information to start with. Furthermore, the competent authorities with the help of Environmental Tribunals may conduct environmental, occupational safety and legal audits to ensure sustainable practices in the area.

References

[1] Aggarwal, A. (2006). Special economic zones: revisiting the policy debate. *Economic and Political Weekly*, 4533-4536.

[2] Bashir, K., Husnain, T., Fatima, T., Riaz, N., Makhdoom, R., & Riazuddin, S. (2005). Novel indica basmati line (B-370) expressing two unrelated genes of Bacillus thuringiensis is highly resistant to two lepidopteran insects in the field. *Crop protection, 24*(10), 870-879.

[3] Chung, S.-W., & Murakami-Suzuki, R. (2008). A comparative study of e-waste recycling systems in Japan, South Korea and Taiwan from the EPR perspective: implications for developing countries. *Kojima. Chiba.*

[4] Cronin, M. T., Jaworska, J. S., Walker, J. D., Comber, M. H., Watts, C. D., & Worth, A. P. (2003). Use of QSARs in international decision-making frameworks to predict health effects of chemical substances. *Environ. Health Perspect., 111*(10), 1391.

[5] EPA, P. (1997). Pakistan Environmental Protection Act, 1997. *Government of Pakistan, Ministry of Environment. 25p.*

[6] Fasih, T. (2007). *Analyzing the impact of legislation on child labor in Pakistan* (Vol. 4399): World Bank Publications.

[7] Godwin, D. L. (1993). Basel Convention on Transboundary Movements of Hazardous Wastes: An Opportunity for Industrialized Nations to Clean up Their Acts, The. *Denv. J. Int'l L. & Pol'y, 22*, 193.

[8] Joseph, K. (2007). *Electronic waste management in India–issues and strategies.* Paper presented at the Eleventh International Waste Management and Landfill Symposium, Sardinia.

[9] Leung, A., Cai, Z. W., & Wong, M. H. (2006). Environmental contamination from electronic waste recycling at Guiyu, southeast China. *Journal of Material Cycles and Waste Management, 8*(1), 21-33.

[10] Mittal, D., Goel, N., & Rani, R. (2012). E-Waste: A Hidden Threat to Global Environment and Health. *VSRD International Journal of Science and Technology, 2*(3), 271-275.

[11] Pirzada, M. D. S., & Pirzada, F. N. (2010). E-Waste: An Impending Challenge. *Pakistan Institute of Nuclear Science & Technology.*

[12] Saoji, A. (2012). E-Waste Management: An Emerging Environmental and Health Issue in India. *National Journal of Medical Research, 2*(1), 107-110.

[13] Sawhney, P., Henzler, M., Melnitzky, S., & Lung, A. (2008). Best practices for E-waste Management in Developed Countries. *Adelphi Research, Austria.*

[14] Skinner, A., Dinter, Y., Lloyd, A., & Strothmann, P. (2010). The Challenges of E-Waste Management in India: Can India draw lessons from the EU and the USA. *ASIEN, 117*, 7-26.

[15] Umair, S., Björklund, A., & Petersen, E. E. (2013). *Social life cycle inventory and impact assessment of informal recycling of electronic ICT waste in Pakistan.* Paper presented at the Hilty L, Aebischer E, Andersson G, Lohmann W, Proceedings of the First International Conference on Information and Communication Technologies for Sustainability ETH Zurich.

[16] Umair, S., Björklund, A., & Petersen, E. E. (2015). Social impact assessment of informal recycling of electronic ICT waste in Pakistan using UNEP SETAC guidelines. *Resources, Conservation and Recycling, 95*, 46-57.

3

Causes of Biodiversity Depletion in Bangladesh and Their Consequences on Ecosystem Services

Md. Redwanur Rahman

Ecology and Biodiversity Research Laboratary, Institute of Environmental Science, University of Rajshahi, Rajshahi, Bangladesh

Email address:
redwan_rahman@lycos.com

Abstract: Bangladesh is exceptionally endowed with a vast variety of flora and fauna due to its unique geophysical location which are maintaining balance ecosystem. There are recorded 5,700 angiosperms, 68 legumes, 500 medicinal plants, 29 orchids and 1,700 pteridophytes. Sundarban is a rich faunal diversity in Bangladesh which mammals 113, birds 628, reptiles 126, amphibians 22, 708 fishes, insects 2,493, mites 19, alges 164 and 4 species of echinoderms. Bangladesh's inland water bodies are known to be the habitat of 266 species of indigenous fish, 13 exotic fish, 56 prawns, about 26 freshwater molluscs, and 150 birds. The marine water bodies (200 nautical miles along the coast) are also remarkable for being habitat of 442 species of fish. There are at least 36 species of marine shrimps. About 336 species of molluscs, covering 151 genera have been identified from the Bay of Bengal. In addition, several species of crabs, and 31 species of turtles and tortoises, of which 24 live in freshwater, are found in Bangladesh. A total of 168 seaweeds, 3 sponges, 15 crabs, 3 lobsters,10 frogs, 3 crocodiles, 24 snakes, 3 otters, 1porcupine, 9 dolphins, and 3 species of whale found in Bangladesh. About 70 bees, many species of wasps and 8-10 per cent of the land area is under good canopy cover. A total of 81 fish species were recorded from Chalan Beel including 72 indigenous fish species and nine exotic species. More than 85 percent of Modhupur forest has been cleared in last 40 years. In most of the areas the age of sal forest is not more than 10 to 15 years. Thousands prawn farms have replaced almost the Chokoria mangrove forest. Thousands prawn farms have replaced almost the entire forest. The Chokoria Sundarbans today- the evergreen forests have been replaced by thousands of shrimp ponds. Two orders of class Oligochaeta of phylum Annelida: five families, nine genera include fifteen species of Tangail District of Bangladesh. A total of 35 species of earthworms have been recorded in Bangladesh. Already, 12 wildlife species have become extinct from Bangladesh. In addition, Mammals 40, birds 41, reptiles 58 and amphibians 8 have listed in red data book of threatened animals of Bangladesh. Anthropogenic pressures, uncontrolled dredging, hydrological intervention, pollution, chemical fertilizers are directly effect on habitat, biodiversity and aquifer. In this article, the causes of depletion of biodiversity and the consequences on ecosystem services were observed.

Keywords: Biodiversity, Depletion, Consequences, Ecosystem, Bangladesh

1. Introduction

"Biological Diversity" means the variability among living organisms from all sources including, inter alia, terrestrial, marine and other aquatic ecosystems, and the ecological complexes of which they are part; it includes diversity within species between species, and of ecosystems (Article II, CBD, 1992). Literal meaning of biodiversity is the diversity of all life forms on earth. This includes the various races and species of all microbes, plants, and animals that live on earth, including their genetic differences, i.e., the gene pool of each species. Bangladesh is a transitional zone of flora and fauna, because of its geographical settings and climatic characteristics. As mentioned in the overview section of the report that there are many rivers and streams existing in the country covering a length of 22,155km. In addition to the regular inland waters, seasonally a large part of the country remains submerged for 3-4 months during monsoon. This country is rich in fish and aquatic resources, and other biodiversity [1]. Sundarban is a largest mangrove forest in the world. It is located in between 21° 32' – 21° 55' N latitude and between 88° 42' – 89° 04' E longitude and 7.5m above sea level. The forest densely covered by sundary (*Heritiera littoralis*) tree. It is also largest reserve for the Royal Bengal Tiger (*Panthera tigiris*) in the world. [1,2] has reported that Bangladesh is exceptionally endowed with a vast variety of flora and fauna due to its unique geophysical location. [3] recorded 5,700 angiosperms species from sundarbans. They also recorded 68 legumes, 500 medicinal plants, 29 orchids and 1,700 pteridophytes from these mangrove forests. Sundarban is a rich faunal diversity in Bangladesh which mammals 113, birds 628, reptiles 126, amphibians 22, 708 fishes (fresh, estuarine and marine), insects 2,493, mites 19, alges 164 and echinoderms 4 species were found in Sundarban [3,4]. Like in other regions around the globe, the biodiversity of Bangladesh is also entering through a critical period. [5] has shown that already, 12 wildlife species have become extinct from Bangladesh. In addition, Mammals 40, birds 41, reptiles 58 and amphibians 8 have listed in red data book of threatened animals of Bangladesh [4]. On the other side, 106 vascular plants under risk of various degree for extinction of Bangladesh [6].

Some of the important relevant literatures about biodiversity in Bangladesh are discussed here. [7] has published an article on Biodiversity Conservation Strategies in Bangladesh: the state of protected areas. He find out some major causes as, rural poverty, ignoring local people, illegal wood extraction, overexploitation, poor forest management, lack of trained personnel and awareness about biodiversity. [8] has published that there is no legal issue related to environment and contains vast information about the affection of human lives by environment and environmental science every day. [9] has published that related to environment as a fusion of theory, basic research, applied research and public education, that is evolving at a peace. Land includes agriculture, irrigation *etc.*, to the study. The writer did not deal with all agriculture and fisheries laws and policies in the light of environmental aspects prevailing in the country.

[10] has recorded a report on the environmental attempts to tell signs of the time with regard to our environment, resources and people of Bangladesh. [11] reported that the biological diversity are rapidly destroying the environments from more than billion years. [12] has described that there is no law resolution nevertheless this book is very important for having informative discussion about extinct wildlife. A complete particulars and solution of some controversial concept about the extinct wildlife are found in this book with information of the studies both nationally and internationally. [13] has mentioned that about the contribution of wildlife with a view to making the people conscious and it is riched with root level information related to wildlife management of Bangladesh. [14] has contributed that the development of our environmental studies from the different grounds. [15] has analysed that about the present and past status of Bangladesh.

The sundarban mangrove forest on the Bay of Bangle formed by Ganges, Brahmaputra and Meghna rivers across the southern part of Bangladesh. According to view point of English Writer Hunter, Sundarbon is erapted in the largest coastal area Hugli to Meghna which is about 81 miles [23]. It is a swamp forest which covers 10,000 sq. km. of which 6,017 sq.m in Bangladesh. The UNESCO in 1997 declared a world heritage as the sundarban. This forest is occupied by 1,874 sq. km water bodies forms of rivers and canals.

Royal Bengal Tiger (*Panthera tigiris*), birds, spotted deer, crocodiles and snakes etc are abundant in Sundarban. Fertile soils have been mostly converted to intensive agriculture. Population pressure, habitat loss, pollution, and poaching are putting tremendous pressures on the existing biodiversity. If the current trend continues, many of the species will be endangered or be even extinct. In case of `Chalan beel" these percentage of lost is near about 3%. The condition of wetlands is even worse than that of forests. Most of the wetlands are being converted into agriculture land and degraded due to development activities.

Throughout the world ecosystems today are threatened with destruction at a rate rarely seen in history. As a result depletion of biodiversity and climate change has gained the foremost importance in the current world order.

But unfortunately, many plant and animal species on Earth are facing severe threats to their survival and are disappearing at alarming rates. And human actions are the cause behind this biological degradation.

Bangladesh is no different story. Here environmental degradation is also severe. But this report will only focus on the Modhupur forest which is one of the oldest forests in this region *i.e.* last deciduous forest in Bangladesh. It was a vast resource of flora, fauna and natural resources. The ecosystem of the forest was very old. But from late 1980's, there have been implemented a number of government and foreign funded projects which cause felling of old diversified species of natural trees and replace them with monotonic exotic trees like rubber, acacia, eucalyptus, mahogany *etc.* which was planted for short term economic benefits. But these practices have adverse effects on biodiversity and ecosystem resulted in destroying several species of flora and fauna for good. As a result, the forest has been shrunken and the livelihood of people depend on forest has also been affected.

Objectives of this chapter

The objectives of the article are as follows:

i. to estimate the present scenario of biodiversity in Bangladesh;
ii. to identify the causes of depletion of biodiversity;
iii. to observe the consequences on Ecosystem Services and
iv. to prepare a set of recommendations that will enunciate the methods of integrating environmental change adaptation into various national actions focused on issues related to eco-zone management.

2. Materials and Methods

This research based on two sources such as primary and secondary source.

Primary Sources: The primary data has been collected through two methods, these are:

- Interview method and
- Field survey with prepared questionnaire
- Interview Method

A pilot survey has been conducted to know the biodiversity depletion in Bangladesh and their consequences on ecosystem services. Primary problems are noticed in this time and then a detail survey with details questionnaire has been conducted. The mental condition, cordiality, willingness, transparency and patience of the respondents are being considered during interview. Because it is important to know the mentality of the respondents it either is positive or not while getting information. The respondents are selected from the head of household based on proportional allocation of stratified random sampling method.

- Field Survey with questionnaire

Data has been also colleted through field survey with a questionnaire to know the real feature of biodiversity condition of the study area. The questionnaire covers all the relevant questions to get the maximum information about the research topic. The authenticity of the data is justified through proper scrutiny. In this regard, cooperation has taken from local peoples.

Secondary Source

It includes related books, national and international journals, Reports, records, publications, government and non-governmental documents, published dissertations, internet, homepage, Bangladesh Statistical Bureau report, from different Institutes and offices *etc*.

The completed research prepared a cross sectional descriptive study. In this work qualitative data collected. Findings of the research presented in an analytical description.

Sample Size and Selection Procedure

Sampling procedure plays an important role of a research. It provides good information about diversity. A sample is a finite part of a statistical population whose properties are studied to gain information about the whole [32].

Methodology for Biodiversity Assessment in Sunderbans

The Sundarban Reserve Forest is situated at the southern part of Khulna, Bagerhat and Satkhira civil district lying in between latitude 21° 27' 30" and 22° 30' 00" North longitude 89° 02' 00" and 90° 00' 00" East. Sundarbans is under the administrative of Khulna, Satkhira and Bagerhat districts.

This case study has been conducted primarily based on the field visit in order to collect raw information from the field, also to get proper understanding and insight in the subject. Along with it, we had to depend on the source of information so that historical analysis and comparison can be made and a suggestive attitude can be developed.

To collect the information we have gone through literature stock from Bangladesh Statistical Bauru, Bangladesh Development Studies and other necessary documents found in other sources. After all, documents collection has also been relied on the internet.

Methodology of Chalan beel Observation

Required data were collected by side observation and interviewing the local people. There is no any specified or appropriate analytical method was used to analyze the collected environmental data. So collected data were tabulated and arranged, according to the researcher choice for achieving the objectives of the study.

3. Results and Discussion

The number of local varieties of varieties crops is steadily declining due to 'the Promotion and economic consideration of only a few selected varieties, especially of the High Yeld Varities (HYV) of rice and wheat and other crops. This is causing of irreparable loss to the genetic diversity of our crop plants and their wild relatives. Introduction of exotic carnivorous fishes like African perch and the African catfish poise a threat to the propagation of many indigenous fishes. It is now a concerned issue in Bangladesh which should take in consideration immediately. There are three prime reasons for which this critical situation has been raised in Bangladesh. These are: i. the lack of adequate conservation efforts to maintain and protect indigenous crop genetic recourses. ii. lack of software for the conservation and propagation of the various local crop varieties. iii. unbalanced competition between the local varieties consequent to the introduction, adoption and promotion of the exotic and locally developed HY varieties.

Table 1. *Extinct wildlife in Bangladesh. According to [12]).*

English Name	Scientific Name	English Name	Scientific Name
Mammals			
Striped Hyena	*Hyaena hyaena*	Wolf	*Canis lupus*
Malayan sun Bear	*Helorctos malayanus*	Great one Horned Rhinoceros	*Rhinoceros unicornis*
Smaller One Horned Rhinoceros	*Rhinoceros sondaicus*	Two Horned Rhinoceros	*Dicerohinus sumatrensis*
Asiatic Gaur	*Bos frontalis*	Banteng	*Bos javanicus*
Wild Buffalo	*Bubalus bubalis*	Black buck	*Antelope cervicapra*
Bulebull	*Bocehpalus tragocamelus*	Swamp Deer/Barasingha	*Cervus duvauceli*
Pygmy Hog	*Sus salvinus*		
Birds			
Bengal pinkheaded Dock	*Rhodonessa caryophyllacea*	Indian Peafowl	*Pavo cristatus*
Bengal Florican	*Eupodotis bengalensis*	Green Peafowl	*Pavo muticus spicifer*
Painted Stork	*Mycteria leucocephala*	Greater Adjutant	*Leptoptilos dubius*
King Vulture	*Sarcogyps calvus*		
Reptiles			
Indian Egg eating snake	*Elachistodon westermanni*	Chittagong Mud Turtle	*Trionyx nigricans*

Table 2. *Extinct plants species in Bangladesh. According to [15].*

Scientific Name	Scientific Name	Scientific Name	Scientific Name
Psilotum triquetrum	*Tectaria chattagramica*	*Aglaonema clarki*	*Aldrovanda vesiculosa*
Aquillaria agallacha	*Cirrhopetalum roxburghii*	*Cymbopogon osmastonii*	*Debregeasia dentata*
Elaeocarpus lucidus	*Hippocratea macrantha*	*Homulium schlchtii Kurz*	*Limnophila cana*
Mantisia spathulata	*Marsdenia thyrsiflora*	*Ophiorrhiza villosa*	*Phrynium imbricatum*
Quercus acuminate	*Rotala sin pliciuscula*	*Semecarpus subpanduriformis*	*Sonneratia griffithii*
Tournefortia roxburghii	*Typhonium listeri*	*Vatica scapula*	*Justicia oreophila*
Knema bengalensis	*Vernonina Thomsoni*		

The population of domestic animals and birds of Bangladesh is insufficient to meet the basic minimum requirements of draft power, meat, milk and eggs. The total number of cattle was estimated at about 2.5 million sheep and goat 14.5 million, poultry 76.5 million heads after 1983-84 agricultural censuses. [16] has shown that the resources base declined steadily since the 40s consequent to the human explosion in Bangladesh.

In addition, Bangladesh is a rich flora and faunal zone. All over the country, 22 amphibians, 400 fishes, 126 reptiles, 628 birds, 113mammals were identified. Approximately, 5700 species of angiosperms were recorded in Bangladesh [4]. Extinct wildlife and plants in Bangladesh are presented in table 1 and 2.

Causes of Biodiversity Depletion in Bangladesh

Many wildlife species have been exterminated in Bangladesh and many more are threatened with extinction. The list of extinct animals of Bangladesh is prepared on the basis of published data and habitat management. The country has lost 10% of its mammalian fauna, 3% avifauna and 4% reptiles over the last 180 years. More than 50 species are critically endangered in Bangladesh of which 23 species are already declared as endangered in the Red Data Book of IUCN (International Union for Conservation of Nature). In addition, 83 species are threatened and are included in the appendices of Convention on International Trade in Endangered Species (CITES). Among them the most endangered species are : Elephant (*Elephas maximus*), Tiger (*Panthera tigris*), Wildcat (*Felis chaus*), Leopard (*Panthera pardus*), Dolphin (*Peponocephala electra*), among mammals; Whitling duck (*Cairina scutulata*), Knobbilled duck (*Sarkidiornis melanotos*), Black partridge (*Perdicula manipurensis*), among birds and Crocodile (*Crocodylus, porosus*) Python (*Python molurus*), Monitor (*Varanus bengalensis*), River Terrapin (*Batagur baska*), Roffed turtle (*Kachuga dhongoka*), Softshell turtle (*Chitra punctata*) and all marine turtle among reptiles. Bird population around the globe is declining at an alarming rate; the primary threats to them are habitat destruction, trade and over hunting [8, 10-22].

Despite the depletion of rivers, canal and flood plains for hundreds of year, Bangladesh still holds the worlds most diverse and abundant inland fisheries. But the catch of many popular local species have drastically reduced and many are no more seen. The fishes face many obstructions and disturbs in the reproduction period for the causes of unplanned construction in the flood plain area. Mistakenly, subsistence fishermen are often blamed for over fishing and thus reducing the fish population.

But the studies and expert opinions do not affirm this blame. The physical loss, shrinkage and modification of aquatic habitats for fish, prawn, turtle and other aquatic organizing are said to be one major factor in depleting the fish varieties.

Grazing, Illegal filling, fuel wood collection uncontrolled and wasteful commercial exploitation are also shown to be causes of depletion of forest resources, with drastically reduced forest, diminishing also are rare wildlife and Biological diversity.

Reasons of Depletion of Biodiversity

Bangladesh is rich flora and fauna diversity which are maintaining balance ecosystem. Presently, biodiversity is affecting due to deforestation, forest exploitations, agriculture and industrial pollution, irrigation and flood control developments, shifting land use and over exploitation of biological resources. Besides, Human beings occupy almost all the earthly land; therefore, the habitats of other lives are seriously hampered.

Human interference for agriculture and settlement such as flood control embankments, drainage systems and conversion of inundated lands into crop land have reduced the water area, prevented or altered the seasonal migratory routes of many fishes and reduced breeding and natural stocking of fishes in many water bodies.

We are facing serious environmental degradation due to unplanned development work and urbanization. Some major causes are as follows:

- Unsustainable use of Natural Resources
- Hill Clearing, Cutting and Destruction
- Industrial and Agro-Chemical Pollution
- Deforestation
- Illegal Hunting
- Habitats Degradation and Destruction
- Wetland Degradation
- Traditional Management and Short Sighted Development Efforts
- Introduction of Modern Agro-Technology
- Rapidly Expanding and Heavy Population Pressure
- Indiscriminate Filling of Trees and thereby reducing tree cover
- Over-exploitation of Particular Species, like Medicinal Plants, Bamboo, canes *etc.*
- Salinity Intrusion due to Reduced Freshwater flow in the Ganges and its Tributaries is affecting the Vegetation of the Sundarbans
- Destruction of the Mangrove forest of Chakaria and Teknaf for Conversion into Shrimp Culture

- Frequent Occurrences of Natural Disasters, like Floods Droughts, Storms, Cyclones, Tornadoes, Tidal Surges *etc.*,
- Industrial, Domestic, Agro-Chemical and other Pollution an Irrational use of Chemical Fertilizers and Pesticides,
- Rapid, Unplanned and Uncontrolled Industrialization and Urbanization
- Changing the Character of the Land and Conversion of Low-lying areas to Build up Areas
- Construction of Upland Roads and Homesteads on Wetland
- Increased Salinity Intrusion in the Coastal areas due Lower Flow in the dry Season
- Pollution from Domestic and Agro-Chemical Waste Products and their runoff
- Hunting, Trapping and Disturbances of Animals
- Management of open water Fisheries Through Auction; Ineffective Legal Measures for the Protection of Undersize Fish
- Lack of Adequate Conservation Efforts to Maintain and Protect Indigenous Crop Genetic Resources
- Lack of Software for the Conservation and Propagation of the Various Local Crop Varieties
- Imbalanced Competition Between the Local Varieties Consequent to the Introduction, Adoption, and Promotion of the Exotic and Locally Developed High Yield Varieties (HYV)
- Gradual Encroachment of the Fallow Grazing Lands for Agricultural Crops and thereby reducing the base for Grass and Fodder Crops
- Regular, Almost Annual, Recurrence of Epizootic Diseases of Cattle and Poultry in some Parts of the Country
- Absence of Trained Manpower, Software and Logistic Support to Contain and Prevent such Recurrent Epizootic
- Long Absence of any Concern for the Production of Fodder Crop and Animal Feed in the Country, Even to the Extent of Exporting Animal Feed Ingredient like Oil Cake
- Wrongful Dependence on Artificial Insemination from Imported Breeds of Bulls for Improving the Cattle Stock since the 1940s without any Visible Success
- Negligence and Apathy in the Systematic, Identification Selection and Breeding of different local races of Cattle for Improvement.

3.1. Case Study: Case Study on Biodiversity of Sundarban and Its Depletion

Sundarbans is the World largest mangrove forest, 62% in Bangladesh and 38% in West Bengal. The area of Sundarbans in Bangladesh is 6017sk.km. (70% land and 30% water). Sundarbans forest is situated on the Southwest coastal areas and it is the largest forest in Bangladesh having 6017sq. miles of areas. The areas 99% covered by the districts of Sathkhira, Khulna and Bagerhat while other areas are in Patuakhali and Barguna districts. Rajmongal and Burishwar stand on the east and west side of Sundarbans respectively.

The Sundarbans Researve Forest (SRF) are situated at the southern part of Khulna, Bagerhat and Satkhira civil district lying in between latitude 21° 27' 30" and 22° 30' 00" North longitude 89° 02' 00" and 90° 00' 00" East (Map 1 & 2).

Map 1. *Map of Sundarbans and surrounding area of Bangladesh.*

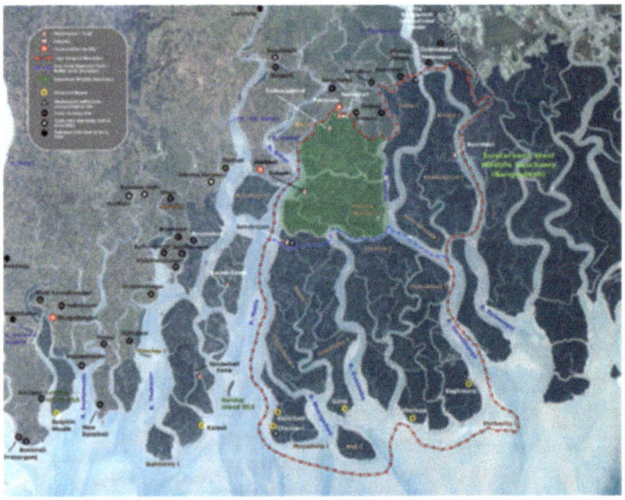

Map 2. *Tiger reserve areas of Sundarban (Dark Green: Biosphere reserve).*

About 450 large and small rivers in Sundarban, Baleswar and Pusur rivers bring fresh water from Padma to Sundarban. There are rivers and canals spread across this forest like a net with their innumerable branches. Nearly 450 large and small rivers occupy about 1 lakh 75 thousands 685 hectares or about 30 (%) of the Sundarbans. The Biggest River is the Pusur. Other rivers worth mention are Baleswar, Sibsa, Arpangasia, Bhola, Horinbhanga, Kalindi, Andharmanik, Raimangal, Kapotaksha, Koira, and Shela Bhadra *etc.* As one proceeds to the south, the rivers widen. Some rivers are so wide that one cannot see one bank form the other. Baleswar and Pusur rivers and their tributaries and disrbutaries are connected with the Ganges. As a result these rivers and their branches have flow of fresh water. The sibsa and other rivers in the western part have their source of fresh water only in the Ganges and the northern portion of the Sundarbans depends upon the rainwater. As a result, the fresh water flow in these rivers decreases during the dry season when there is a massive intrusion of saline water. Moreover, the condition of the rivers and canals in the Sundarbans is deteriorating shoals are forming and navigability is declining. River erosion is taking place at some places.

The rivers, canals, creeks *etc.* spread across of Sundarbans like a net with their innumerable branches. There are neatly 450 large and small rivers that occupy the water areas of the Sundarbans. The biggest river is the pasur. Other rivers worth-mentioning are the Baleswar, the Sibsa, the Arpangasia, the Bhola, the Horinbhanga, the Kalindi, the Andharmanik, the Raimangal, the Kapotaksha, the Koira, the Shela Bhadra *etc.* As they once proceed to the south, the rivers become widen.

There are about 102 Islands in the Sundarhans of which some 58 Islands are inhabited. A large number of these inhabitants are directly or indirectly involved in fishing, fry collection or fish marketing *etc.*

Floral and faunal diversity of Sundarban

Sundarban is a large unique mangrove ecosystem habitat in the world. This mangrove habitat is abundant of many rare and endemic species. Sundarban mangroves are habitats of many endangered animals *viz*., *Batagur baska, Pelochelys bibroni, Chelonia mydas*). World famous Royal Bengal Tiger (*Panthera tigris*), Spectacular Spotted Deer (*Axis axis*), Jungle Fowl (*Gallus lafayetii*) and Rhesus Monkey (*Macaca mulata*) are critically endangered.The forest has a unique biota comprising 334 species of plants, 49 species of mammals, as many as 400 species of fish, 315 species of birds and 53 species of reptiles [24, 25]. Besides numerous species of Phytoplankton, Fungi, Bacteria, Zooplankton, Benthic Invertebrates, Molluscs, Reptiles, Amphibia and Mammals are found in the Sunderbans. Species composition and community structure varies east to west, and along the hydrological and salinity gradients. The forest is particularly important as a barrier to tidal cyclones.

The mangroves of the Sundarban are unique prime species are sundari (*Heritiera littoralis*), from which the Sundarban takes its name. There are four types *viz*., low mangrove forests, tree mangrove forests, salt-water *Heritiera* forests and freshwater *Heritiera* forests. *Ecoecaria agallocha,* a dense understory of Goran (*Ceriops decandra*) and dense patches of Hantal palm (*Phoenix palludosa)* on drier soils. Dhundal (*Xylocarpu granatum*) and Passur *(Xylocarpu mekongennnsis)* and Bakul kankra (*Bruguiera sexangule)* sporadically found throughout the area [12].

Five species, viz., Javan Rhinoceros (*Rhinoceros sondaicus)* (CR), Wild Buffalo (*Bubalus bubalis*) (EN), Swamp Deer (*Cervus duvauceli*) (VU), Gaur (*Bos frontalis)* (VU) and probably Hog Deer (*Axis porcinus)* (LR) have become extinct [12].

A total 315 species of bird have been recorded by [4]. Stork-Billed kingfishers (*Pelargopsis amauropterus*), White-bellied Sea-Eagle (*Haliaeetus zeucogaster),* Grey-Headed Fish Eagle (*Zchthyophaga ichthyaetus*), Pallas's Fish-Eagle (*Haliaeetus zeucoryphus*), Raptors (*Leptoptillos dubius*), Herons (*Ardeola grayii*), Egrets (*Bubulcus ibis*), Storkas (*Cochoa purpurea*), Sandpipers (*Actitis hypoleucos*), whimbrel (*Numenius phaeopus*), Curlew (*Numenius phaeopus*), Woodpeckers (*Micropternus brachyurus),* Mynahs (*Gracula religiosa*), Minivets (*Pricrocotus flammeus),* babblers (*Pomatorhinus horsfieldii*) are found in dry season [4].

Reptile 53 and 8 amphibians species have been recorded from Sundarbans by [4]. Mugger (*Crocodylus palustris*) (Vulnarable) is now extinct. Estuarine crocodile (*Crocodylus porosus)* still now survives but its numbers have been greatly depleted [4].

Four species of marine turtle have been recorded from the Sunderbans by [12], Olive Ridley *(Lepidochelys olivacea)* (EN=Endangered), Green turtle *(Chelonia mydas)* (EN), *Loggerhead (Caretta caretta)* (EN), Hawksbill *(Eretmochelys imbricate)* (Critically Endangered) are found in Sundarban. King Cobra *(Ophiophagus hannah)* and Spectacled Cobra *(Naja naja)* three Vipers and six Sea-snakes are recorded in Sundarban. Shrimps, Prawns and Lobsters are rich in water of canals and rivers [12].

Impact of Climate Change on Sunderbans

Sundarban wetlands, freshwater reaching the mangroves was considerably reduced from the late 19th century due to diversion of freshwater in the upstream area. As a result, the salinity increased of the Sundarban. The sea level rise due to global warming and then the vulnerability to tropical cyclone storm surge flooding increase in the Sundarban areas [36].

SIDR (Tidal Cyclone)

Barguna, Pautuakhali, Jhalokati, Pirojpur, Bagerhat, Barisal, Bhola, Lakshmipur, Chandpur. Madaripur, Shariatpur, Munshigong, Narayanganj. Dhaka, Comilla, Bramanbaria, Narsingdi, Habigang. Kishoregonj, Maulvi Bazar, Sunamganj and Sylhet district of Bangladesh were affected by the tidal cyclone SIDR. In 2007 super cyclone SIDR, originating in the Bay, ravaged the entire south and southwestern coast with peaking winds over 220km on November 15. This coastline is homeless of nearly 12 million people. At least 5,000 people have been died caused by cyclone SIDR. The flora and fauna of the Sundarbans, became a major casualty of the devastating cyclone. The Forest Department is yet to make an assessment of the colossal losses of wildlife in the Sundarbans.

SIDR damaged 25% of the south eastern part of the Sundarbans. Out the damaged area 7% (ca 30,000ha) damage was severe 18% (ca.77000ha) damage was partial. SIDR uprooted and broke severely damaged crown of still standing trees, foliage have been striped off from branches of trees in nearly one third of the Sundarbans mainly including the World heritage Sites, many forest stations, boats, jetties etc have been severely damaged, fresh water ponds have been contaminated by saline water, six out posts have been totally destroyed. Two tigers, 23 deer, I forest staff have been found dead because of tidal upsurge. Residential accommodation of nearly 1000 persons has been badly damaged or washed away. Sundarbans will require years to recover from the damage. Forest Department's official apprehension that the total damage of all kind might amount to US$145 million while Governments official source supports suggest the damage to be US$150 million Forest Departments strategies are to rebuild damaged outpost, revive the safe drinking fresh water, mark and estimate fallen trees and record them by species make fill scale estimation of damage, assist in natural regeneration and plant where necessary.

AILA (Tidal Cyclone)

At least 300 people have been died, 8,208 more missing, 1million people homeless, 20 million people at risk of post disaster diseases due to Aila. The majority portion of the biosphere reserve was destroyed by the Aila. The gradual climatic change gave back a heavy return which not only had a great impact on the life of the dwellers in the mentioned region but also made them aware of the benefits of the mangrove forests [http://www.surfat10.com/climateicare]. Salinity and Sea Salinity Rise Threaten on Sundarban.

Many diseases, like 'Agamora' (top dying), affect different species of plants and trees, including the Sundari tree (*Heritiera littoralis*), in the Satkhira district due to the salinity intrusion in Range of the Sundarbans. Rising salinity in rivers, canals and other water bodies in the mangrove forest was also one of the reasons for the death of trees. Sundari (*Heritiera littoralis*), Passur (*Xylocarpu mekongennnsis*), Keora (*Sonneratia apetala*), Gewa (*Excoecaria agallocha*), Baen (*Avcennia officinalis*), Goran (*Ceriops decandra*), Hantal palm (*Phoenix palludosa),* Dhundal (*Xylocarpu granatum*), Bet (*Calamus tenuis*), Jhau (*Tamarix galica)* and Hogla (*Typha elephantiana*) affected by 'Agamora' (top dying) disease due to salinity intrusion in the sundarban area [14]. Many species of animals and birds have been facing extinction due to salinity on the soil and water in the sundarban area.

Endangered and Extinct Species of Sunderbans

Axis porcinus, Wild buffalo *(Bubalus bubalis),* Swamp Deer *(Cervus duvauceli),* Javan Rhinoceros *(Rhinoceros sondaicus)* and the Mugger Crocodile or Marsh Crocodile *(Crocodylus palustris)* have been extinct in the Sundarbans recorded by [14]. Royal Bengal Tiger *(Panthera tigris)* Estuarine Crocodile *(Crocodylus prosus),* River Terrapin *(Batagur baska),* Olive Ridley Turtle *(Lepidochelys olivacea) (EN)* Green turtle *(Chelonia mydas) (EN),* Gangetic Dolphin *(Platanista gangetica),* Hawks Bill Turtle *(Eretnochelys imbricate)* and King Crabs *(Horse shoe)* are endangered species of Sundarban [4]. Capped Langur *(Semnopithecus pileatus),* Smooth-Coated Otter *(Lutrogale perspicillata),* Oriental Small-Clawed Otter *(Aonyx cinerea)* and Great Indian Civet *(Viverra zibetha)* are threatened [14].

Illegal Practice of Sundarban Biodiversity

The total area of Sundarban in Bangladesh is calculated about 5800 sq.km including rivers and Canals. Wild species such as wild buffalo, Bengal Rhinoceros, Swamp deer have already become extinct from this forest. About two species of amphibians, fourteen species of reptiles, twenty five species of birds, five species of mammals have now become extinct from the Sundarban. Some natural phenomena and human interventions are the main causes in this regard. There are about 334 species of trees including vegetation, creepers, and harbs are available in the forest. Peoples are authorized with legal endorsement in the forest to catch fish, collect honey, Golpata and Gewa tree and there are frequent illegal trespassers among them who collect the natural resources indiscriminately.

There are many valuable trees in the Sundarban which should be preserved to save the ecosystem and biodiversity of Sundarban. But unfortunately the ecosystem and biodiversity of Sundarban is damaging day by day because of illegal activities trough the Sundarban. Sundarban is being a crime zone day by day and it is using as safe zone to the criminals. A few groups of terrorists are placed in Sundarban and they are doing crime in the surrounding area. Criminals loot the locals and hand over the valuable trees to the black market. Stealing is a common practice in Sundarban by which Sundarban is losing its valuable assets day by day. Anthropogenetic impacts like reclamation, human encroachment and influence plays a vital role to increase the depletion of biodiversity of Sundarban. High salinity, low water table and acidity problem, loss of soil fertility, coastal erosion and a steep fall in fishery resources, conversion of mangrove tracts for aquaculture and agriculture Extension of other non forestry land use into mangrove forest, uncontrolled collection of prawn seedling, pollution from both the landward and seaward sides are also the illegal practice in Sundarban. Hunting different kinds of valuable animal like tiger, deer, crocodile, birds, snakes *etc* is the most affective illegal practice in Sundarban which are seriously damaging the ecosystem of Sundarban.

Causes of Depletion in Biodiversity of Sunderbans

The sundarban bio-diversity is threatened by unabated poaching, increasing conversion of the forest land for agriculture, shrimp farming, hunting, illegal felling of trees, exotic plant plantation, unplanned and over extraction of resources and insufficient conservation effort.

Natural calamities like storms, cyclones and tornadoes pose a big threat to trees. More than a dozen of cyclones have hit the Sundarbans during the last decade. Some ecological changes like increase in salinity, diseases like “top dying” of the Sundari trees a serious threat to rich biodiversity of the Sundarbans.

Diversion of the natural courses of rivers, construction of embankments, dams and bridges in the upstream as well as decreased flow of freshwater in the rivers resulted in the increase in salinity level and over-silting in many places and caused subsequent changes in the mangrove ecosystem. In the Sundarbans, for bringing non-commercial cover areas under tree cover in different areas of Chandpai, Sarankhola and Khulna forest ranges, plantation of exotic species, like *Acacia nilotica* and *Alibizia procera* have taken place. Though at a limited scale, many other exotic species have been introduced in the Sundarbans mangrove ecosystems, particularly in areas around the forest offices. It is apprehended that these exotic species may pose a threat to the native variety of mangrove species.

Wild Buffalo, 2 species of Deer, Javan Rhinoceros extinct

and presently 2 species of amphibians, 14 species of reptiles, 25 species of birds and 5 species of mammals are considered as endangered species reported by [25].

Causes of Depletion of Biodiversity of Sundarban

1) Anthropogenic activities
2) Causes of neo-tectonic movement of the Bengal Basin
3) Global Warming and Sea Level Raise
4) Huge silt Deposition of river and canal in the sundarban area
5) Salinity intrusion
6) Reduction in Periodicity and Quantity of Freshwater Reaching the Mangrove Environment due to Diversion of Freshwater in the Upstream Areas (especially due Farakka Barrage Constructed by India) and Change in Course of main rivers
7) Conversion of Mangrove Tracts for Aquaculture and Agriculture
8) Over wood extraction.
9) Illegal hunting of Tiger, Deer, Wild Boar, Marine Turtles and Horse Shoe Crab *etc.*
10) Illegal practices of prawn farming.
11) Illegal fishing and overfishing
12) Uncontrolled Prawn Seed Collecton
13) Pollution from usage of Excessive Pesticides, Chemicals for Agricultures, Industries, Exploitation of Mineral Gas and Oil *etc.*
14) Organizational and Infrastructure Deficiencies;
15) Lack of Public Awareness
16) Lack of trained manpower

Study on the search of appropriate cause of biodiversity depletion in Sunderban region has provided an insight which is in need of immediate implementation. Such as the conservation attempt by the government must be accentuated by the local people. This can be materialized by managing alternative employment opportunities of those whose livelihood rests on Sunderban and by developing sustainable management procedure which will replenish the loss of resources for the local livelihood sustenance.

Secondly, the most troublesome cause that has been identified by our study is climate change impact. Recent cyclone like SIDR, AILA has devastated the entire region by striking on the long trees and by depleting the wild animals, even the tiger has been found dead after the disaster. Hence, an intensive and immediate measure is to be taken so that a proper strategy can be put in place before any such calamity strikes soon.

Consequences of Mangrove deplation

The mangroves plantations are the breeding grounds for thousands of species which thrive in a combination of saline and freshwater reservoirs. Due to illegal cutting and grazing the mangroves are rapidly vanishing.

If current trends persist then entire biodiversity of the area would be lost, also leading to serious social and economic repercussions. Over 90 percent of the population of sunderbans are illiterate and lives well below the poverty line. If the sunderbans forest is exhausted not only that area would be vulnerable to cyclones, but also innumerable fishermen would loose their livelihood and their way of life hence it's vital for the ecosystem without which this fragile habitat would fall victim to extinction.

3.2. Case Study: Causes of Biodiversity Depletion of Chalan Beel Area

Chalan beel is the largest wetland, water body and natural water reserver in Bangladesh. It covers the area of eight upzilas of three districts, namely Natore, Sirajgonj and Pabna. Singra, Boraigram and Gurudaspur of Natore, Taras, Raigonj and a part of Ullapara of Sirajgonj, Chatmohor and Vangura of Pabna districts come within the range of Chalanbeel. It is some 2000 square kilometers in its size, 51 kilometers in length (east-west) and 39 kilometers in its width (north-south).

Once it stretched from Raninagor, Atrai of Naogan to Faridpur, Bera and Sador of Pabna, as well as Nandigram and Adamdighi of Bogra district. But now it has diminished in its size owing to siltation. The construction of highway has also had a negative impact on its size.

The topography of the land is low. The area is about 25 meters high from the sea levels. The area is under Agro Ecological Zone (AEZ) named High Ganges River Flood plain. There is an overall pattern of olive-brown silt loams and silty clay loams calcareous soils. Organic matter content in the soil is low. Soils are slightly alkaline in reaction. General fertility level is very high.

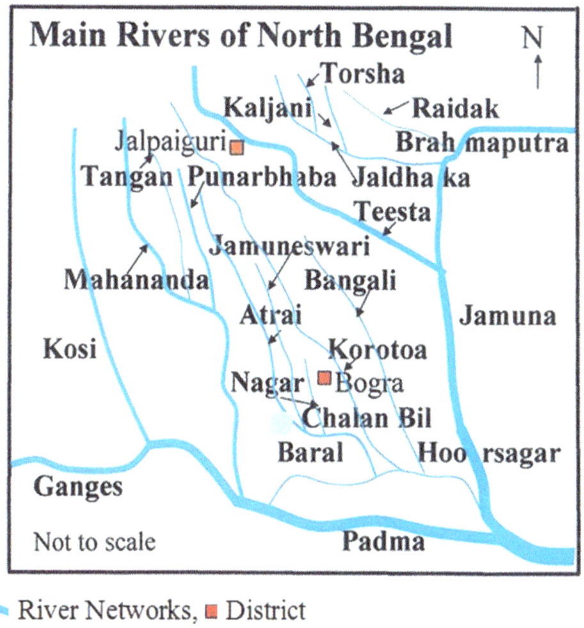

Map 3. *Overview map showing Chalan Beel among the main rivers in North Bengal of Bangladesh.*

Climate and Temperature of Chalan beel Area

The climate of this area is mainly warm and moist also. In terms of moisture chalanbeel resembles Rajshahi. The temperature of this region is on average from maximum 35°C to minimum 25°C. In winter it is from 12°C to 15°C.

Sometimes it increases to 45°C in summer and decreases 14°C.

In 1909: The original area of the chalan beel was above 1088 km² [29].

Current area of beel is 375 km². The study area is less populated than other parts of Bangladesh. About 1749484 people live here in 64 union councils, 7 municipalities, and 1878 villages. Average family size is 6, literacy rate 40 and 37.36 male and female respectively. About 82% population is Muslim, 16% Hindus and others 2%.

Purba Maddhanagar, Piprul, Dangapara, Laror, Tajpur, Niala, Chalan, Majhagaon, Briasho, Chonmohan, Satail, Khardaha, Darikushi, Kajipara, Gajna, Bara, Sonapatila, Ghugudaha, Kuralia, Chiral, Dikshi and Gurka are main depression of Chalan beel. The big-size depressions are mostly in Pabna district which are Gajna, Bara, Sonapatila, Ghugudaha, Chiral and Gurka. Gajna beel cover an area of 123 sq km. Bara beel covers an area of 31 sq km. Sonapatila beel in Pabna which an area of 35 sq km [29].

About 10,000 families live on fishing and this number increases to the number of 30,000 in the rainy season. A great number of families depend on farming, business and other occupation.

Road and Communication (Table 3)

Table 3. *Constructions on Boral River in several times.*

Locations	Construction Year	Construction Agency
Charghat Sluis Gate, Rajshahi	1985	Water Development Board
Ramnagar Dam, Chatmahar	1986	Up-Zilla Parishad
Bodhar Dam, Chatmahar	1987	Up-Zilla Parishad
Natun Kheya Ghat, Chatmahar	1988	Up-Zilla Parishad
Dohopara Sluis Gate, Chatmahar	1988	Up-Zilla Parishad

As it is a lower wetland, the modern communication system has not developed even in this scientific age. Boats and carts are still the main mode of transport and communication. Though a number of upzilas and districts are linked with metalled and semi-metalled roads and streets, no particular improvement could be found in communication. Natore-Nagorbari highway, Bogra-Natore Bogra-SirajgonjI highway and recently built Hatikumrul, Bonpara-Dhaka highway has contributed greatly to the development of the people of this area. Atrai, and Gour, Karatoa and Fulzhor, Boral, Barnoi, Tulsi, Chechua, Vadai, Chiknai, Bangonga and Khubjipur Telkupi are of the main rivers of this area. Different waterbodies under Chalan beel and their areas are presented in table 4.

Table 4. *Different waterbodies under Chalan beel and their area.*

Type of water body	Number	Dry season Area (ha)	Monsoon Area (ha)
Rivers	21	709	3300
Beels	93	2227	9164
Floodplain	-	-	22 369
Ponds	12 817	2293	2617
Borrow pits	214	–	50
Total	-	5229	37 500

(Sources: [29] and Upazila Fisheries Offices (2006).)Fishes of Chalan Beel

Chalan beel is the heaven of fish. The number of fishermen family in this area is 10000 in dry season and 30000 in the rainy season. The fish hunters freely collect their fish in the marshes and flooded land of the area.

Every year 12,255 metric tons of fish are caught only in the rainy season and thereby some 72 crores 2 lacs 55 thousand tk at tk. 60 per kg is earned every year. The amount of fishes can meet the demands of the local people and can be exported to other regions of the country as well. The availability of the agricultural and fishing land of Chalan beel and the surrounding is decreasing day by day due to increased demands of settlements.

A total of 81 fish species were recorded from Chalan Beel including 72 indigenous fish species and nine exotic species [26]. Small freshwater prawns, crabs and some other fisheries items were very common in Chalan Beel. The available fish were recorded in the Chalan Beel Area. The available fish were recorded in the Chalan Beel Area (Table 5).

Majority drying were done by spreading raw fishes on bamboo rack without any protection measure from insects or dust. Among birds are as follows in table 6 [27].

Migratory Birds of the Chalan beel Area

Titihas (*Anas platyrhynchos*), Pamtamukhy (*Anas plapeata*), Lenja has (*Anas acuta*), Chiti has (*Anas platyrhynchos*) etc.

List of Plant Diversity in the Study Area

Paddy is the only crop because the depth of water in the rainy season is 6/7 meter.

Though the region is farmed only once, 51 thousand 30 metric ton is produced in 1 lac 20 thousand hector lands, 50% of the total products are transport to other regions of the country. Once Amon, Aus and Boro were the main type of paddy which was produced in this area but at the advent of IRRI (International Rice Research Institute) the farmers are getting benefited. In the relatively highland of this region some other crops are also produced. These include Wheat (*Triticum aestivum*), Onion (*Allium cepa*), Garlic (*Allium stivum*), Black gram (*Vigna mungo)* and vegetables. Among the fruits are as follows in table 7.

Table 5. *Fish fauna in Chalan beel.*

Local Name	Scientific Name	Local Name	Scientific Name
Mola	*Amblypharyngodon mola*	Bele	*Awaous grammepomus*
Bailla	*Awaous guamensis*	Katol	*Catla catla*
Baghair	*Bagarius bagarius*	Guratengra	*Chandramara chandramara*
Bata	*Bangana. Ariza*	Tengra	*Batasio tengana*
Tengra	*Batasio batasio*	Lombachanda	*Chanda nama*
Gozar	*Channa marulius*	Hatchetfish	*Chela cachius*
Gachua	*C. orientalis*	Chital	*Chitala chitala*
Taki	*C. punctata*	Mrigal	*Cirrhinus cirrhosus*
Shol	*C. striata*	Koi	*Clarias batrachus*
Kachki	*Corica soborna*	Muri bacha	*Eutropiichthys muriu*
Olive danio	*Danio dangila*	Rivercatfish	*E. vacha*
Banspata	*Devario devario*	Gang tengra	*Gagata cenia*
Bhut bele	*Eleotris fusca*	Catfish	*G. youssoufi*
Darkina	*Esomus danricus*	Ghor poa	*Garra annandalei*
Bele	*Glossogobius giuris*	Angra labeo	*Labeo angra*
Kani tengra	*Glyptothorax cavia*	Bata	*L. bata*
Chapila	*Gonialosa manmina*	Ghonia	*L. boggut*
Shingi	*Heteropneustes fossilis*	Kalibaus	*L. calbasu*
Ghora mach	*Labeo dyocheilus*	Baitka	*L. pangusia*
Rui	*L. rohita*	Madhu pabda	*Ompok pabda*
Chep chela	*Laubuca laubuca*	Kani pabda	*O. bimaculatus*
Gutum	*Lepidocephalichthys annandalei*	Chital	*Notopterus notopterus*
Puiya	*L. berdmorei*	Batasi	*N. atherinoides*
Tara baim	*Macrognathus aculeatus*	Nodoi	*Nandus nandus*
Baim	*Mastacembelus armatus*	Gangatic mystus	*Mystus cavasius*
Along	*Megarasbora elanga*	Kuchia	*Monopterus cuchia*
Pangas	*Pangasius pangasius*	Bhol	*Raiamas bola*
Ranga chanda	*Parambassis ranga*	Darkina	*Rasbora rasbora*
Snake eel	*Pisodonophis cancrivorus*	Khorsula	*Rhinomugil corsula*
Gang magur	*Plotosus canius*	Tit punti	*Puntius ticto*
Punti	*Puntius chola*	Bhadi punti	*P. sophore*
Kanchan punti	*P. conchonius*	Shorpunti	*P. sarana*
Rita	*Rita rita*	Chela	*Salmophasia acinaces*
Chela	*Salmophasia phulo*	Chela	*S. bacaila*
Savon khorka	*Schistura Savona*	Phasa	*Setipinna phasa*
Ayre	*Sperata aor*	Tepa	*Tetraodon cutcutia*
Potka	*Tetraodon fluviatilis*	Khailsha	*Trichogaster fasciata*
Lal kholisha	*Trichogaster lalius*	Boal	*Wallago attu*
Kakila	*Xenentodon cancila*		

Table 6. *Native Birds of Chalan beel.*

Local Name	Scientific Name	Local Name	Scientific Name
Balihas	*Nettapus coromandelianus*	Chokha Chokhi	*Tadorna ferruginea*
Manikjor	*Ciconia episcopus*	King Fisher	*Alcedo atthis*
Tiya	*Psittacuua krameri*	Kokil	*Eudynamys scolopacea*
Bokkodal	*Platalea leucorodia*	Tilagugu	*Streptopelia chinensis*
Gang Salic	*Acridotheres ginginianus*	Fishing Eagole	*Haliaetus leucoryphus*
Babui	*Ploceus philippimus*	Kana Bok	*Ardeola grayii*
Pankori	*Phalacrocorax niger*	Crane	*Grus antigone*
Dahuk	*Amaurornis phoenicurus*	Kora	*Gallicrex cinerea*
Sada Bok	*Egretta garzetta*	Choroi	*Passer domesticus, etc.*

Table 7. Available fruit gardens in Chalan beel areas.

Local Name	Scientific Name	Local Name	Scientific Name
Mango	*Mangifera indica*	Jujube	*Ziziphus zizyphus*
Elephant apple	*Dillenia indica*	Wood apple	*Aegle marmelos*
Golden apple	*Spondias cytherea,*	Black berry	*Rubus fruticosus*
Amloky	*Phyllanthus emblica*	Dalim,	*Punica granatum*
Jamrul	*Syzygium samarangense*	Papaya	*Carica papaya*
Jackfruits	*Artocarpus heterophyllus*	Lichi	*Litchi chi0nensis*
Cucumbers	*Cucumis sativus*	Banana	*Musa spp.*
Coconut	*Cocos nucifera*	Makhna	*Euryale ferox salis*
Guava	*Psidium guajava*	Saluk	*Nymphaea capensis*
Sada Sapla	*Nymphaea nouchalli*	Paddo	*Nelumbo nucifera*

At present in different regions of Raigonj, some parts of Boriagram, and west part of Gurudaspur a huge amount of Garlic (*Allium stivum*), are being grown. Lichi (*Litchi chinensis),* is also being farmed in Nazipur of Gurudaspur nowadays.

Sustainability of agriculture is now a major global concern. In the study area the Sustainability of Chalan beel and the its surrounding is serious concern in the fact of lac of proper designing and planning of land use as a result, declining yield of agricultural production, fishing for increasing population.

The availability of the agricultural and fishing land of Chalan beel and the surrounding is decreasing day by day due to increased demands of settlements, roads and other non-agricultural uses. The crop yield has to be increased by using scientific land use planning, suitable crops selection, cropping patterns, used balanced fertilizers, pesticides, and other essential management practices. With the introduction of modern production systems, farmers disregard the traditional cropping practices such as mixed cropping, crop rotation and intercropping.

Depletion of Biodiversity in Chalan beel

As is the case with other parts of the country, the dam of Farakka is inevitably reducing the inflow of water in Chalan beel. As a result the diverse animal life in Chalan beel is heel is being obstructed in their movement.

Construction of railroads and a number of highways has greatly barred the free movement of the fishes and other water animal. The reproduction process of certain fishes and water animal are being hindered due to the railroads and highways. As a result some of the species of fish like Nodoi (*Nandus nandus*), Sarpoti (*Puntius sarana*), Bacha (*Eutropiichthys muriu*), Bata (*Bangana ariza*), Pangas (*Pangasius pangasius*) and Pabda (*Ompok pabda)* are becoming extinct. Even the water animals like Ud Biral (*Lutra lutra*), Mesobagh (*Felis viverrina*) have also become non-existent.

Closure of water movement use to Farakka and its resultant negative effects on the fertility compels the farmer of this area to use chemical fertilizers on their lands to increase the production. But though the practice of fertilizing leads to a greater production of crops, it ultimately affects the biodiversity negatively. Life of the fishes and other animal's dependent on the water of this region becomes a victim to the large amount of pesticides used by the farmers in this area.

Land Degradation of Chalan beel (Map 4 & 5)

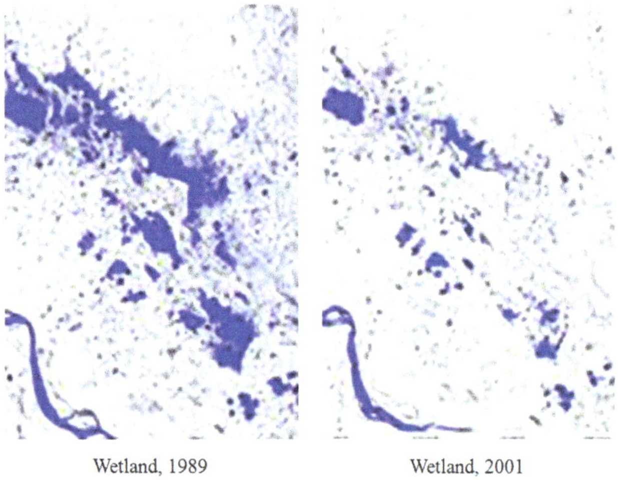

Map 4. *Chalan beel Wetland Decreasing day by day in every year.*

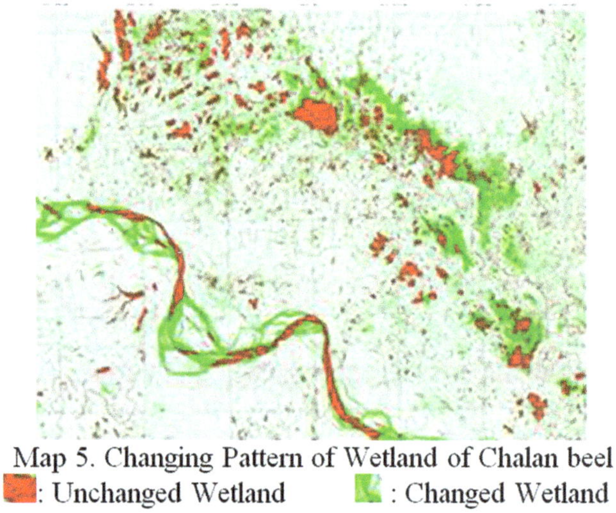

Map 5. *Changing Pattern of Wetland of Chalan beel.*

Poverty and population growth, absence of a proper land use policy, over-exploitation of natural resources are major causes for land degradation of Chalan beel area. Soil degradation in the Chalan beel area results from unplanned land management. The extent of land degradation varies according to region, season, and year due to the diverse nature of the driving forces and causes [28, 29]. Land degradation in the Chalan beel is occurring mainly due to rapid changes in demography, traditional shifting cultivation practices, construction of roads and high ways and other physical infrastructure. Land degradation in the Chalan beel is caused mainly due to over exploitation of biomass from agricultural lands, and cultivation of HYV (High Yield Varity) rice.

Chalan beel was a riched very potential source of water resources. There was large green land in Chalan beel for which huge number of wild buffalo came here from Japaiguri of India. But for the misuse of that land the green field of Chalan beel is no more and the primary shape of the historical Chalan beel also got a dramatically change. Initially the Chalan beel was extended about eight hundred square kilometers [30] but today the major part of that area is reduced which are not possible to identify as the part of Chalan beel because of establishing village, constructing roads and high ways *etc.* [29]. In 1914 the Chalan beel is divided into two by the construction of Dhaka-Rajshahi Railway and finally the Sirajganj-Bonpara high way stopped its water flow. At once different kinds of dangerous animal like Mesobagh (*Felis viverrina*), Wild buffalo (*Bubalus bubalis*), Crocodile (*Crocodylus prosus*), Snake (*Naja naja*), Elephant (*Elephas maximus*) *etc.* were found openly in the area of Chalan beel but for the indiscriminate aggression to that area those animals are not found even in search. Chalan beel contributes the rich biodiversity of Bangladesh which is a valuable heritage of our present and future generations. Its value goes beyond the local context; it has a global significance. Over the last hundreds years, the country has lost about 10 percent of its biodiversity. Population pressure, habitat loss, pollution, and poaching are putting tremendous pressures on the existing biodiversity. But unfortunately, unplanned social development and illegal capturing of the water resources Chalan beel can not contribute globally even to our local biodiversity. If the current trend continues, many of the species will be endangered or be even extinct. In case of `Chalan beel" these percentage of lost is near about 3%. The condition of wetlands is even worse than that of forests. Most of the wetlands are being converted into agriculture land and degraded due to development activities. Chalan beel was a great source of fisheries but for collecting fish from Chalan beel indiscriminately damaging this source of fisheries day by day.

Negative Impact On Chalan Beel Ecosystem

Negative Impact of Fertilizer and Pesticide Uses (Table 8 & 9)

Chemical fertilizer, pesticides, insecticides, herbicides, fungicides and rodenticides are commonly used in Bangladesh. The trends of irrigated land and use of chemical fertilizers and pesticides from 1991 to1995 in Bangladesh are presented in fig-2. In 1991, the use of nitrogenous fertilizer along accounted for about 67% of total fertilizer use, which rose to 88% in 1995. Per hectare use of urea for HYVs rice cultivation in the study area. It is found that more than 90% of the farmers use 298 to 334 kg of urea per hectare, which is much below the average use of urea for HYVs rice cultivation at national level (457 kg per hectare) [29].

Table 8. *Consumption of Fertilizer in Singra Beel from 1985 to 2005 (per hectare in quintal).*

Years	Compost		Chemical		Grand total	
	Total	% of total	Total	% of total	Total	Percent (%)
1985	14.28	65.56%	7.5	34.44%	21.78	100%
1995	6.31	17.75%	29.23	82.25%	35.54	100%
2005	2.59	3.09%	81.2	96.91%	83.79	100%

A part of land of Singra *Upazila* is being used for settlements and non-agricultural purpose, so landuse of *Upazila* center is not as in the rural areas. The main reason is enhanced of population. Agricultural activities have been developed in wetland area. Therefore, the landuse for agriculture has been developed in rural area and non-agricultural landuse has been developed in *Upazila* center. The population is increasing but not agricultural land. So the farmers are using chemical fertilizers to meet the demand of food for excess population and to take economic benefit from the land. The peasants are using chemical fertilizers and insecticides to increase production. Excess use of chemical fertilizer declines soil fertility.

The data of chemical fertilizer use in Singra *Upazila* is as below (Table 6) [28].

Table 9. *Net Fertilized of Cultivated Area (hectare) in Singra Upazila (Sub District).*

	Net cultivable land	Net fertilized area	
		Total	(%)
1985	113,312	37959.5	33.5
1995	115,807	75981	65.61
2005	120,168	110518	91.9696

Main data source: *Upazila* Agricultural Office, Singra *Upazila*, Natore and Field Survey' 2005.

There has been no significant increase of total chemical fertilizer use at national level. But, significant increase in pesticide use has been observed which has serious implications to land and ecosystem. Application of pesticides in the study area still remains much below the national average (0.35kg per hectare). The local people of the study area claimed that fallow and grazing lands were turned into the lands of rice monoculture and thereby adversely affect the production of green cattle feed. In addition, they also mentioned that, the domestic animals eating pesticides affected grasses also suffer from health problems.

With the increased production of paddy there would be concomitant increase in paddy straw. But the palatability and digestibility of the straw has declined due to excessive use of chemical pesticides. The deep-water aman cultivation was a valuable source of supply of fresh cattle fodder. Moreover, farmers spraying pesticides suffer from heart and skin diseases because of ignorance.

Extensive monopoly of HYVs rice cultivation is removing other crops needed for nutritional welfare of the people. Such monoculture is depleting some specific soil nutrients very rapidly. Deficiency of different micro and macro elements is prevalent in the farming soils. With such complicated adversities, grain yield may be reduced. About 28% of the farmers blamed on the HYVs rice monoculure for the loss of soil fertility of the area. [29] in their study claimed that up to 40% of the fertilizers used are leached to low-lying areas where weeds proliferate and an unknown amount of pesticides end up in the irrigation and drainage system. It is also encourages a proliferation of weeds and algae growth which depletes the oxygen content of the water. They also claimed that, in the water of Chalan Beel, the contention of oxygen is only 3mg/Lin the dry season and this is not enough to sustain aquatic life, which needs a minimum of 5mg/L. This may promote the formation of poisonous nitrates in the water. These conditions are responsible for the reduction in the fish population. There are frequent complaints from the Department of Fisheries about the agrochemical toxicity to the fishes. The high level of pesticides leached to the water sources has also been blamed for causing regular outbreaks of epidemic diseases in fishes and is said to have decimated the fish population in certain areas [30, 31]. There is however, no detailed and specific study on the effects of pesticides on fisheries or on wildlife.

Ecological Characteristics of Chalan Beel

Water temperature (11.5–40°C), air temperature (8–40°C), water transparency (18–55cm), pH (7.2–8.5), dissolved oxygen concentration (5.1–9.6ppm), free carbon dioxide (1.3–12.3ppm), carbonate alkalinity (8.3–15ppm) and bicarbonate alkalinity (82–117ppm) of chalan beel were recorded by [29]. Rivers, floodplains, beels and ponds constitute distinct habitat types in the Chalan beel. During inundation, however, floodplains also become open-access resources for fishing. When most of the beels and floodplains dry up, several large rivers retain deeper pools. Most owners drain their ponds (*kuas*) late in the dry season in order to harvest the remaining fish. No aquaculture in the rivers, beels and floodplain, although pond owners culture carps over the 6 to 9 month period when the grow-out conditions are optimal [29].

Illegal fishing were very common feature in Chalan Beel. Fishermen were harvesting fish without considering the natural broods tock or fry. Fishing nets were used in small meshed, capable of catching almost all types of fishes. During dry season, fishermen harvest fish from the beel by complete dewatering.

Impact of Irrigation on Chalan Beel Area (Figure 1 and Table 10)

The tremendous increase is made in the installation of shallow tube wells and Deep Tube wells for ground water in the Chalan Beel area during the last 30 years. In these areas the cropping pattern is mostly transplanted HYVs Boro/ Aus followed by the rain-fed transplanted Aman. As a result the land remains inundated in most of the seasons (about 6 to 8 months), which keeps adverse effects on soils because of continued oxygen deprivation in the sub-soils.

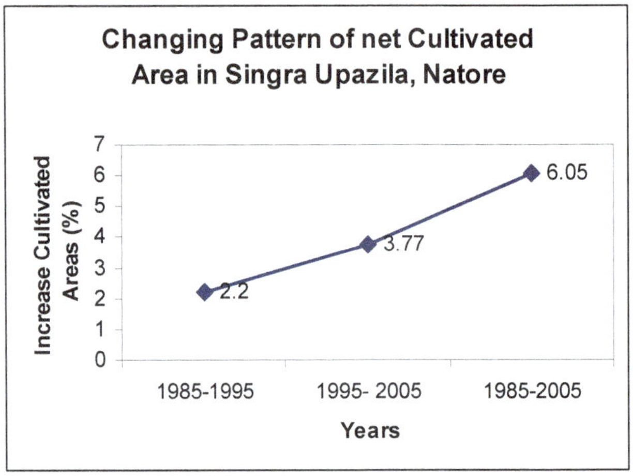

***Figure 1.** Changing Pattern of net Cultivated Area in Singra Upazila from 1985 to 2005.*

Main data Source: Agricultural Census, 1997, *Upazila* Agricultural Office, Singra *Upazila*, Natore and Field Survey' 2005.

Above figure 1 shows that the land use for cultivation in Singra has been increased to 2.2% in 1995. The cultivable land use has been increased to 3.77% from 1995 to 2005 and it has been increased to 6.05% from 1985 to 2005 (Agricultural Census, 1996 and Field survey, 2005). So the land use for agriculture is increasing for increase of population. So the density of crops is also increasing.

***Table 10.** Land Use for Different Crops in Different Years (in hector).*

Crops	2005-06	04-05	03-04	02-03	01-02	00-01	99-00	98-99
Ropa Aman	32300	32300	33100	-	32150	31000	30150	29500
Aus	3300	4150	3160	2860	3600	2860	2170	1934
Bona Aman	10100	10400	11150	9150	11000	9150	10830	10950
Mastered	1900	2000	1650	1800	2300	0	-	1550
Wheat	2250	2300	1895	1610	2000	0	-	1670
Boro	31500	32100	32000	32100	32050	31310	30930	30660

Singra Upazila Agricultural Office, Natore.

One cropland has been turned into two crops land, two crops land has been turned into three. The above table 10 shows that the maximum land is using for paddy cultivation. The excess cultivation makes land degradation. One crop is sustainable in Chalan beel area. It is not sustainable to cultivate many crops in Chalan beel area. During the rainy season there remains water in the beel area. When the beel area is remaining under floodwater then the silt and sand deposition occurs. Even floodwater causes damages to the crops. These types of land use-changing causes land degradation.

Impacts of Different Factors on Chalan Beel Fisheries

Siltation, construction of flood control embankments and roads, uncontrolled use of pesticides, insecticides and chemical fertilizers, excessive removal of surface water and extraction of groundwater for irrigation, diversion of water courses *etc* are main causes for degradation. These factors, which also contribute to already significant reductions of Chalan beel area.

Recommendations for Chalan beel Conservation

Modifying Existing Infrastructure on Chalan *Beel*

Bridges, culverts, sluice gates, flood control embankments, *etc.*, has disrupted the spawning, breeding and feeding migration of fish and other aquatic animals due to unplanned development in the beel area.

Rational use of Chemical Fertilizers, Pesticides and implement of Integrated pest management (IPM) program.

a) Maintenance of Minimum Water Depth in the beel
b) Regulation of Selective Fish Gears, Mesh Sizes, and Fishing Dewatering
c) Establishment of Natural Beel Nurseries and Fish Sanctuaries
d) Strict Application of Existing Biodiversity Rules and Regulations
e) Establishment of Community-based Organizations (CBO)
f) Awareness Building
g) Construction of Eco-friendly Raver Dam
h) Construction Water Reserver
i) Training Programme for Stakeholder
j) Cannel and Water Bodies Link to each others
k) Established Permanent Fish Sanctuary Developed
l) Established Integrated Pest Management System
m) Proper Management of Industrial Effluent of Chalan Beel Area
n) Regulation of Selected Fish Gears, Mesh Sizes and Fishing Dewatering and Sanctuary
o) Maintenance Gene Tank
p) Established Community Based Organization (CBO)
q) Formation of Paying Ecosystem Service (PES)
r) Reconstruction of Ecosystem

i. Charghat Sludge Gate
ii. Atgori Sludge Gate
iii. Mini bridge and Culvert Reconstruction
iv. Water Bodies Reconstruction
v. Reconstruction of Existing 23000 Ponds
vi. Reconstruction of Boral River and to Link Every Water Bodies of Chalan beel

This section does not attempt to provide an extensive analysis of complex balance of environment and the interaction between the population growth and the use of land resource. Part of the reason is the difficulty in covering such a wide topic and dimension of the problems, as well as the inadequacy and range of data available that can facilitate critical analysis. However, it is found that the intensification of land use and it's associated externalities and spillover have produce environmental problems to a certain level in the area under study. The major types of environmental problems are as follows:

a. Degradation of Soil Quality Due to the Inundation of Land in most of the Seasons Round the year;
b. Decline of Soil Quality due to Increasing Incidence of Unscientific HYVs rice Monoculture and Unbalanced use of Chemical Fertilizer;
c. Increased Pesticide use Particularly of Insecticides with Adverse Environmental Consequences;
d. Increased Health Problems due to Unbalance Use of Pesticides;
e. Substantial Reduction in the Supply of Cattle Fodder due to Over Harvesting;
f. Substantial Reductions in the Stock of Island Fish due to Use of Pesticides;
g. Increased Deforestation Due to Extension of Crop Production and Human Interventions;

The environmental impacts of land use in Chalan beel area may not be considered serious enough in the present context. Balanced and timely application of chemical fertilizers and pesticides will be helpful to improve the environmental quality at this stage. Application of strict regulative measures and creation of farmer's awareness should be considered as top priority in this respect. Integrated pest management and use of traditional knowledge (TKs) which are considered environmentally suitable may also reduce the use of chemical pesticides. To minimize soil nutrient depletion, monoculture of rice needs to be discouraged.

All the above measures may be effective for the present context of the area concerned. But the situation is not easy in the national concept. Bangladesh will have to ensure the supply of food for an additional 40-50 million people during the next twenty years. To meet the increased demand, it is essential that food grain output will have to expand 90 to 100% during the period. In order to meet the challenge the country may face serious environmental hazards in near future. The present knowledge about the process of land degradation because of unplanned efforts of production maximization is inadequate. The potentiality of preventing further degradation in future is also uncertain as the country suffers from lack of innovative technology friendly to environment.

Consequences of Chalan Beel Ecosystem

Chalan beel ecosystem comprise the increasing biotic and abiotic pressures.

Biotic

Uncontrolled siltation, infestation, waste water, industrial effluents, *etc.* resulting in increasing of aquatic weeds. Habitat destruction leading to loss of fish and decrease in number of animals in the beel area.

Abiotic

Anthropogenic pressures, uncontrolled dredging, hydrological intervention, pollution, chemical fertilizers are directly effect on habitat, biodiversity and aquifer in beel and there surrounding area.

3.3. Case Study: Effects of Exotic Plantation and Its Implication on Biological Diversity: A Case Study on Modhupur Forest

About Modhupur Forest (Map 6)

Among the different types of forests that exist the Modhupur Deciduous Forests (Dhaka, Tangail and Mymensingh) in Bangladesh stands out for its uniqueness. It is valuable in ecological aspects of natural forest. This tract is a large upland area in the central part of Bangladesh. This part is known as Modhupur Garh. It is a terrace from 1 to 10 metres above the adjacent floodplains areas. The total area of this forest is 4,244 sq km. The soils of the forest have developed largely on Modhupur clays. Which are nutrient poor and somewhat acidic and red or brown in colour. Average temperatures vary from 28 to 32°C in summer, falling to 20°C in winter, with extreme lows of 10°C. Rainfall ranges between 1,000mm and 1,500mm annually. This tract is well known for jackfruit gardens. Kaliakair and Savar are also known for their production of mango, Bengal Olive, Pomelo and Star Apple, Pineapples are available in Modhupur.

Exotic Plantation in Modhupur Forest and Its Effects on Biological Diversity

Exotic species are harmful to native biodiversity such as, competitors, predators, parasites, and spreading disease. Consequently, the greatest threats to biodiversity. The Department of Forest is apparently depleting the Modhupur Sal Forest through implementation of different donor-funded aforestation projects, which mainly plant exotic trees of commercial value. The majority of the donor-funded projects in the past or the projects the government and development partners plan to implement shortly are about planting alien species of wood trees or constructing infrastructures inside the national park area of the forest. They are actually encouraged the officials to strip off natural trees and plant alien species including acacia, eucalyptus, goran, gamari and rubber. The indigenous Garos people were unwilling to take part, the department encouraged the Bangla-speaking people to resettle in the forestland.

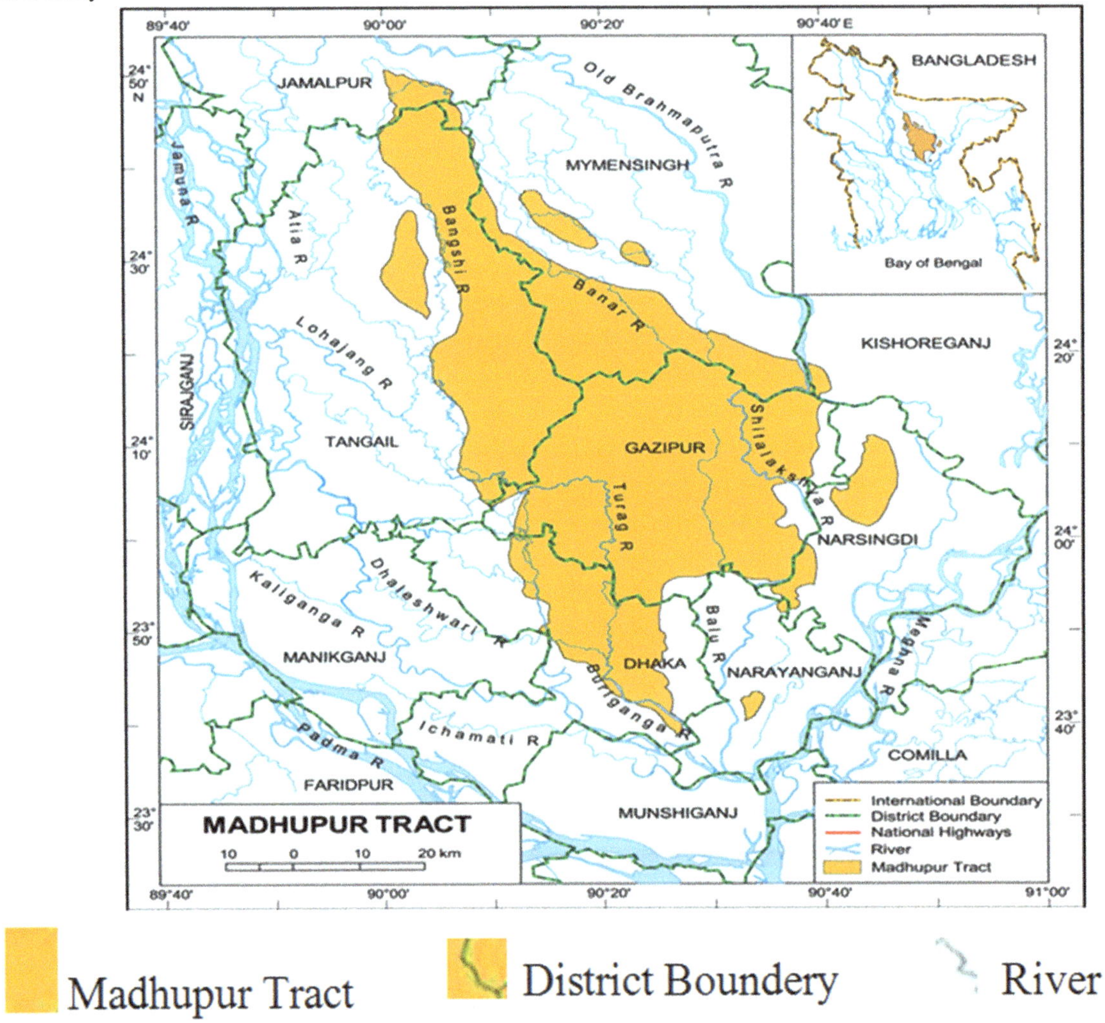

Map 6. Map of Modhupur Forest Area.

Natural Forest vs Anthropogenic Activities (Map 7)

Sal (*Shorea robusta*) was the dominant species of this track. About 600 sq km of forest remains and new woodlands planted with exotic species such as Akashi (*Acacia auriculiformis*) and eucalyptus (*Eucalyptus camaldulensis*) have transformed the ecosystem in many areas due to illegal forestation. [3] analysed a series of satellite images taken in between 1963 to 2003 and detected how the government officials cleared the forest and let others do it. They mentions in the summary that remote sensing techniques have shown quality degradation of the forest, signs of areas cleared for air force firing range and social forestry, conversion of forestland into rubber plantation, evidence of engulfing forests with settlement expansion, and indications of the influence of infrastructure. This study reveals that more than 85 percent of the Modhupur forest has been cleared in last 40 years. Whatever remains, in most of the areas the age of sal forest is not more than 10 to 15 years. Map 8 a comparison of satellite images of 1962 and 2003 clearly shows 85 percent greenery of Modhupur has disappeared in last 40 years. Now noise of vehicles and bombings and shootings of the armed forces overwhelm the chirping of birds.

Forest Map Distribution in Madhupur tract Area

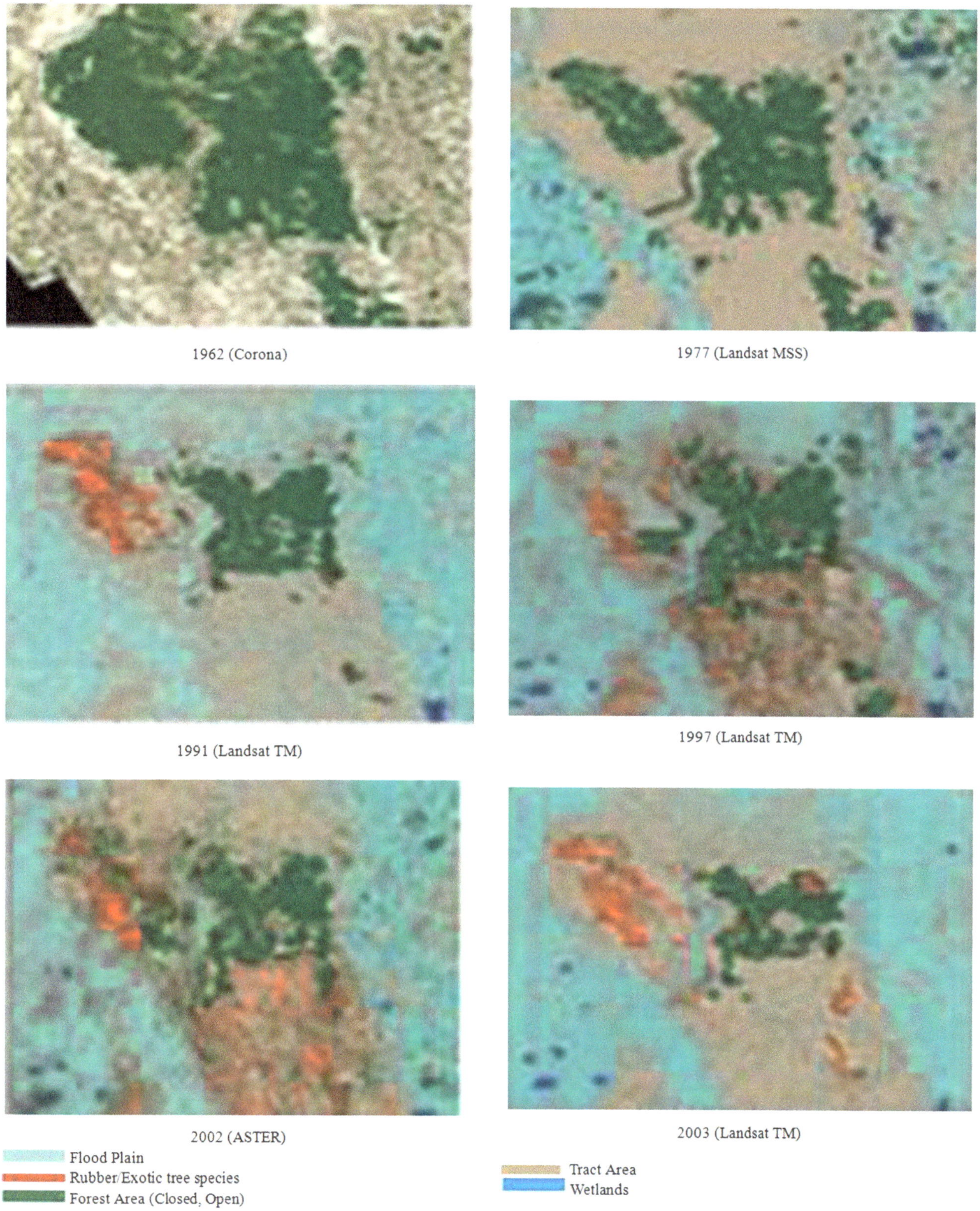

Map 7. *Forest areas are Decrising day by day for the Anthropogenic Activities.*

Illegal extraction and poaching of wildlife has seriously reduced their potential. Two centuries ago elephant and rhinoceros have been reported in these forests but became extinct in the late nineteenth century. Tigers and leopards were numerous, but tigers dwindled in the early twentieth century and extinct fifty years ago. Leopards are feared to be close to extinction. The Leopard-Cat, Fishing-Cat, Jungle Cat and small Indian Civet are still to be found. The Peacock was at one time quite plentiful but became extinct thirty years ago. Tribal people, Koch and Mandi (garo) are dependent on the Modhupur Tract. In the past thirty years the landscape and ecosystems of most of the Tract has changed drastically. Bengalis has developed stable agro-horticultural systems from Savar to Kaliakair. Urbanisation is extended dramatically in this Tract. Bangladesh Agriculture Research Institute (BARI), Bangladesh Rice Research Institute (BRRI), Bangladesh National University, Bangladesh Open University, Bangladesh Citrus Research Institute, Bangabandhu Agricultural University, *etc.* are situated in Gizipur. Moreover the government has set up an export processing zone at Ashulia in Savar upazila which is continually expanding.

However illegal extraction and poaching of wildlife has seriously reduced their potential. Two centuries ago elephant and rhinoceros have been reported in these forests but became extinct in the late nineteenth century. Tigers and leopards were numerous, but tigers dwindled in the early twentieth century and extinct fifty years ago. Leopards are feared to be close to extinction. The Leopard-Cat, Fishing-Cat, Jungle Cat and small Indian Civet are still to be found. The Peacock was at one time quite plentiful but became extinct thirty years ago.

Two tribal people, the Koch and the Mandi (garo), who are dependent on these forests, live in the Modhupur Tract. The Kochs are among the earliest people of Bangladesh, while the Mande have their main centre of dispersal in the Garo Hills in India.

Findings in Field Visit to Modhupur Forest

On October 23, 2010 the team comprising the members of assignment submitter visited Modhupur forest. The visit was one day stay in the forest (areas *Jalachatra*, telki, Rasulpur) with discussion with local people including people of Garo community, rickshaw/van pullers, and shopkeepers. The findings of discussion with Garo community as perceived by the team are as follow:

Garo Community

They are Dissatisfied with Government Projects of Exotic Plantation.

People of Garo Community are transforming their Profession from Forest based Gatherers to Common Professions like Rickshaw Pulling, Shop Keeping, *etc.*

Many people of Garo Community have been Migrated to Different Places of Country for the sake of Livelihood Sustenance.

There is no Properly Documented Land Record System and Land Ownership is based on *de facto* Land Ownership System.

Repression against them for Protesting Projects like Eco Park (killing of Choles Ritchil,) created grievance to them.

There is working a Religious Organization which Convincing Non-Muslim Garo People to be Convert to Muslim with Material Benefit. But one Religious Leader of the Organization termed it Resettlement plan for socially displaced people for religion convert.

The findings of discussion with Bengali community people (Rickshaw/van puller) as perceived by the team are as follow:

Bengali Community

1. Economic pull factor working for coming Bengali people from surrounding area of Modhupur forest.
2. There are presences of *forest thieves* who are felling trees for timbers.
3. There is growing groups of pirates snatching valuables from tourists.
4. The forest is eroding day by day.
5. There is no major conflict between Garo and Bengali people.

Overall Observation

1. Government agriculture department is working to develop man made rubber plantation in order for economic development. Biological Diversity has been ignored even by the very organization who is supposed to protect and preserve the natural forest.
2. There are depleting image of the forest.
3. There is less presence of wild animals in the forest.
4. There is a Bangladesh Air Force firing Range at Rasulpur inside the forest which is creating threats to wild animals as there are heavy sounds of firing in firing sessions which frightens wild animal and disturbs their reproductive activities.
5. The existence of the forest as there is sight of depletion in every part of the forest.

Biodiversity Depletion in Modhupur Forest in Bangladesh and Their Consequences on Ecosystem Services

Exotic species are damaging to biodiversity of the area and transform the indigenous ecosystem into drought through their increased water uptake characteristics. Moreover, climate change might increase the temperature in this region which infect alters the ecological processes in this tracts. Increased evapo-transpiration causing moisture stress (water deficiency) in winter is affecting the survival of sal forest ecosystem.

3.4. Case Study: Chokoria Mangrove Forest (Map 8)

Chokoria Sundarban in a mangrove patch is situated in the southern part of the country. Once it was a habitat of diverse plants and animals but it is completely replaced by brackish shrimp cultivation. The forest was home of tigers, deer's, wild cats, wild boars, foxes *etc.* These all animals are now rarely seen in the region. The forest was unique with 20 species of trees, some as high as 15 meters. Sundari, Geoa, Keora, Bain were some of the major species. Today the lush green vegetation is gone and also gone is the unique, diverse life its environment sustained. In the coastal belt the mangrove forest serves as a shield against cyclones and that tidal surge and gives much-needed protection to the people who live close to the Bay of Bengal. The Chokoria Sundarban provided protection to the people who live in the locality. Today no trees in the Chokoria Sundarbans and no such safety bulb for a few thousand people who live close to the forest that was. Today there is no such forest is seen in that area. The Chokoria Sundarban enriched the local environment with the diverse animal and plant life nurtured. Chokoria Mangrove Forest must be damages within a short period.

This disappearance of Chakoria Sundarban first started in 1929, when British Governments leased out 3910 acres of forests land to 262 local landless families by forming an association of settlers for cooperative cultivation for their subsistence income. That was the beginning of forest clearing activities by human beings (Bangladesh Environment Facing the 21st Century Edited by Philip Gain). Gias Uddin Chawdhuri, a promonent local of Chokoria Sundarban is leased 563 acres land of Chakoria Sundarban in 1977 with a view to establishing Shrimp farm, duck farm and agro fisheries (Bangladesher Biponno Bon by Philip Gain). The major assault on the area came in 1980 when so called development organization has come to develop shrimp firm clearing the forest. The statistics from the Department of Environment reveals that Asian Development Bank took major portion of the forest (around 7000 hectors) land for shrimp cultivation and then it is occupied by the ministry of Land only to perpetuate the development of shrimp firming in the area. The Asian Development Bank begans to fund shrimp cultivation in 1982. Its loan helped set up over a hundred shirmo firms (each eleven acres in size) and build a 16 kilometers embankment. Then came US$ 26.5 million dollar shrimp culture project funded by the World Bank. The project included a grant from the United Nations Development Program (UNDP). The World Bank's loan [Under Structural Adjustment Program] was used to develop 468 plots (each ten acres in size) and infrastructure (Bangladesh Environment Facing the 21st Century Edited by [10]. By this impractical and unscientific investment of the many authorized organization the disappearance of a most potential forest of Bangladesh is started.

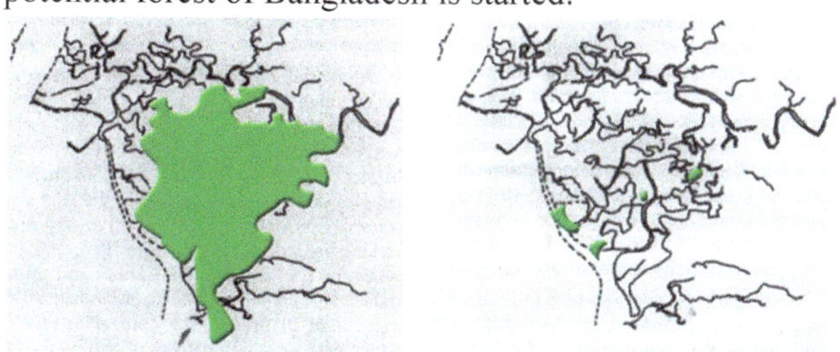

1976 Chokoria Sundarban-19,617 acres

1995 Chokoria Sundarban-466 acres

Forest Area

***Map 8.** Chokoria Sundarban Forest (1976-1995).*

Total disappearance of the 21,000 acres Chokoria Mangroves in Cox's Bazar, largely due to Shrimp cultivation, is an example of how shrimp cultivation has caused unprecedented harm to the unique mangrove ecosystem. The unique mangrove patch in the southeast of Bangladesh has lost all but its name today. Thousands prawn farms have replaced almost the entire forest.

Biodiversity deplation on Chokoria Sunderbans Forest and its concequences on ecosystem services:

The Chokoria Sundarbans today- the evergreen forests have been replaced by thousands of shrimp ponds. An ecocide indeed!

Impact on the environment due to the disappearance of natural forest is alarming. Change like Soil turned into salty, human health at risk, animal husbandry appeared tough in want of grazing land, Sea water in absence of forestry often flooded the adjacent area, has abolished the natural ecosystem. Furthermore, natural protection from disaster like SIDR by the forest has been demolished due to the disappearance of the chokoria forest. An attempt has taken to reverse the process by the World Bank but did not stop the degradation. Thus the emergence of development effort has jeopardized the normal dynamics of environment.

3.5. Case Study: Hilsa Fish (Hilsa ilisa)

There are many rivers throughout the Bangladesh. Some of them are very potential source of Hilsa Fish. Hilsa is a national fish of Bangladesh and it is very popular over the world. Hilsa gives us both protein and foreign currency. Twelve percent of total fish production of Bangladesh comes from Hilsa. Bangladesh contributes 50-60 percent of world's Hilsa production. Almost 20-25 lac peoples are engaged with the Hilsa production directly or indirectly (Jatio Motso Saptah Sonkolon-2010). It is also known as Hilsa Shad fish. Hilsa or Hilsa Shad fish is now a most cherished treasure of Bangladesh.

May to October is generally considered the hilsa catching season in Bangladesh. But unfortunately the production of Hilsa fish is decreasing day by day for violating the laws imposed with a view to ensuring high production. As a result, dispute and frustration are arising among fishermen and traders in the coastal region.

The prime reason of this crisis is not to have proper implementation of the relevant law. Many fishermen were using current net and catching Jhatca illegally but law enforcement authority did not play their responsibility properly.

According to my visiting experience of Meghnaghat (Chadpur), I went to Meghnaghat (Chadpur) in Mach 2010. Though it was not a seasonal period of Hilsa fish nevertheless a largest amount of hilsa is caught by the fishermen.

However, it is happy news for us that, If the law enforcement authority and other empowered authority are been able to implement the laws related hilsa as it should be done then the production of hilsa will be increased. From the table given above, it is clear to us that after coming the government to the power and implementing the laws properly the production of Hilsa are continuously increased.

3.6. Case Study: Wildlife Consumption and Trade

Different type of people in Bangladesh used to consume some amount of wildlife species or their products because food, fuel, tibe, fertilizer comes from biodiversity or wildlife species.

Different type of people in Bangladesh used to consume some amount of wildlife species or their products because food, fuel, tibe, fertilizer comes from biodiversity or wildlife species.

Frogs:

The frog legs, live turtles and other biological resources are legally for exported from Bangladesh due to inappropriate policies and guidelines. Most of the frogs used to be harvested from the wild and exported as a frozen food item. As a consequence, the population of some wildlife species has become commercially threatened. To manage this situation the government has forced a ban on trade in wildlife. To control this situation the government has imposed a ban on trade in wildlife without addressing the issue for a long term sustainable management. The frog legs and turtle meats have been allegedly traded under the camouflage of frozen fish/foods even after frequent ban on commercial trade in wildlife.

Frogs have a high demand in the international market. This led the Bangladeshi traders to export frog legs in large quantity until the early eighties. Most of the frogs used to be harvested from the wild and exported as a frozen food item.

Consequences on Ecosystem Services:

Indiscriminately harvest of frogs-main predators in crop fields resulted in enormous use of insecticides in the agriculture fields to control the massive attack of the pests. Uncontrolled harvest of frogs from the wild adversely impacted not only the food-chain, but also had an enormous impact on the population of predators such as monitor lizards, snakes and birds of prey. The environmental activities from home and abroad strongly demanded a ban on frogs collected from the nature. Under pressure of the wildlife scientists and environmental pressure groups, the Bangladesh Government proclaimed some laws to control the trade in frog legs and imposed a ban on the export of frogs. Importing countries of the frog legs from Bangladesh are USA, United Kingdom, Hong Kong, Netherlands, Canada, Belgium, Korea, Bulgaria *etc.*

Turtles and Tortoises (Geoclenys hamitonil, Melanochelys trijuga)

Turtles, tortoises and terrapins are locally known as Kachim/Kachamp/Kaitta (*Batagur baska*) Bangladesh supports 25 species of Chelonia of which 18 species are freshwater turtles, five are marine turtles *Cuora amboinensis, Geocle,nys hamitonil, Hardella thurjii, Kachuga dhongoka, Kachuga kachuga* and two are land tortoises *Melanochelys trijuga, Melanochelys tricarinata*. Besides their natural ecological role, the Chelonia population in Bangladesh provides meat protein to a percentage of people, especially the ethnic communities. Turtle has been one of the major non-traditional export earners or the country for a long time. Other than use as table food, different byproducts extracted from turtle fats, carapace, plastron and eggs are widely used for treatment, making handicrafts *etc*. Bostami turtle named after a shrine, Bayezid Bostami in Chittagong is the only endemic wildlife of Bangladesh out of over 1000 species of wildlife so far recorded. This giant turtle (*Aspideretes nigricans*) is found only in a lone pond of the shrine in Chittagong and nowhere else in the world.

Turtles are considered as religious symbols. According to Hindu mythology, the universe is supported by four elephants standing on a turtle's back. People show special reverence to the turtle as they consider the turtle as one of the ten incarnations of `Vishnu' the supreme God. River Jamuna, mythological used turtle as her vahan (carriage).

The soft-shell turtles are omnivorous and they are scavengers to the dead bodies of other animals. A huge quantity of turtles is exported from Bangladesh. Turtles are exported mainly to Singapore, Hong-Kong and China from Bangladesh. All the specimens exported are collected from nature. There is no farm for procreating turtles in captivity. Harvesting of turtles from the wild has drastically reduced their population and established their habitats. To save the turtles from going extinct in Bangladesh, the present unsustainable harvest must be stopped and must implementation of biodiversity laws.

Consequences on Ecosystem Services:

Turtles and Tortoises are an important role in the ecosystem of their surroundings as both predator and prey. Mostly they are carnivorous. As an integral component of the food chain they often fulfill the role of 'recycler.' They are also scavengers, feeding on dead and decaying matter. As a result Turtles and Tortoises remove much of the debris from the ecosystem, thereby ensuring that the stored energy can be reused. Sea turtles play an important role in ocean ecosystems by maintaining healthy sea grass beds and coral reefs, providing key habitat for other marine life, helping to balance marine food webs and facilitating nutrient cycling from water to land.

Molluscs in Bangladesh

Subtropical monsoon climate prevails throughout the year in Bangladesh, with the hilly region in the North-Eastern and South-

Eastern part of the country possesses low, flat and fertile flood plains. Ample rainfall and fertile flood plains with rich vegetation as well as the natural rivers, streams, swamps and artificial lakes and water reservoirs offer good habitats for the land and freshwater molluscs. Freshwater gastropods occupy important trophic levels in the aquatic ecosystems. Carnivorous freshwater fishes consume some of the snails. Snails also supplement animal protein for prawn farms, ducks and people of low earning groups, especially the tribal people. On the other hand, certain freshwater snails are the registered intermediate hosts of digenetic trematodes, which cause diseases of man and animals are burning problems throughout the world. People belonging to different tribes in

Rajshahi, Naogaon, Dinajpur, Chittagong, Rangamati viz, Chakma, Pahan, Orao, Buna, Coch, Garo, Hazong, Khasia, Tripura, Marma, Murang, Pahari, Saontal, Rakhain, Buna are used to consume the different type of snails such as *Pila globosa, Bellamya, bengalensis, Lamellidens marginalis etc.* regularly to meet up their animal protein requirments. They are used to prepare their curries, with the flesh and viscera of snails in various forms such as ghati, soup, wetfried, fried [33]. A total of 6 species of freshwater mussels under the family Unionidae, order- Unionoida and a slug were collected, which are new records for Bangladesh [34, 35].

At the same time flesh of snails is being expensively utilized in freshwater prawn (*Macrodrachium rossenbergii*) and the brekish water prawn (*Peaneus monodon)* culture in south western part and of their region of the country. In addition to that domestic duck poultry, and diffirents cat fishes are also feed with flesh of the snails species. As a result Madaripur, Gopalgonj, Sariatpur, Faridpur, Rajbari, Magura, Norail, Jhenaidoha, Kustia, Chuadanga, Meherpur, Jessore and Bagerhat the significant wetland on south west Bangladesh are under threat of over exploitation area of many kind of snail.

The natural snail populations in the wetland areas of the country are the exclusive source to meet up such high demands. As a result over exploitation without any scientific management of the snail population is causing the bioderversity depletion. A lot of evidences to indicated that the population of the species, *P. globosa* is decreasing day by day from Bangladesh. Consequently, a study of induced breeding of *P. globosa* was found to be essential in the sense that it may provide information and exploitation of this freshwater resource. In spite of being densely populated Bangladesh possesses vast suitable habitats for freshwater edible molluscs which can supplement the animal protein for the people especially for the lower income group and various types of aquaculture such as prawn farming and cat fish farming.

Consequences on Ecosystem Services:

The snail is an important component of the biodiversity playing vary vital role for the maintenance of the aquatic ecosystem in one hand and on the other hand recently the animal is being utilized as the supplementary feed for prawn, cat fish and poultry farming. It is being eaten by numerous predators, fish, birds and mammals such as various tribal peoples in Bangladesh.

Earthworms: So far knowledge goes till now there is in total 35 species of earthworms have been described by [37] from Bangladesh. Earthworms are widely distributed throughout the world and their population contributes 80% of the total biomass of the soil ecosystem. They play role to neutralize the soil. The food they ingest, pass through the gut with calcium carbonate particles coated mucus substance by calciferous glands and possibly influence pH control [38]. *Drawida limella, Drawid anepalensis, Glyphidrilus tuberosus, Amynthas alexandri, Lampito mauritii, Metaphire houlleti, Metaphire posthuma, Perionyx excavatus, Perionyx horai, Perionyx modestus, Perionyx simlaensis, Dichogaster modiglianii, Dichogaster saliens, Eutyphoeus gigas, Eutyphoeus orientalis* were recorded by [38] in tangail district in Bangladesh.

Gharial *(Gavialis ganjeticus)* and Crocodile *(Crocodylus prosus)*

Once this animals were found in the padma, jamuna tista, Mohnanda including Rajshahi area. But now a day it has been very rare. Experts opine that there are only 20-22 pairs of Gharial still exist in our country. But most disappointing news is, even after searching for years all throughout the country, their existence has not been located. Moreover, Crocodile in southern region is not seen as much as it was before.

Why the number is decreasing day by day is sorted out as: i. Smuggle to abroad since its flesh and leather is widely demanded and hence earn a lot of money, ii. Loss of Habitat, iii. Water Pollution, iv. Cleavage of Natural Food Chain, v. Obstacles to normal Reproduction.

Consequences on Ecosystem Services:

Gharial (*Gavialis ganjeticus*) and Crocodile (*Crocodylus prosus*) had played a vital role in a river's ecosystem and their presence is important to the environment. They are important components of the food webs in most aquatic ecosystem. They are also filling a critical role both as predator and prey species.

Royal Bengal Tiger (Panthera tigris)

Royal Bengal Tiger is the national animal of Bangladesh. It selects her accommodation considering with mainly food, shelter, risk less, mutual activities facilities *etc.* Mainly it lives in hilly forest, ever green forest, free forest and mangrove forest through the country. But now, it is not found without sundarbon at anywhere in Bangladesh. It's alive in sundarban with very crisis. Its found also except Bangladesh, in India, Bhutan, Nepal, Myanmar, Indonesia, Malaysia, China and Russia. According to the red list of IUCN, there are 2500 tiger in the world. In Bangladesh, there are 200 but in 2004, in a survey completed by Bangladesh government, the number of tigers is 440 by the conversation with the bearable villager of the sundarbon, the information is given by them, the number tiger is about 200-250. The number of the tiger will be reduce continuously, if we do not take mutual and effective step, to continue and increase also this number of tiger in sundarbon. The causes of reduction of tigers as under follows;

i. weather changing
ii. to increase natural disaster
iii. minimize sweet and maximize salty water
iv. setup new accommodation,
v. Increasing Blakers

Consequences on Ecosystem Services:

Royal Bengal Tiger (*Panthera tigris*) live in the sunderbans forest ecosystems. They are the top predators in their ecosystems. This means that while they have many prey species, they have no naturally occurring predators.

Top predators control the populations of their prey animals. By feeding on wild deer, for example, tigers keep the deer population within certain boundaries. If the population of deer were to exceed these boundaries, overgrazing of vegetation would likely occur.

Mongoose (*Herpestes edwards*)

These mammals live throughout Bangladesh. It likes to reside beside the paddy fields. It hides in the hole of the bank of the river. There are three kinds of Mongoose But those found more often are smaller in size. These animals preserve the balance of environment and protect crops. Nevertheless, these animals are decreasing gradually. On the other side, it fall prey to the people of the neighboring locality while it tries to catch its prey such as hen, duck, *etc.*

Killing this animal is increasing at a massive scale. If this continues, this will lead to extinction by short time. Massive killing is committed in the tribal areas in Chittagong, Rajshahi, and Moymensingh.

Eventually it is found after study that the key animals of Bangladesh like Gharial and Crocodile, Kuchia, Snails, Monitor Lizards, Turtles, Tortoises, Frogs, Snails, Monitor Lizards, Turtles, Tortoises, Royal Bengal Tiger, Mongoose *etc.* which play vital role in our environment are decreasing day by day for lack of appropriate residence, security and necessary food of them. Though there are few laws to protect this massive decreasing of animals but not properly enacted and effective.

Consequences on Ecosystem Services:

Mongooses have a large impact on their prey population, including snakes and rodents. They are important in eliminating rats, mice and especially snakes.

Chronological Reduction of Biodiversity in Bangladesh

From above mentioned case studies it is understood that there are many reasons behind the reduction to endanger, critical endanger, extinct of biodiversity in Bangladesh. The main reasons of them are lack of enforcing laws properly, damaging accommodation, over exploitation, and uncontrolled negative impact in the field of biodiversity throughout Bangladesh.

In water sources, chalan beel displays the causes which are so diverse as to range from the over fishing in the area to the innocuous, protective regulation, unable to deter the people.

Moreover, human beings are dependent on biodiversity for different fundamental demand such as food, cloth, fuel, treatment *etc.* If the existence of biodiversity is been endangered then the existence of human being will be destroyed. So the protection of biodiversity is highly essential for the interest of human living sound. After that human beings are hunting many useful animals from the natural sources for their unconsciousness. Be short of proper management and sustainable development, natural plants and animals are being lost day by day.

Overall Recommendations for Biodiversity Conservation

The major problem was the sluggish realization of the UN Convention on Biological Diversity (CBD). There is still no adequate data basis for recording the effects of use and over-exploitation on biodiversity. Research can make an important contribution here – in monitoring activities and in developing integrated approaches to biodiversity use. Biodiversity can be managed in a sustainable way only if the management plans take proper account of the socio-economic context. It may include a requirement to develop jobs based on the sustainable exploitation of biodiversity.

The Following Suggestions can be made which can Provide Sustainable Biodiversity

i. Biodiversity Rules and Regulations must be Updated and Enforced Strictly. Must be planned to Formulate Appropriate Biodiversity Law and its Competent Implementation.
ii. For sustainable management must be implement traditional knowledge.
iii. The Ministry of Environment and Forestry should have a separate Wildlife Cell or Department.
iv. Human activities often conflict with the conservation of biodiversity.
v. Illegal Capture of Wild Fauna and Aquatic Endangered or Threatened Species must be Stopped and Over Harvesting Stopped.
vi. At least some Major Forest Area, Wetlands and portion of River should be marked as Sanctuary for Bio-diversity Conservation under the Direct Supervision of Government.
vii. Unsustainable development and consumption lead to biodiversity loss and ecosystem damage
viii. Water Pollution by any means should be Minimized and Existing Rules and Regulations on Water Pollution should be Effectively Enforced. Effluent Treatment Plant has to set up to the Outlet of the every Industry.
ix. Use of Pesticide in the Agricultural Fields should be reduced and Biological Control of Pest should be introduced.
x. The Height of Chimney of the Brickfield must by beyond 120ft with a Scrubber in its Outlet. Brickfield must be Established Maintaining a Reasonable Distance and Position.
xi. Cutting Hill in the name of Development and Burning the Trees to cultivate *joom* is to be prohibited.
xii. Mass Media like Radio, Television, Leaflets and Announcement *etc.* must be used in Wider Extent for Awareness Building among the Common people about the Consequence of Bio-diversity and its Conservation.
xiii. Appropriate Research and Non-governmental Organizations (NGOs) can play an important role for biodiversity conservation.

4. Conclusion

The real effect of biodiversity depletion on ecosystem services is defined by the interactions between genetic diversity and species richness and identity effects in combination with the influence of the environmental context. For future research on the subject of biodiversity- ecosystem functioning, better understanding of the composition of natural food webs and the traits and strength of the interactions among species is needed, as well as empirical evidence of how biodiversity loss drivers affect ecosystem function. As long as we ignore the real impacts of human activities on ecosystem services, conservative management policies should be applied in order to minimize the effects of these activities on the remaining biodiversity. Biological resources and biological diversity form the basis of both the ecology and economy of Bangladesh. The country's agriculture, fisheries, and livestock, along with a number of other sectors are heavily depended, directly or indirectly, on biological resources. There is a great potential in Bangladesh for biodiversity-based sustainable development. In spite of threatened wild fauna and flora, there are nearly10,000 species of plants, animals, and microbial organisms - a good percentage of which are found still in abundance. All out efforts are needed to be taken. so that these biological resources are not over-exploited and sustainable management plans need to be formulated and implemented.

References

[1] Nishat A, Huq SM, Barua ISP, Reza AHMA, Khan ASM. Bio-ecological Zones of Bangladesh. The World Conservation Union (IUCN), Dhaka, Bangladesh. 2002, 141pp.

[2] Hossain MK. Overview of the forest biodiversity in Bangladesh. In: Assessment, conservation and sustainable use of forest biodiversity (CBD Technical Series no. 3). Secretariat of the Convention on Biological Diversity, Montreal, 2001, 33-35pp.

[3] Islam MM, Amin ASMR, Sarker SK. National Report on Alien Invasive Species of Bangladesh. In: Pallewatta, N., Reaser, J.K. and Gutierrez, A. T. (Eds) Invasive Alien Species in South-Southeast Asia: National Reports & Directory of Resources. Global Invasive Species Programme, Cape Town, South Africa, 2003, 7-24pp.

[4] IUCN. Red List of Threatened Animals of Bangladesh. The World Conservation Union (IUCN), Dhaka, Bangladesh. 2000, 54 pp.

[5] Rahman MM. Forest Resources of Bangladesh with Reference to conservation of biodiversity and wildlife in particular for poverty alleviation. In: Sim, H. C. Appanah, S. and Youn, Y. C. (Eds), Forests for poverty Reduction: opportunities with clean development mechanism, environmental services and biodiversity. Proceedings of the workshop held at Seoul, Korea in 27–29 august 2003. FAO Regional Office for Asia and the Pacific (FAO-RAP), Bangkok. 2004, 139-148pp.

[6] Khan MS, Rahman MM, Ali MA (Eds). Red Data Book of Vascular Plants of Bangladesh. Bangladesh National Herbarium, Dhaka. 2001, 179 pp.

[7] Mukul SA. Biodiversity Conservation Strategies in Bangladesh: the state of protected areas. 2007; 34(3): 28-32.

[8] Bashar MA. Instant Basics of Environment wide approach to nature lover. Positron publication, 38/2/ka, Banglabazar, Dhaka-1000, 2004, 176pp.

[9] Primack BR. Essentials of Conservation Biology Sinauer associate inc. Publisher, sunderland, massachusetts USA 1993, 22pp.

[10] Gain P. Bangladesher Biponna Bon (The Endangered Forest of Bangladesh). Society for Environment and Human Development (SEHD). Lalmatia Dhaka, Bangladesh, 1997, 276pp.

[11] Wilson EO. Biodiversity, National Academy Press, Washington DC 1990, 521pp.

[12] Asmat GMS. Bangladesher Bilupta Bannyaprani (Extinct Wildlife of Bangladesh). Published by Bangla Academy, Dhaka, 2001, 184pp.

[13] Aziz MA, Sultana, S., Begum, S. Bonnyaprani O. Tar Babosthapona (Wildlife and its Management), Kabir publication House, 38/3, Bangla Bazar, Dhaka-1100, 2010, 272pp.

[14] Khan MAR. Bangladesher Bannyaprani. Published by-Bangla Academy, Dhaka 1996, 148pp.

[15] Paul KN., Sarker MAM. Biodiversity. Eureka Book Agency, New Market, Rajshahi 2001, 244pp.

[16] Karim MA. An Introduction to Environmental Law. Isamoto Prokashoni, 38 Banglabazar, Dhaka-1000, 2004., 79p

[17] Pocock RI. The Fauna of British India, Including Ceylon of Burma. 2nd. edn. Memmals.Vols. 1 & 2 Taylor of Francis. London 1939, 1941pp.

[18] Husain KZ. An Introduction to the Wildlife of Bangladesh, Dhaka. 1974.

[19] Prater SH. The book of Indian Mammals. 3rd. Revised Reprint edn. Bombay, Nat. Hist. Soc. India. 1980.

[20] Khan MAR. Mammals of Bangladesh: A Field Guide. Nazma Reza, Dhaka Bangladesh 1985, 92p.

[21] Khan MAR. Wildlife of Bangladesh: A Check list Dhaka University. Bangladesh, 1982, 173p.

[22] Gazi ASM. Bangladesher Bilupto Bannyaprani (Extinct Wildlife of Bangladesh). Bangla Academy Dhaka. 2001, 184pp.

[23] Gain P. Bangladesh Environment, Facing the 21st Century, Dhaka: Society for Environment and Human Development (SHED), 4/4/1 (B), Lalmatia, Dhaka-1207, Bangladesh 2005, 340pp.

[24] Rahman MR., Asaduzzaman M. Ecology of Sundarban, Bangladesh. J. Sci. Foundation 2010; 8(1&2): 35-47.

[25] Mannan MA. Environmental Impact Assessment of Development Activities In The Sundarbans. Ph.D. Thesis, University of Rajshahi, Rajshahi, Bangladesh 2012. 423pp.

[26] Kostori FA, Parween S, Islam MN. Availability of Small Indigenous Species (SIS) of Fish in the Chalan Beel: the largest wetland of Bangladesh. Univ. j. zool. Rajshahi Univ. 2011, 30: 67-72.

[27] Samad MA, Galib SM, Flowra FA. Fish Drying in Chalan Beel Areas, Bangladesh. J. Sci. Ind. Res. 2009; 44(4): 461-466.

[28] Rahman MR. Wetland Environment of Chalan beel: A Case Study on Singra Beel. J. Environ. Sci. & Natural Resources 2009; 2(1) 89-95.

[29] Rahman MR. The Plan for the Chalan Beel Restoration. As a key note paper, Organized by Bangladesh Environmental Lawyer Association (BELA) 21st October 2012, 25pp.

[30] Hamid AM. Chalon Beeler Iti Katha. Magnum Opus, 12, Aziz Supper Market, Shahbag Dhaka, 1967, pp124-143.

[31] Hamid MA. Chalanbiler Eti Katha (A book based on the History of Chalan beel Territory). Published by- Aziz Super Market, Shahbag, Dhaka-1000, Bangladesh 2002, 342pp.

[32] Webster M. Webster`s ninth new collegiate dictionary. Meriam - Webster Inc. 1985.

[33] Bhuiyan MMS, Jahan MS, Rahman MR, Haque M. Freshwater Molluscs in the Northeastern Region of Bangladesh. Proceedings of the International Conference on Environmental Technology and Construction Engineering Sustainable Development, ICETCESD-2012, SUST, Sylhet, Bangladesh, pp510-521.

[34] Bhuiyan MMS, Rahman MR, Jahan MS, Haque M. New Record of Some Freshwater Mussels (Bivalvia: Uniodae) From North-East Bangladesh. Bangladesh J. Zool. 2013, 41(1): 29-42.

[35] Bhuiyan MMS, Rahman MR, Jahan MS, Haque M. Ecology of Slug, (*Girasia crocea*) (Godwin-Austen) (Stylommatophora : Ariophantidae). J. Environ. Sci. & Natural Resources, 2013, 6(1): 227-232.

[36] Rahman MR. Environmental Impact Assessment of Sundarbans in Bangladesh. International Journal of Environmental Monitoring and Analysis. doi: 10.11648/j.ijema.20140203.17, 2014, 2(3): 175-184.

[37] Pramanik MN. Annelida. In: Encyclopedia of Flora and Fauna of Bangladesh. Vol. 16 Annelida, Asiatic Society of Bangladesh, Dhaka, 2009, 97-106.

[38] Bahar I, Jahan MS, Rahman MR. Distribution of Earthworms at Different Habitats in Tangail, Bangladesh and Significantly Impacts on Soil pH, Organic Carbonand Nitrogen. American Journal of Life Sciences. Vol. 3, No. 3, 2015, pp. 238-246. doi:10.11648/j.ajls.20150303.26.

4

Research of Soil Resources Degradation Processes in Georgia

George Kordzakhia, Lali Shavliashvili, Gulchina Kuchava, Nugzar Buachidze

The Institute of Hydrometeorology, Georgian Technical University, Tbilisi, Georgia

Email address:

giakordzakhia@gmail.com (G. Kordzakhia), shavliashvililali@yahoo.com (L. Shavliashvili), gkuchava08@gmail.com (G. Kuchava), emc.buachidze@yahoo.com (N. Buachidze)

Abstract: Proper exploitation of soil resourses is one of the significant problem of the adverse impact of climate change. Soil formation is a long process, while its degradation develops much faster. Soil degradation means the process of deterioration of soil resources fertility, as a result of both natural and anthropogenic impact. Significant impact on soil resources take place in East and West Georgia. The results of the present study show that the main reasons of soil resources degradation in Georgia are: salinization, erosion processes and natural disasters impact. In the work the following issues are overviewed, namely: i. The climate components regime and their connection with modern ecological processes. In particular, the temperature and precipitation trends and their possible impact on chemical data are researched. The regression relation between global warming intensity and soil chemical data are revealed. ii. In selected degraded soils total amount of easily soluble salts, the main ions composition for the determination of saline soil's quality, humus and soluble forms of nutrient elements aiming soil fertility definition. iii. The mechanical analysis of the soil referring to research soil physical properties and salt migration in the ground. Natural disasters impact on the soil degradation is considered. Research results defined the territories and types of soils degradation in Georgia. The recommendations to improve the soil fertility are created. The received results could promote the socio-economic development of the districts/ regions in Georgia and introduce considerable material and economic effects, assist the sustainable management of soil resources and alleviation of poverty.

Keywords: Soil Degradation, Sustainable Development, Agriculture, Climate Change, Mitigation

1. Introduction

The area of productive soils in the world is limited and faces increasing pressure from competing uses such as cropping, forestry and pastures/range land, urbanization, as well as energy production and mineral extraction.

The soils represent at least a quarter of global biodiversity, and play a key role in the supply of clean water and resilience to floods and drought. Crucially, plant and animal life depend on primary nutrient recycling through soil processes.

In human terms, soils are a finite resource, of which about 24 billion tons of fertile soil per year is being lost on agricultural lands. Land degradation often occurs in areas that are already characterized by land scarcity and increasing population [1].

The inventory of soil resources for their decsription would be performed [2]. Based on these the urgent actions to improve the health of the world's limited soil resources and to stop land degradation [3] are required, so as to ensure that future generations have enough supplies of food, water, energy and raw materials.

Some stabilization mechanizms are overviewed and studied in [4].

Georgia is a transcontinental country, located along the dividing lines of Asia and Europe in the South Caucasus. Georgia is situated east of the Black Sea, south of Russia, north of Turkey and Armenia and northwest of Azerbaijan. Georgia covers an area of 69.700 km^2.

The climate of Georgia is extremely diverse, there are two main climatic zones, roughly separating eastern and western parts of the country. The Greater Caucasus Mountain range plays an important role in moderating Georgia's climate and protects the area from the invasion of colder air masses from the north.

The Lesser Caucasus Mountains partially protect the region from the influence of dry and hot air masses from the south as well.

Georgia is an agrarian country with small territory, which has the potential to produce high quality and neat agricultural products. Therefore, effective management of land resources is important from the environmental and socio-economic points of view.

Approximately 54% of the territory of Georgia is mountainous. The high mountains and plains occupies respectively 33% and 13% of country's territory. Correspondingly, the 70% of the territory is below the altitude of 1,700 meters [5]. These altitudes contributes to the development of agriculture, and at the greater heights are mainly located only pastures.

Geography of Georgia is diverse: humid, subtropical lowlands and swamps; lakes and numerous rivers and soils; valleys; semi-deserts; hills; mountains covered by forests and glaciers [6]. Some characteristics about Georgia are presented in Table 1.

Georgian population is equal to 4.4 million. According to 2010 data [7] approximately 47% of the population lives in rural areas, 52.3% of this amount deal with agricultural sector. The proportion of the population living in rural areas is more or less stable with respect to the city residents. Agriculture has an important role in Georgia. 43% of the land resources is suitable for cultivation, but certain amount of this soil is degraded and no longer used for agriculture [8].

Table 1. *Some Data about Geography of Georgia.*

Total territory (km^2)	Agricultural lands (km^2)	Forests (km^2)	Mountains (Above 1000 m) (km^2)	Waters (Lakes, Rivers, Reservoirs) (km^2)	Swamps (km^2)	Glaciers (km^2)	Semi Deserts (km^2)
69, 700	30, 258	24, 562	37, 640	8, 351	600	511	100

The soil resources distribution in Georgia is presented in Fig. 1.

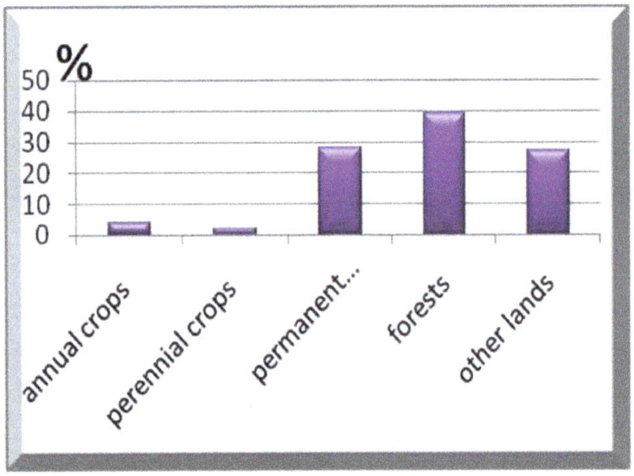

Fig. 1. *The Soil ResourcesDistribution in Georgia.*

According to this Figure soil resources allocation are as follows: annual crops make up 4%, correspondingly perennial crops - 2%; permanent meadows and pastures - 28%; forests - 39%; other lands - 27%. In 2005, 1.07 million hectares of land were processed. From this amount annual crops were planted in the area of 75%, while the rest were occupied by long-term cultures.

2. Study Area & Methods

The study area is the territory of Georgia allocated for the agriculture. Historically soil fertility specified the importance of the agriculture sector as an important part of the Georgian economy. Agricultural productivity is currently limited by soil degradation, which is an important problem. The soil formation is a long process, while its degradation develops much faster. Soil degradation means the process of deterioration of soil resources fertility as a result of both natural and anthropogenic impact.

In Georgia the soil degradation is mainly caused by inefficient management of ground resources and ignoring the principles of sustainable development. It is stipulated by various aspects, namely:

- Significant increase of soil pollution level;
- Substantial decrease of the pastures and forests;
- Expansion of soil erosion;
- Increasing of saline soils.

The tendency concerning deterioration of the quality of soil resources is defined by the change of the economic relations i.e. transition of the country to free economy.

The main reasons are [9]:

i. Considerable shortage of activities to increase soil productivity;
ii. Interruption of the work for soil quality improvement and erosion reduction;
iii. Minimized use of organic and chemical fertilizers;
iv. Full or partial reduction of projects against pollution of soil resources.

As a result of inefficient management we receive:

- Degraded soils due to ground salinization. The typical examples are Georgian districts: Kakheti and Kvemo Kartli;
- Soil erosion caused by water impact. This is well expressed in Lentekhi district;
- Land desertification due to wind erosion. This is spread mainly in Dedoplistkaro district.

Due to these problems the complex study of the soil degradation in abovementioned districts of Georgia is very important.

Modern climate change has significant impact on the development of agriculture. Climate change has significant influence on natural disasters connected with climate, weather and water. The frequency and intensity of these disasters grow considerably. This leads to the soil degradation in large areas.

The ultimate outputs of all above-mentioned are the reduction of agricultural lands and arable areas [10]. That significantly worsens social and economic conditions of life of the population.

Integrated study of abovementioned issues is the scientific novelty of the research.

Research areas and problems has a broad variation to variety of materials and methods being used. In a monography [9] methodologies and data sets utilized are overviewed. In order to carry out corresponding actions for accomplishment of tasks scheduled by the research it is necessary to carry necessary actions according to standards elaborated by International Standards Organization (acronym ISO). These measures with corresponding standards are:

- Inventory, taking and transportation of soil analysis samples in the districts/regions under investigation implemented according to ISO 10381-1 and 10381-2;
- Preparatory treatment and appropriate preparation (drying, disintegration, grinding, sieving etc.) of soil samples for chemical analysis carried based on ISO 11464 and ISO 565;
- Sampling and transportation of water analysis samples implemented according to ISO 5667-4:1987 standards;
- Determination of turbidity of water samples and suspended load using sedimentation method carried according to ISO 5667-17;
- Determination of humus in soil samples implemented using colorimetric method (at a depths of 0-10,10-20 and 20-30 cm);
- Determination of assimilated forms (N, P, K) in soil carried according to ISO 103041: 2007 and 9964-3:2010.

3. Results & Discussions

In 2009-2010 the production of annual and perennial crops was reduced. This is preconditioned by: a reduction in the sowing area, unfavorable weather conditions, pests and limited use of agricultural practices [7].

The decline in agricultural production are defined due to: water and wind erosion, other anthropogenic and natural agricultural processes that led to the degradation of 35% of agricultural lands [7].

The landscapes of East Georgia are vulnerable towards the climate change [8, 9]. For the last century in most of the territory of East Georgia increase of average air temperature in the range of 0.5 – 0.7 ^{0}C is observed. At the same time the frequency of the droughts grow considerably. The amount of the precipitation in the mentioned territory does not exceed 200-250 mm and the existing reserve of the productive moisture in a meter of thick soil is equal to 50 – 200 mm only. On the background of the global warming due to more rapid draughts, the transformation of the natural landscapes is anticipated.

In South-East of Georgia almost 3 th. sq. km territory located in semi-arid zone is damaged with droughts and wind-erosion and is liable to degradation and desertification. The process of desertification is clearly pronounced in Kiziki, Gare Kakheti and some districts of Kvemo Kartli. The territory of 120 th. ha of Dedoplistskaro district, the territory of 47 th. ha of Sighnaghi and Sagarejo districts, the territory of 32 th. ha of Gardabani district, the territory of 30 th. ha of Marneuli district are liable to the impact of desertification.

Table 2. *The Saline Soils of Georgia in 2012 (Th. ha).*

Region	Saline Soils					
	Arable land	**Pasture / hayfield**	**Pasture / Shrubbery**	**Forest**	**Perennials**	**Total**
Signaghi	1.68	5.5	2.25	0.09	0.15	9.67
Gurjaani	0.24	0.47	0.005	-	-	0.72
Sagarejo	2.51	3.8	0.63	0.05	0.06	7.05
Gardabani	5.1	3.29	2.1	0.51	1.1	12.1
Dedoplistskaro	2.21	4.17	0.51	0.67	0.44	8.00
Lagodekhi	2.12	-	-	-	-	2.12
Marneuli	1.59	7.3	1.0	0.04	0.75	10.68
Bolnisi	0.7	-	-	-	-	0.7
Qareli	0.06	0.08	-	0.005	-	0.15
Total	13.39	24.61	6.495	1.365	2.50	48.37

One of the types of degraded lands are saline and solonetz soils. They are mainly spread in Alazani valley. The medium and heavily saline and solonetz soils occupy more than 40 % of the total territory of Georgia. (tables 2-3).

Saline and solonetz soils are spread in the following districts: Sighnaghi (54 th. ha), Marneuli (53 th. ha), Dedoplistskaro (48th.ha), Gardabani (40 th. ha), Sagarejo (23 th. ha), Gurjaani (8 th.ha). In Georgia total area of saline soils is more than 205 th. ha and from this amount more than 84 th. ha are used. Therefore, they are characterized by low productivity.

It is necessary to research erosion processes activated due to temporary climate change. The basic factors that are stipulating the development of erosion processes are climate, relief, surface waters, soils and geological content of the Earth interior.

The problems of water erosion are strongly pronounced in mountainous territories that are mainly stipulated by over pasturing and unsystematic cutting of the forests. At present 50-60 % of the mountainous territory of Georgia are exposed to adverse impact of the water erosion.

The increase of water erosion leads to development of the accompanying processes such as landslides, mudflows, creation of the ravines.

The wind erosion occurs in the cold season of the year in winter-spring months (December-April).

That time North-West winds are predominant and their speed reaches 25-30 m/s. The main factors providing the wind erosion are: climate, dry up, wind speed, absence of the wind-proof belts, absence of the plants in winter-spring period and irregularity of the precipitation.

Dedoplistkaro district is one of the most vulnerable in Georgia regarding desertification. Natural disasters, such as drought and strong winds have a substantial damage to agriculture.

Table 3. *The Solonetz Soils of Georgia in 2012 (Th. ha).*

Region	Solonetz Soils					
	Arable land	Pasture / hayfield	Pasture / Shrubbery	Forest	Perennials	Total
Signaghi	17.79	20.15	4.6	0.33	1.77	44.64
Gurjaani	2.46	1.89	0.019	0.017	0.51	4.896
Sagarejo	5.87	8.37	0.69	0.11	0.067	15.11
Gardabani	19.9	5.49	1.46	0.65	0.50	28.0
Dedoplistskaro	4.3	24.18	7.02	2.14	0.92	38,56
Lagodekhi	1.7	0.16	0.2	0.012	0.05	2.12
Marneuli	11.3	9.85	0.34	0.03	0.56	22.08
Bolnisi	0.26	0.11	0.32	0.011	0.003	0.70
Qareli	0.34	0.02	0.008	0.01	0.13	0.51
Total	63.92	70.22	14.66	3.31	4.51	156.62

3.1. Land Resources Degradation Due to Salinization

The most saline soils are observed in Kakheti region. The complex investigation of such type of soil degradation was carried for this region [9]. Agricultural soils of 606 thousand ha of Kakheti region the area of 133 thousand ha are saline soils that makes 22 % of total.

The necessary researches on the saline soils conditions were carried. The received results are listed in [9].

Thus:

- The climatic components changes in the Alazani valley and their influence on saline soils;
- The dynamic of the chemical composition changes of saline and alkaline soils;
- The migration of easily soluble salts and the soil productivity alteration;
- The dynamic of the processes of soil salinization - desalinization against: the seasonality, the irrigation, the underground water level, the soil physical characteristics and the climate components;
- The assessment of migration and accumulation in soil of the salts that have the toxic impact on the plants in connection with the climate conditions;
- The chemical content of collector-drainage, underground and irrigation waters are determined.

On Fig. 2- 5 the dry residue or total amount of easily soluble salts in 0-200 cm depths are given in dynamics during 2012 and 2013 for various plots. The analysis of presented graphs shows that in the soils of Dzveli Anaga vineyards salt content is increased in the depth during 2013. The dry residue has a growth tendency in grass soils in a depths of 60-80 cm in 2013.

The total amount of easily soluble salts amount is more during the 2012 in comparison with 2013.

However, total salinization has the growth tendency for plots with drainage and without drainage.

It would be mentioned that total salinization content in drainage plots are more than in the plots without drainage. This clearly shows the positive role of the drainage.

The negative impact of the climate change causes the decrease of some food products production. After caring out necessary activities on the saline lands it is possible to receive crop plants and food grasses that are necessary for the stock-breeding development.

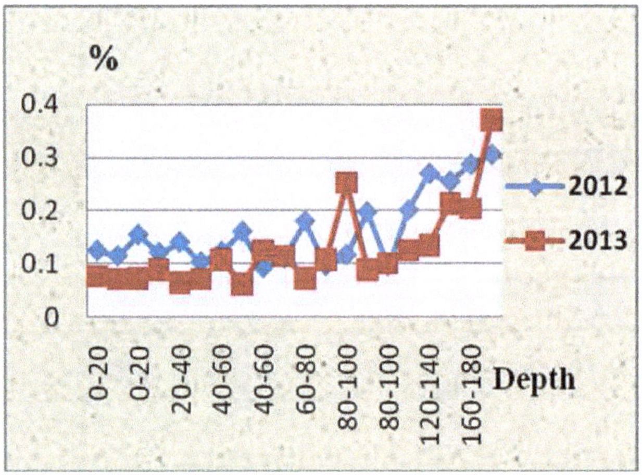

Fig. 2. *DryResidue Total Amount, (%) (Dzveli Anaga, vineyards) for 2012 and 2013.*

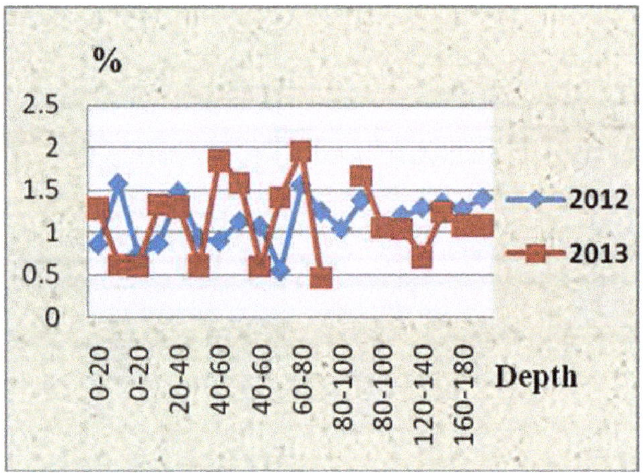

Fig. 3. *Dry Residue Total Amount, (%) (DzveliAnaga, Grass Soils) for 2012 and 2013.*

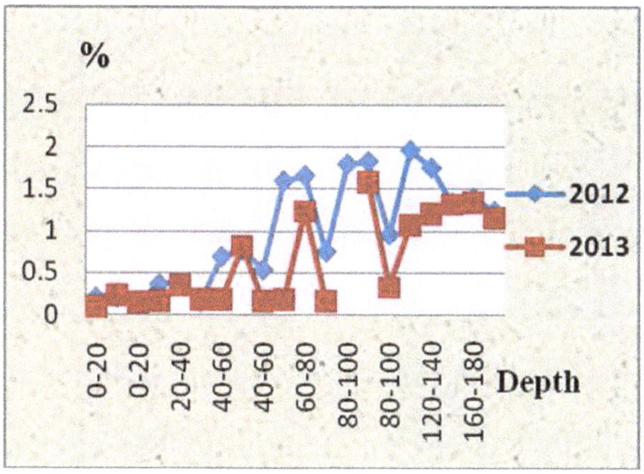

Fig. 4. Total Amount (%) of Dry Residue Tsnori, drainage plot) for 2012 and 2013.

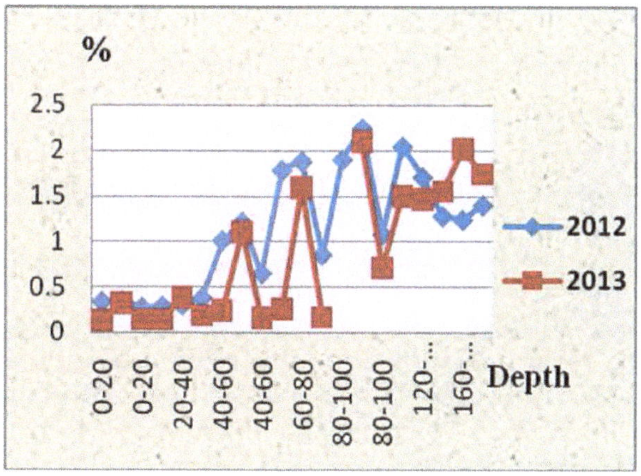

Fig. 5. Total Amount (%) of Dry Residue (Tsnori, plot without drainage) for 2012 and 2013.

The nonlinear character of climate change makes the creation of the adaptation measures and their implementation topical. In a new ecological situation characterized by big emmissions of technological pollutants in environment, plants and microorganisms exhibited new capabilities for the absorption and metabolic degradation of toxicants [4, 11, 12].

The analysis of the measures revealed necessity of the following activities fulfillment:

- Development of irrigation systems;
- Introduction of the biotechnologies (selection of drought and salt resistant plants);
- Enhancement of the scientific research in the agriculture.

3.2. Land Resources Degradation Due to Erosion Processes

Research of erosion processes were activated due to temporary climate change. The basic factors that are stipulating the development of erosion processes are relief, climate, surface waters, soils and geological content of the Earth interior. The following land-erosion zones were defined: water erosion in East and West Georgia, the wind erosion in East Georgia and Kolkheti lowland (table 4).

As mentioned above the problems of water erosion are strongly pronounced in mountainous regions that are mainly stipulated by over pasturing and unsystematically cutting of the forests. At present 50-60 % of the mountainous territory of Georgia are exposed to adverse impact of the water erosion. The increase of water erosion caused development of the accompanying processes: landslides, mudflows and creation of the ravines. At present in Georgia 10 th. landslides and 2 th. mudflows causing ecological equilibrium disturbance are detected. Top layer losses make 50-90 t/ha due to water erosion in various land-climate conditions of East Georgia and the same value for West Georgia is in the range of 120-150 t/ha [13].

Table 4. The Area of Eroded Arable Soils in Georgia.

Territory	Arable lands		Degradation Level					
			Weak		Middle		Strong	
	100 ha	%	100 ha	%	100 ha	%	100 ha	%
West Georgia	141.9	100	24.0	16.9	15.5	10.9	6.1	4.3 (32.1)
Abkhaseti	36.1	100	3.7	10.2	5.3	14.7	3.1	8.6 (33.5)
Adjara	8.8	100	2.7	30.7	2.5	28.4	0.1	1.1 (60.2)
Total West Georgia	186.8	100	30.4	16.3	23.3	12.5	9.3	5.0 (33.8)
East Georgia	467.7	100	74.6	15.9	45.0	9.6	9.8	2.1 (27.6)
South Osetia	18.7	100	5.5	29.4	6.1	32.6	1.7	9.1 (71.1)
Total East Georgia	486.4	100	80.1	16.5	51.1	10.5	11.5	2.4 (29..4)
Total Georgia	673.2	100	110.5	16.4	74.4	11.0	20.8	3.1 (30.5)
Total eroded soils	205.7 thousand ha - 30.6%							

In West Georgia the erosion processes caused a high magnification of the alluvium amount– up to 30 mln. Metric Ton annually. From this amount approximately 9 mln. Metric Ton fall on r. Rioni basin, correspondingly 9 mln. Metric Ton - on r. Chorokhi basin and 3 mln. Metric Ton - on r. Enguri basin.

As mentioned above the wind erosion occurs in the cold season of the year in winter-spring months (December- April). That time North-West winds are predominant and their speed reaches 25-30 m/s. The main factors providing the wind erosion are: climate, dry up, wind speed, absence of the wind-proof belts, lack of the plants in winter-spring period and irregularity of the precipitation.

Dedoplistkaro district is one of the most vulnerable regarding desertification in Georgia. Natural disasters, such as drought and strong winds have a substantial damage to agriculture. For the last 50 years the severity of these events had increased from 54 to 72 days and the frequency doubled.

The recurrence of strong winds (30 m/s and more) from the beginning of 1980 s increased 5 times.

In the second half of the 20-th century the average annual air temperature increased by 0.6^0 C and annual amount of precipitation grew by 6%. For 2100 the forecasted precipitation would be decreased by 14%. The aridity of the local landscapes would be transformed into semi desert and desert ones.

In Dedoplistskaro district the soil degradation is one of the most topical problems. The thickness of humus layer in arable lands, earlier characterized with high productivity, noticeably decreased due to wind erosion. In Shiraki area the humus amount of black earth soils decreased from 7.5% till 3.2 %. The soil productivity decreased almost twice respectively.

At present degraded soils are spread over 25 300 ha in the Dedoplistskaro district. From this amount the area of 20 000 ha is eroded by the wind. The wind negative impact is spread over 100 000 ha of the territory of East Georgia. More than 80 % of winter pastures are damaged due to over pasturing and partially because of climatic conditions [8]. This process is spread on some areas (Shiraki, Eldari, Iori, Taribana, Natbeuri, Naomari, Ole) fields, sublime localities and significant part of south slopes of Kakheti range [14].

3.3. Land Resources Degradation Due to Natural Disasters

Lentekhi region is one of the most vulnerable regarding natural disasters in Georgia. This is due to the fact that climate change consequences are expressed in increasing frequency and intensity of natural disasters. In the region for the last 50 years the average annual temperature and precipitation increased by $0.4\ ^0$ C and 106 mm (8%) respectively. For 2100 the forecasted precipitation would be decreased by 14%. The aridity of the local landscapes would be transformed into semi desert and desert ones.

At present degraded soils are spread over 25 300 ha in the Dedoplistskaro district. From this amount the area of 20 000 ha is eroded by the wind. More than 80 % of winter pastures are damaged due to over pasturing and partially because of climatic conditions [8]. This process is spread in some areas (Shiraki, Eldari, Iori, Taribana, Natbeuri, Naomari, Ole), sublime localities and significant part of south slopes of Kakheti range [11]. The main reason for these facts are that due to the climate change the frequency and intensity of natural disasters are increased considerably.

The time period 1967-1989 is divided into two equal intervals (1967-1978 and 1978-1989) and the analysis of the observation data on flash floods for these time intervals define that for the second period the recurrence of the flash floods doubled. The maximal value of water discharge is increased by 9%. At the same time the duration of flash floods decreased by 25%. That is explaining significant increase of the intensity of flash floods.

Since 1980 the amount of land slides increased by 43% and the slumped sites number reached 117 [15]. The landslides activated since anomalous rich snow winter that took place in Georgia in winter of 1986/87. For the last 2 decades in Lentekhi district increase of plentiful precipitation caused that the frequency of the mudflows doubled. For the last decade the land erosion is increasing due to the increase of frequency and intensity of natural disasters (flash floods, landslides, mudflows, snow avalanches, glacial catastrophes). The land erosion has significant negative impact on the agriculture and damage forests, spoil roads and infrastructure.

Since 1986 the population of Lentekhi district was decreased by 40% due to stirring up of the landslides and flash floods. The Adaptation Fund supported the project “Developing Climate Resilient Flood and Flash Flood Management Practices to Protect Vulnerable Communities of Georgia” (2012-2016). In frames of this project disasters risks are assessed and corresponding adaptation measures are created and carried out for mitigation land erosion in the region.

In addition to the impact of mudflows on degradation processes having place in most vulnerable Lentekhi district it would be useful to overview the heavy catastrophic events that happened in Georgia for the last two years. They are activated mainly by the modern climate change [13, 14].

In 13-14 June 2015 the severe mudflow took place in the river Vere basin. The disaster was connected with long lasted heavy rain. The extreme rain continued 4 hours. The precipitation amount made 49 mm (67 % of the monthly norm). The water level was substantially raised (more than 9 meters). It was accompanied with catastrophic mudflow. The consequences were terrible (see fig. 6). Due to torrential flow (the determined discharge was equal to 259 m^3/s) the human casualties took place, the infrastructure was destroyed, the soil resources were fully degraded. The material losses amounted to several hundred USD.

Fig. 6. Consequences of Catastrophic Flash flood/ Mudflow in r. Vere Basin.

The severe catastrophic events (heavy rain and strong mudflows were caused by the ice avalanches occurred in May and August, 2014 from Devdoraki glacier.

High mountains (Big Caucasus Range) are spread at the North of the Georgia. The main problem here are the glacial and hydrological disasters such as glacier falls, the river bed blockage and natural dam formation, with consequent break and passing of catastrophic flash floods and/or mudflows.

Consequently the large areas of land are degraded. To study

these processes the most effective and easy way is assessment of the degraded soil resources from the space i.e. the use of satellite remote sensing [16].

The main reason of the glacial mudflows genesis are consequences of the glaciers subduction. Most of the mountainous glaciers are characterized by rapid pulsing movement. This is a result of the processes going inside the glacier, i.e., fast change of glacier bed conditions and distribution of substances in accumulation and ablation area. In Caucasus there are pulsating glaciers (e.g. Kolka and Devdoraki), which are characterized by sharp advancing of the glacier body the so-called surge accompanied by glacier falls.

When glacier retreats the glacier tongue, due to restricted feeding (result of decreasing of snow precipitation and reduction of its accumulation in firn field) is grinded. The crevices are created on the glacier tongue that make prerequisites for moraine material accumulation (rocks and their debris that falls on glacier surface). In these places due to high temperatures and frequent precipitation on the surface of glacier the certain amount of water is accumulated. It will break the created dams capturing all the remaining ice and moraine material. As a result the destructive mudflows are created.

Obviously, the mudflows that took place in the past were the result of the glaciers such pulsation. At that time periods the characteristics of the Georgian glaciers were mainly increasing.

Last catastrophic events occurred in May and August, 2014 during the period when the characteristics of the Georgian glaciers were steadily diminishing [17].

Based on video, photo material analysis, it can be concluded that the creation of the May 17 and August 20, 2014 catastrophic mudflows are not fully connected with the pulsation of the Devdoraki glacier. In May 17, 2014 from Kazbegi slope at 4, 700 m elevation was cut ice block, with the volume of 5 million cubic meters. This mass fell down along 80-85 degrees inclined slope below 2, 500 m. Then it jumped over the Devdoraki glacier tongue and stroke the left side of the narrow gorge. The ice mass from the mountain slope was accompanied with the volcanic slag. Vice versa can also be considered, i.e., volcanic slag was off first and then it was accompanied with ice.

The red trace of the volcanic slag is well reflected in the area's of the strike of the gorge sides and debris flow trace from the source to the confluence (fig.7). Mudflow material surface that arrived in the r. The Terek valley also was red due to the volcanic slag [17].

The torrential stream generated due to the ice block collapse into pieces. Water/ice masses mixed with moraine material and vulcanic slag (red slag pieces, including andesite & dacite rock material) generated semi-viscous mass. This material passed through narrow Devdoraki-Amali gorge with the speed of 100 km/h in 4-5 min.

Semi-viscous mass stream blocked r.Terek riverbed and artificial reservoir was established (fig. 8).

Fig. 7. *The red trace of the volcanic slag.*

Fig. 8. *The Terek river bed blockage.*

The consequences were heavy. Outgoing pipeline was exploded. Casualties amounted to 9 persons. Established artificial reservoir occupied the area about 2 hectares and the pond was 15 m deep. Luckily water was discharged by under construction derivation tunnel of hydropower station. The outgoing road was closed for a long-term time. The soil resources were fully degraded.

It should be noted that throughout the history of the survey of Devdorak glacier such big scale and strong catastrophe was not observed twice a year.

4. Conclusion

Soils represent at least a quarter of global biodiversity, and play a key role in the supply of clean water and resilience to floods and drought. Crucially, plant and animal life depend on primary nutrient recycling through soil processes.

Proper exploitation of land resourses is one of the main problem of the adverse impact of climate change. Soil formation is a long process, while its degradation develops much faster. Soil degradation means the process of deterioration of soil resources fertility as a result of both natural and anthropogenic impact. The current escalating rate of soil degradation threatens the capacity of future generations to meet their needs.

Research of soil degradation processes were activated due to climate change adverse impact worldwide. The basic factors that are stipulating the development of degradation processes are relief, climate, surface waters, soils, geological content of the Earth interior.

In Georgia the soil degradation is mainly caused by inefficient management of ground resources and ignoring the principles of sustainable development. The following physical processes are stipulating degradation: salinization, erosion processes, natural disasters impact, substantial decrease of the pastures and forests.

Due to abovementioned the complex study of the soil degradation in connection with climate change tendencies in Georgia were done.

Modern climate change has significant impact on the management of soil resources and development of agriculture.

The landscapes of East Georgia are vulnerable towards the climate change. In most of the territory of East Georgia increase of average air temperature in the range of 0.5 – 0.7 ^{0}C is observed in last century. At the same time the frequency of the droughts grow considerably. The amount of the precipitation in the mentioned territory does not exceed 200-250 mm and the existing reserve of the productive moisture in a meter of thick soil is equal to 50 – 200 mm only. On the background of the global warming due to more rapid draughts, the transformation of the natural landscapes is noted. In South-East of Georgia almost 3 th. sq. km territory located in semi-arid zone is damaged with droughts and wind erosion processes and is liable to degradation and desertification. The process of desertification is clearly pronounced in some districts of East Georgia. The territory of 120 th. ha of Dedoplistskaro district, the territory of 47 th. ha of Sighnaghi and Sagarejo districts, the territory of 32 th. ha of Gardabani district, the territory of 30 th. ha of Marneuli district are liable to the impact of desertification.

Research areas and problems has a broad variation to variety of materials and methods being used. In order to carry out up-to-date accomplishment of tasks scheduled by the research, the necessary measures and actions are carried according to standards elaborated by International Standards Organization.

One of the types of degraded lands is saline soils. They are mainly spread in Alazani valley. The medium and heavily saline soils occupy more than 40% of the total territory of Georgia. Saline soils are spread in the following districts: Sighnaghi (54 th. ha), Marneuli (53 th. ha), Dedoplistskaro (48th.ha), Gardabani (th. ha), Sagarejo (23 th. ha), Gurjaani (8 th.ha). In Georgia total area of saline soils is more than 205 th. ha and more than 84 th. ha are used. Therefore, they are characterized by low productivity.

Soil resources degradation due to erosion processes activated because of temporary climate change is researched. The basic factors that are stipulating the development of erosion processes are climate, relief, surface waters, soils etc.

The wind erosion occurs in the cold season of the year in winter-spring months (December-April). That time North-West winds are predominant and their speed reaches 25-30 m/s. The main factors providing the wind erosion are: climate, dry up, wind speed, absence of the wind-proof belts, absence of the plants in winter-spring period and irregularity of the precipitation. Natural disasters, such as drought and strong winds have a substantial damage to agriculture.

In Dedoplistskaro district the soil degradation is one of the most topical problems. The thickness of humus layer in arable lands earlier characterized with high productivity noticeably decreased due to wind erosion. The content of humus of black earth soils in Shiraki area decreased from 7.5% till 3.2 %. The land productivity decreased almost twice respectively.

At present degraded soils are spread over 25 300 ha in the most vulnerable Dedoplistskaro district. From this amount the area of 20 000 ha is eroded by the wind. The wind negative impact is spread over 100 000 ha of the territory of East Georgia. More than 80 % of winter pastures are damaged due to over pasturing and partially because of climatic conditions.

Modern climate change has significant impact on the management of soil resources and development of agriculture. Climate change has significant influence on natural disasters connected with climate, weather and water. The frequency and intensity of these disasters grow considerably. This causes the soil degradation on large areas.

Lentekhi region is one of the most vulnerable regarding natural disasters in Georgia. This is due to the fact that climate change consequences are expressed in increasing of the frequency and intensity of natural disasters. In the region for the last 50 years the value of average annual temperature and the amount of precipitation increased by 0.4 ^{0}C and 106 mm (8%) respectively.

The analysis of the observation data on flash floods showed that for the active anthropogenic influence period (time period from 1978) the recurrence of the flash floods doubled in comparison with the previous period.

The maximal value of water discharge is increased by 9%. At the same time the duration of flash floods decreases by 25%. That can explain significant increase of the intensity of flash floods.

Since 1980 the amount of land slides increased by 43% and the slumped sites number reached 117. The landslides activated since anomalous rich snow winter (1986/87). For the last 2 decades in Lentekhi district increase of plentiful precipitation caused that the frequency of the mudflows doubled. For the last decade the land erosion is increasing due to the raise of frequency and intensity of natural disasters (flash floods, landslides, mudflows, snow avalanches, glacial catastrophes). The land erosion has significant negative impact on the agriculture and damage forests, spoil roads and infrastructure.

The nonlinear character of climate change makes the creation of the adaptation measures and their implementation topical. In a new ecological situation characterized by big emissions of technological pollutants in environment, plants and microorganisms exhibited new capabilities for the absorption and metabolic degradation of toxicants.

The analysis of the measures revealed necessity of the following activities fulfillment:

- Development of irrigation systems;
- Introduction of the biotechnologies (selection of drought and salt resistant plants);
- Enhancement of the scientific research in the agriculture.

Research results allow the study of soil degradation, improve the soil fertility and create effective methodology and corresponding measures for prevention or/and mitigation of soil degradation.

References

[1] A. Müller and J. Weigelt (2013) "Governace for a land Degradation Neutral World" in IISD Land Policy and Practice Knowledge Database. http://nr.iisd.org/guest-articles/governance-for-a-land-degradation-neutral-world/.

[2] FAO, Guidelines for Soil Description (2012). Rome, 97 p.

[3] Tsao D. (2013). Phytoremediation: Advances in Biochemical Engineering and Biotechnology. Springer, Berlin-Heidelberg-New York, p.207.

[4] G. Khasitashvili, L. Machavariani, R. Gakhokidze (2015). Improving Phytoremediation of Soil Polluted with Hydrocarbons in Georgia, Soil remediation and Plants. Chapter 19. New-York-Paris-Tokyo, Elsevier, 547-569.

[5] L. Maruashvili (1971) – Physical Geography of Georgia. Tbilisi. 260 p. (in Georgian).

[6] T. Urushadze (1999) - The map of Georgian soils. (scale 1:500 000), Tbilisi. (in Georgian).

[7] FAO and Ministry of Agriculture of Georgia (Winter 2011). Georgia: Agriculture Sector Bulletin.http://www.fao.org /fileadmin/templates/tc/ tce/pdf/ Georgia Agriculture Sector Bulletin Winter 2011.

[8] UNFCC (2009). The Second National Communication under the UNFCC, Tbilisi, 230 p.

[9] Shavliashvili L., Kordzakhia G., Elizabrashvili E., Kuchava G., Tugushi N. (2014). Degradation of Alazani Valley Soils on Background of Modern Climate Change. Publisher" Universal". P. 181 (in Georgian).

[10] Materials of the Statistic Department of Georgia, (2009). Land Use and Its Structure According to Various Forms of Landownership and Land Tenure. Tbilisi, pp.33-50.

[11] G. Khasitashvili, L. Machavariani, R. Gakhokidze (2015). Improving Phytoremediation of Soil Polluted with Hydrocarbons in Georgia, Soil remediation and Plants. Chapter 19. New-York-Paris-Tokyo, Elsevier, 547-569.

[12] Kadjaia G., 2008. The Principles of Environment Protection//"Intelekt", Tbilisi. 272 pp. (in Georgian).

[13] Climate Change Impact on Dedoplistskaro Region (2008). Tbilisi, pp.47-110, (in Georgian).

[14] Emil Tsereteli, George Gaprindashvili, Tsetsili Donadze, Tinatin Nanobashvili, Otar Kutsikidze (2015). Some Aspects of the Methodology of Disaster Geological Processes Hazard and Risk Mapping on the Example of Georgia, International Conference "Applied Ecology: Problems, Innovations". Proceedings ICAE - 2015, pp 30 - 36. http//icae-2015.tsu.ge/.

[15] Von Ina Keggenhoff, Tatjana Keller, Mariam Elizbarashvili, Ramin Gobejishvili und Lorenz

[16] King (2011). Naturkatastrophen durch Klimawandel im Kaukasus? Hochwasser und Hannstabilitaten in Georgien immer haufiger. Spiegel der Forschung, N2, 16-23.

[17] Lubomir Prokes, Michaela Vanharova and Josef Havel (2014) Image analysis in remote sensing of soils. The materials of the International Conference on Research and Innovation for Sustainable Soil Management.

[18] G. Kordzakhia L., Shengelia G., Tvauri V., Tsomaia, M. Dzadzamia (2015). Satellite remote sensing outputs of the certain glaciers on the territory of East Georgia

[19] G. Kordzakhia, L. Shengelia, G. Tvauri, M. Dzadzamia, Research of Devdoraki Glacier Based on Satellite Remote Sensing Data and Devdoraki Glacier Falls in Historical Context, American Journal of Environmental Protection. Special Issue: Applied Ecology: Problems, Innovations. Vol. 4, No. 3-1, 2015, pp. 14-21. doi: 10.11648/j.ajep.s.2015040301.13.

5

Sprouting Value Index: A Mathematical Approach for Evaluation of Vegetative Propagation in Tree Species

Justin R. Nayagam*, Mani Varghese K. I.

Department of Botany, Union Christian College, Aluva -2, Kerala, India

Email address:
gigirn79@gmail.com (Justin R. Nayagam)

Abstract: Rooting of *Saraca asoca* stem cuttings using three IBA (Indole 3-butyric acid) concentrations 300 ppm, 500 ppm and 1000 ppm IBA, have been carried out and the data obtained was evaluated with SVI (sprouting value index) method in order to interpret the vigor in rooting/sprouting and the suitable medium, which gives maximum results. Field studies were carried out using three rooting medium at a location in Central Kerala, Peninsular India, for three years at four months regular interval. The control cuttings do not recorded rooting. The results obtained indicate high SVI (sprouting value index), when prop root cuttings planted in root trainers with coir pith compost (RTCP) for all the three concentration of IBA applied. Sprouting percentage was increased and the delay in completion of sprouting/rooting initiation decreased by the use of IBA treatment. Increased CWR and SUP percentages indicate defects in management practices, dormancy and or the genotype of the cultivar.

Keywords: Stem Cuttings, Sprouting Value Index, *Saraca asoca,* Rooting Media

1. Introduction

Plant propagation has a global effect, as it is a fundamental occupation of human kind and its discovery dates back to the origin of civilization and ornamental gardening attained high level during the period between 500 B.C to A.D. 1000 [1], [2]. The vegetative propagation of herbaceous plants can be performed easily as many of them produces bulbs, rhizomes, corms, tubers, offsets, suckers and stolons which are natural vegetative propagating structures. But vegetative propagation in tree species is a difficult process and it is attained through hormone application mostly in stem cuttings, which eventually initiates adventitious root and shoots [3], [4], [5]. Again, there are several mathematical expressions and explanations to measure seed germination and viability in sexually reproducing plants [6], [7], [8], [9], [10] but mathematical expressions for determining efficiency of vegetative planting material or for clonal propagation is scanty. The data obtained through the present investigation are being interpreted using sprouting value index (SVI).

Saraca asoca (Roxb.) de Wilde is an enlisted medicinal tree species of the tropics with astringent bark, which is reported to have stimulating effect on the endometrium and ovarian tissue, and is useful for treating menstrual pain and menorrhagia due to uterine fibroids, leucorrhoea, and internal bleeding, haemorrhoids and haemorrhagic dysentery. In Ayurvedic medicine, it is also used for treating dyspepsia, diseases of the blood, biliousness, tumours, abdominal enlargement, colic, piles, ulcers and bone fractures. The flowers pounded in water, are used to treat haemorrhagic dysentery and the dried flowers for diabetes; they are considered to be an excellent tonic, and useful for the treatment of syphilis and biliousness. The edible seeds are diuretic. [11], [12], [13], [14] [15]. The lack of technology for regeneration from cuttings was considered for selecting the plant for the experiment trials.

2. Materials and Methods

2.1. Specimen Collection and Conduct of Field Trials

For the present investigation plant samples were collected from T.C Joseph Memorial Botanical Garden, of the Department of Botany, Union Christian College, Aluva, (+10° 7' 30.65", +76° 20' 3.32") Ernakulam district, Kerala State, India. Plant was identified by literature scrutiny, as the plant is not recorded in any local and national herbaria in South India. All field trials was conducted in the plant nursery of T.C Joseph Memorial Botanical Garden, of the Department of Botany, Union Christian College, Aluva, (+10°7'30.65", +76°20'3.32") Ernakulam district, Kerala State, India.

For the sake of large scale planting material production in mechanized gardens, trials were conducted in 100cc root-trainer blocks (24 celled) using potting mixture (RTPM), root trainers with vermin compost (RTVC) and root trainers with coir pith RTCP as rooting medium.

2.2. Experiment Design

For all field trials, average of the three replicates during the months of February (P1), and June (P2), October (P3) from Feb 2009 to October 2012 were conducted in three different rooting medium such as root-trainers with potting mixture (RTPM), root trainers with vermi compost mixture (RTVC) and root-trainers with coir pith (RTCP). Stem cuttings with an average size of 7.5 to 10 cm were used. The sample size was kept twenty-four for each trials separately as the 100cc root-trainer block contains 24 cells. A non-auxin control and three Indole 3-butyric acid (IBA) concentrations were designed in this experiment with 300ppm, 500ppm and 1000ppm (parts per million) in order to detect the rooting/sprouting ability by quick dip method. A randomized complete block design was employed. After 45 days, the cuttings were evaluated for rooting/sprouting percentages, mortality percentage and viability percentage. The data obtained were subjected to one factor analysis, employing analysis of variance (ANOVA) and two-way ANOVA.

2.3. Calculation of SVI

In order to develop the new idea of calculating sprouting value index (SVI), data regarding sprouting and successful rooting percentages (SP – sprouting percentages), percentage of planting material with callus production but without rooting (CWR – callus without rooting), percentage of sound unsprouted propagule without callus production (SUP) (was determined by vertical cut test: planting material with more than three nodes living tissues was considered viable) was prepared. Viability percentage can be prepared using the formula (VP = SP + CWR + SUP), peak value (PV = maximum mean sprouting recorded at any time during the test), final mean sprouting (final MDS = cumulative percentage of full sprouting at the end of the test divided by number of days to finish sprouting) were calculated. SVI index method developed through the present study is calculated by the equation, SVI = PV * MDS.

3. Results

Sprouting and rooting studies of stem cuttings were carried out in three different rooting medium during three seasons in an year (June, October and February), for three consecutive year and the various attributes obtained are given in table 1 to table 6. Stem cuttings of seven-year-old *Saraca asoca* plants was used as mother plants in order to obtain stem cuttings at four months interval.

Table 1, table 2 and table 3 gives the VP of prop root cuttings in three concentrations used against the control whereas table 4, table 5 and table 6 gives SVI in different concentrations. Callus production as well as sprouting and rooting was very poor in control. High VP was obtained with all the three concentrations of IBA used (ranging between 83.05% to 95.84%) in the rooting/sprouting medium RTCP.

Table 1. *VP of IBA 300ppm treated stem cuttings of Saraca asoca.*

Rooting medium	Period	CWR	SP	SUP	VP
	P 1	0	0	8.33	8.33
RTPM	P 2	4.17	0	5.55	9.72
	P 3	2.72	0	5.55	8.27
	P 1	2.72	8.33	5.55	16.6
RTVC	P 2	9.72	4.17	8.33	22.22
	P 3	4.17	5.55	9.72	19.44
	P 1	5.55	8.33	9.72	23.6
RTCP	P2	8.33	12.5	12.5	33.05
	P3	6.94	13.89	9.72	30.55

Table 2. *VP of IBA 500ppm treated stem cuttings of Saraca asoca.*

Rooting medium	Period	CWR	SP	SUP	VP
	P 1	4.17	0	9.72	13.89
RTPM	P 2	5.55	0	8.33	13.88
	P 3	6.94	0	9.72	16.66
	P 1	6.94	34.72	9.72	51.38
RTVC	P 2	8.33	37.5	12.5	58.33
	P 3	5.55	41.66	8.33	55.54
	P 1	11.11	44.45	12.5	68.06
RTCP	P 2	8.33	45.83	13.89	68.05
	P 3	6.94	50	9.72	66.66

Table 3. *VP of IBA 1000ppm treated stem cuttings of Saraca asoca.*

Rooting medium	Period	CWR	SP	SUP	VP
	P 1	5.55	0	5.55	11.1
RTPM	P 2	8.33	0	4.17	12.5
	P 3	6.94	0	6.94	13.88
	P 1	8.33	58.33	13.89	80.55
RTVC	P 2	5.55	44.45	12.5	62.50
	P 3	6.94	50	9.72	66.66
	P 1	8.33	63.89	13.89	86.11
RTCP	P 2	8.33	75	13.89	97.22
	P 3	9.72	66.89	8.33	84.94

Table 4. *SVI results of IBA 300ppm treated stem cuttings of Saraca asoca.*

Rooting medium	Period	PV	MDS Final	SVI
	P 1	4.17	0.19	0.79
RTVC	P 2	4.17	0.09	0.38
	P 3	4.17	0.12	0.50
	P 1	4.17	0.19	0.79
RTCP	P 2	4.17	0.28	1.17
	P 3	4.17	0.31	1.29

Table 5. *SVI results of BA 500ppm treated stem cuttings of Saraca asoca.*

Rooting medium	Period	PV	MDS Final	SVI
	P 1	6,94	0.77	5.34
RTVC	P 2	8.33	0.83	6.91
	P 3	4.17	0.93	3.88
	P 1	8.33	0.99	8.25
RTCP	P 2	6.94	1.02	7.08
	P 3	6.94	1.11	7.70

Table 6. *SVI results of IBA 1000ppm treated stem cuttings of Saraca asoca.*

Rooting medium	Period	PV	MDS Final	SVI
	P 1	8.33	1.30	10.83
RTVC	P 2	6.94	0.99	6.87
	P 3	8.33	1.11	9.25
	P 1	9.72	1.42	13.8
RTCP	P 2	11.11	1.67	18.55
	P 3	9.72	1.49	14.48

With stem cuttings of *Saraca asoca,* maximum SVI was obtained in RTCP 1000ppm treated cuttings (29.17). The speed of completion of sprouting/rooting in RTVC was found lower than in RTCP.

The ANOVA results on callus production show significance at 1% level between concentrations of IBA and significance at 5% level between media of study. The ANOVA results on callus formation and callus with root formation show significance at 1% level between concentrations of IBA and between mediums. The results was non significant at 5% level between years and month of study.

The ANOVA results on callus formation and callus with root formation show significance at 1% level between concentrations of IBA and significance 5% level between media of study for callus formation root production show significance at 1% level between medium of study.

4. Discussion

The SVI studies worked out in the present study is a mathematical approach to determine the quality of rooting material, suitability of the medium used for rooting and the optimum concentration of IBA to be used. Production of elite genotypes of any plant species can be generated by vegetative method of propagation round the year. Successful propagation using stem cuttings has been reported by several studies in various plants, which uses stem cuttings, rhizomes or other vegetative parts [16], [17], [18], [19], [20] but rooting of stem cuttings of tree species is difficult to the cultivation field.

In field trials, all planting materials gave elite performance in RTCP planting medium. In the trials using 1000ppm IBA (Indole 3-butyric acid) treated cuttings also SVI was highest in RTCP (29.17). Significant differences in rooting were found between various rooting media when effect of rooting, in sheanut stem cuttings were performed [21]. Sprouting percentage and speed of completion of sprouting/rooting initiation increased by the use of IBA treatment [14]. However, it is shown by other workers [21] that very high IBA concentration have negative effect in rooting. Comparing the CWR and SUP percentages one can also assess the defects in management practices, dormancy and the genotype of the cultivar.

Sprouting index value (SVI) proposed through the present study is a modified form of Germination value (GV), proposed by Czabator [6] for seed germination studies. It is also suitable in field and nursery trials for vegetative propagation and IBA treated rooting of cuttings. The incorporation of CWR (callus production without rooting) along with SUP (sound unsprouted propagule) in calculating VP (Viability percentage) is effective in finding the field oriented defects. SVI and VP is an integrated measure of planting material quality. The speed of sprouting/rooting ability along with the completeness of sprouting can also be determined vegetative planting materials. Suitable rooting media hold considerably high rooting ability [14] [21].

Even though SVI for control were found zero, the VP for all the IBA treated trials are > 90% (table 1, table 2 and table 3) which indicates that by using alternate methods like pretreatments and management practices, SVI can be increased. The internal physiology of the planting material may be the reason for the same. The change in planting material and IBA treatment used affects the rooting process as in the present study; SP was increased in higher concentration (1000ppm IBA). Methods for reducing CWR and SUP values can increase SP and the sample in rooting medium with least difference in SP and VP value will give maximum performance. Results of the rooting experiments showed that with very high IBA concentration in sheanut tree cuttings, rooting ability decreased [21] and hence three concentrations of IBA were used in the present experiment.

5. Conclusions

The present study focuses on regeneration of plantlets from stem cuttings of *Saraca asoca* and the data obtained were interpreted using sprouting value index (SVI) to establish suitable vegetative planting material and suitable rooting medium through field trials. Trials conducted in different rooting medium reveal the sprouting efficiency and vigor in different medium. SVI is a statistically treated data, which is obviously a modification of germination value for seed germination proposed by Czabator [6] the incorporation of CWR (callus production without rooting) along with SUP (sound unsprouted propagule) in calculating VP (Viability percentage) makes it suitable for vegetative cultivation practices and it interprets the quality of planting material, failure due to management practices and the selection of suitable rooting medium. This method of planting stock preparation is valuable in large-scale cultivation and much promising in producing quality clonal planting material production in economically important plants in future.

Acknowledgement

The authors expresses their heartfelt gratitude to Dr. Benny Cherian, Principal and Dr Thara K. Simon Head of the Botany Department, Union Christian College, Aluva, for providing space to conduct field trials in the Dr. T.C Joseph Memorial Botanical Garden, Department of Botany, Union Christian College, Aluva. Thanks to Mr. Thomachen, Gardener, Dr. T.C Joseph Memorial Botanical Garden, Department of Botany, Union Christian College, Aluva, for maintaining the field specimens throughout the study period. Extending a word of thanks to Mr. Jabir who helped in statistical correlation.

References

[1] Harlan, J. R. 1992. *Crops and man.* 2nd ed. Madison, Wia. Amer. Soc. Of Agron., Inc. Crop Science of America. America.

[2] Solbrig, O.T., D.J Solbrig., 1994. So shall you reap. Farming and crops in human affairs. Island Press. Washington, D.C.

[3] Davies, F. T., T.M. Davies, D.E. Kester, 1994. Commercial importance of adventitious rooting to horticulture. In T. M. Davis and B. E. Hassing, eds. *Biology of adventitious rooting.* Plenum Press, New York and London. pp.53-60.

[4] MacDonald, A. B., 1986. Propagation facilities-past and present. *Comb. Proc. Intl. Plant Prop.* Soc. 35: 170-75.

[5] Ritchie, G. A., 1994. Commercial application of adventitious rooting to forestry. In T. M. Davis and B. E. Haissing, eds. *Biology of adventitious root formation.* Plenum Press, New York and London. pp. 37-52.

[6] Czabator, F.J., 1962. Germination value: an index combining speed and completeness of pine seed germination. Forest Science 8: 386-396.

[7] Nichols, M. A. and Heydecker, W., 1968. Two approaches to the study of germination data. Proceedings of the International Seed Testing Association 33: 531-540.

[8] Brown, R.F., Mayer, D.G., 1988. Representing cumulative germination. 1. A critical analysis of single-value germination indices. Annals of Botany 61: 117-125.

[9] Bewley, J.D., Black, M., 1994. Seeds: physiology of development and germination. 2nd ed. Plenum Press, New York.

[10] Santana, D.G., Ranal, M.A., 2004. Análise da germinação: um enfoque estatístico. Editora Un B, Brasília.

[11] FRI. 1983. Troup's The Silvicultureof Indian Trees. Vol. IV. The Controller of Publications, Delhi.

[12] Sasidharan, N. 2004. Biodiversity documentation for Keralaa Part 6: Flowering Plants. KFRI publication, KFRI, Peechi, Thrissur.

[13] Nair, K. K. N., 2000. Manual of non-wood forest produce plants of Kerala. KFRI publication, KFRI, Peechi, Thrissur.

[14] Nayagam, J.R., 2015. Plantation Technology for Seven Tropical Tree Species. LAP Lambert Academic Publishing, OmniScriptum GmbH & Co. KG, Saarbrücken, Germany.

[15] Praddhan, P., L. Joseph, V. Joseph, V. Gupta, R. Chulet, H. Arya, R. Verma, A. Bajpai. 2009. Saraca asoca (Ashoka): A Review. Journal of Chemical and Pharmaceutical Research, 2009, 1 (1): 62-71.

[16] Sharma, S.D., Aier, N.B., 1989. Seasonal rooting behaviour of cuttings of plum cultivars as in.uenced by IBA treatments. Scientia Horticulturae, 40: 297–303.

[17] Howard, B.H., 1996. Relationships between shoot growth and rooting of cuttings in three contrasting species of ornamental shrub. *J. Hortic. Sci.* 71, 591–605.

[18] Rosier, C.L., Frampton, J., Goldfarb, B., Blazich, F.A., Wise, F.C., 2004. Growth stage, auxin type, and concentration in uence rooting of stem cuttings of Fraser r. Hort Science, 39: 1392–1396.

[19] Hartmann, H.T., Kester, D.E., Davies, F.T., Geneve, R.L. 1997. Plant propagation principles and practices. 6th edition, Asoke K. Ghosh, Prentice-Hall of India. Private Limited, M-97, Connaught Circus, New Delhi – 110 001.

[20] Hambrick, C.E., Davies, F.T., Pemberton, H.B. 1991. Seasonal changes in carbohydrate/ nitrogen levels during field rooting of *Rosa* multi.ora 'Brooks 56' hardwood cuttings, Scientia Horticulturae, 46: 137–146. 1991.

[21] Akakpo, D.B., Amissah, N., Yeboah, J., Blay, E., 2014. Effect of Indole 3-Butyric Acid and Media Type on Adventitious Root Formation in Sheanut Tree (*Vitellaria paradoxa* C. F. Gaertn.) Stem Cuttings. American Journal of Plant Sciences, 5, 313-318.

6

Carbon Stock Variations Along Altitudinal and Slope Gradient in the Forest Belt of Simen Mountains National Park, Ethiopia

Tibebu Yelemfrhat Simegn[1,*], Teshome Soromessa[2]

[1]Ethiopian Wildlife conservation Authority, Addis Ababa, Ethiopia
[2]Center of Environmental Science, Addis Ababa University, Addis Ababa, Ethiopia

Email address:
tibes27@yahoo.com (Simegn T. Y.)

Abstract: Forests play a significant role in climate change mitigation by sequestering and storing more carbon from the atmosphere which is released by anthropogenic causes. The overall objective of this study was to measure Variations of Carbon Stock along Altitudinal and Slope Gradient in the forest belt of Simen Mountains National Park. And it aimed to add values of the lowland forest belt of the park for climate change mitigation contribution in Ethiopia. The work was accomplished properly using random sampling to estimate the forest carbon in above and below ground biomass by considered each trees and shrubs which had DBH ≥5 cm. Above ground biomass was estimated by using allometric models equation while below ground biomass was determined based on the ratio of below ground biomass to above ground biomass factors. Dead wood, leaf litter, herb and grass (LHG) and soil organic carbon were conducted according to sampling quadrates data and laboratory result. The results shown that, there were twenty species with a density of 2334 trees and shrubs in the study sites which had DBH ≥5 cm. The mean above ground and below ground biomass carbon stock were 270.89±154.50 and 54.18±30.81 t ha^{-1} respectively. The mean above ground biomass carbon per species was 20.42±17.99 t ha^{-1}. The mean carbon in dead wood, LHG and soil carbon were 0.7258±1.0479, 0.019±0.008 and 242.51±46.42 t ha^{-1} respectively. The total Carbone stock variation along altitudinal gradient was 542.6, 550.73 and 627.01 t ha^{-1} for upper, mid and lower altitude respectively by which higher amount of carbon was stored in lower altitude. The total carbon stock variation along slope gradient was 487.3, 557.00 and 625.877 t ha^{-1} for upper, mid and lower slope respectively by which higher amount of carbon was stored in lower altitude. So that in this study both altitude and slope had the same impact on carbon stock potential of the species in the areas of study.

Keywords: Allometric models, Anthropogenic causes, Biomass carbon, Mitigation, Simen Mountains National Park

1. Introduction

Carbon sequestration is the process of removing excess carbon dioxide (CO_2) from the atmosphere and depositing it in a reservoir [1]. It is a way to mitigate the accumulation of greenhouse gases in the atmosphere released by the burning of fossil fuel and other anthropogenic activities. Through biological, chemical or physical processes, CO_2, which is one of the greenhouse gases, can be captured from the atmosphere. While a carbon sink is a reservoir that collects and stores carbon containing chemical compound, it removes CO_2 from the atmosphere through absorption. Forest and soil are potential sinks for elevated CO_2 emissions and are being considered in the list of acceptable offsets [1]. Sustainable forest development and forested landscape expansion is one of the key approaches for reducing atmospheric carbon concentration. It is a safe, environmentally acceptable, and cost-effective way to capture and store substantial amounts of atmospheric carbon. The concurrent development of tradable carbon credits provides financial incentives for considering carbon storage in forest management decisions [2].

Carbon sequestration from atmosphere can be advantageous from both environmental and socio-economic perspectives. There are evidences from several studies in Ethiopia and other countries. The environmental perspective includes the removal of CO_2 from the atmosphere [3], the improvement of soil quality [4], and the increase in biodiversity[5]; while socioeconomic benefits include increased yields [6], monetary incomes from potential carbon trading schemes [7], normalizing droughts through its potential for creating atmospheric condensation making cloud seeding, as well as reducing flood hazards and increasing ground water recharge by increasing water infiltration through soil columns. Globally, forests act as a natural storage for carbon, contributing approximately 80% of terrestrial above-ground, and 40% of terrestrial belowground biomass carbon storage [8]. So biomass is an important element in the carbon cycle, specifically carbon sequestration. It is used to help to quantify pools and fluxes of greenhouse gases (GHG) from the terrestrial biosphere to the atmosphere associated with land use land cover changes [9]. There are many conventional methods for quantification of sequestered carbon. Many of these methods are complicated, expensive and limited in their coverage. Such limitations impede sound quantification and monitoring of carbon [10]. One of such approaches is forest inventory data sets which often provide the required base line data to enable the large area mapping of biomass and subsequent carbon accounting over a range of spatial and temporal scales. However, spatially explicit estimates of biomass over large areas may be limited by the spatial extent of the forest inventory relative to the area of interest (*i.e.*, inventories not spatially exhaustive), or by the omission of inventory attributes required for biomass estimation. These spatial and attribution gaps in the forest inventory may result in an underestimation of large area biomass [11].

This research was conducted with the aim of understanding carbon stock variations along altitudinal and slope gradient in different carbon pools of the forest ecosystem Simen Mountains National Park, Ethiopia

2. Materials and Methods

2.1. Description of the Study Area

This study was undertaken in lowland forest of Simen mountains national park forest, Amhara National Regional State, North Gondar, Ethiopia located at about 846 km North of Addis Ababa. The forest has an altitudinal gradient ranging from 1900 to 3000m above sea level with the highest peak at Ras dashin Mountain. The forest covers 171 hectares. The study area is characterized by moderate climate, locally known as woina dega and it has a mono modal rainfall distribution and the rainy season is from June to August with The annual average rain fall of SMNP 1367mm and the mean annual maximum and minimum temperatures are 11°C and 19°C, respectively.

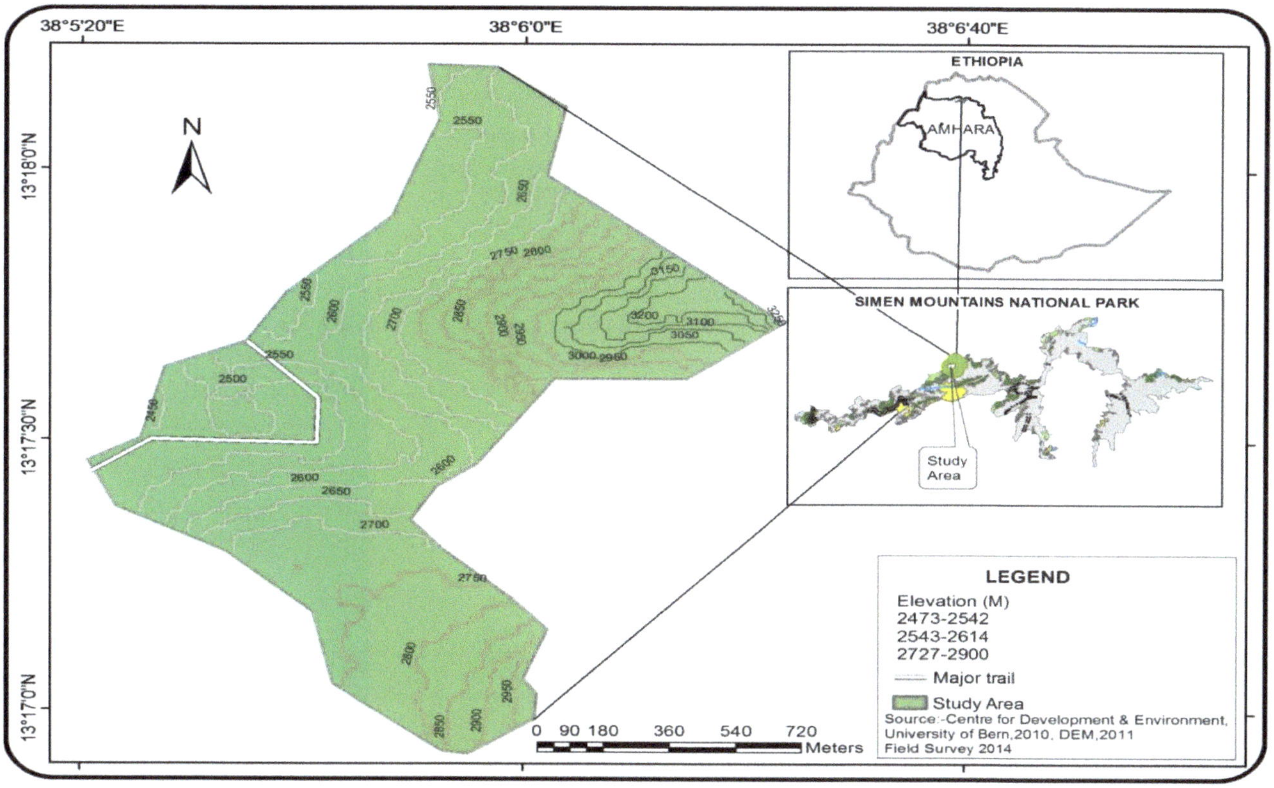

Figure 1. *Map of the study area*

2.2. Delineation of the Study Site

Delineation of the forest boundaries was the first step in forest carbon stock measurement. The boundary of the study forest area was delineated by taking geographic coordinates with GPS at each turning point. The GPS points that were taken from the study site to indicate each sample plots were recorded.

2.3. Sampling Techniques on the Field

Simple random sampling method was used to take samples. Sample plots were laid along line transects based on altitudinal variation of the study area. A randomly sampling plot of (10 m x 20 m) in each site was established. To reveal the tree composition and biomass, all live trees with a diameter ≥ 5 cm were recorded as indicated by [12]. The diameter was measured at breast height (DBH, 1.3 m height from the ground) to estimate biomass and the size class distribution of trees in a sampling plot. DBH was measured by using tree measuring tape. Trees with multiple stems connected near the ground were counted as single individuals and bole circumference was measured separately. Tree height was recorded by using measuring clinometers. The methodology and procedures used to estimate carbon stocks were simple step by step procedures using standard carbon inventory principles and techniques [12]. Procedures were based on data collection and analysis of carbon accumulating in the above ground biomass, below-ground biomass, leaf litter, and soil carbon of forests using verifiable modern methods.

2.4. Stratification of the Study Area

Stratification was done in the forest in order to take accurate data from the field as well as to maintain the homogeneity of the area. Altitude was the major parameter to classify the study area. The strata were defined at each elevation, starting from the bottom to the top of the mountain. Based on altitudinal variation, the study site was stratified into three zones namely: lower (2473-2542 m), middle (2543-2614 m) and higher (2615-2900m). Slope gradient was the second parameter to classify the area. Therefore, slope classes classified into lower (0-10%), middle (>10-20%) and higher (>20%). Aspect was also another parameter that was considered in the study forest and classified in to four classes: N (North), S (South), E (East), W (West).

2.5. Field Measurements

Sample plots (10m x20m) were laid through stratified random sampling method with nine transect lines in the various qualitatively classified biomass levels to account for the largest variability in the biomass range. Ground inventory data of tree parameters i.e., DBH and height of the trees were collected.

2.5.1. Vegetation Data Collection and Identification

The estimations of above and below ground carbon depend on the above ground biomass of living tree species. To estimate the above ground biomass all tree species within selected sample plots DBH ≥ 5cm were identified and recorded. Trees with multiple stems at 1.3 m height were treated as a single individual and DBH of the largest stem was taken. Plant identification was done by using Flora of Ethiopia and Eritrea.

2.5.2. Field Carbon Stock Measurement

The major activities of carbon measurement during the field data collection were above-ground tree biomass, below-ground biomass, leaf litter, and soil organic carbon measurements. Detailed methods are explained under the following sub headings.

(i). Above Ground Biomass (AGB)

The above ground biomass consists of all living vegetation above the soil, inclusive of stems, stumps, branches, bark, seeds and foliage. The DBH (at 1.3m) and height of individual trees greater than or equal to 5cm and DBH were measured in each sampling plots.

(ii). Litter Biomass (LB)

The leaf litter is defined as all dead organic surface material on top of the mineral soil. A quadrate with a size of 1 m × 1 m was established to sample litters. In each sample plots a total of five small quadrates were laid four at the corner and one in the center to minimize heterogeneity. The litter samples were taken in sub quadrate of (1 m × 1 m) along diagonal from one corner to the other and then the leaf litters within the $1m^2$ sub plots were collected. Laboratory analysis: The 100 gram sub sample fresh weights were sampled from the five sub-samples collected from each quadrant which were mixed homogenously and then taken to laboratory and oven dried at 105^oc.

(iii). Soil Organic Carbon (SOC)

Soil samples were collected from the field with five sub-plots within each major plot. The Samples were dug using core sampler with a diameter of 5 cm and the depth of the soil in which that took the sample was 30 cm. Mixing of soils was done properly by taking equal amount of soil from each sub plots to make a composite in order to make homogeneity. After organizing the samples in such a way, the samples were taken to Wondo Genet College of forestry and Natural resources for laboratory analysis. In the laboratory, soils are prepared and oven dried at 105^oc for 24 hours to remove the soil moisture so as to determine the percentage of organic carbon. Finally, the bulk density and soil organic carbon were determined after getting percentage of organic carbon.

2.5.3. Estimation of Carbon Stocks in Different Carbon Pools

(i). Estimation of Above Ground Carbon Stock (AGC)

As defined by [13]; allometric equation as a statistical relationship between key characteristic dimensions of trees that are fairly easy to measure, such as DBH or height, and other properties that are more difficult to assess, such as above ground biomass. The equation developed by [14], was used to calculate the above ground biomass as given below:

$$AGB = 34.4703 - 8.0671(DBH) + 0.6589(DBH2) \quad (1)$$

Where, AGB is above ground biomass, DBH is diameter at breast height.

(ii). Estimation of Below Ground Carbon Stock (BGC)

Below ground biomass estimation is much more difficult and time consuming than estimating aboveground biomass [15]. Roots play an important role in the carbon cycle as they transfer considerable amounts of Carbon to the ground, where it may be stored for a relatively long period of time. As indicated by [10], standard method for estimation of below ground biomass can be obtained as 20% of above ground tree biomass i.e., root to shoot ratio value of 1:5 is used. The equation is given below:

$$BGB = AGB \times 0.2 \quad (2)$$

Where, BGB is below ground biomass, AGB is above ground biomass, 0.2 is conversion factor (or 20% of AGB).

(iii). Estimation of Carbon Stocks in Dead Wood

The allometric equation confirmed in REDD methodology (2009) was used to estimate the amount of biomass in standing dead wood.

$$BSDW = \Sigma_{i=0}^{n} 1/3\ (D/200)^2\ h*s \quad (3)$$

Where, biomass is expressed in kg,

h = length (m), D = tree diameter (cm) and

s = specific gravity (g cm-3) of wood. The specific density is estimated at 0.5 g cm^{-3} as default value, but can be around 0.8 for dense hard woods and around 0.3 for very light species in tropical regions [16].

The carbon content in dead wood is calculated by multiplying total biomass of dead wood with the [17] default carbon fraction of 0.47.

(iv). Estimation of Carbon Stocks in the Litter Biomass

According to Pearson *et al.* (2005), estimation of the amount of biomass in the leaf litter can be calculated by:

$$LHGB = \frac{Wfield}{A} * \frac{Wsub_{sample(dry)}}{Wsub_{sample(fresh)}} * \frac{1}{10{,}000} \quad (4)$$

Where: LB = Litter biomass (ha^{-1})

W field = Weight of wet field sample of litter sampled within an area of size 1 m^2 (g); A = Size of the area in which litter were collected (ha);

W sub-sample, dry = Weight of the oven-dry sub-sample of litter taken to the laboratory to determine moisture content (g), and

W sub-sample, fresh = Weight of the fresh sub-sample of litter taken to the laboratory to determine moisture content (g). The carbon content of vegetation is surprisingly constant across a wide variety of tissue types and species. [18] noted that Carbon content of biomass is almost always found to be between 45 and 50% (by oven dry mass). In many applications, the carbon content of vegetation is estimated by simply taking a fraction of the biomass by multiplying 0.5.

C= 0.5* LB Where C= is carbon content by mass, and LB= is oven-dry biomass.

Therefore, total carbon content of litter (ton/ha) =Total dry litter biomass* carbon fraction

$$CL = LBM \times \%\ C \quad (5)$$

Where, CL is total carbon stocks in the litter in ton/ha,

%C is carbon fraction determined in the laboratory (Pearson *et al.,* 2005).

(v). Estimation of Soil Organic Carbon (SOC)

The carbon stock of soil was done by using the following formula which is recommended by [12] from the volume and bulk density of the soil.

$$V = h \times \prod r^2 \quad (6)$$

Where, V is volume of the soil in the core sampler in cm^3, h is the height of core sampler in cm, and r is the radius of core sampler in cm [12]. More over the bulk density of a soil sample was calculated as follows:

$$BD = Wav, dry/V \quad (7)$$

Where, BD is bulk density of the soil sample per, Wav, dry is average air dry weight of soil sample per the quadrant, V is volume of the soil sample in the core sampler auger in cm^3 [12].

$$SOC = BD * D * \%\ C \quad (8)$$

Where, SOC= soil organic carbon stock per unit area (t ha-1) BD = soil bulk density (g cm^{-3}),

D = the total depth at which the sample was taken (30 cm), and %C = Carbon concentration (%).

2.5.4. Estimation of Total Carbon Stock Density

The total carbon stock is calculated by summing the carbon stock densities of the individual carbon pools of the stratum using the [12] formula. Carbon stock density of a study area:

$$CT = AGC + BGC + LC + SOC \quad (9)$$

Where, CT = Total Carbon stock for all pools (ton/ha), AGC=above ground carbon stock (ton/ha), BGC= below ground carbon stock (ton/ha), LC=litter carbon stock (ton/ha) and SOC= soil organic carbon (ton/ha). The total carbon stock was then converted to tons of CO^2 equivalent by multiplying it by 44/12, or 3.67 as indicated by [19].

3. Data Analysis

After the data collection was completed, data analysis of various carbon pools measured in the forests were accomplished by organizing and recording on the excel data sheet. The data obtained from DBH, diameter, height of each species, field weight (W_w), fresh weight-(FW) and dry weight (W_{dry}) of LHG and soil were organized by excel 2007 and analyzed using and MINITAB software version 16. DBH data was arranged in classes ≤10, >10-20, >20-30, >30-40 and >40 for applying appropriate model of biomass estimation equation. The relationship between each parameter was tested by descriptive statistics. Differences at the 95% (α=0.05) confidence interval was used to see the significance differences.

4. Results

4.1. Carbon Stocks of Different Pools Along Altitudinal Variation

The presence of variation in altitudinal gradient affects the carbon stock of different pools in the forest. The lower parts of altitude is high in above ground carbon stocks while the upper and middle parts of altitude have low to moderate carbon stocks in above ground biomass. 318.8174, 258.247 and 240.6356 ton/ha carbon stocks were recorded at the lower, middle and upper altitude respectively.

Similar trend was shown in below ground biomass in which 63.76, 51.65 and 51.64 ton/ha carbon stocks were recorded in the lower, middle and upper altitude respectively with highest value found at the lower part of altitudinal classes followed by the middle and upper parts. But this is not very much significance at 95% confidence interval (F= 0.618, P= 0.543) in AGC and (F=0.618, P=0.543) in BGC stocks. It is also the same in the case of litter carbon stock and SOC. With significance difference at α=0.059 (F=0.991, P=0.378) in litter carbon stock and (F=0.653, P=0.525) SOC stocks. The litter and SOC carbon stock were higher in the lower parts of altitude and low in the upper altitude. 243.943, 240.3293 and 240.1949 ton/ha stocks of carbon were recorded in the lower, middle and upper altitude respectively in the soil pool. In general, the lower part of the altitude contains more carbon stocks (243.943 ton/ha), followed by the middle (240.329 ton/ha) and the upper altitudinal gradient (240.194 ton/ha).

4.2. Carbon Stocks of Different Pools Along Slope Gradient

The slope gradient was also a second factor which affects the carbon stocks of different pools in the studied forest. Above ground biomass and below ground biomass and their consecutive carbon stocks were found to be low in hilly areas of the forest due to the fact that no more vegetation cover were found there. The carbon stock of the middle slope gradient was moderate in both above ground carbon and below ground carbon stocks and higher in the upper slope gradient in both pools. 316.172, 260.355 and 187.128 ton/ha carbon stocks were recorded at the lower, middle and upper slope gradient respectively in above ground carbon stocks. Similar trend was also shown in below ground biomass in which 63.234, 52.071 and 37.426 ton/ha carbon stocks were recorded in the lower, middle and higher slope classes respectively with highest value found at the lower of slope classes followed by the middle and upper slope classes. But this was not statistically significance at α=0.05 (F= 0.169, P= 0.845) in above ground biomass and (F=0.169, P=0.845) in below ground biomass. At the same in the case of dead wood, litter carbon stock and SOC, there was no significance difference at α=0.05. The litter carbon stock was higher in the lower slope classes and low in the higher slope classes with its value arranged 0.020, 0.0171 and 0.017 ton/ha in the lower, middle and higher slope gradient respectively. The carbon stock in the soil pool was higher in lower slope classes and lower in the upper slope classes with moderate carbon stocks in the middle slope classes. 246.417, 244.453 and 240.974 ton/ha stocks of carbon were recorded in the lower, middle and upper slope classes respectively in the soil pool. However for dead wood carbon stock higher carbon stock in higher slope class (1.466 ton/ha) with lower in the lower slope class (0.290) and moderate stock in the middle slope class (0.421ton/ha).

Table 1. *Mean biomass and carbon stocks (t ha-1) in different carbon pools along altitudinal range*

Altitude class	Altitude range (m)	No of Plots	AGC	BGC	LC	DWC	SOC	Total carbon (ton/ha)
Upper	2615-2900	17	240.635	51.6494	0.459	0.016	240.195	542.582
Middle	2543-2614	19	258.247	48.1271	0.255	0.017	240.329	550.73
Lower	2473-2542	18	318.817	63.7635	0.896	0.021	243.943	627.014

Table 2. *Carbon stocks in different pools with respect to slope gradient*

Slope Class	Slope Range (%)	No of Plots	AGC	BGC	LC	DWC	SOC	Total carbon (ton/ha)
Upper	>20	22	187.128	37.426	0.017	1.466	240.974	487.396
Middle	>10-20	23	260.355	52.071	0.017	0.421	244.453	557.005
Lower	0-10	9	316.172	63.234	0.02	0.29	246.417	625.877

Table 3. *One-way ANOVA results with altitude gradients*

gradient	Carbon pools	F-value	P-value
Altitude	AGC	0.618	0.543
	BGC	0.618	0.543
	DWC	3.339	0.043
	LC	0.991	0.378
	SOC	0.653	0.525

**Bold values are significant at α=0.05 (95%)

Table 4. One-way ANOVA results with altitude gradients

gradient	Carbon pools	F-value	P-value
	AGC	0.169	0.845
	BGC	0.169	0.845
Slope	DWC	1.107	0.338
	LHGC	0.54	0.586
	SOC	0.077	0.926

5. Discussion

5.1. Storage of Biomass in Different Pools

The maximum above ground biomass per plot was 1795.64 ton/ha and the minimum was 162.10 ton/ha. The average biomass stock recorded in above ground biomass was 576.36 ton/ha. The results are more or less similar to the previous researches of above ground biomass of afromontane forest which were 567.2 ton/ ha [20] and 591.00 ton/ ha [21] while it was highly larger than Humbo forest of Wolaita Zone [22]. The average above ground biomass observed from these three location was 574.89 ton/ha. The global above ground biomass in tropical dry and wet forests ranged between 30-275 ton/ha and 213-1173 ton/ha respectively as indicated by [23], which is lower than the above ground biomass of Tara Gedam forest.

In this study, the differences in biomass and carbon accumulation among plots could be largely due to differences in the growth rates of plants as indicated by [24]. Litter constitutes an important flux of soil organic Carbon. The forest litter consisted of a relatively high number of trees, although the density varies among samples and species; in densely populated trees few litters were found due to the closeness of plants each other makes their litter not fall down [25]. The Carbon stocks in the litter of the study forest ranged from 2.6–3.8 ton/ha which were comparable to those reported for tropical seasonal rainforests (1.4 ton/ha) carbon [26] and tropical secondary forest at the Makiling Forest Reserve in the Philippines (1.9 ton/ha) [27]. On the other hand, the litter carbon varies on other tropical forests (2.6-3.8 ton/ha) as reported by [28]. The relatively low quantities of Carbon stored in litter carbon stock in the studied forest may be due to the high decomposition rate as reported in a 10-year study by [29] and Seasonal Wild fire distraction in some part of the forest and caring number of cattle during the winter that affect the litter carbon stock directly in the area.

The average values of soil organic carbon in the study area was 242.507 ton/ha, which was similar to the Carbon density estimates of Afromontane Rain Forests of the Eastern Arc Mountains which were found to be between 252 and 581 ton/ha as indicated by [30, 31]. The distribution of Carbon stocks in each sample plot of the study forest is known to vary due to the presence of different tree species, soil nutrient availability, climate, and topography and disturbance regime [32]. This indicates that, the higher soil organic carbon in the soil could sequester more CO_2. The bulk density of the soil in this study was found to be 0.78 g/cm^3 minimum value and 0.999 g/cm^3 maximum value with an average value of 0.93 g/cm^3. The presence of low bulk density in the soil indicates that the soil has high potentials to accumulate large amount of organic matter in it [32].

5.2. Carbon Stocks of Different Pools Along Altitudinal and Slope Gradient

Most of the time altitudinal and slope gradient were the factors that affect the storages of carbon in different pools. As observed, [33] a strong effect of slope on the SOC stock of a subalpine forest in the Olympic Mountains of Washington state. The carbon stock of the study forest was highest at the lower altitudinal range followed by the medium and lowest at higher altitude of the mountain. This is due to the decreasing of layer of large DBH trees at higher altitudinal range of the forest site and the layer of large DBH trees increases towards the lower altitudinal ranges naturally and trees and shrub deforestation was also higher in upper altitude. Especially, altitudinal variation has an impact on AGC, BGC and SOC stock because of its influence on soil water regime [34]. Similarly the carbon stock of the study forest was increased as the degree of slope gradient decreased. This is due to the inclined arrangement of the study forest that affects both forest biomass and soil nature of the area. Most steep areas were covered by lower plants (grasses, herbs and bushes) which have lower biomass and carbon relatively.

6. Conclusion

The study conducted in low land forest of Simen Mountain National Park showed that the forest contains moderately diversified plant species. A total of twenty different species of plants were collected, of which *Erica arborea* was the dominant in density and *Buddeleia polystachya* was the least dominant in the study site. Based on DBH and height class distribution the forest has a bit different trends observed in both classes. The densities of tree species decreases as the height class increases but For DBH, it has an irregular pattern. This implies that, the predominance of small sized tree and shrub species in the lower classes than in the upper classes. The analysis of these two parameters in the study forest indicated that higher percentage of number of tree species in the lower than in the higher height frequency classes. The carbon stocks of the study site shows a variation among the plots due to the presence of higher biomass in some plots and small biomass in other plots. The average carbon stock of the different carbon pools of this study was almost similar with most researches done in tropics and Ethiopia. The ANOVA result showed that at 95% confidence interval, the carbon stocks in the different carbon pools (AGC, BGC, DWC, LHGC and SOC) were different due to environmental factors (altitude and slope). The lower parts of altitude class carbon stock potential were higher in all pools (except for the DWC pool which is significant for most carbon pools) while the upper and medium parts of altitude had low to moderate carbon stock. This is due to the fact that there were dense vegetation cover in the lower altitudinal range and the presence of favorable conditions in this part. DWC was higher in higher altitude and low to moderate in medium and lower altitude. This is because the higher altitude part of the forest was under less nutrient and different environmental factors like landslide and erosions that facilitate dead wood accumulation. At same time the litter and soil carbon stocks were higher in the lower parts of altitude and slope with lower in the higher altitude due to the presence of large decides forest and by the nature of the lower part to receive the medium and higher feeds to down. The carbon stock of the higher slope gradient was lower in all four pools (AGC, BGC, LHGC, and SOC) and higher to moderate stock in lower and medium slope gradients. This is because the better density of the study forest was found in lower and medium slope area due to the presence of stable environmental conditions better relative disturbance. But DWC was higher in higher slope because environmental pressure is high in this range which affects plant life span.

References

[1] UNFCCC,"Kyoto Protocol to the United Nation Framework Convention on Climate Change". Retrieved 18/09/2010, from http://unfccc.int. University Press, Cambridge,UK 1997.

[2] Siry, J., Bettinger, P., Borders, B., Cieszewski, C., Clutter, M., Izlar, B., Markewitz, D. and Teskey, R, "Forest Carbon Estimation Protocol for the State of Georgia", USA 2006.

[3] Yetebitu Moges, Zewdu Eshetu and Sisay Nune, "Manual for assessment and monitoring of carbon in forest and other land uses in Ethiopia" Ethiopian Forest Research Center, Addis Ababa, Ethiopia 2010.

[4] Zewedu Eshetu "Forest soils of Ethiopian highlands: Their characteristics in relation to site history". Doctorial thesis, Swedish University of Agricultural Sciences, Sweden 2000.

[5] Batjes, N.H. and Sombroek, W.G." Possibilities for carbon sequestration in tropical and subtropical soils." Global Change Biology 1997 3: pp,161–173.

[6] Sombroek, W.G., Nachtergaele, F.O. and Hebel, A,"Amounts, dynamics and sequestering of carbon in tropical and subtropical soils". 1993, 22: 417–426.

[7] McDowell, N, "Developing countries to gain from carbon trading fund", Climate Change Mitigation: Cam bridge, U.K.: Cambridge University Press 2002.

[8] Kirschbaum, M.U.F. "The carbon sequestration potential of tree plantations in Australia" In: (eds.), Environmental Management: The Role of Eucalypts and Other Fast Growing Species, Eldridge, K.G., Crowe, M.P., Old, K.M. Forestry and Forest Products, 1996 ,20:77–89.

[9] Cairns, M. A., Brown, S., Helmer, E. H., Baumgardner, G. A,"Root biomass allocation in the world's upland forests", Oecologia 111: 1-11, 2003.

[10] MacDicken, K. G, "A Guide to Monitoring Carbon Storage in Forestry and Agro- forestry Projects. In", Forest Carbon Monitoring Program. Winrock International Insti-tute for Agricultural Development, Arlington, Virginia 1997.

[11] Wulder, M.A., White, J.C., Fournier, R.A., Luther, J.E., and Magnussen, S, "Spatially Explicit Large Area Biomass Estimation", Three Approaches Using Forest Inventory and Remotely Sensed Imagery in a GIS 2008, 8:529-560.

[12] Pearson, T., Walker, S. and Brown, S, "Sourcebook for land-use, land-use change and forestry projects": Win rock International and the Bio-carbon fund of the World Bank. Arlington, USA, pp. 19-35, 2005.

[13] Bhishma, P.S., Shiva, S.P., Ajay, P., Eak, B.R., Sanjeeb, B., Tibendra, R.B., Shambhu, C. and Rijan, T, "Forest Carbon Stock Measurement: Guidelines for Measuring Carbon Stocks in Community Managed Forests": 2nd ed. International Centre for Integrated Mountain Development, Asia Network for Sustainable Agriculture and Bioresearches, Federation of Community Forest Users, Kathmandu, Nepal, pp. 15-36, 2011.

[14] Brown, S.A.J., Gillespie, J.R. & Lugo, A.E, "Biomass estimation methods for tropical forests with application to forest inventory data". *For. Sci.*, 35(4): 881–902, 1989.

[15] Russell AE, Raich JW, Valverde OJ, Fisher RF "Tree species effects on soil properties in experimental plantations in tropical moist forest" Soil Sci Soc Am J 71:1389–1397, 2007.

[16] Hairiah, K., Sitompul, S.M. van Noordwijk M. and Palm C, "Methods for sampling carbon stocks above and below ground" International Centre for Research in Agroforestry; Bogor, Indonesia 2001.

[17] Intergovernmental Panel on Climate Change (IPCC). "Guidelines for National Greenhouse Gas Inventories" National Greenhouse Gas Inventories Program, IGES, Japan 2006.

[18] Schlesinger, W.H, "Biogeochemistry, an Analysis of Global Change", New York, USA, Academic Press 1991.

[19] Pearson, T. R., Brown, S. L. and Birdsey, R. A, "Measurement guidelines for the sequestration of forest carbon", Northern research Station, Department of Agriculture, 2007, Washington, D.C, pp. 6-15.

[20] Getachew Tesfaye, "Structure, biomass and net primary production in a dry tropical afromontane forest in Ethiopia". Addis Ababa University, Ethiopia.

[21] Muluken, N.B., Teshome S., Eyale B., "Above- and Below-Ground Reserved Carbon in Danaba Community Forest of Oromia Region, Ethiopia: Implications for CO2 Emission Balance", American Journal of Environmental Protection. 2015, 4(2): 75-82. doi: 10.11648/j.ajep.20150402.11

[22] Alefu C., Teshome S., Eyale B., "Carbon Stock in Woody Plants of Humbo Forest and its Variation along Altitudinal Gradients: The Case of Humbo District, Wolaita Zone, Southern Ethiopia", International Journal of Environmental Protection and Policy. 2015, 3(4): 97-103. doi: 10.11648/j.ijepp.20150304.13

[23] Murphy, P. G. and Lugo, A. E," Ecology of tropical dry forest", Ann. Rev. Ecol. Syst. 17:67-88, 1986.

[24] Brown, S. "Estimating biomass and biomass change of tropical forests" a Primer *FAO Forestry Paper* 1997,134:120-130.

[25] Demel Teketay "Germination ecology of 12 indigenous and 8 exotic multipurpose legu minous species from Ethiopia". *Forest Ecology and Management* 1996, 80: 209-223.

[26] Lüet, Yin JX, Jepsen MR and TangJW, "Ecosystem carbon storage and partitioning in a tropical seasonal forest in Southwestern China". *Forest Ecology and Management* 260: 1798–1803, 2010.

[27] Lasco RD, Guillermo IQ, CruzRVO, Bantayan NC & Pulhin FB, "Carbon stocks assess ment of a secondary forest in Mount Makiling Forest Reserve", Philippines. *Journal of Tropical Forest Science 2004,* 16: 35–45.

[28] Lugo, A. E; Parrotta, J. A; and Brown, S, "Loss in species caused by tropical defore station and their recovery through management" 1993, 22: 106-109.

[29] Moges, Y., Eshetu, Z., and Nune, S, "Current Status and Future Management Options In View of Access to Carbon Finances". Ethiopian Climate Research and Networking and the United Nations Development Programme (UNDP) Addis Ababa, Ethiopia 2010, PP.56.

[30] Munishi PKT, The Eastern Arc Mountains Forests of Tanzania: "Their Role in Biodiversity and Water Resources Conservation and Net Contribution to Atmospheric Carbon", PhD Thesis 2001, North Carolina State University, Raleigh, NC USA. pp.128.

[31] Munishi PKT & Shear TH, "Carbon storage in afromontane rain forests of the Eastern Arc Mountains of Tanzania": their net contribution to atmospheric carbon. Journal of Tropical Forest Science 2004, 16: 78–93.

[32] Houghton RA,"Aboveground forest biomass and the global carbon balance" Global Change Biology 11: 945–958, 2005.

[33] Brady, N. C, "The Nature and Properties of Soils", 8th ed., Macmillan, New York 1974.

[34] Powers, J.S., Schlesinger, W.H. "Relationships among soil carbon distributions and bio physical factors at nested spatial scales in rainforests of northeastern Costa Rica", 109:165–190, 2002.

7

Role of Enset (*Ensete ventricosum* (Welw.) Cheesman) in Soil Rehabilitation in Different Agro-ecological Zones of Hadiya, Southern Ethiopia

Chakoro Tamire[1], Mekuria Argaw[2]

[1]Environemntal Management Team, Ethiopian Railways Corporation/ERC, Addis Ababa, Ethiopia
[2]Center for Environmental Science, Addis Ababa University, Addis Ababa, Ethiopia

Email address:
oronurebo@gmail.com (C. Tamire), m_argaw@yahoo.com (M. Argaw)

Abstract: Soil degradation is a problem both in developed and developing countries. Its impact is of great concern to farmers that depend on rain fed agriculture. The objective of this research was to assess the role of enset in soil rehabilitation. Household survey and soil quality analysis were conducted to collect primary data from the fields. Soil laboratory analysis result revealed that enset field had significantly ($p<0.05$) higher level of TN (0.3%), OC (3.6%), Av. P (66.7ppm) and lower level of BD (1.2) than annual crop fields. This might be due to the materials added to the soil from the plant parts and organic farming (application of manure and domestic waste). Enset has highly degradable parts such as leaf midribs, pseudostem sheath, pseudostem core/ which are used to maintain soil fertility. Laboratory analysis of soil has shown that an average size of clay, silt and sand fraction under enset cultivation was 31.2%, 49.3% and 19.5%, respectively, while in the annual crop land the proportion of clay, silt and sand were 37.5%, 40.3% and 22.2%, respectively. The mean value of TN and OC were found to be greater in *dega* than the *w/dega* and *dry w/dega* zones of agro-ecology. The overall output from the research showed that enset has contribution in environmental rehabilitation.

Keywords: Agro-ecology, *Enset ventricosum*, Soil Rehabilitation

1. Introduction

The rehabilitation of degraded soil and water resources will be critical in meeting future global food demand in the context of future climate variability and in addressing associated negative impacts on the environment that are a consequence of current agricultural systems. At the global level, combating soil degradation (soil rehabilitation) will help offset greenhouse gas emissions, provide a better environment, guarantee more food to an increasing population, and contribute to the economic progress of future generations [2]. However, in Ethiopia there is practice of rehabilitating the degraded low-potential areas mainly by construction of physical soil and water conservation structures along with tree planting on hillsides [5].

Ensete venticosum is a monocot perennial crop that belongs to order S*chistaminae*, family *Musaceae* and genus *Enset*. Genus enset has different species, but there is some ambiguity regarding their numbers. However, all authors agree that *Ensete ventricosum* is the only cultivated species in Ethiopia. It looks like a large, thick single stemmed banana like plant [14]. Usually it is larger than banana and 6-12 meters tall. The leaves are 5-7 meters tall and 1 meter in diameter and are more erect than a banana plant [18]. The stem has three parts. The pseudo-stem, which is made of tightly clasping leaf sheaths, is 2-3 meters in height and which are average of 1 meter diameter. The underground corm is an enlarged lower portion of the stem with an average of 0.7 meter length. The true stem is between the pseudo-stem corms near the ground. Usually it grows up during maturity and initiates a single flower head, which forms multiple flower fruits and seeds [18].

In land point of view, fertility is suitable for an environment for production and in which the climate and the soil are the main components [13]. Soil fertility is made up of three important properties (physical, chemical and biological) that greatly influence the soil's suitability for production. According to this, some parameters of soil fertility are total nitrogen, phosphorous, organic carbon, pH, bulk density and some others. According to this, some parameters of soil fertility are discussed below which are used to study soil rehabilitation.

Total nitrogen measures the total amount of nitrogen present in the soil, much of which is held in organic matter and is not immediately available to plants. It may be mineralised to available forms. It composed of inorganic (NH_4^+, NO_3^- and NO_2^-) and organic forms (OM) are subject to change due to various factors. Management (cropping, fertilization, erosion and leaching) and climate (temperature and moisture) determine its level and dynamics [8]. Climatic conditions, especially temperature and rainfall generate dominant influence on amounts of nitrogen and organic matter found in soils. As one moves from warmer to cooler zones, the organic matter and nitrogen contents of comparable soils tend to increase [14].

Phosphorus forms the most critical essential element in influencing plant growth and production across the world. Unlike nitrogen, it is not supplied through biochemical fixation but must come from other sources to meet plant requirements [14]. The sources include commercial fertilizer, animal manure, plant residues, human, industrial and domestic wastes and native compounds of phosphorus. Hence, its deficiency is directly related to food security issues, especially in the tropics where severe soil degradation is responsible for serious deterioration in soil quality [7, 12].

Soil organic carbon is important for the function of ecosystem and agro-ecosystems having a major influence on the physical structure of the soil, the soil's ability to store water (water holding capacity), and the soil's ability to form complexes with metal ions and supply nutrients. Loss of SOC can, therefore, lead to a reduction in soil fertility, land degradation and even desertification. Hence, the purpose of this research was to asses the role of enset in soil rehabilitation at the time of climate related hazard has occurred.

2. Materials and Methods

2.1. The Study Area

Hadiya zone is one of the zones of southern nations, nationalities and peoples region. It is situated roughly at the margin of the great Ethiopian Rift Valley at western margin in north western part of SNNPR. It located between $7^{0}07'$- $7^{0}92'$N and $37^{0}29'$- $38^{0}13'$E. The zone bordered in East and North East with Alaba Special Wereda and Siliti Zone, respectively. In the North it bordered with Kembata Tembaro, in the South East it is bordered with Oromiya region. Hossana is the capital city of the zone and it is 232km away from Addis Ababa.

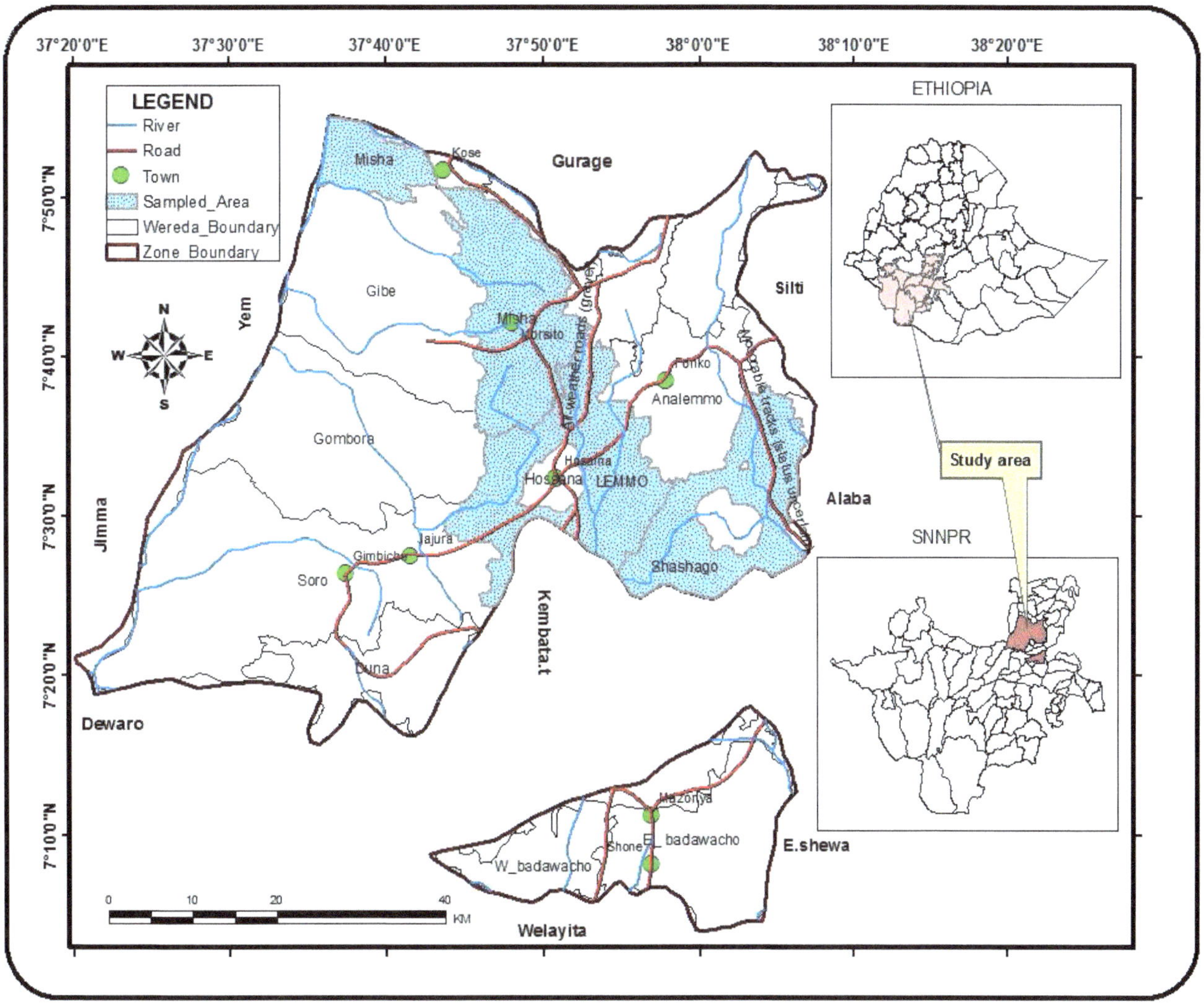

Figure 1. Map of the study area.

Topography of the zone is sloppy (54%) and flat (46%) out of the total land [22]. The altitudinal variation between the average highest peak and the lowest point ranges from 3200m at Tulla in Misha district and 860m at Gortancho and Gibe valley in Soro district, respectively.

2.2. Soil Sampling and Laboratory Analysis

In order to see whether enset contributes towards the rehabilitation of degraded land through soil fertility improvement, soil samples were taken and analysed.

The soil samples were taken from annual crop fields and enset field from *dega, woina dega* and *dry/w/dega* AEZs. Equal numbers of soil sample (i.e.15) were taken from each field. At each sampling field five sub samples were collected by using soil auger at depth of 0-30cm. The five soil samples were mixed of which 1kg of composite soil sample was taken for laboratory analysis.

The collected samples were analysed at Holeta Agricultural Research Center. Finally, the following soil parameters analysed: soil reaction (pH), organic carbon (OC), total nitrogen (TN), available P (Av. P), bulk density (BD) and soil texture. These physicho-chemical soil parameters were analysed using a standard methods. The total nitrogen content in soils was determined using the Kjeldahl procedure by oxidizing the organic matter with sulphuric acid and converting the nitrogen into ammonium. Then ammonium was determination in the digest. Soil pH was determined by potentiometric method at soil to water ratio of 1:2. Organic carbon (OC) was determined by Walkley Black Method.

Depending on soil pH of the available phosphorous was determined by Olsen method. The soil texture was determined by the Bouyoucos Hydrometer method it was calibrated at 67^0F to read directly in grams or percentage material remaining in suspension with a fixed amount of soil. The bulk densities of the soil sampled were estimated by taking undisturbed soil by core sampler after drying the soil samples in an oven at 105^0C to constant weights. In addition to this the researchers were used house hold survey, key informants interview and field observation.

3. Results and Discussion

The field survey report and soil laboratory analysis showed that enset has a role in soil rehabilitation. What was observed in the study area is that, continuous cultivation of crop without using any conservation measure in annual crop land; and climate related hazards (drought and runoff) has negative effect on the land. This practice might lead to soil erosion and soil nutrient depletion hazard, which further enhanced the land for declined crop productivity and food insecurity.

People of Hadiya have rehabilitated the soil by cultivating enset crops near the households. Because of its farming system (organic farming), protecting of soil from evapotranspiration by broad leaves (canopy), water holding and releasing properties to the soil, that increase the fertility of the soil.

3.1. Effect of Enset in Chemical Properties of Soil

Some chemical parameters such as total nitrogen (TN), organic carbon (OC), available phosphorus (Av. P) and soil reaction (pH) of soil were used to describe the contribution of enset by comparing with annual crop fields, and with the general standards. The results obtained to evaluate the effect of enset on soil chemical properties are presented in table 1 below.

Table 1. *Mean values ± SEM of soil chemical properties under the enset and annual crops.*

No	Soil parameters	Mean value and rank			
		Enset		Annual crop	
		Mean value	Rating	Mean value	Rating
1	pH H_2O (1: 2)	6.94±0.153	Neutral	6.17±0.130	Slightly acidic
2	OC %	3.60±0.280	Very high	2.58±0.256	High
3	Av. P (ppm)	66.76±12.092	Very high	15.16±3.653	Medium
4	Total N (%)	0.30±0.020	High	0.23±0.021	Medium

In addition to the above table the effect of enset in chemical properties of soil in different agro-ecological zones are shown in table 2 below.

Table 2. *Mean value ± SEM of soil chemical properties under the enset in different AEZs.*

No	Soil parameters	Mean value		
		Dega	w/dega	Dry w/dega
1	pH H_2O (1: 2)	7.18±0.138	7.27±0.25	6.36±0.20
2	OC %	4.33±0.26	4.18±0.23	2.28±0.22
3	Av. P (ppm)	46.20±8.24	86.28±30.0	67.80±17.45
4	Total N (%)	0.36±0.23	0.30±0.36	0.23±0.02

Total Nitrogen/TN

Total nitrogen is one of the indicators of the fertility status of the soil. According to the soil analysis result, the average value of total nitrogen under enset and annual crop land was 0.3% and 0.23%, respectively (Table 1). Further 'Levene's Test for Equality of variances' in t-test showed significant difference ($t = 2.200$ and $sig = 0.036$) at ($p < 0.05$) level between the two groups (under enset and annual crop land) of total nitrogen. Literally the amount of total nitrogen in enset field was greater than annual crop field; this shows that enset has an effect to maintain the amount of nitrogen in the soil.

The laboratory results confirm that enset could rehabilitate the degraded land by increasing important soil nutrients. This might be due to the materials added to the soil from the plant parts and organic farming (application of manure and domestic waste). Enset has highly degradable parts such as leaf midribs, pseudostem sheath, pseudostem core/ which are used to maintain soil fertility.

Their degradability is higher than green chloris gayana, setaria grass, elephant grass and guatemala grass [3]. Therefore, there is a great opportunity to expand enset agriculture, and by it environmental rehabilitation is high [19], leaching and losses of plant nutrient, particularly nitrogen, may be reduced by enset as compared to annual crops. This should be possible because of the continuous soil occupation by roots.

Some literature has rate the content (%) of nitrogen as very high (>0.5), high (0.25-0.5), medium (0.15-0.25), low (0.05-0.15) and very low (<0.05). Based on this, enset land could be classified in the higher rank (Table 1).

The mean value of total nitrogen (TN) was found to be greater in *dega* than the *w/dega* and *dry w/dega* zones of agro-ecology (Figure 2; Table 2). The mean TN content in the zones were 0.36%, 0.30%, and 0.23% for *dega*, *w/dega* and dry *w/dega*, respectively (Table 2). The mean values of TN content from enset land of the different zones of agro-ecology are shown in figure 2 below.

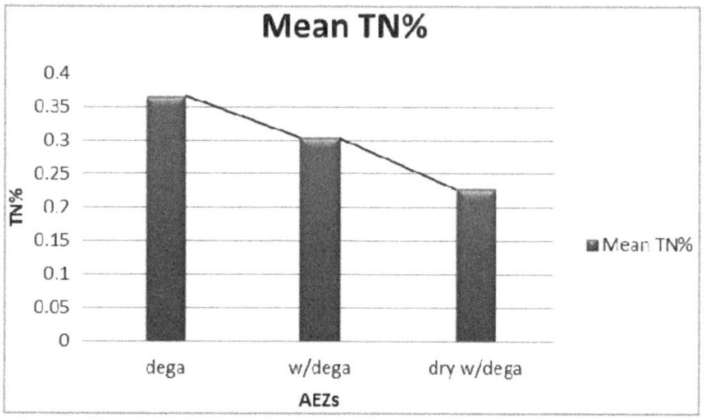

Figure 2. *Total nitrogen amount in different AEZs of the study area.*

ANOVA result proves that significant difference was observed at ($P < 0.05$) under the agro- ecological zones from the enset cultivated field. Based on the ANOVA analysis, TN was significantly increasing in highlands (rainfall and elevation is high) than other AEZs. Further the statistical test of LSD verifies that the mean TN content of enset field in *dega* show positive significant difference from *dry w/dega*; but no significant difference was found from *w/dega*. Even the *w/dega* had high mean value of TN from *dry w/dega*; it was not significantly different. However, the mean difference in each consecutive agro-ecological zone was not significant.

The results also confirm that enset has an effect on total nitrogen; as the elevation and coolness increase the coverage and cultivation of enset increase and in turn the fertility also increases. As altitude increases the suitability to enset cultivation also increases and total nitrogen availability increases in a given enset cultivation field. As one moves from warmer to cooler zones, the organic matter and nitrogen contents of comparable soils tend to increase [14].

This might be because of climatic conditions; especially temperature and rainfall which generate dominant influence on amounts of nitrogen and organic matter found in the soils.

Organic Carbon/Organic matter

The mean value of organic carbon was 3.60% and 2.58% for enset and annual crop land, respectively (Table 1).The t-test for equality of means verifies that difference between the two mean values were also statistically significant at ($P<0.05$). Organic carbon content of the soil under enset was higher which may be related to organic materials incorporated to soil from the plant and application of cattle manure. Soil organic matter consists of roots, plant residues, and soil living or dead organisms [4]. Further enset cultivation attracts most of the organic waste whereby farmers allocate about 80% of the organic manure and crop residue to the enset and banana fields, including the nutrients coming from outfield in terms of feed and mulch [11, 21].

Organic matter has been termed the "life blood" of soils. Pearson correlation analysis in table 19 shows that organic carbon has significantly strong correlation with TN (0.913) and pH (0.57). It has insignificantly strong correlation with BD (0.69), but week correlation with Av.P (0.273).

It has a tremendous impact and influence upon the chemical, physical and biological properties of the soil, soil fertility status, plant nutrition and biological activity in the soil [4, 14]. The same to organic carbon the level of the organic matter evaluated using soil organic carbon was higher under enset. Organic matter is calculated from the levels of organic carbon in the soil by multiplying by 1.72 [9, 16]. The reason behind the difference of organic matter may be due to framers' management practices in enset plot, where they enrich it with manure and crop residues.

Again the reason for soil organic matter accumulation could be high enset cover in enset land which result litter input and thus higher accumulation of organic matter in the soil. An improvement in organic matter content follow the enset cultivation is important sign of soil restoration. By which both physical and chemical properties of soils can be improved. Some research note that lowering of organic matter may increase soil erosion and degradation, and enhanced greenhouse effect from released CO_2. According to [16], organic carbon was classified in rat of very high and high respectively for enset and annual crop land (Table 1). The mean value of organic carbon (OC) was found to be greater in *dega* than the *w/dega* and *kola* or *dry w/dega* zones of agro-ecology (Table 2).

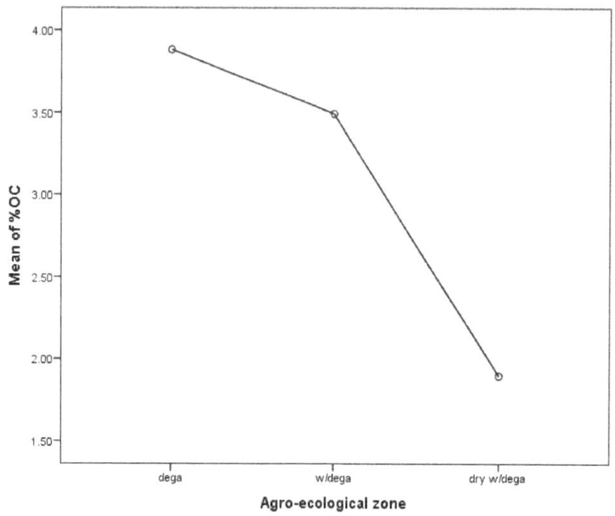

Figure 3. *The mean means plots of organic carbon in different AEZs.*

ANOVA result proves that significant difference was observed under the agro-ecological zones from enset cultivated field at ($P < 0.05$). This may be because of climatic conditions; especially temperature and rainfall that generate dominant influence on amounts of nitrogen and organic matter found in soils. The mean values of organic carbon content of the different AEZs from enset cultivated fields are shown below in means plot (Figure 3).

The statistical test of LSD verifies that the mean OC content of enset field in *dega* shows positive significant difference from dry *w/dega*; but no significant difference was found from *w/dega*. However, *w/dega* had significantly high mean value of OC than *dry w/dega* of enset field. The amount of organic carbon from enset land at *dega* and *w/dega* has been grouped in to very high (>0.3) (Table 2). According to [16], this kind of soil has good structural condition and high stability, and probably water is repellent.

Soil Reaction (pH)

Soil reaction (pH) is one of the soil chemical properties that indicate soil quality in this study. The mean pH value obtained underneath the enset was 6.94 while that obtained from annual crop land was 6.17. The t-test for equality of means verifies that difference between the two mean values was also statistically significant at ($P<0.05$).

According to [15], pH of the soil from the enset and annual crop land was nearly neutral and slightly acidic, respectively. This pH range is preferable for many crops [16]. Therefore, enset cultivation is important to moderate the soil to the crops.

Available Phosphorus

Next to nitrogen, phosphorus is the most critical essential element in influencing plant growth and production across the world. Laboratory result of soil analysis further revealed that the mean available phosphorus was 66.76ppm and 15.16ppm for enset land and annual crop land, respectively (Table 1). Independent sample t-test proves that enset land has significantly high Av. P mean value than annual crop land at ($P<0.05$) level. The increased mean Av.P value at enset land was compared to that of the annual crop cultivated land probably due to the applications of organic materials such as manure, plant residues. This might show that enset crop cultivation has an effect to increase Av. P in its field.

Mulching and the addition of organic matter to the soil can keep phosphorus from becoming bound to aluminium and other ions in acid soils, thus making it more available for plants [10]. Its sources include commercial fertilizer, animal manure, plant residues, human, industrial and domestic wastes and native compounds of phosphorus. Hence, its deficiency is directly related to food security issues, especially in the tropics where severe soil degradation is responsible for serious deterioration in soil quality [7, 12].

According to study conducted by [6], whole farm nutrient flow analysis by farmers, and yield variability across the farm units in the area indicated that there was a positive nitrogen and phosphorus balance, while the outfields indicated extremely low nitrogen and phosphorus balances for all farmers of various social categories.

Table 17 shows that the mean values of Av.P for *dega, w/dega* and *dry w/dega* were 46.2ppm, 86.28ppm, 67.8ppm, respectively.

However, multiple comparison tests using the tamhane showes that, no significant difference in each agro-ecological zone. This could shows that phosphorus in the *w/dega* enset field is more available for crop than other AEZs.

3.2. Effect of Enset in Physical Properties of Soil

The soil physical properties are one of the soil quality indicators that may be changed through enset cultivation. For this study bulk density and soil texture were taken as a physical indicator of soil fertility. The results obtained to evaluate the effect of enset on soil physical properties are presented in table 3 below.

Table 3. *Mean values ± SEM of soil physical properties under the enset and annual crops.*

No	Soil parameters	Mean value and Rank			
		Enset		Annual crop	
		Mean value	Rating	Mean value	Rating
1	BD	1.2±0.03	Low	1.3±0.03	Low
2	Texture	-	Silt clay loam	-	Clay loam

In addition to the above table the effect of enset in physical properties of soil in different agro-ecological zones are shown in table 4 below.

Table 4. *Mean value ± SEM of BD and soil texture under the enset in different AEZs.*

No	Soil parameters	Mean value		
		Dega	w/dega	Dry w/dega
1	BD g/cm^3	1.26±0.24	1.12±0.07	1.22±0.03
1	Clay %	34.5±2.76	28.0±1.23	40.5±2.62
2	Silt%	50.25±3.16	45.5±1.39	38.75±4.05
3	Sand%	15.25±1.91	26.5±1.0	20.75±4.28
	Class	Silty loam	Silty clay loam	Silty clay

Bulk Density/BD

The soil lab analysis showed that, the mean amount of BD of enset field and annual crop field were 1.2 g/cm^3 and 1.3 g/cm^3, respectively (Table 3). Enset field has low (1.2g/cm^3) BD with no significant difference at ($P<0.05$) level. BD describes the potential for leaching, productivity, and erosion.

Any practice of cultivating enset could improve soil structure. However, in some cases these improvements may not be sustainable at the time of scraping it, because of labour movement which may compress the soil. However, according to the [16], all the sampled soils were categorized in to the low rank (1.0 to 1.3). It indicates that these soils are structurally good because it has low bulk density and high pore space to aeration and infiltration.

In the beginning of the rainy season and after maturation, annual crops have little root proliferation and little effect on nutrient leaching. For established enset, roots already proliferate in the soil profile at the beginning of the rainy season [18]. It makes the environment favourable for plant and crop roots, which in turn increases soil fertility in general. According to [23], loose porous soils and those rich in organic matter have lower bulk densities which agree with this study.

The large mass of the plant should serve as a storage reserve, reducing the availability of the nutrients in the soil for leaching [18].

The mean value of BD was 1.26 g/cm^3, 1.12 g/cm^3, and 1.22 g/cm^3 for *dega, w/dega* and *dry w/dega* AEZs (Table 4). ANOVA revealed that no significant difference in amount of BD in different AEZs. Multiple comparison test using the tamhane also used showed that, no significant difference in each AEZs.

Soil Texture

Soil texture is another important physical property of soil. It indicates the potential rate of infiltration, water-holding capacity, degree of leaching and erodibility of the soil [20]. Laboratory analysis of soil has shown that average size of clay, silt and sand fraction under enset cultivation was 31.2%, 49.3% and 19.5%, respectively, while in the annual crop land the proportion of clay, silt and sand were 37.5%, 40.3% and 22.2%, respectively (Figure 4).

The mean value of textural classes was different on the enset and annual crop lands that is given in figure 4.

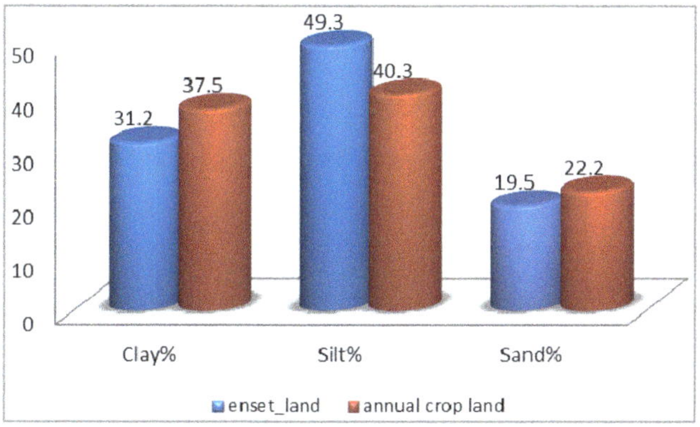

Figure 4. *The mean value of textural classes on the enset and annual crop lands.*

Based on the USDA soil texture triangle chart reading the above soil textural classes in the enset and annual crop land were silt clay loam and clay loam, respectively (Table 3). Based on this it is possible to suggest that both plots (enset and annual crop land) have good soil textural class. These are highly desirable for cultivation of many crops because it can stand drought and unsuitable condition and/or it has low evapotranspiration. Sometime organic substances are applied to the modified soil and have better physical characteristics, such as water holding capacity [18]. When compared to the sandy soils both have the ability of water holding capacity.

Figure 17 show that annual crop land (37.5%) has high amount of clay than enset (31.2%). However, clay cannot easily release water to the crops. Even clay soil has an ability to hold water so tightly that it is not easily available for plants [1]. By using soil textural triangle the soil class of *dega, w/dega* and *dry w/dega* were silty loam, silty clay loam and silty clay, respectively.

3.3. Conservation of Soil and Water by Using Enset Crop

According to the survey, about 86.8% of respondents indicated that enset is used to conserve soil and water. From all sampled household almost all were the users of enset to protect and reduce land degradation and soil erosion in addition to food source. This could be due to conservation tillage by enset canopy leaf, litter and long roots. Conservation tillage is one of the components of sustainable agriculture, which is environmentally friendly. US Soil Conservation Service, defines conservation tillage as any tillage system that leaves at least 30% of surface covered by plant residues for control erosion by water; for controlling erosion by wind [17]. Furthermore, farmers favour mulching crops like banana and enset because they know that it suppresses weed growth, maintains soil fertility and conserves moisture for those shallow rooted crops [11, 21].

Based on the field observation and response of farmers, enset can reduce soil erosion and land degradation. This is similar with the work of [15] and he noted that enset protects the soil from erosion and degradation of soil because of its canopy leaves, abundant accumulation of litter and continuous application of manure. The root system and non-fertilizer cultivation are other important characteristics to protect soil erosion and land degradation. Some respondents noted that they cultivate enset in the face of flood and drain it in to enset farm in order to reduce further soil erosion.

Literatures shares the above concept as enset is an erosion barrier particularly when it is planted on the top of the farm [21]. About 85.5% of respondents revealed that enset crop can reduce soil erosion with its canopy leaves and long roots. Its funnel like leaves and spongy root systems form mate-like structure in the root zone that minimize soil erosion and run-off, which ultimately improve the water and nutrient budget of the sub unit [21].

In order to understand enset's role in protection of soil erosion and land degradation according to different parameters, the researcher forwarded different questions for farmers. According to this, soil fertility, other crop yield, food security of the household and sustainability of spring and borehole water could be increasing due to enset cultivation in their homestead. At the time of the survey, it has been commonly observed that species like enset, with deep roots and leaf canopies of long duration, improve the water status of an area.

Further other literature confirm that, as the proportion of these species increases with respect to annual species, water infiltration increases and surface runoff decreases, resulting in more water in the soil and aquifers. Enset crop could increase water availability and greater volume and duration of discharge to springs, decreasing the effective length of the dry season of the area [18].

4. Conclusion

Soil quality like total nitrogen, soil organic carbon and available phosphorus in enset field has a higher status than other crop fields. It indicates that enset cultivation is sustainable system with regard to maintain soil fertility significantly higher in the enset field. This might be due to the materials added to the soil from the plant parts and organic farming (application of manure and domestic waste). Enset has highly degradable parts such as leaf midribs, pseudostem sheath, pseudostem core/ which are used to maintain soil fertility.

Any practice of cultivating enset could improve soil structure. However, in some cases these improvements may not be sustainable at the time of scraping it, because of labour movement which may compress the soil.

Enset can reduce soil erosion and land degradation. The root system and non-fertilizer cultivation are other important characteristics to protect soil erosion and land degradation. Some farmers cultivate enset in the face of flood and drain it in to enset farm in order to reduce further soil erosion.

References

[1] Alemayehu Assefa, "Impact of Terrace Development and Management on Soil Properties in Anjeni area, Eest Gojam." MSc. Thesis. Addis Ababa University, Addis Ababa, 2007.

[2] C. García, A. Zsolnay, and M. Kuderna, "Soil Rehabilitation: Indicators and Thresholds for Desertification, Soil Quality, and Remediation." Sixth framework programme (global change in ecosystems), 2004.

[3] D. Fekadu and I. Ledin, "Weight and chemical composition of plant parts of enset (*Enset ventricosums*) and the intake and degradability of enset by cattle." *Livest. Prod. Sci.* 49(3): 249-257, 1997.

[4] E. Akinrinde, "Soils: Nature, Fertility Conservation and Management." (V. Chude, M. Amakiri,..eds). AMS Publishing, Ibadan, 2004.

[5] EFAP, "The Challenge for Development." Vol. II, Final Report. Ministry of Natural Resources Development and Environmental Protection. Addis Ababa, 1997.

[6] Elias Eyasu, "Soil enrichment and deletion in southern Ethiopia." In: Nutrients on the Move: Soil fertility dynamics in African farming systems, Pp 65-82 (T. Hillhorst, and F. Muchena., eds), 2000.

[7] H. Zhang and G. Zhang, "Landscape-scale soil quality change under different farming systems of a tropical farm in Hainan, China." *Soil U. Manag.* 21: 58-64, 2005.

[8] ICARDA, "soil and plant analysis laboratory manual." www.icard.org/publication/lab:Data accessed: March 05, 2013, 2005.

[9] J. R. Landon, "Tropical Soil Manual." A handbook of soil survey and agricultural land evaluation in the tropical and subtropical. Longman. Broaker, Pp. 447, 1991.

[10] K.J. Schlather, "The Dynamics and Cycling of Phosphorus in Mulched and un Mulched Bean Production Systems Indigenous to the Humid Tropics of Central America." Ph.D. Dissertation, Cornell University, New York, 1998.

[11] M. Bekunda, "Farmers' response to soil fertility decline in banana-based cropping systems of Uganda." *Man. Afri. Soils* 4:1-17, 1997.

[12] M. A. Stocking, "Tropical Soils and Food Security: the next 50 years." *Science* 302: 1356-1359, 2003.

[13] Mekuria Argaw, "Forest conversion - soil degradation - farmers' perception nexus: Implications for sustainable land use in the southwest of Ethiopia." In: Ecology and Development. (Vlek, P. L. G; Denich, M.; Martius, C.; Rodgers, C.; Giesen, N. V. D., eds.), 2005.

[14] N. Brady, and R. Weil, "The Nature and Properties of Soils, 13rd edn. Prentice Hall. Upper saddle river, New Jersey. pp960, 2002

[15] P. Peveri, "Enset, the Tree of the Poor." Nutrition and identity in Hadiya zone (south-central Ethiopia).University of Bologna, Bologna, 2011.

[16] P. A. Hazelton, and B. Murphy, "Interpreting Soil Test Results." What do all the numbers mean? 2nd edn, CSIRO publishing, Australia, 2007.

[17] R. Miller and R. Donahue, "Soils in Environment." 7th edn. Prentice Hall of India, New Delhi, 1997

[18] S. A. Brandt, A. Spring, C. Hiebsch, S. T. McCabe, T. Endale, D, Mulugeta, W/M. Gizachew, Y. Gebre, M. Shigeta, and T. Shiferaw, "The 'Tree Against Hunger'. Enset-based Agricultural Systems in Ethiopia." American association for the advancement of science, Washington, DC, 1997.

[19] Tadesse Kippie, "Five Thousand Years of Sustainability? A case study on Gedeo land use, Southern Ethiopia." Treemail publishers, Heelsum, Netherlands, 2002.

[20] Teshome Yirgu, "Assessing Soil Nutrient Depletion to Household Food Insecurity in the Smallholders Farming System in the Western Hills of Lake Abaya." *Ethio. J. Env and E. Sci.* 2(4), 2012.

[21] Tilahun Amede and Mulugeta Diro, "Optimizing soil fertility gradients in the Enset (*Ensete ventricosum*) systems of the Ethiopian highlands: Trade-offs and Local Innovations." In: Improving Human Welfare and Environmental Conservation by Empowering Farmers to Combat Soil Fertility Degradation (Bationo *et al.,* eds). African highlands initiative (AHI), working papers No 15, 2005.

[22] UNEP, "Environment and Biodiversity Impact assessment." Field survey report, Hossana, 2012.

[23] USDA, "Soil quality indicators." USDA, Natural Resources Conservation Service, Nyle, 2008.

8

Evaluation of the Extent and Potential Impacts of Noise Pollution inside Public Service Vehicles (PSVs) in Nairobi City, Kenya

Dancan Otieno Onyango[1, *], Chritopher L. Kanali[2], James Wambua Kaluli[1]

[1]Department of Soil, Water and Environmental Engineering, Jomo Kenyatta University of Agriculture and Technology, Nairobi, Kenya
[2]Department of Biomechanical and Processing Engineering, Jomo Kenyatta University of Agriculture and Technology, Nairobi, Kenya

Email Address:
dancanthomas@gmail.com (D. O. Onyango), c_kanali@yahoo.com (C. L. Kanali), wambuak@gmail.com (J. W. Kaluli)

Abstract: Public service vehicles (PSVs) in Nairobi City, Kenya, have a tendency of playing loud music inside the vehicles resulting in disturbing noise. The objective of this study was to evaluate the extent of noise pollution inside PSVs in Nairobi City and predict possible impacts of the noise on PSV users. A sample of 60 randomly selected PSVs were analyzed. Posing as passengers inside the vehicles, noise measurements were taken using a digital integrating sound level meter (SVANTEK 971) inside these matatus. The measured noise levels were compared with corresponding legislated limits and the deviation thereof used to quantify the extent of pollution and probe the effectiveness of existing noise laws. Additionally, potential effects of the noise were predicted by comparing measured noise levels with respective guidelines recommended to protect public health and welfare from various adverse impacts of noise. The study results showed that PSV users in Nairobi are exposed to 86.3±9.6 dB A of noise inside PSVs and all the sampled vehicles did not comply with Kenyan noise regulations. PSVs plying northern Nairobi area were found to be the noisiest at 92.2±9.4 dBA, followed by Nairobi West PSVs at 88.7±9.6 dBA; Nairobi South PSVs at 83.1±6.3 dBA and Nairobi East PSVs at 81.2±7.7 dBA. The growing habit of PSVs playing extremely loud music/radio using sound amplifying equipment, was found to be a major contributor to noise pollution inside the vehicles, in addition to normal vehicular traffic sounds. The high noise levels put city dwellers at risk of such adverse effects of noise as noise-induced hearing loss (NIHL); physiological and psychological noise effects; speech interference, sleep disturbance; annoyance, social and behavioral effects. This calls not only for intensified noise law enforcement in the city, but also puts to question the effectiveness of existing noise laws to control noise in the city.

Keywords: Noise, Noise Pollution, Public Service Vehicles, Noise Impacts

1. Introduction

Noise is defined in Kenyan law as "any undesirable sound that is intrinsically objectionable or that may cause adverse effects on human health or damage to the environment". The law goes further to define noise pollution as "the emission of uncontrolled noise that is likely to cause danger to human health or damage the environment" [1, p.3]. Generally, noise pollution is recognized in three categories: community noise, industrial noise and airport noise. Industrial noise and airport noise refer to noise emanating from industrial processes and airport operations, respectively, whereas community noise (or environmental noise) is noise emanating from human activities such as transportation, recreation, entertainment, worship, businesses, animal rearing, construction and internal domestic activities [2]

In Kenya, not much has been done to study noise pollution. However, in 2009, the noise problem was recognized and the Noise and Excessive Vibration (Pollution) Control Regulations, 2009 (hereafter referred to as the Noise Act or Legal Notice Number 61) enacted. This legislation, annexed to the Environmental Management and Coordination Act, EMCA, 1999, provides for relevant noise criteria for different environments and empowers the National Environment Management Authority (NEMA) to enforce the noise regulations. NEMA has singled out noise pollution as an issue of grave environmental concern [3] in Kenya, particularly in major urban centers such as Nairobi, Kisumu,

Mombasa and Eldoret, among others. In addition to NEMA, the Directorate of Occupational Safety and Health Services (DOSHS) is tasked with protecting workers from occupational noise.

Despite attempts by NEMA and DOSHS to control noise in Kenya, community noise remains a nuisance and a major health risk factor in Nairobi City. This is due to rapid industrialization, population pressure and growth in commercial, entertainment, transportation and religious activities in the city. The public transport sector, specifically, is a notorious source of noise. As a matter of fact, a walk through the city streets and estates reveals alarming noise levels from moving PSVs (popularly known as *Matatus)*. The situation is exacerbated by a growing habit of PSVs playing extremely loud music/radio in the vehicles. This is aided by the use of sophisticated sound amplifying equipment and numerous loud speakers installed everywhere inside the PSVs. In essence, the resulting high noise renders telephone or speech communication practically impossible while traveling in the PSVs. Additionally, the PSVs are accustomed to uncontrolled hooting and touting (PSV crew shouting and banging vehicle bodies), particularly as they approach bus termini to attract passengers. This further worsens the noise situation. Nairobi City is prone to traffic congestion that tremendously slows down traffic speeds. This implies that PSVs take longer to reach their destinations and this increases the exposure time of PSV users to the high noise. The high noise may be subjecting PSV users to risks of suffering noise-induced impacts on their health and welfare.

Unfortunately, as compared to other environmental issues concerning soil and water, for instance, noise has not been given much attention in Kenya, even by scholars. In fact, Kenyan noise data is generally missing. This study sought to evaluate the extent and potential impacts of noise pollution inside PSVs in Nairobi, the capital city of Kenya. The results thereof would be used to probe the effectiveness of existing laws to control noise pollution and formulate possible technical and policy recommendations for noise control.

2. Literature Review

Noise pollution poses a wide range of risks to public health and welfare. Adverse effects of noise pollution include annoyance, speech interference, sleep disturbance and hearing loss. Recent studies have also associated noise with cardiovascular diseases such as hypertension and myocardial infarction [4, 26]. Traffic noise-related annoyance and sleep disturbance have become common problems throughout the world, with adverse effects on life quality and health [5, 27]. Another adverse effect of noise pollution that has been documented since the 16th century, and is still a problem today, is occupation-related hearing loss. World records estimate that about US$835 million was paid to workers between 1978 and 1987 in compensation for occupation-related hearing loss. It has also been proven that noise-induced sonic boom can result in physical damage of structures [6].

All over the world, noise levels have reached alarming levels, particularly in urban areas. The impact of noise pollution on communities is fast escalating especially in densely populated urban areas near very busy roads [7, 29, 32]. In Varanasi City, India, about 85% of the people are disturbed by traffic noise, 90% of whom consider noise as the main cause of headaches, hypertension, giddiness and lethargy [8, 25]. In Canada, a nationwide survey indicates that half of Canadians are bothered by noise emanating outside their homes, mainly from road traffic [9]. In Denmark, a one (1) dB increase in noise levels results in a corresponding 1% decrease in house prices in main residential areas, indicating how reactive people there are to noise. In Rawalpindi and Islamabad, the environment management agency has raised concerns that the daily maximum and daily equivalent noise levels are significantly higher than the maximum permissible level [10]. In most towns/cities of Nigeria, noise emanating from traffic, industry, worship and commercial activities has been listed as an issue of key environmental concern [11, 12, 30]. In Uganda, has been noted as the main source of environmental pollution in the city of Kampala [13]. Additionally, in a study of noise pollution in restaurants of Morogoro town, Tanzania, [14] recorded between 61 and 64 dBA of noise, which is way above the legislated maximum permissible limit of 55 dBA and is associated with music in the restaurants, customer conversations as well as activity on adjacent streets. In sum, rapid urbanization, population growth, industrialization and associated socio-economic activities is fast escalating the level of noise pollution in most world cities and urban centers.

3. Theory

This study was based on various theoretical concepts of Acoustics. These include: sound characterisation, sound measurement, and human sound perception.

3.1. Sound Characterization

Sound is characterized in terms of three important parameters namely, sound pressure, sound power and sound intensity [15, 31]. Sound pressure (denoted as ΔP_m) is the maximum pressure differential, above and below the existing atmospheric pressure, which is responsible for sound propagation. This is based on the basic fact that sound results from a differential/disturbance in air pressures caused by the vibration of a sound source. The disturbance leads to formation of two pressure bands, the compression band, characterized by high pressures, and the rarefaction band of low pressures [16, 28, 31]. In the absence of sound pressure, no sound transmission occurs and the ear perceives no sound. Sound pressure is expressed as in equation 1 in which U is the velocity of propagation of sound; ρ is the mass density of air; ω is the angular frequency of the sound wave; S_m is the maximum displacement and β is the bulk modulus of elasticity of air [15].

$$\Delta P_m = (U\ \rho\omega)\ Sm;\ U = \sqrt{\frac{\beta}{\rho}} \quad (1)$$

Like a compressed spring, the air that propagates a sound wave possesses potential energy, in its compressed state, and kinetic energy in the rarefaction state [16]. Neglecting any energy losses, the total energy content of a sound wave equals the sum of the potential and kinetic energy. Sound power (P_w) is the rate of change of the energy (from potential to kinetic) with respect to time as given in equation 2 in which A is the cross-sectional area of propagation of the sound wave [15]. The rate of transmission of sound power per unit area of cross section (A) in the direction of travel of the sound is known as sound intensity (I). Sound intensity is commonly expressed in terms of the root mean square of the sound wave as shown in equation (3) [15].

$$P_w = \frac{1}{2}\rho A v \omega^2 S_m^2 \quad (2)$$

$$I = \frac{\Delta P_{rms}^2}{\rho v} \quad (3)$$

3.2. Sound Measurement

3.2.1. The A-Weighted Decibel Scale

Although sound is characterized in terms of pressure, power and intensity, noise measurements are never reported in the units thereof. This is because sound that a human ear can perceive comprises a wide range of pressures and intensities. Thus, to accommodate these wide variations, a logarithmic scale of zero to 140 units, known as decibels (dB), is used [17, 34]. A decibel is defined by equation 4 in which: x is sound pressure/sound power/sound intensity; y refers to some arbitrary reference values which depend on the type of decibel being defined. The arbitrary reference values are 10^{-12} W for sound power; $2(10^{-5})$ Pa for intensity and 10^{-12} W/m^2 for sound pressure as shown in equations 5, 6 and 7 [18].

$$dB = 10\log\frac{x}{y} \quad (4)$$

$$L_w = 10\ log\frac{P_w}{10^{-12}}\ dB \quad (5)$$

$$L_I = 10\log\frac{I}{10^{-12}}\ dB \quad (6)$$

$$L_p = 10\log\frac{\Delta P_{rms}^2}{(2\times10^{-5})^2}\ dB \quad (7)$$

Most sounds comprise of a wide range of frequencies which determine human ability to hear them. Humans readily hear sounds whose frequencies range from 1000 to 6000 Hz. Sounds dominated by frequencies below 100 Hz and above 10,000 Hz are difficult to hear [19]. Sound measurements need to be done in a way that approximates human response by giving more weight to readily heard frequencies. In order to achieve this, several weighting scales have been developed for sound measurement. These include A, B, C, D and E-weighted scales [18]. The A-weighted scale has been conventionally adopted for environmental noise measurements due to its convenience and accuracy for most purposes. A-weighted sound level measurements are reported in units of dBA [15].

3.2.2. Decibel Addition

A given sound spectrum may consist of various decibel readings. These must be added up to give rise to an overall reading. The addition is performed logarithmically such that the overall/total decibel readings (L_{pT}) for a sound spectrum comprising of sound pressures ranging from L_{p1}, L_{p2}, ... to L_{pn} is given as shown in equation (8). Similarly, the total decibel readings corresponding to sound power level and sound intensity levels are computed as shown in equations 9 and 10, respectively.

$$L_{pT} = 10\log\left(\sum_{i=1}^{i=n}\sqrt{10^{\frac{L_{pi}}{10}}}\right)^2 \quad (8)$$

$$L_{WT} = 10\log\sum_{i=1}^{1=n}10^{L_{wi}/10} \quad (9)$$

$$L_{IT} = 10\log\sum_{i=1}^{1=n}10^{L_{Ii}/10} \quad (10)$$

The logarithmic additions described above have been summarized as shown in Table 1 [19];

Table 1. Decibel addition.

$L_1 - L_2$, dB	Add to L_1
0 or 1	3 dB
2 or 3	2 dB
4 - 8	1 dB
9 or more	0 dB

3.2.3. Miscellaneous Measures of Noise

In practice, three miscellaneous measures of noise have evolved. This enables noise pollution to be regarded in three dimensions namely, community noise, industrial noise and airport noise. Industrial noise and airport noise refer to noise emanating from industrial processes and airport operations respectively [20]. Community noise (also known as environmental noise) refers to noise emanating from common day-to-day human activities such as transportation, recreation, entertainment, worship, business, animal rearing, construction and domestic activities. Most often, noise from a combination of these sources add up to give rise to overall community/environmental noise and it may not be easy to discriminate and quantify the individual point sources. Thus, as a measure of noise, community/environmental noise seeks to conceptualize the noise environment in which human beings are exposed in their day-to-day life.

Various measures have been developed for quantifying community noise. These include: the statistical measure (L_N), continuous equivalent sound level (L_{eq}) day-night average sound level (L_{dn}) and noise pollution level (L_{NP}).Out of these measures, the continuous equivalent sound level is the most widely used for community noise assessment. L_{eq} is preferred for describing sound levels which vary over time to result in a single decibel reading. It accounts for the total sound energy involved over the given period of time and gives an

A-weighted average equivalent for the period. L_{eq} is expressed as shown in equation 11 in which: p_0 is the reference pressure level (typically taken as 20 µPa); p_A is the acquired sound pressure; t_1 and t_2 are the start time and end time for measurement, respectively [15].

$$L_{eq} = 10 \log\left[\frac{1}{t_1-t_2}\int_{t_1}^{t_2}\frac{p_A^2}{p_0^2}dt\right] \tag{11}$$

The statistical measure refers to a sound level that is either equaled or exceeded a given percent of the time, based on a probability distribution analysis. For instance, L_{90}, L_{20} and L_{10} refer to noise levels that are equaled/exceeded 90%, 20% and 10% of the time, respectively. The day-night average sound level is the equivalent sound level over a 24 hour period with an additional 10 dB (A) during the night (10.00 P.M. to 7.00 A.M.). Community noise levels can also be described in terms of another measure known as the community noise equivalent level (CNEL). CNEL is related to L_{DN} except that an additional 5 dB (A) is imposed between 7.00 P.M. and 10.00 P.M. and 10 dB (A) is added between 10.00 P.M. and 7.00 A.M. Noise pollution level is the equivalent sound level with a probability of exceedance of 0.5% [8, 15].

3.2.4. The Sound Level Meter

The instrument used to measure noise is the sound level meter. The meter should be located at a reasonable distance from the sound source to avoid obstruction [21]. The sound level meter is first calibrated by a *calibrator.* When attached to the sound level meter, the calibrator produces a reference sound that calibrates the meter. Basically, a sound level meter consists of four parts namely, a microphone, sound amplifiers, weighting network and a display which shows readings in decibels [22, 33]. The sound pressure of an incoming sound is received by the microphone where the pressure pattern of the sound is mimicked by a pressure transducer (such as a diaphragm) and the signal is converted to a small current of electricity. The resulting electrical signal if amplified by the amplifiers and an equivalent reading shown on the screen [19].

It is important to note that the sound pressure received by the meter is not the same as that which the ear perceives. As a result, a weighting network (A, B, C, D and E) – simply an electrical circuitry – is incorporated in the meter to help approximate human response and give a reading as close to it as possible [23]. The work of the weighting network circuitry is to electronically subtract the actual sound pressure level, at a particular frequency, received at the microphone with respect to the network upon which the meter is set [19]. A sample plot of weighting networks for scale A, B and C is shown in the Figure 1. The difference between the actual pressure level and a given frequency-dependent value, from the weighting network plot, is displayed as the sound level reading. The units of this reading depend on the scale used thus it may be reported as dBA, dBB or dBC for scales A, B and C, respectively. While the A-network is recommended for most community noise measurements, the D network is best suited for approximating human response to airport noise [15].

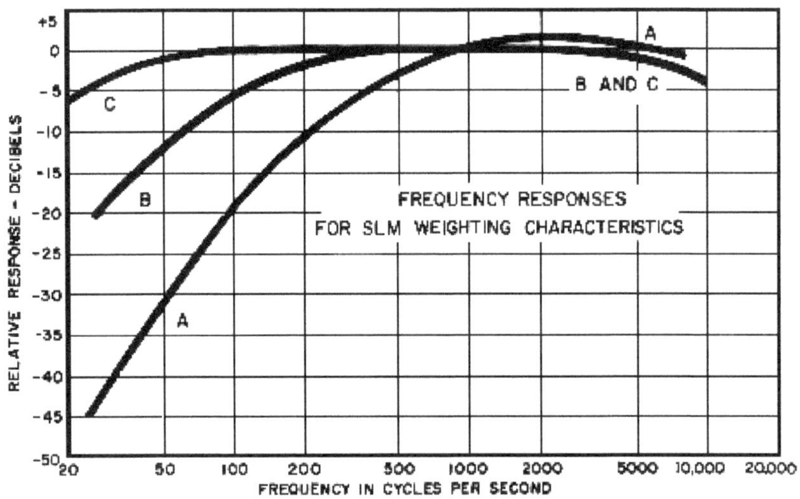

Adapted from: (http://www.sfu.ca/sonicstudio/handbook/Graphics/Sound_Level_Meter.gif)

Figure 1. *Plot of weighting networks.*

4. Materials and Methods

The study area was divided into four geographical routes, namely, Nairobi CBD-Nairobi East, Nairobi CBD-Nairobi West, Nairobi CBD-Nairobi North and Nairobi CBD-Nairobi South as shown in Figure 2. On each route, fifteen (15) PSVs were selected by simple random sampling for analysis. These comprised buses, mini-buses and 14-seater vans which are the commonest means of public road transport in Kenya. Posing as passengers in every PSV, continuous equivalent sound measurements (LAeq) were taken using a digital integrating sound level meter (SLM). The SLM used (SVANTEK 971) is a Type 1 SLM which gives measurements to the IEC 61672-1:2002 standards. The meter is also capable of computing LAeq levels directly thus minimizing computations and enhancing the accuracy of resulting measurements. In order to enhance the integrity of measured noise levels, PSVs were assessed for noise pollution anonymously, without the consent of their crew. This was due to fear that some may deliberately regulate, particularly, the level of music playing in their vehicles, if they knew they were under study.

Prior to its use, the SLM was calibrated, as per the manufacturer's manual, using a calibrator model SV34 and set to the "fast" and "A-Weighting network" modes. Measurements were taken with the SLM placed at a convenient sitting position. In each PSV, measurements were taken at five minutes intervals for the entire duration of travel, which ranged from 15 minutes to one hour. An average of the measurements taken was then computed and taken as the average LAeq for each PSV. The fifteen measurements taken in every route were further averaged to obtain a representative value for each route.

The measured LAeq were compared with recommended (legislated) limits and the extent of pollution quantified by the percentage deviation from (above) the limits [1]. Additionally, in order to assess the potential effects of this noise on PSV users, the measured LAeq were compared with the World Health Organization and the Environmental Protection Agency's guidelines recommended to protect public health and welfare from adverse impacts of noise. The degree of deviation from the protective guidelines was used to determine the level of risk posed to victims by noise inside PSVs.

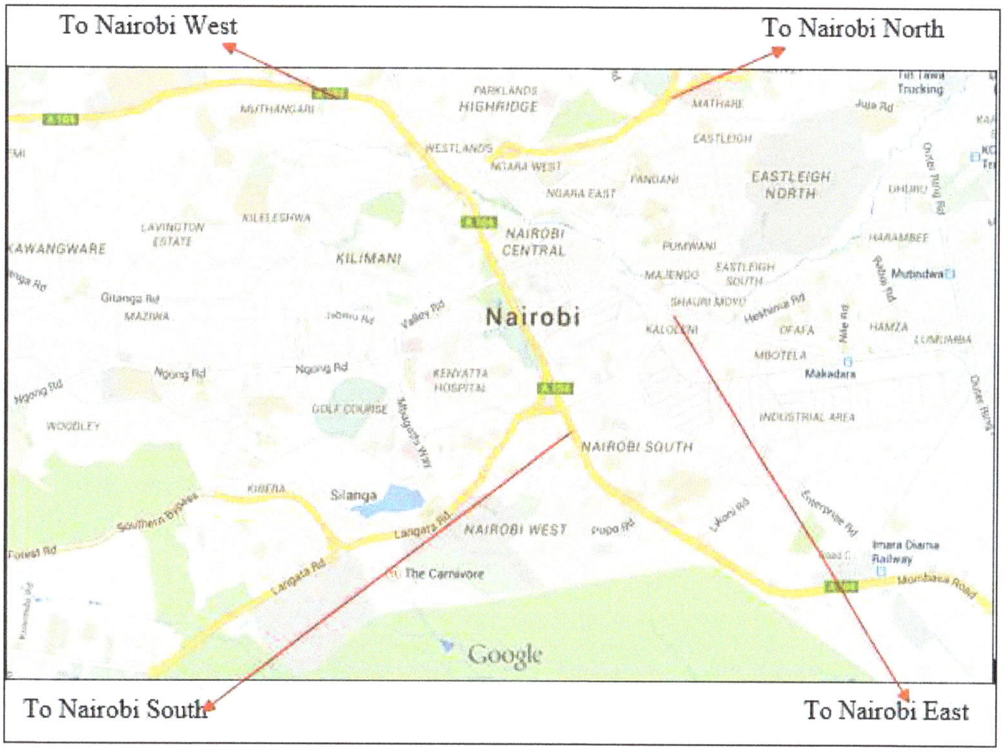

(Source: Google Earth)

Figure 2. *A map showing the main PSV routes in Nairobi City.*

5. Results and Discussion

5.1. The Extent of Noise Pollution inside PSVs

The extent of noise pollution inside PSVs in Nairobi City was established as follows.

5.1.1. Nairobi CBD–Nairobi North PSVs

Figure 3 shows the noise levels observed among PSVs operating between Nairobi CBD and various destinations north of the CBD such as *Githurai 45, Kahawa, Kenyatta University (KU), Kasarani, Lucky Summer, Mwiki and Thika (46;237),* among others. The main road serving this route is the Nairobi-Thika Superhighway which connects to various feeder roads. On average, the noise level inside these PSVs was found to be 92.2±9.4 dB (A). Two Githurai 45 buses exhibited the highest noise level in this route, 108.8 and 108.2 dB (A), respectively. These were followed closely by a Thika-bound minibus (Route 237) in which 107.3 dB (A) was recorded. Other significantly high noise levels were measured inside a KU 45 Minibus (98.8 dB A) and a Mwiki 17 bus (96.6 dB A). In contrast, a 14-seater KU-bound Matatu and a Thika 237 minibus exhibited the lowest noise levels in this route: 76.6 dB (A) and 84.4 dB (A), respectively.

Going by these results, noise levels in all Nairobi North PSVs are higher than the maximum permissible level of 60 dBA. Even the lowest noise levels recorded in some PSVs in this route are still significantly higher than the permissible limit, with 54% average rate of non-compliance with the Noise Act.

This study further revealed that the high noise level in northern Nairobi PSVs is a direct result of loud music/radio played inside the vehicles using sound amplifying equipment. Touting (banging of the vehicle bodies by touts; whistling; shouting; and uncontrolled hooting) were seen to be another major source of noise, especially among Githurai 45 buses and minibuses plying the Nairobi CBD-Nairobi North route.

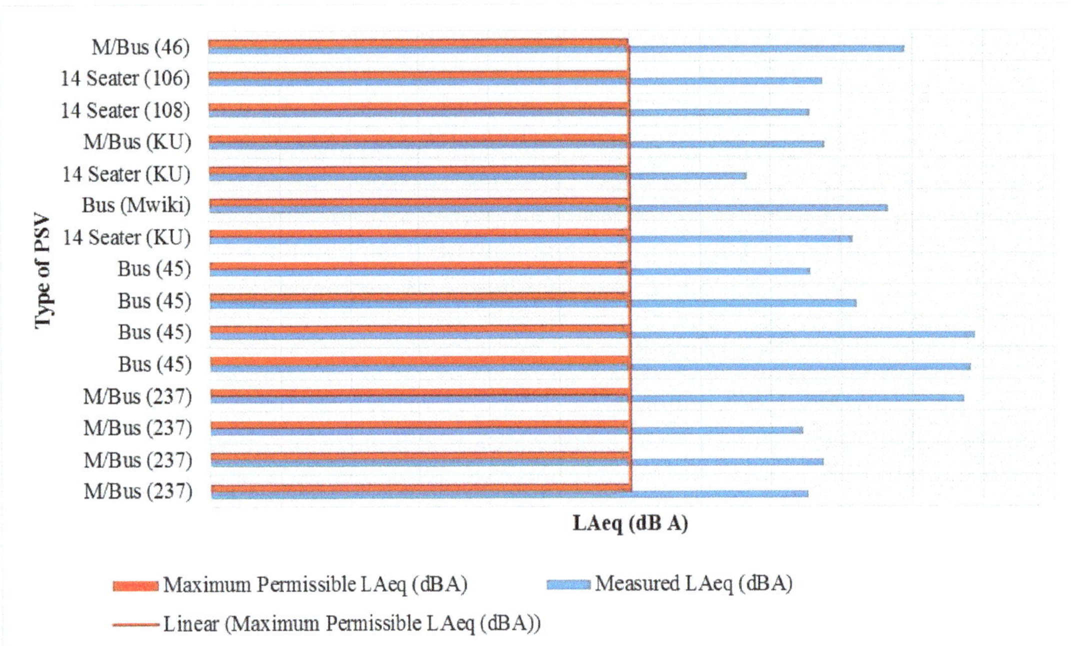

Figure 3. *Noise inside Nairobi CBD-Nairobi North route PSVs.*

5.1.2. Nairobi CBD-Nairobi South PSVs

The extent of noise pollution observed inside Matatus operating between Nairobi CBD and various destinations south of the city (such as: *South B, South C, Hazina, Langata, Ngong, Karen, Kibera, Kiserian, Ongata Rongai* and *Bomas*, among others) is presented in Figure 4. The main roads serving the southern area are Langata Road and Ngong Road and their associated feeder roads. Like in the northern region, all Matatus sampled in the south exceeded the maximum permissible noise level of 60 dB (A). On average, a noise level of 83.1±6.3 dB (A) was recorded in these Matatus. This represents a 39% extent of non-compliance with the Noise Act. A South B (Route 11B) minibus and Ngong Minibus exhibited the highest level of noise (96.5 dBA) in this route. This signifies a high of 61% non-compliance with the Noise Act. The high noise levels are directly attributable to loud music played in the Matatus using sound amplifying equipment. To a little extent, touting could also be pointed out as a noise source in Nairobi South Route Matatus, though not as extensive as witnessed in the north.

In contrast, the least noise level recorded in this region was 75.2 dBA. This was observed in a mini bus plying Nairobi CBD–South B (route 11B). Unlike the matatus that exhibited the highest noise levels, this minibus played radio without the aid of sound amplifying equipment. Thus, the 75.2 dBA noise level could be attributed to other factors observed in the matatus such as: touting, normal vehicle engine sounds, human (passenger) sounds and possible sounds from the outside environment. This shows that the use of sound-amplifying equipment inside PSVs greatly increases the noise pollution situation therein. Other matatus, in this region, that exhibited relatively lower noise levels include: an 11B bus (76.6 dB A); an 11B Minibus (77.7 dB A); and an 11B 14-seater (78.1 dB A).

5.1.3. Nairobi CBD-Nairobi East PSVs

The Nairobi CBD-Nairobi East route serves the vast eastern Nairobi region (popularly known as East-lands). The main roads traversing this region include: *Jogoo* Road, Mombasa Road and *Juja* Road and their associated feeder roads. Major estates in these region include: *Buruburu, Doonholm, Kariobangi, Umoja, Makadara, Pipeline, Kayole* and *City-Kabanas*, among others. East-lands' route matatus recorded an average internal noise level of 81.2±7.7 dB (A), signifying 35% average non-compliance with the Noise Act (Figure 5). Some three minibuses exhibited the highest noise levels in this region, one plying Buruburu Route 58 (94.6 dB A); the second plying Kayole Route (94.6 dB A) and another Buruburu Route 58 minibus (91.2 dB A). Like in the north and south, the high noise is a direct result of amplified music in the vehicles. This is in contrast to matatus without amplified music such as Double M (Route 33/34) which exhibited the lowest noise level of 68.8 dB (A). This represents a low of 15% non-compliance with the Noise Act and is attributed mainly to normal vehicular traffic sounds such as engine sounds, human sounds and possible sounds from the outside environment. While inside these less noisy matatus, it was observed that, unlike inside the noisy ones, speech/telephone communication could go on uninterrupted.

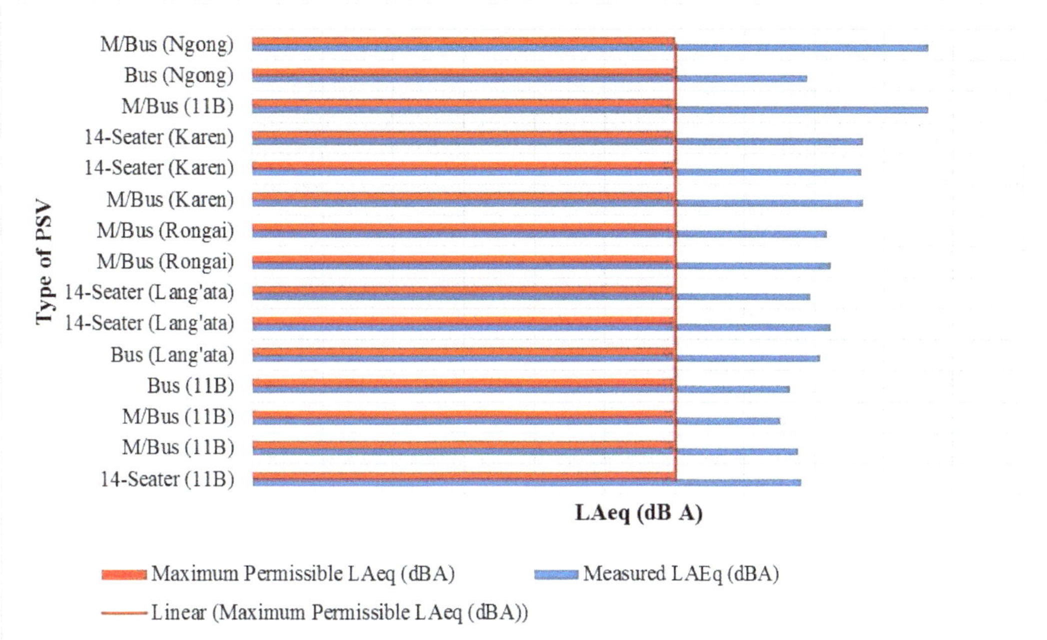

Figure 4. *Noise inside Nairobi CBD-Nairobi South route PSVs.*

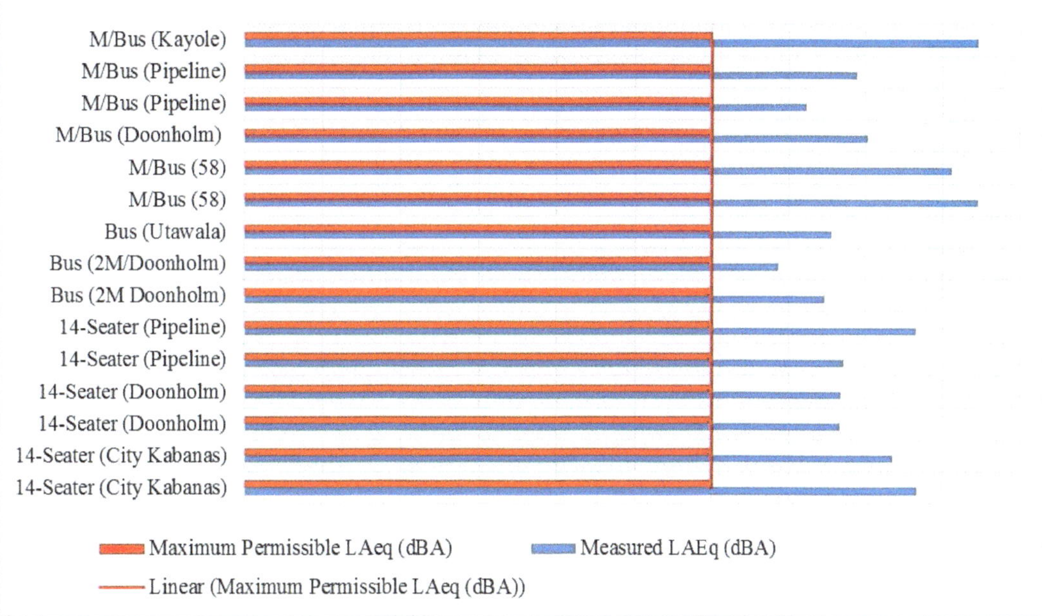

Figure 5. *Noise inside Nairobi CBD-Nairobi East route PSVs.*

5.1.4. Nairobi CBD-Nairobi West PSVs

The Western region of Nairobi includes areas such as *Westlands, Kabete, Uthiru, Kinoo, Parklands, Mountain View, Satellite and Kikuyu*, among others. The main road serving this area is *Waiyaki Way* and its associated feeder roads. The average noise level measured inside matatus in this region was 88.7±9.6 dB (A), representing 48% non-compliance with the Noise Act (Figure 6). Two minibuses (Star Bus) plying this route exhibited the highest noise level of 107.2 and 104.1 dB (A), respectively. These were followed closely by a third minibus in which 98.9 dB (A) was recorded. Like the other regions, it was observed that the high noise levels in these matatus is a direct result of the use of sound-amplifying equipment to play extremely loud music in the vehicles. In fact, the loud music overpowered all the other noise sources such as engine and human sounds and it was practically impossible to make speech/telephone communication in the vehicles. On the other hand, some PSVs in the Western region exhibited relatively low noise levels. These were buses operating in the Nairobi-Kabete route and their observed noise levels varied from 75-77 dB (A). They played either no music at all or very low music/radio without using sound-amplifying devices. As opposed to their counterparts with loud noise, it was possible to communicate easily and make calls in these matatus without shouting.

5.1.5. Comparison of Various Routes

Figure 7 compares the mean noise levels measured in PSVs plying routes in the four regions of Nairobi City. The results show that the Nairobi CBD-Nairobi North PSVs are the noisiest, with an average noise level of 92.2±9.4 dBA. The second noisiest matatus are those in the Nairobi-Nairobi West route at 88.7±9.6 dBA. These are in turn followed by matatus in the southern region which recorded an average of 83.1±6.3 dBA. The lowest noise levels were observed in the eastern region whose matatus average 81.2±7.7 dBA. This implies that, in terms of compliance to the Noise Act, the most non-compliant route is Nairobi North at 54% followed by: Nairobi West (48%); Nairobi South (38%); and Nairobi East (35%). As shown in Table 2 and Table 3, these deviations from legislated noise limits are statistically significant, as per one-sample t-tests conducted at 95 and 99% confidence intervals.

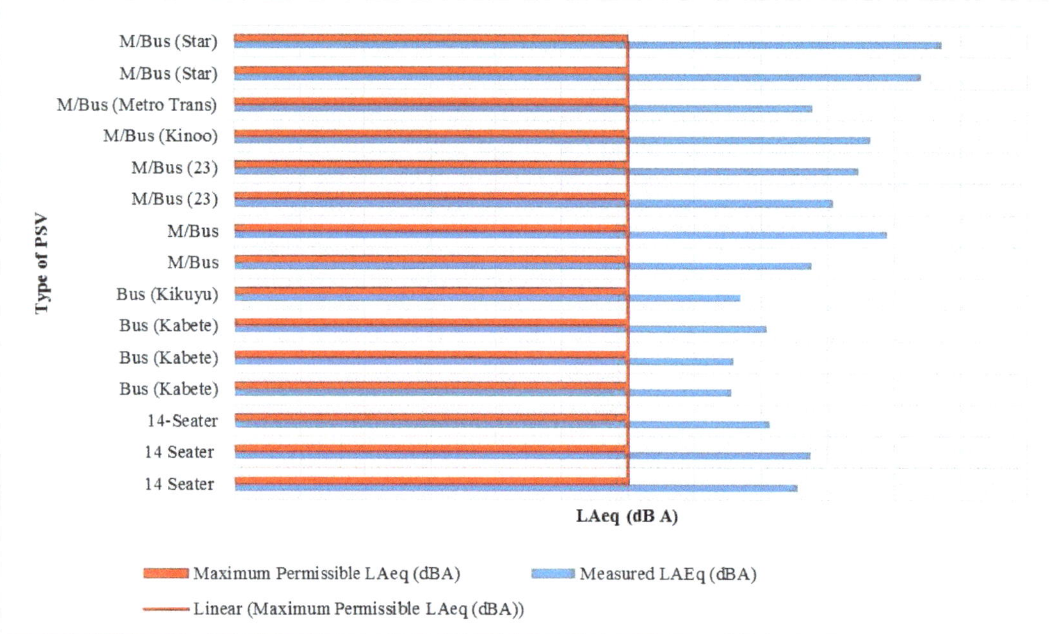

Figure 6. *Noise inside Nairobi CBD-Nairobi West route PSVs.*

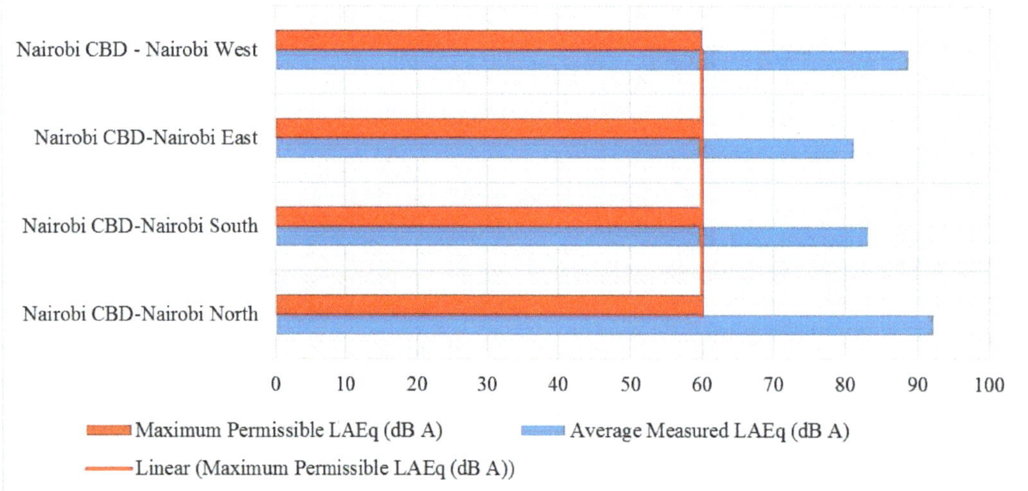

Figure 7. *Comparison of noise pollution levels for different PSV routes in Nairobi City.*

Table 2. *One-sample t-test at 95% confidence interval (CI).*

	t	df	Sig. (2-tailed)	Mean Difference	95% Confidence Interval of the Difference	
					Lower	Upper
PSV Measured Values	10.395	3	.000	26.300	18.25	34.35

Table 3. *One-sample t-test at 99% CI.*

	t	df	Sig. (2-tailed)	Mean Difference	99% Confidence Interval of the Difference	
					Lower	Upper
PSV Measured Values	10.395	3	.002	26.300	11.52	41.08

5.2. Causes of Noise Pollution inside PSVs

Regarding the causes of noise inside PSVs, this study further revealed that the high noise level in northern Nairobi PSVs is a direct result of loud music/radio played inside the vehicles using sound amplifying equipment. This renders communication endeavors (speech or telephone communication) virtually impossible while travelling in the PSVs. In addition, touting (banging of the vehicle bodies by touts; whistling; shouting; uncontrolled and uncontrolled hooting) was also seen to be another major source of noise, especially among Githurai 45 buses and minibuses plying the Nairobi CBD-Nairobi North route. Other than communication interference, the high noise levels in PSVs put passengers at risk of suffering noise-induced hearing loss, physiological and psychological defects [24].

5.3. Potential Impacts of Noise Pollution

As already indicated, 100% of PSVs and bus stations in Nairobi city violate the legislated noise regulations. This poses a wide range of risks to public health and welfare, particularly among passengers. Also, most of the PSV roads and bus stations are located in/near residential estates, offices, hospitals or schools. It is also important to note that Nairobi faces a serious problem of traffic congestion thereby increasing the duration for which passengers, pedestrians and near-road users/dwellers are exposed to traffic noise. Considering the average noise levels established in this study, an average Nairobi resident is exposed to 79.7±6.0 dB (A) of noise while awaiting to board a PSV in a bus station; followed by 86.3±9.6 dB (A) while travelling in a matatu and as the PSV maneuvers through slow traffic to and from work, daily.

To predict the possible noise impacts, an overall average value was computed of all the measured noise levels in this study. The value was found to be 83 dB (A). To predict possible effects of noise, Table 4, compares this overall average noise level with various international standards recommended to protect public health and welfare with an adequate margin of safety [24, 25, 26]. From the table, it is evident that Nairobi faces a high risk of various adverse noise effects. These include: NIHL, physiological effects (e.g. hypertension and cardiovascular disease), psychological effects, annoyance and behavioral effects. This is because the overall average value (83 dB A) exceeds the maximum permissible levels of 70; 65-70; 70; and 80 dB (A), respectively. In addition, the study established that there is a very high risk of other effects of noise such as speech interference, sleep disturbance, and activity interference whose recommended limits are 35-55; 30; and 45-55, respectively.

Table 4. *Potential impacts of noise pollution inside PSVs in Nairobi City.*

Effect of noise	Protective noise limit (LAEq, 24h, dB A)	Duration of exposure (Years)	Measured noise level (LAEq, 24h, dB A)	Inference (level of risk)
Noise Induced Hearing Loss (NIHL)	70	20-40	83	High Risk
Physiological Effects (Hypertension, Cardiovascular Disease, etc.)	65-70	5-30	83	High Risk
Psychological/Mental Illness/Stress	70	ST-LT	83	High Risk
Speech Interference (Indoors)	35	ST	83	Very High Risk
Speech Interference (Outdoors)	55	ST	83	Very High Risk
Sleep Disturbance	30	ST	83	Very High Risk
Activity Interference	45-55	ST	83	Very High Risk
Annoyance/Social/Behavioural Effect	80	ST	83	High Risk

Key: ST – Short-term (Instant) Effects LT – Long-term Effects

6. Conclusions and Recommendations

Kenyans are exposed to dangerously high noise inside PSVs. An average noise level of 86.3±9.5 dB (A) was recorded inside PSVs. These levels are way above the maximum permissible limit of 60 dB (A) and are mainly associated with the use of sound amplifying equipment to play loud music/radio inside PSVs. The high noise pollution level raises two main concerns. Firstly, there is inadequate enforcement of environmental laws in Nairobi City. Secondly, existing laws for noise control are inadequate to effectively control noise.

This calls on law-enforcement agencies to intensify environmental law enforcement in the city. In addition, the laws governing noise pollution in the country need to be reconsidered to ascertain not only their ability to control noise pollution, but also whether they are attainable to law keepers. It is also imperative that more attention is given to the study and monitoring of noise pollution in Nairobi City. Additionally, other mechanisms of creating public awareness of noise pollution should be explored to help reduce irresponsible generation of and exposure to excessive noise among members of the public as is the case with PSV crew in Nairobi.

Acknowledgement

Sincere gratitude is expressed to the Division of Research, Production and Extension, Jomo Kenyatta University of Agriculture and Technology, Nairobi Kenya, for funding the study. The authors also thank the Environmental management agencies in Kenya (NEMA and DOSHS) for their support and advice; and the Nairobi County Environment Office, for approving and facilitating access to the study area.

Appendices

Measured Noise Levels

Nairobi-Nairobi North Route PSVs

S/N	Type	LAEq (dBA)	L90 (dBA)	Remarks
1	M/Bus (237)	85.2	78.8	Radio
2	M/Bus (237)	87.4	80.1	Music
3	M/Bus (237)	84.4	78.9	Music
4	M/Bus (237)	107.3	85.3	Music/Touting
5	Bus (45)	108.2	85.7	Music/Touting
6	Bus (45)	108.8	79.5	Music/Touting
7	Bus (45)	92.1	86.2	Music/Touting
8	Bus (45)	85.6	76.2	Radio/Old Vehicle
9	14 Seater (KU)	91.6	84.6	Music/Old Vehicle
10	Bus (Mwiki)	96.6	87.2	Music
11	14 Seater (KU)	76.6	73.1	Low Radio
12	M/Bus (KU)	87.6	81.3	Music
13	14 Seater (108)	85.6	79.9	No Music/Old
14	14 Seater (106)	87.4	82.1	Music
15	M/Bus (46)	98.9	94.6	Music

Nairobi-Nairobi West Route PSVs

S/N	Type	LAEq (dBA)	L90 (dBA)	Remarks
1	14 Seater	85.6	79.9	
2	14 Seater	87.4	82.1	
3	14-Seater	81.3	76.2	Music/Old
4	Bus (Kabete)	75.6	67.4	Low Music/Old
5	Bus (Kabete)	75.9	69.9	No Music
6	Bus (Kabete)	80.9	77.8	Music
7	Bus (Kikuyu)	76.9	70.9	Low Music/Old
8	M/Bus	87.6	81.3	
9	M/Bus	98.9	94.6	Music
10	M/Bus (23)	90.9	84.7	
11	M/Bus (23)	94.6	85.1	
12	M/Bus (Kinoo)	96.4	90.1	
13	M/Bus (Metro Trans)	87.7	76.2	Music
14	M/Bus (Star)	104.1	79.5	Music/Touting
15	M/Bus (Star)	107.2	85.6	Music/Touting

Nairobi-Nairobi East Route PSVs

S/N	Type	LAEq (dBA)	L90 (dBA)	Remarks
1	14-Seater (City Kabanas)	86.7	81.3	Music

Nairobi-Nairobi North Route PSVs

S/N	Type	LAEq (dBA)	L90 (dBA)	Remarks
2	14-Seater (City Kabanas)	83.6	78.8	No Music
3	14-Seater (Doonholm)	76.6	74.6	No Music
4	14-Seater (Doonholm)	76.7	71.5	No Music
5	14-Seater (Pipeline)	77.1	71.3	Music
6	14-Seater (Pipeline)	86.6	81.6	Music/Old/Touting
7	Bus (2M Doonholm)	74.6	64.6	Low Music
8	Bus (2M/Doonholm)	68.8	61.1	Music
9	Bus (City Shuttle/Utawala)	75.5	68.8	Music
10	M/Bus (58)	94.6	86.6	Moderate Music
11	M/Bus (58)	91.2	76.9	Low Music
12	M/Bus (Embassava/Doonholm)	80.3	73.9	Music
13	M/Bus (Embassava/Pipeline)	72.2	63.2	No Music
14	M/Bus (Embassava/Pipeline)	78.8	74.8	Moderate Music
15	M/Bus (Kayole)	94.6	88.6	Music

Nairobi-Nairobi South Route PSVs

S/N	Type	LAEq (dBA)	L90 (dBA)	Remarks
1	14-Seater (11B)	78.1	72.3	Low Music
2	M/Bus (11B)	77.7	74.1	Low Radio/Old
3	M/Bus (11B)	75.2	54.8	No Music/Old
4	Bus (City Shuttle/11B)	76.6	70.9	Low Radio
5	Bus (Lang'ata)	80.7	66.9	Low Music/Old
6	14-Seater (Lang'ata)	82.2	75.5	Music/No Passengers
7	14-Seater (Lang'ata)	79.4	74.1	Moderate Music
8	M/Bus (Rongai)	82.3	77.4	Music
9	M/Bus (Rongai)	81.7	76.2	Music
10	M/Bus (Karen)	86.9	81.3	Music
11	14-Seater (Karen)	86.7	81.3	Music
12	14-Seater (Karen)	86.9	81.4	Music
13	M/Bus (11B)	96.5	90.3	Music
14	Bus (Ngong)	78.9	75.3	Low Music
15	M/Bus (Ngong)	96.5	90.1	Music/Touting

References

[1] Government of Kenya (2009). The Environmental Management and Coordination (Noise and Excessive Vibration Pollution) (Control) Regulations, 2009. Legal Notice No. 61.

[2] Ahmad, K., (1998). A study of noise pollution in Dhaka City, Department of Environment (DOE), Bangladesh.

[3] National Environmental Management Authority, Kenya, NEMA (2011). *Assessment of Compliance to Noise and Excessive Vibration Regulations, 2009 in Nairobi Province.* Nairobi, Kenya: NEMA.

[4] Sincero, A.P. & Sincero, A.G. (1995). *Environmental Engineering: A Design Approach.* Prentice Hall.

[5] Boldin, *et al,* (2012). Survey context and question wording affects self-reported annoyance due to road traffic noise: a comparison between two cross-sectional studies. *Environmental Health,* 11(14), pp. 234-239.

[6] Salvato, E.L. (2003). *Environmental Engineering (Fifth Edition).* John Wiley & Sons, Inc: Hoboken, New Jersey.

[7] Alam, J. B. (2006). Study on traffic noise level of sylhet by multiple regression analysis associated with health hazards, *Iranian Journal of Environmental Health, Science and Engineering,* 3(2), pp.71-78.

[8] Ehrampoush, M.H., Halvani, H.G., Barkhordari, A. & Zare, M. (2012). Pollution in urban environments: A study in Yazd City, *Iran. Pol. J. Environ. Stud.* 21(4), pp.1095-1100.

[9] Goines, L. & Hagler, L. (2007). Noise Pollution: A modern plague. *Southern Medical Journal,* 100(3), pp. 2345-2356.

[10] Pakistan Environmental Program (2006). Measurement of Noise levels at different locations of Rawalpindi and Islamabad [Online] Available at <www.environment.gov.pk/pub-pdf/noise%20%study%20isb-rpindi.pdf> [Accessed October 23 2013].

[11] Ityavyar, E.M. & Tyav, T.T. (2013). Environmental Pollution in Nigeria: The Need for Awareness Creation for Sustainable Development. *Journal of Research in Forestry, Wildlife and Environment,* 4(2), 1-14.

[12] Ochuko, A. (2013). Evaluation of environmental noise pollution in Abuja, the capital city of Nigeria. *IJRRAS*, 4(2), pp. 470-477.

[13] Matagi,S.V.(2002).Some issues of environmental concern in Kampala, the capital city of Uganda. *Environmental Monitoring and Assessment*, 77(1), 121–138.

[14] Samagwa, D., Mkoma, S.L. & Tungaraza, C. (2009). Investigation of Noise Pollution in Restaurants in Morogoro Municipality,Tanzania, East Africa. *Journal of Applied Sciences and Environmental Management*, 13(4), 29 – 33.

[15] Liu, D. (1999). Environmental Engineers Handbook: Noise Pollution. CRC Press LLC: New Jersey.Piccolo, A., Plutino, D. & Cannistraro, G. (2005). Evaluation and analysis of the environmental noise of Messina, Italy. *Applied Acoustics,* 66(1), 447-465.

[16] Richard, L., Pierre, S.T. & Maguire, D.J. (2004). The impact of A-weighting Sound Pressure Level measurements during the evaluation of noise exposure. MARC Inc: UK.

[17] Bies, D.A. & Hansen, C.H. (1996). Engineering Noise Control: Theory and Practice (2nd Edition). E and FN SPON: London.

[18] Bruel, A. & Kajer, C. (1984). Measuring Sound. Larson and Son Printers: Denmark.

[19] EPA South Australia (2009). Guidelines for the use of the Environment Protection (Noise) Policy, 2007. Government of Australia.

[20] Gary, W.E., Lercher, P., Meis, M., Ising, Harmut & Kofler, W.W. (2001). Community noise exposure and stress in children. Journal of Acoustical Society of America, 109(3), pp.1023-1027.

[21] I-INCE (2011). Guidelines for community noise impact assessment and mitigation. Final report of the I-INCE Technical Study Group on Community Noise: Environmental Noise Impact Assessment and mitigation, March 2011.

[22] World Health Organization (1999) *Guidelines for Community Noise.* United Nations: Geneva, Switzerland.

[23] World Health Organization WHO (2009). *Night Noise Guidelines for Europe.* WHO Regional Office in Europe: UK.

[24] United States Environmental Protection Agency, EPA (1974). *Information on Levels of environmental noise requisite to protect public health and welfare with an adequate margin of safety.* US Government Printing Office: Washington DC.

[25] Jamir, L., Nongkynrih, B., & Kumar, S.G. (2014) Community noise pollution in urban India: Need for public health action. *Indian Journal of Community Medicine*. 39(1), pp. 8–12.

[26] Olayinka, O.S. (2012). Noise map: Tool for abating noise pollution in urban areas. *Open Access Scientific Reports.* 1(3), pp. 1-7.

[27] Vijay, E.L. (2015). Assessment of honking impact on traffic noise in urban traffic environment of Nagpur, India. Journal of Environmental Health Science & Engineering, 13(10), pp. 1-9.

[28] Balashanmugam, P., Ramanathan, A.R., Nehrukumar, V. & Balasubramaniyan, K. (2013). Assessment of noise pollution in Chidambaram town. *International Journal of Research in Engineering and Technology*, 2(10), pp. 85-93.

[29] Freire, S. & Gomes, N. (2013). Advancing environmental noise pollution analysis in urban areas by considering the variation of population exposure in space and time. *International Archives of the Photogrammetry, Remote Sensing and Spatial Information Sciences,* XL-4(W1), pp.155-160.

[30] Fadairo, G. (2013). Urban center's noise pollution: A case study of Akure, Nigeria. British. *Journal of Arts and Social Sciences,* 14(2), pp. 125-132.

[31] Rosenhouse, G. (2012): *The essence of noise in nature with reference to acoustics. In Hernandez and Brebbia (Eds) Design and Nature VI.* TheWITpress: United Kingdom.

[32] Momir, R.P., & Darko, I.M. & Dragan, S.C. Measurement and evaluation of the environmental noise levels in the urban areas of the city of Nis (Serbia). *Environmental Monitoring & Assessment,* 186(1), pp. 1157–1165.

[33] Tandel, B., Macwan, J. & Ruparel, P.N. (2012). Urban Corridor Noise Pollution: A case study of Surat city, India. 2011 International Conference on Environment and Industrial Innovation IPCBEE vol.12 (2011) IACSIT Press, Singapore.

[34] Kim, M., Seo, I.C., Jeong C.S., Holt, B.J., Park, T.H., Ko, J.H. & Croft, J.B. (2012). Road traffic noise: annoyance, sleep disturbance, and public health implications. *American Journal of Preventive Medicine*, 43(4), pp. 1211-1221.

9

Studies on Physicochemical Properties of Biofield Treated 2,4-Dichlorophenol

Mahendra Kumar Trivedi[1], Alice Branton[1], Dahryn Trivedi[1], Gopal Nayak[1], Ragini Singh[2], Snehasis Jana[2,*]

[1]Trivedi Global Inc., Henderson, NV, USA
[2]Trivedi Science Research Laboratory Pvt. Ltd., Bhopal, Madhya Pradesh, India

Email address:
publication@trivedisrl.com (S. Jana)

Abstract: The chlorinated phenols are widely used in chemical industries for the manufacturing of herbicides, insecticides, *etc.* However, due to consistent use they create hazards to the environment. This study was designed to use an alternative method *i.e.* biofield energy treatment and analyse its impact on the physicochemical properties of 2,4-dichlorophenol (2,4-DCP), which are the important factors related to its degradation. The 2,4-DCP sample was treated with Mr. Trivedi's biofield energy and analyzed as compared to the untreated 2,4-DCP sample (control) using various analytical techniques. The X-ray diffraction studies revealed up to 19.4% alteration in the lattice parameters along with approximately 1.8% alteration in the molecular weight, unit cell volume and density of the treated sample. The crystallite size of treated sample was increased and found as 215.24 nm as compared to 84.08 nm in the control sample. Besides, the thermal study results showed an alteration in the thermal stability profile of the treated sample as compared to the control. The differential scanning calorimetry studies revealed the decrease in the thermal decomposition temperature from 137.9°C (control) to 131.94°C in the treated sample along with 92.19% alteration in the quantity of heat absorbed during the process. Moreover, the thermogravimetric analysis showed that onset temperature of degradation was decreased, while the percent weight loss of the sample was increased from 59.12% to 71.74% in the treated sample as compared to the control. However, the Fourier transform infrared and UV-visible spectroscopic studies did not show any significant alteration in the spectra of the treated sample as compared to the control. Hence, the overall studies revealed the impact of biofield energy treatment on the physical and thermal properties of the 2,4-DCP sample.

Keywords: 2,4-Dichlorophenol, Biofield Energy Treatment, Pollutants, Complementary and Alternative Medicine

1. Introduction

The chlorinated derivatives of phenols are an important class of pollutants that originate from the industrial chemicals [1]. They are non-biodegradable and hence remain in the environment for a longer duration. 2,4-dichlorophenol (2,4-DCP) is a widely used chlorinated phenol that comes under the category of industrial toxic compounds and listed by United State Environmental Protection Agency (US EPA) as primary pollutants [2, 3]. 2,4-DCP is used as a precursor in the synthesis of 2,4-dichlorophenoxy acetic acid (2,4-D), an herbicide. Other uses include as a wood preservative, insecticide, plant growth regulator, *etc.* [4, 5]. However, due to the consistent use, it poses hazards to the environment hence, these toxic and bio-resistant compounds need to be removed from the environment or transformed into some other form that are less harmful.

When a chemical enters the environment, it distributes itself in the different compartments *viz.* air, water, soil and living organisms. Moreover, the distribution, as well as degradation of the chemical from these compartments, were decided by the physicochemical properties of that chemical. These properties include molecular weight, solubility, volatilization, and polarity, *etc.* [6, 7]. The environmental fate of 2,4-DCP from the aquatic surface and the soil surface is expected through the volatilization process [8, 9]. Its molecules reach in the atmosphere through the process of volatilization where they degraded by reacting with the photochemically produced free radicals [10]. However, this process is slow and not suitable for high concentration of pollutants. Recently, several processes are designed for making this process fast that include advanced oxidation processes [11], but they are not very much cost- effective. Therefore, it is of great significance to search an alternative strategy that can help in the fast degradation of these chemicals from the environment in a cost effective manner. Biofield energy treatment is recently known for its ability in altering the physicochemical properties of several metals and organic compounds [12, 13]. Biofield energy is scientifically related to the human energy field that is formed by the body in the surrounding space. This energy field is generated by the physical and thought processes of the human body and its interaction with the environment [14]. The living organisms are regularly exchanging this energy with the surroundings to maintain themselves [15]. The biofield energy and its interaction with the environment are used by the practitioners as the healing therapies. Its effectiveness was reported in the reduction of tension, anxiety and pain. Moreover, these therapies are considered as complementary and alternative medicines (CAM) by National Institute of Health (NIH)/National Centre for Complementary and Alternative Medicine (NCCAM). It was believed that the healers/practitioners channel the energy from the environment and direct it towards certain target [16-18]. Thus, the human has the ability to harness the energy from the environment and can transmit it to any living or non-living object(s) around the Globe. By absorbing this energy, the object will respond in a useful way. This process is known as biofield energy treatment. Mr. Trivedi is well known to possess a unique biofield energy treatment (The Trivedi Effect®). The impact of his biofield energy treatment has been reported for causing alterations in various research field *viz.* agriculture, microbiology, biotechnology [19, 20], and materials science [12, 13]. Hence, the present study was designed to analyse the impact of the biofield energy treatment on the various physicochemical properties of the 2,4-DCP using different analytical methods *viz.* X-ray diffraction (XRD), differential scanning calorimetry (DSC), thermogravimetric analysis (TGA), Fourier transform infrared (FT-IR) and UV-visible (UV-Vis) spectroscopy.

2. Materials and Methods

2,4-Dichlorophenol (2,4-DCP) was procured from Loba Chemie Pvt. Ltd., India. The procured sample was divided into two parts and named as control and treated. Mr. Trivedi's biofield energy treatment was provided to the treated part while no treatment was given to the control part. For treatment, the treated part was handed over to Mr. Trivedi in sealed pack under standard laboratory conditions. Mr. Trivedi provided the treatment to the treated part through his unique energy transmission process. The control and biofield treated samples were further characterized using various analytical techniques.

2.1. X-ray Diffraction (XRD) Study

The Phillips Holland PW 1710 X-ray diffractometer system was used to obtain the X-ray powder diffractogram of control and treated samples. The X-ray generator was equipped with a copper anode with nickel filter operating at 35kV and 20 mA. The XRD system used 1.54056 Å wavelength of radiation. The data were collected from the 2θ range of 10°-99.99° and a counting time of 0.5 seconds per step along with a step size of 0.02°.

The crystallite size (G) was calculated from the Scherrer equation:

$$G = k\lambda/(b\cos\theta) \quad (1)$$

Where, k is the equipment constant (0.94), λ is the X-ray wavelength (0.154 nm), b in radians is the full-width at half of the peak and θ is the corresponding Bragg angle. The PowderX software was used to calculate the other parameters *viz.* lattice parameter, unit cell volume, molecular weight and density of the control and treated samples.

Moreover, the percent change in crystallite size was calculated using the following equation:

$$\text{Percent change in crystallite size} = [(G_t-G_c)/G_c] \times 100 \quad (2)$$

Here, G_c and G_t denotes the crystallite size of control and treated powder samples, respectively. Similarly, the percent change in lattice parameter, unit cell volume, molecular weight, and density was calculated to analyse the impact of biofield treatment on crystal parameters of treated sample as compared to the control.

2.2. Thermal Analysis

The effect of temperature on the thermal stability profile of 2,4-DCP was analyzed using DSC and TGA/DTG studies. The impact of biofield treatment was analyzed by comparing the results of treated sample with the control sample.

2.2.1. Differential Scanning Calorimetry (DSC) Study

The DSC analysis of control and treated sample was carried out using Perkin Elmer/Pyris-1. The control and treated samples were accurately weighed and hermetically sealed in aluminium pans. Further, the samples were heated at a rate of 10°C/min under air atmosphere (5 mL/min). The thermograms were collected over the temperature range of 50°C to 250°C. An empty pan sealed with cover pan was used as a reference sample.

2.2.2. Thermogravimetric Analysis / Derivative Thermogravimetry (TGA/DTG)

The effect of temperature on the stability of the control and treated samples of 2,4-DCP was analyzed using Mettler Toledo simultaneous thermogravimetric analyser (TGA/DTG). The samples were heated from room temperature to 350°C with a heating rate of 5°C/min under air atmosphere. From TGA/DTG curve, the onset temperature, T_{onset} (temperature at which sample start losing weight), the end set temperature (temperature at which thermal degradation was completed), T_{max} (maximum thermal degradation temperature), and percentage weight loss of samples were recorded.

2.3. Fourier Transform-Infrared (FT-IR) Spectroscopic Characterization

For FT-IR characterization, the samples were crushed into fine powder and mixed in spectroscopic grade KBr in an agate mortar. Then the mixture was pressed into pellets with a hydraulic press. The FT-IR spectra were recorded on Shimadzu's Fourier transform infrared spectrometer (Japan) in the frequency range 4000-350 cm^{-1}. The FT-IR spectral analysis was helpful in determining the effect of biofield energy on the strength of bonds and stability of the samples.

2.4. UV-Vis Spectroscopic Analysis

The UV-Vis spectral analysis was measured using Shimadzu UV-2400 PC series spectrophotometer. The spectrum was recorded with 1 cm quartz cell having a slit width of 2.0 nm over a wavelength range of 200-400 nm. This study helps to investigate any change in the electron transfers behavior between the orbitals or bands of atoms, ions and molecules from the ground state to the first excited state.

3. Results and Discussion

3.1. X-ray Diffraction (XRD)

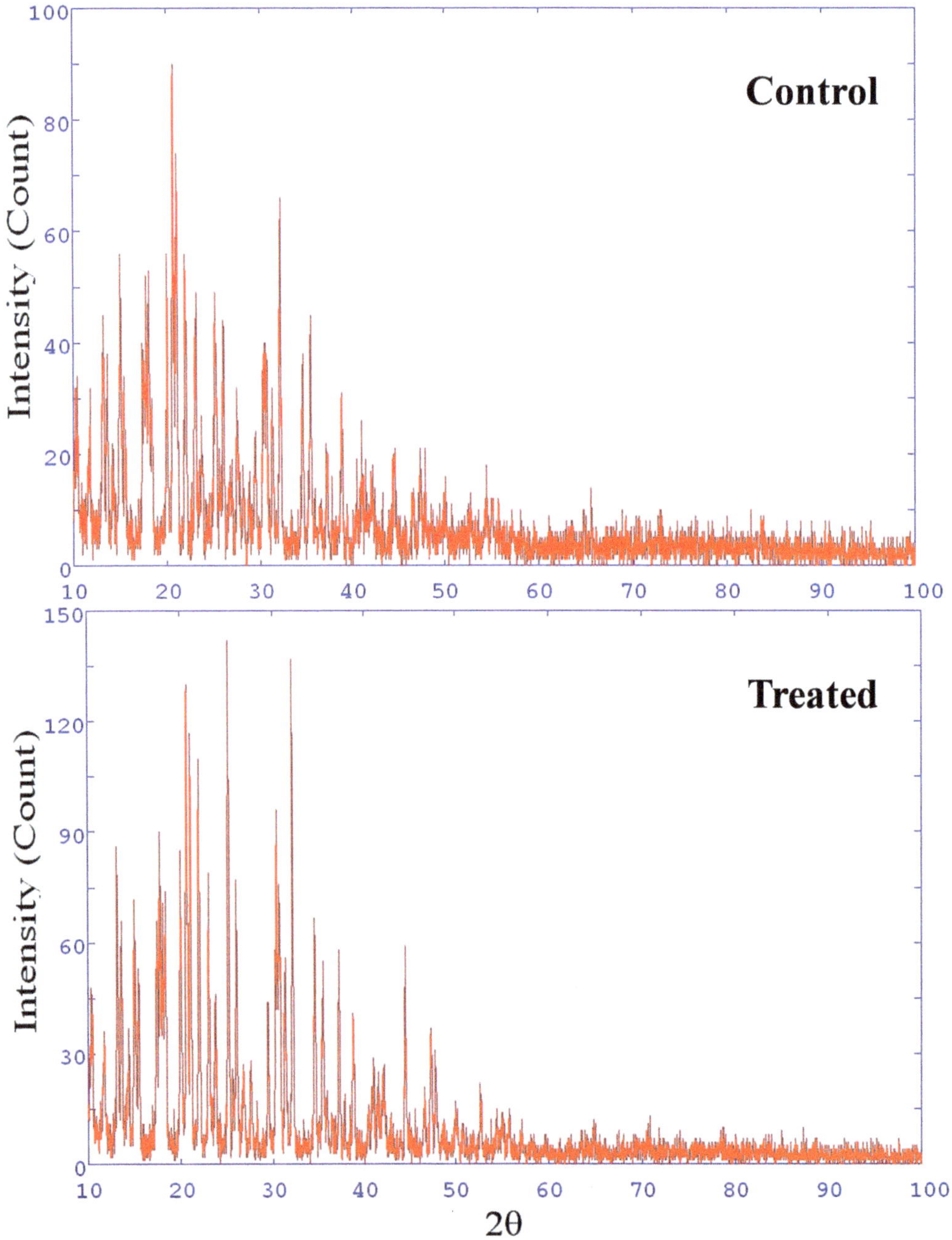

Fig. 1. *XRD diffractograms of control and treated samples of 2,4-dichlorophenol.*

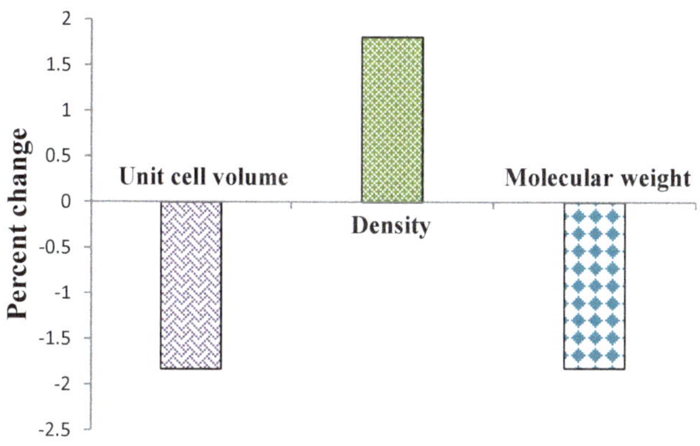

Fig. 2. Percent change in crystal parameters of treated 2,4-dichlorophenol.

The X-ray powder diffractograms of control and treated samples of 2,4-DCP are presented in Fig. 1. The diffractograms showed a series of sharp peaks in the regions of 10°<2θ>50°. The sharp peaks in the diffractograms showed the crystalline nature of 2,4-DCP [21]. In addition, the most intense peak in control sample was observed at 2θ equal to 20.62°; however, in treated sample it was observed at 32.11°. It indicated that the relative intensities of XRD peaks were altered in the treated 2,4-DCP as compared to the control sample. Moreover, the crystal structure parameters such as lattice parameter, unit cell volume, density, and molecular weight were computed using PowderX software. The results are presented in Table 1, and the data showed that the lattice parameter 'a' was increased by 19.4%, whereas 'c' was decreased by 2.44% in the treated sample as compared to the control. Similarly, the unit cell volume, molecular weight, and density were also altered in the treated sample as compared to the control (Fig. 2). It was reported that the temperature has a significant impact on the crystal parameter, and it is directly related to the lattice parameter [22]. Hence, it is assumed that the biofield treatment provided some energy to the treated sample due to which its lattice parameter was increased as compared to the control. Besides, the crystallite size of the control sample was found as 84.08 nm whereas; in the treated sample it was found as 215.24 nm. It suggested that crystallite size of the treated sample was significantly increased by 155.99% as compared to the control. Furthermore, it is assumed that biofield treatment might induce some internal strain in the treated sample that resulted in the re-orientation of neighboring planes in the same planes. This may lead to increased crystallite size in the treated sample of 2,4-DCP [12]. The decrease in the crystallite size may results in decreasing the lattice parameter and *vice versa* [23]. Hence, the alteration in crystallite size was supported by the result of lattice parameter. Besides, the process of degradation through volatilization increases with the decrease in the molecular weight. Also, the less water soluble chemicals tend to be on the water surface and hence increased volatilization [24]. It was observed that the biofield treated 2,4-DCP had decreased molecular weight as compared to the control sample. Moreover, the crystallite size was increased in the treated sample, which can be further correlated with increased particle size, and hence decreased solubility [25]. Both parameters of treated sample suggested the possible increase in decomposition of the biofield treated 2,4-DCP through the process of volatilization in the environment as compared to the control.

***Table 1.** X-ray diffraction analysis of control and treated 2,4-dichlorophenol.*

Parameter	Control	Treated
Lattice parameter		
a (Å)	30.05	35.88
b (Å)	24.23	23.64
Unit cell volume ($\times 10^{-21}$ cm^3)	3.175	3.117
Density (w/cm^3)	1.709	1.740
Crystallite size (nm)	84.08	215.24

3.2. Thermal Analysis

The DSC analysis of control and treated samples of 2,4-DCP are presented in Table 2. The peak width of samples revealed that both samples exhibited a broad endothermic peak, in which the control sample exhibited it at 137.90°C, whereas the treated sample exhibited the similar broad peak at 131.94°C. The peaks are due to the decomposition of the control and treated samples of 2,4-DCP that occurs through the process of vaporization as suggested by the literature [26]. The result suggested about 6°C decrease in the decomposition temperature of the treated sample as compared to the control. However, a very slight alteration was observed in the onset and end set temperature of the decomposition as well as the peak width in the control and treated samples. The data also showed that the control sample absorbed 90.83 J/g of heat in this process, whereas the treated sample absorbed 174.57 J/g of heat. It indicated that amount of heat absorbed was significantly increased by 92.19% in the treated sample as compared to the control. The heat needed to decompose/vaporize the sample is related to the internal energy of the molecules. It is presumed that biofield energy might alter the internal energy stored in the molecules of the treated sample.

***Table 2.** Differential scanning calorimetry analysis of 2,4-dichlorophenol.*

Parameter	Control	Treated	Percent change
Onset temperature (°C)	126.40	127.65	0.99
Peak temperature (°C)	137.90	131.94	-4.32
Endset temperature (°C)	143.73	145.95	1.54
Peak width (°C)	17.33	18.30	0.06
ΔH (J/g)	90.83	174.57	92.19

Besides, the TGA/DTG studies revealed the change in weight of sample during thermal decomposition. The TGA/DTG thermograms of the control and treated samples of 2,4-DCP are presented in Fig. 3, and data are presented in Table 3. The control sample started to decompose around 101.5°C (onset), and the process of decomposition was completed around 135.5°C (end set). However, the treated sample started losing weight around 95°C (onset) and terminated around 142°C (end set). It indicated that onset temperature was decreased while the end set temperature was increased in the treated sample as compared to the control. In this process, the control sample lost 59.12% of its initial weight; however the treated sample lost 71.74% of its initial weight. The weight loss mainly occurred due to the volatilization/vaporization of the sample after heating. From this data, it was observed that the total weight loss in the treated sample was increased during its decomposition as compared to the control. Besides, DTG thermogram data showed that T_{max} was observed at 120.54°C in the control sample and 119.31°C in the treated 2,4-DCP sample. It indicated that T_{max} was slightly decreased in treated sample as compared to the control. Furthermore, the decrease in onset temperature of decomposition, T_{max}, and increased weight loss in the treated sample of 2,4-DCP with respect to the control sample (Fig. 4) may be correlated with the decreased thermal stability and increase in volatilization of the treated sample after biofield treatment. It was reported that in 2,4-DCP molecules, the phenolic H atom possess the intermolecular hydrogen bonding [27]. Moreover, the bonding between the neighboring molecules may results in packed molecules resulting in higher thermal stability [28]. It is assumed that biofield energy was absorbed by the treated sample that might break the hydrogen bonding between the molecules. Due to this the thermal stability of the treated sample was reduced as compared to the control sample. The treated sample with increased volatility suggested the increased degradation of the treated 2,4-DCP in the environment through the process of vaporization as compared to the control sample.

Table 3. *Thermogravimetric analysis of 2,4-dichlorophenol.*

Parameter	Control	Treated
Onset temperature (°C)	101.50	95.00
Peak temperature (°C)	120.54	119.31
Endset temperature (°C)	135.50	142.00
Peak width (°C)	34.00	47.00
Change in weight (%)	-59.12	-71.74

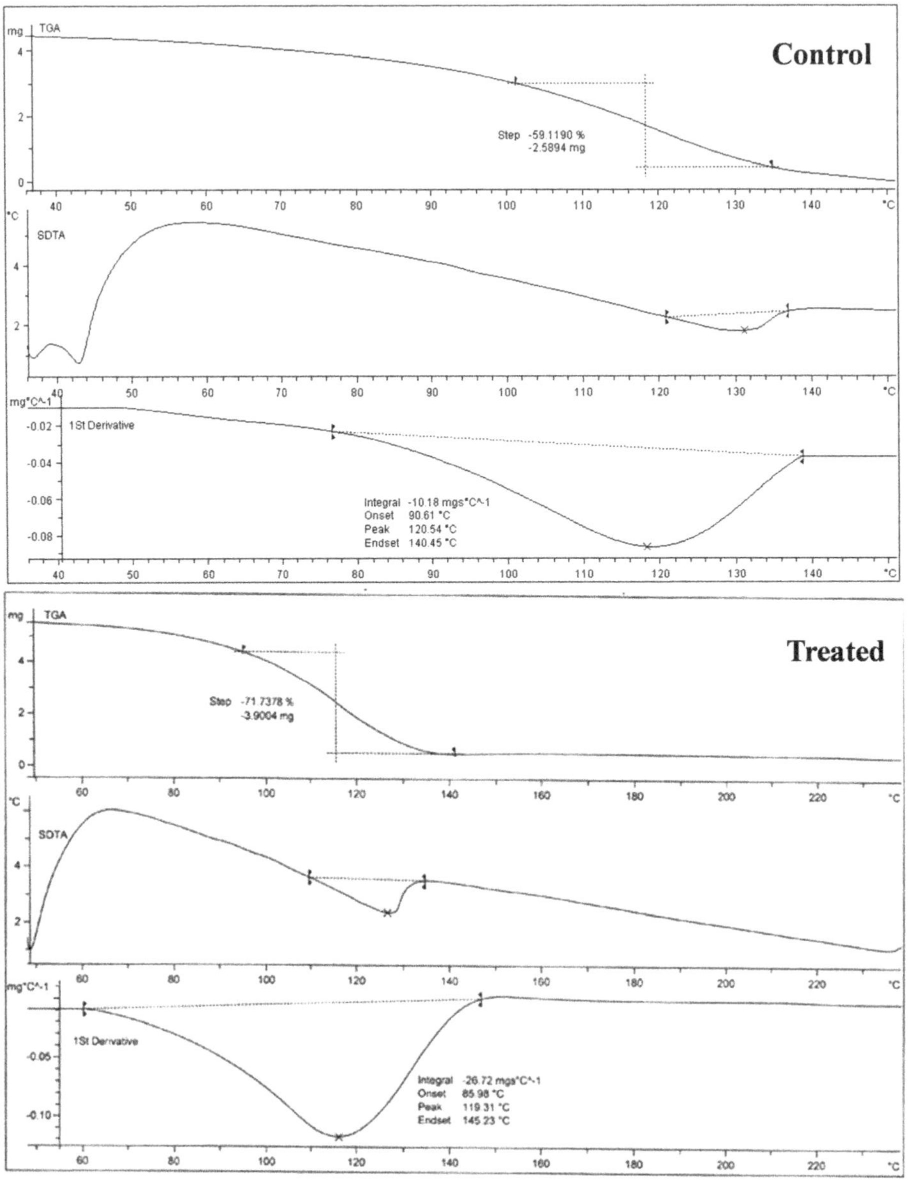

Fig. 3. *TGA/DTG thermograms of control and treated samples of 2,4-dichlorophenol.*

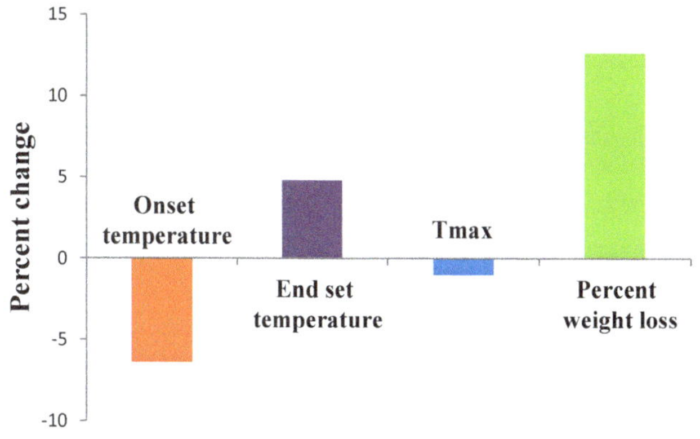

Fig. 4. Percent change in thermal analysis parameters of treated sample.

3.3. FT-IR Spectroscopic Analysis

The FT-IR spectra of control and treated samples of 2,4-DCP are shown in Fig. 5. The major vibration peaks observed were as follows:

O-H vibrations

The OH group gives rise to three vibrations: stretching, in-plane bending and out-of-plane deformations. In the present study, the strong band observed at 3437 cm^{-1} in control and 3435 cm^{-1} in the treated sample was assigned to O-H stretching mode of vibration. The peaks observed at 1251 and 653 cm^{-1} in both, control and treated samples were assigned to OH in-plane and out-of-plane bending vibration, respectively.

C-O vibrations

The C-O stretching vibration peak was observed at 1184 cm^{-1} in the control sample and 1186 cm^{-1} in the treated sample. Moreover, the peak due to C-O in-plane bending was appeared as merged with the O-H out-of-plane bending at 653 cm^{-1} in both, control and treated samples.

Aromatic C-H vibrations

The aromatic C-H stretching vibration was observed as a weak band at 3091 cm^{-1} in both, control and treated samples. Similarly, the in-plane CH bending peaks were observed at 1238 and 1093 cm^{-1} in both, the control and treated samples.

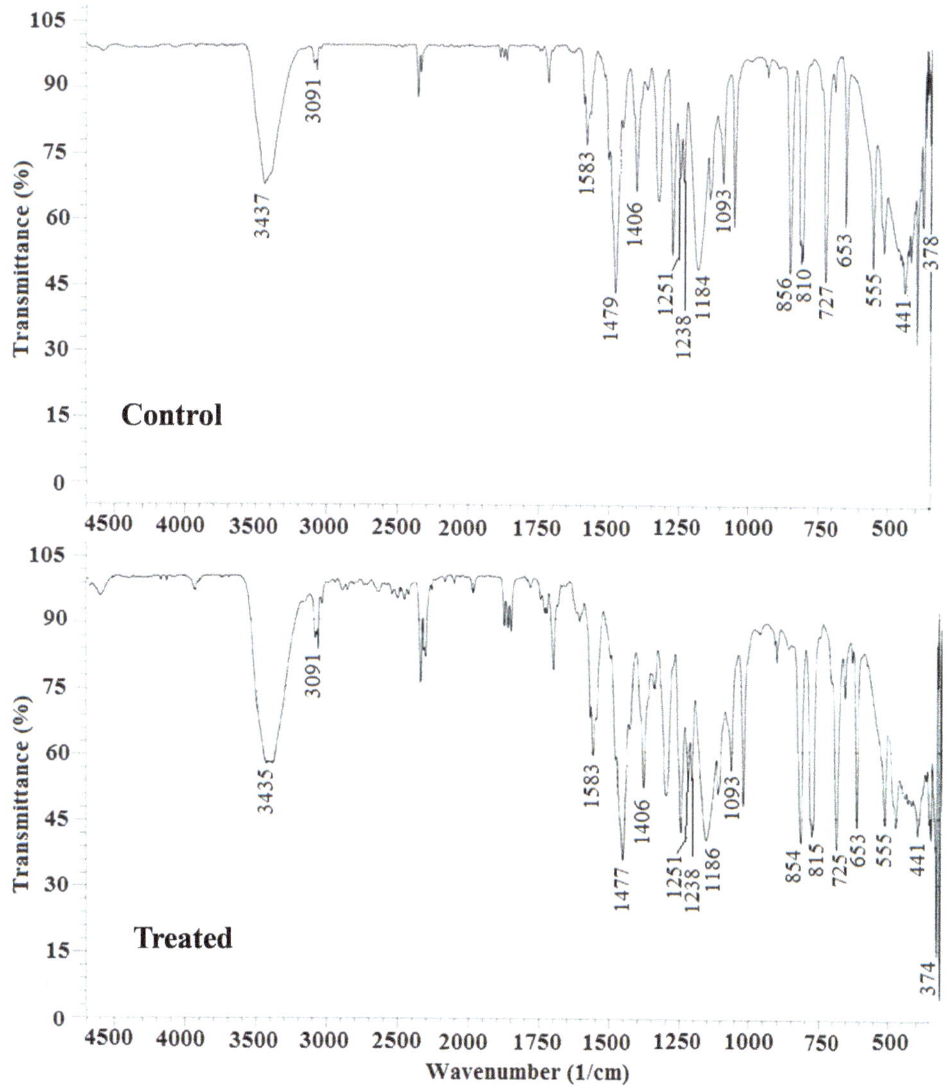

Fig. 5. FT-IR spectra of control and treated samples of 2,4-dichlorophenol.

C-C vibrations

The C-C aromatic stretching peaks were observed at 1479 and 1406 cm^{-1} in the control sample; whereas 1477 and 1406 cm^{-1} in the treated sample. Moreover, the C-C-C in-plane bending peak (441 cm^{-1}) and out-of-plane bending peak (555 cm^{-1}) were observed at the same frequency in both, control and treated samples.

C-Cl vibrations

The control and treated 2,4-DCP showed a very strong band of C-Cl stretching due to the presence of two chlorine atoms. In the FT-IR spectra, C-Cl stretching was observed at 727 and 725 cm^{-1} as the strong, sharp bands in control and treated samples, respectively. The C-Cl in-plane bending peak has appeared at 378 cm^{-1} in the control and 374 cm^{-1} in the treated sample.

Ring vibrations

The ring stretching vibration peak was observed at 1583 cm^{-1} in both control and treated samples. Moreover, the 1,2,4-trisubstituted benzene ring showed a doublet peak in the region of 750-850 cm^{-1} [29]. In the present study, the peaks were observed at 810 and 856 cm^{-1} in the control sample, while 815 and 854 cm^{-1} in the treated sample.

The FT-IR spectrum of the control sample of 2,4-DCP was well supported by literature [29,30]. The FT-IR spectra of treated 2,4-DCP sample showed the similar pattern of IR absorption peaks as the control sample. It suggested that biofield energy treatment did not cause any significant alteration in the bond strength, dipole strength, *etc.* of the treated sample.

3.4. UV-Vis Spectroscopic Analysis

The UV spectra of control and treated samples of 2,4-DCP are shown in Fig. 6. The UV spectrum of control sample showed absorption peaks at λ_{max} equal to 205, 220, 228, and 287 nm. The spectrum of control sample was well supported by the literature [31, 32].

However, the biofield treated sample showed absorption peaks at 206, 229, and 287 nm. The λ_{max} was found similar in both, the control and treated samples, except the peak at 220 nm in the control sample that was not observed in the treated sample. The peak at 220 nm was probably merged in the peak at 229 nm in the treated sample.

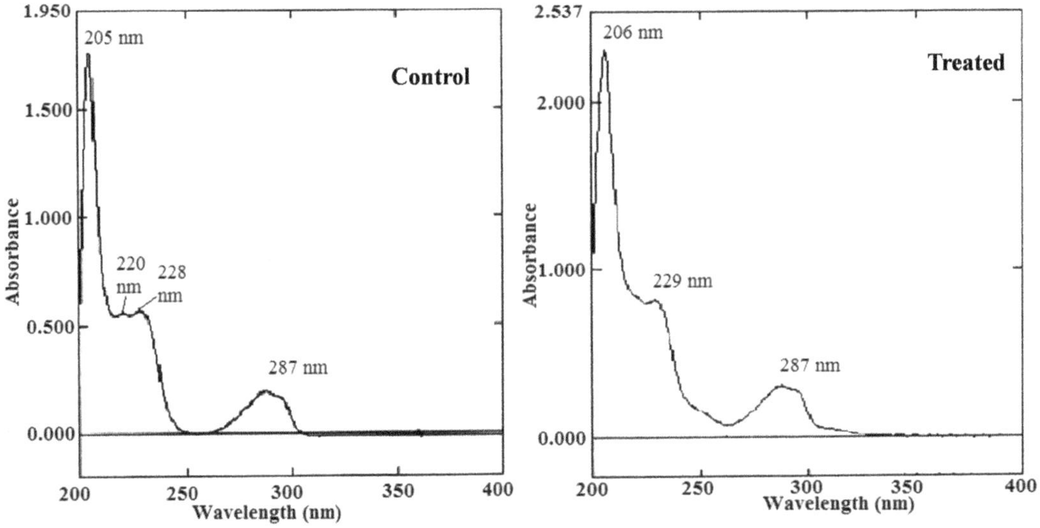

Fig. 6. *UV-Vis spectra of control and treated samples of 2,4-dichlorophenol.*

4. Conclusions

The overall study was accomplished to determine the impact of biofield treatment on various physicochemical properties of the sample of 2,4-DCP. The XRD results showed an alteration in the lattice parameters and molecular weight along with 156% increase in crystallite size of the treated sample as compared to the control. It suggested the presence of some internal strain due to the biofield energy treatment. Moreover, the increased crystallite size may ensure the less solubility of the treated sample; thereby the sample remained on the water surface and hence might result in more volatilization. Besides, the thermal analysis data revealed the decreased thermal stability and increased vaporization of the treated sample as compared to the control. It probably occurred due to the breaking of intermolecular hydrogen bonding after biofield treatment. The decreased thermal stability of the treated sample suggests its fast vaporization from ground and water surface to the atmosphere where it can further degrade by reacting with photo-chemically produced free radicals. Hence, it was concluded that the biofield treated 2,4-DCP could be less harmful to the environment due to their fast degradation process.

Acknowledgements

Authors are very grateful for the support of Trivedi Science, Trivedi Master Wellness and Trivedi Testimonials in this research work. The authors would also like to acknowledge the whole team from the Sophisticated Analytical Instrument Facility (SAIF), Nagpur and MGV Pharmacy College, Nashik for providing the instrumental facility.

References

[1] Pera-Titus M, Garcia-Molina V, Banos MA, Gimenez J, Esplugas S (2004) Degradation of chlorophenols by means of advanced oxidation processes: A general review. Appl Catal B Environ 47: 219-256.

[2] Sabhi S, Kiwi J (2001) Degradation of 2,4-dichlorophenol by immobilized iron catalysts. Water Res 35: 1994-2002.

[3] Sponza DT, Ulukoy A (2008) Kinetic of carbonaceous substrate in an upflow anaerobic sludge blanket (UASB) reactor treating 2,4-dichlorophenol (2,4 DCP). J Environ Manage 86: 121-131.

[4] Temel NK, Sokmen M (2011) New catalyst systems for the degradation of chlorophenols. Desalination 281: 209-214.

[5] Hendricks NR, Waryo TT, Arotiba O, Jahed N, Baker PGL, et al. (2009) Microsomal cytochrome P450-3A4 (CYP3A4) nanobiosensor for the determination of 2,4-dichlorophenol-an endocrine disruptor compound. Electrochim Acta 54: 1925-1931.

[6] Rathore HS, Nollet LML (2012) Pesticides: Evaluation of environmental pollution. CRC press, Florida.

[7] Domb AJ, Kost J, Wiseman D (1998) Handbook of biodegradable polymers. CRC Press, Florida.

[8] Shiu WY, Ma KC, Varhanickova D, Mackay D (1994) Chlorophenols and alkylphenols: A review and correlation of environmentally relevant properties and fate in an evaluative environment. Chemosphere 29: 1155-1224.

[9] Lyman WJ, Reehl WF, Rosenblatt DH (1982) Handbook of chemical property estimation methods: Environmental behaviour of organic compounds. McGraw-Hill, Washington, DC.

[10] Altschuh J, Bruggemann R, Santl H, Eichinger G, Piringer OG (1999) Henry's law constant for a diverse set of organic chemicals: Experimental determination and comparison of estimation methods. Chemosphere 39: 1871-1887.

[11] Gupta SS, Stadler M, Noser CA, Ghosh A, Steinhoff B, et al. (2002) Rapid total destruction of chlorophenols by activated hydrogen peroxide. Science 296: 326-328.

[12] Trivedi MK, Nayak G, Patil S, Tallapragada RM, Latiyal O, et al. (2015) Characterization of physical and structural properties of brass powder after biofield treatment. J Powder Metall Min 4: 134.

[13] Trivedi MK, Patil S, Mishra RK, Jana S (2015) Structural and physical properties of biofield treated thymol and menthol. J Mol Pharm Org Process Res 3: 127.

[14] Prakash S, Chowdhury AR, Gupta A (2015) Monitoring the human health by measuring the biofield "aura": An overview. Int J Appl Eng Res 10: 27654-27658.

[15] Rubik B (2002) The biofield hypothesis: Its biophysical basis and role in medicine. J Altern Complement Med 8: 703-717.

[16] Mager J, Moore D, Bendl D, Wong B, Rachlin K, et al. (2007) Evaluating biofield treatments in a cell culture model of oxidative stress. Explore 3: 386-390.

[17] Uchida S, Iha T, Yamaoka K, Nitta K, Sugano H (2012) Effect of biofield therapy in the human brain. J Altern Complement Med 18: 875-879.

[18] NIH, National Center for Complementary and Alternative Medicine. CAM Basics. Publication 347. [October2008]. Available at: http://nccam.nih.gov/health/whatiscam/.

[19] Trivedi MK, Patil S, Shettigar H, Gangwar M, Jana S (2015) An effect of biofield treatment on multidrug-resistant *Burkholderia cepacia*: A multihost pathogen. J Trop Dis 3: 167.

[20] Nayak G, Altekar N (2015) Effect of biofield treatment on plant growth and adaptation. J Environ Health Sci 1: 1-9.

[21] Rudrangi SR, Bhomia R, Trivedi V, Vine GJ, Mitchell JC, et al. (2015) Influence of the preparation method on the physicochemical properties of indomethacin and methyl-β-cyclodextrin complexes. Int J Pharm 479: 381-390.

[22] Muniraju NKC (2012) Crystal and spin structure and their relation to the physical properties in some geometrical and spin spiral multiferroics. Forschungszentrum Julich, Germany.

[23] Sui ML, Lu K (1994) Variation in lattice parameters with grain size of nanophase Ni_3P compound. Mater Sci Eng A Struct Mater 179-180: 541-544.

[24] Linde CD (1994) Physico-chemical properties and environmental fate of pesticides. Environmental monitoring and pest management branch, California.

[25] Hefter GT, Tomkins RPT (2003) The experimental determination of solubilities. Wiley series in solution chemistry. John Wiley and Sons.

[26] http://www.ilo.org/dyn/icsc/showcard.display?p_card_id=0438.

[27] Cai XQ, Jin ZM (2014) A hydrogen-bonded three-component complex: bis (dicyclohexylammonium) 2,4-dichlorophenolate 2,4,6-trichlorophenolate 2,4-dichlorophenol. Acta Crystallogr C Struct Chem 70: 207-209.

[28] Poletto M, Zattera AJ, Forte MM, Santana RM (2012) Thermal decomposition of wood: Influence of wood components and cellulose crystallite size. Bioresour Technol 109: 148-153.

[29] Sundaraganesan N, Anand B, Dominic Joshua B (2006) FTIR, FT-raman spectra and *ab initio* DFT vibrational analysis of 2,4-dichloro-6-nitrophenol. Spectrochim Acta A Mol Biomol Spectrosc 65: 1053-1062.

[30] Lambert JB (1987) Introduction to organic spectroscopy. Macmillan, New York, USA.

[31] Zhang YZ, Zhang NX, Ren AQ, Zhang J, Dai J, et al. (2010) Spectroscopic studies on the interaction of 2,4-dichlorophenol with bovine serum albumin. J Solution Chem 39: 495-510.

[32] Es'haghi Z (2011) Extraction and determination of three chlorophenols by hollow fiber liquid phase microextraction-Spectrophotometric analysis, and evaluation procedures using mean centering of ratio spectra method. Am J Analyt Chem 2: 1-8.

10

Carbon Stock Analysis Along Altitudinal Gradient in Gedo Forest: Implications for Forest Management and Climate Change Mitigation

Hamere Yohannes[1, *], Teshome Soromessa[2], Mekuria Argaw[2]

[1]Department of Natural Resource Management, College of Agriculture and Natural Resource Science, Debre Berhan University, Debre Berhan, Ethiopia

[2]Center for Environmental Science, College of Natural Science, Addis Ababa University, Addis Ababa, Ethiopia

Email address:

hamerey@gmail.com (H. Yohannes)

Abstract: Forests provide important ecological and environmental benefits. They serve as natural sinker of atmospheric CO_2 to mitigate climate change. In Ethiopia although, there is significant forest resource, the studies on carbon stock potential and factors that affect this potential have not been well studied. This study was done with the aim of estimating carbon stock potential and related factors that affect carbon sequestration in Gedo forest. Data was collected from 10m x 20m plot along transect in systematically stratified forest part. The forest had total mean carbon stock of 523.64 ± 29 ton ha^{-1} with aboveground biomass (281 ± 23.34 t C ha^{-1})and belowground biomass 56.1 ± 4.66 t C ha^{-1}), litter biomass (0.41 ± 0.008 t C ha^{-1}), deadwood biomass (2.37 ± 1.33 t C ha^{-1}) and soil organic carbon (183.69 ± 6.17 t C ha^{-1}). Spatial distribution of the carbon stock varied along environmental gradient. Altitude has inverse relation with aboveground biomass, belowground biomass, deadwood carbon and total carbon density. Altitude also has significant effect on all carbon pool except litter biomass and soil organic carbon. More aboveground biomass, belowground biomass and total carbon were found in the middle altitude and lower carbon was found in the upper altitude. Soil organic carbon and litter biomass carbon decreases with altitude. Deadwood biomass carbon pool was found only in lower altitude. Based on overall result it is concluded carbon sequestration in a forest ecosystem is determined by altitudinal gradient.

Keywords: Altitudinal Gradient, Biomass Carbon, Climate Change, Gedo Forest, Soil Organic Carbon

1. Introduction

One of the greatest environmental issues facing modern society is global climate change. This change is very likely a result of rising atmospheric concentrations of anthropogenic greenhouse gases (GHGs) such as carbon dioxide (CO_2), methane (CH_4), nitrous oxide (N_2O), and certain fluoride compounds [1]. There is strong evidence that warming of the earth over the last half-century has been caused largely by human activities for instance burning of fossil fuels, urbanization and change in land use, including agriculture and deforestation [2].

Forests play an important role in global carbon (C) cycle because they store large quantities of C in vegetation and soil, exchange C with the atmosphere through photosynthesis and respiration particularly primary forest [3] However, the vegetation has undergone destruction and degradation in the modern times due to human intervention like deforestation and degradation. This intervention has resulted in emission of carbon in the ecosystem. Therefore, there is a need to address environmental issues related to this. According to [4], despite role of forest, little attempts have been made to organized study that shows forest potential in carbon stock in developing countries. This is true in Ethiopia, where there is a significant forest resource but these resources are not well addressed and managed. Ethiopia forest cover was about 420,000 square kilometers (35% of Ethiopia's land) during twentieth century. These forest cover have declined to 16% in 1952, 3.6% by 1980, 2.6% by 1987, and an estimated 2.4% in 1992 [5]. The recent study shows less than 14.2% of the forest of the entire country is covered with trees. The current forests resources of Ethiopia store 2.76 billion tons of carbon in the aboveground biomass, which will be released to the atmosphere in 50 years if the deforestation continues at the present rate of about 2% [6]. Forest management, therefore, has the potential to mitigate climate change [7].

In Ethiopia there has been very limited forest carbon stock study by considering environmental factors that affect carbon stock. Therefore, this study was done to assess and correlate the environmental variable with carbon amount. And this gives basic information for the management of the forest to mitigate climate change.According to [8], In Gedo forest there is severe human intervention problem and it was intact but now it is depleted and the existing conditions call for a critical mitigating means. To get considerable attention, one way could be show its role in different perspective like its potential to sequester carbon. This has a contribution to control global warming as well as it will have a benefit for the community and country through carbon trade if it is well protected. And in Gedo forest no such study has been conducted with the aim of assessing its carbon stock amount and its dynamics along altitudinal gradient.

2. Materials and Methods

2.1. Description of the Study Area

This study was conducted in Cheliya District, West Shewa Zone of Oromia National Regional State. The district has altitudinal range of 1300-3060m a.s.l [9]. Gedo Forest lies approximately between latitudes 9^0 01` and $9^0$ 09` North and longitudes 37^0 15` and $37^0$ 27` East (Figure 1). The natural forest of the area is estimated to be the continuation of Jibat regional forest priority area in the southwest which is part of tropical rain forest. Gedo forest is legally demarcated in 1994. It is one of the national forest priority areas of Ethiopia. It covers the total area of above 10,000 hectare [10]. On the other hand, according to Oromia Forest and Wildlife Enterprise Office, based on recent survey the natural forest area is estimated about 5,000 ha excluding settlement, farmland, grazing land and bare land.

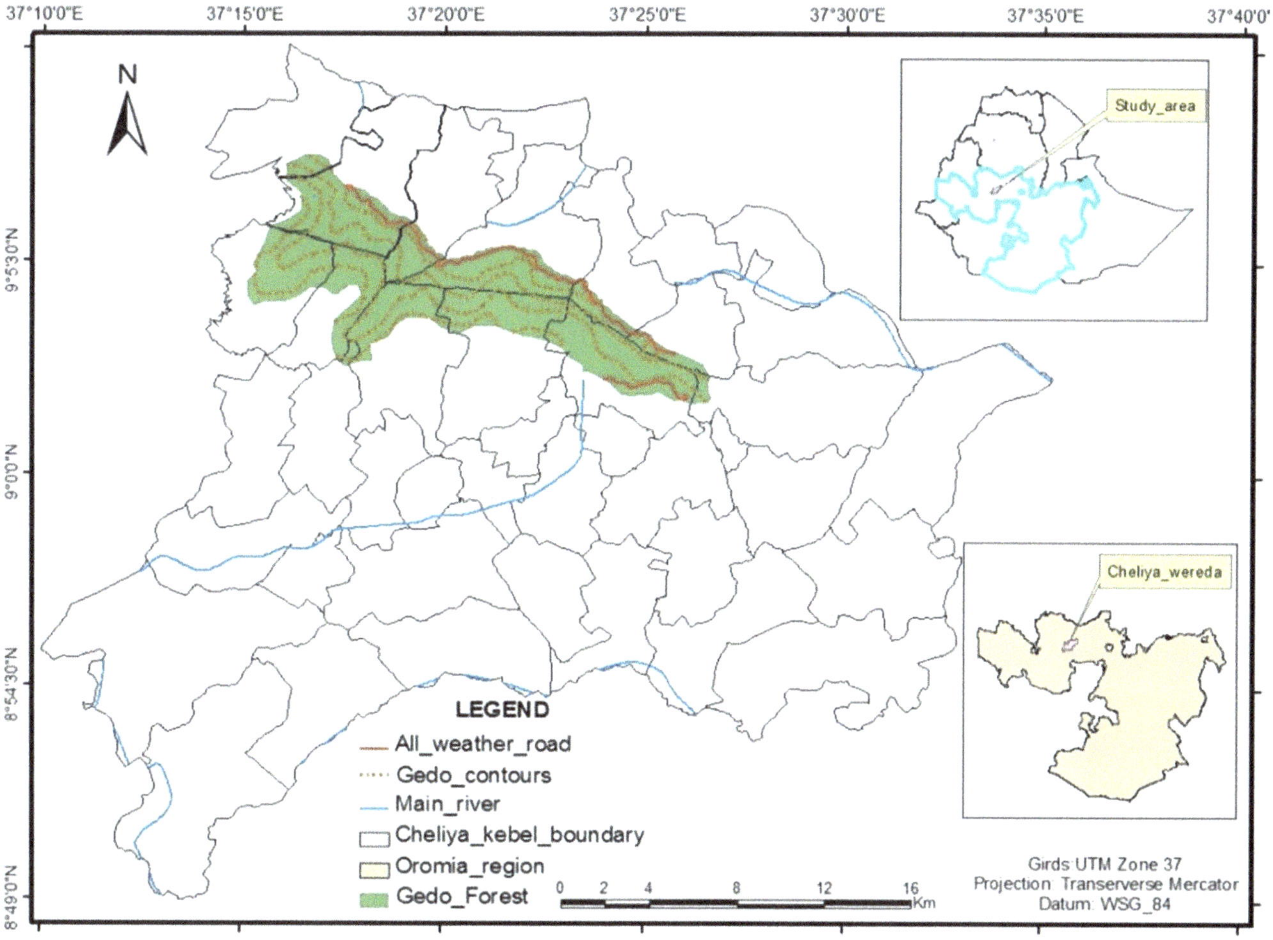

Figure 1. *Map of the study area.*

The annual mean temperature of study area ranges from 11°C to 19°C with a mean of 15°C. According to National Meteorological Services Agency data obtained for Gedo area, annual rainfall in the forest area is between 500mm and 1500mm and has mean annual rainfall of 1054 mm.Gedo forest include under Dry evergreen montane forest vegetation classification of Ethiopia. The study site dominated by *Oliniarochetiana, Oleaeuropaea subsp. cuspidata, Prunus africana, Ekebergia capensis,, Syzygium guineese subsp.Afromontanum, Ficu ssur, Podocarpusfalcatus.* And also the forest includes about five hundred hectare of plantation such as *Acacia melanoxylon, Eucalyptus spp, Grevillea robusta, Juniperus procera, Jacaranda mimosifolia and Pinus patula* [8].

2.2. Methodology

2.2.1. Delineation and Stratification of Study Area

The spatial boundaries of the study area were clearly defined and properly recognized to facilitate accurate measuring, accounting and verification. GPS tracking were used for boundary delineation. Then stratification was done based on altitudinal variation to obtain homogenous units, this increase the precision of measuring and estimating carbon stock without increasing the cost improperly. At the end, the study site classify into three stratum: Lower altitude (2279m - 2420m); Middle altitude (2421m - 2549m) and higher altitude (>2550m).

2.2.2. Sampling Techniques and Field Measurement

In this study, transect approach was applied for tree sampling and measurement. Sampling sites fromthe forest were arranged by the line transects at an interval from the bottom, middle and at thetop of the forest. A plot of 10m x 20m (200 m^2) was systematically set 200 meter gap betweeneach plot and the location of the plot were recorded.Trees above and equal to 5 cm in DBH within sample plots were measured by using diameter tape. Five smaller sub plot of 1 square meter in size were established at the center and at corner of each plot to collect leaf litter and soil. Together with field tree vegetation measurement the location of each plot including altitude was recorded by GPS. In order to identify measured tree species; a complete list of trees in each plot was done. Plant Specimens were collected, pressed, dried then identified at National Herbarium of Addis Ababa University using specimens in the Herbarium and published of Flora of Ethiopia and Eritrea.

Field Carbon Stock Measurement

Aboveground Tree Biomass (AGB): The DBH (at 1.3m) of individual trees greater than or equal to 5cm DBH were measured in each rectangular plot, and marking each tree to prevent repetition.

Leaf litter biomass: Five 1 square meter sub plot of in size were established at the center and each corner of eachplot. The leaf litter within the 1m^2 sub plots were collected and weighed. Hundred grams of evenly mixed sub-samples were brought to the laboratory placing in a sample plastic bag to determine moisture content, from which total dry mass and organic matter can then be calculated [11].

Dead Wood Biomass: According to [12],if a stump is taller than 1.3 m and branched, it wasmeasured in the same way as standing live trees. If it is less than 1.3 m tall, the diameter andheight was measured as close as possible to the top and considered as logged trees. Therefore, deadwood trees were measured and recorded on separate data sheet.

Soil Organic Carbon (SOC): Soil samples for soil carbon determination were collected at the same sampling sub-quadratsrecommended for litter sampling. From the center and corner of each plot a pits of up to 30cm in depth were dug to best represent forest. Hundred grams of composite sample werecollected from one plot from three depths (0-10 cm, 10-20 cm and 20-30 cm) by digging thesoil with the help of standardized 100 or 300 cm^3 metal soil sampling auger. The soil samplescollected from plot were brought to the laboratory placing in a sample paper bags with itslabel. Individual soil sample was taken by core sampler to estimate bulk density.

Carbon Estimation at Each Pool

Above Ground Biomass (AGB) carbon stock

The equation to be used to calculate the above ground biomass is given below:

$$Y = 34.4703 - 8.0671(DBH) + 0.6589(DBH^2) \qquad \text{(equ.1)}$$

Where, Y is above ground biomass in kg., DBH is diameter at breast height in cm.

The carbon content in the biomass were estimated by multiplying 0.47 while multiplication factor 3.67 needs to be used to estimate CO_2 equivalent [12].

Below Ground Biomass (BGB) carbon stock

The equation developed by [13], to estimate below-ground biomass was used.

The equation is given below:

$$BGB = AGB \times 0.2 \qquad \text{(equ.2)}$$

Where, BGB is below ground biomass, AGB is above ground biomass, 0.2 is conversion factor (or 20% of AGB).

Leaf litter Biomass carbon stock

According to [12], estimation of the amount of biomass in the leaf litter can be calculated by:

$$LB = \frac{Wfield}{A} \times \frac{Wsub\text{-}sample(dry)}{Wsub\text{-}sample\ (before\ dry)} \times \frac{1}{10{,}000} \qquad \text{(equ. 3)}$$

Where: LB = Litter (biomass of litter t ha-1);

W field = weight of wet field sample of litter sampled within an area of size (g);

A = size of the area in which litter were collected (ha);

W sub-sample, dry = weight of the oven-dry sub-sample, and

W sub-sample, before dry = weight of the fresh sub-sample of litter taken to the laboratory to determine moisture content (g).

Then Carbon stocks in dead litter biomass calculated by:

$$CL = LB \times \%\,C \qquad \text{(equ.4)}$$

Where, CL is total carbon stocks in the dead litter in t ha^{-1}, % C is carbon fraction determined in the laboratory [12].

Dead Wood biomass carbon stock

Standing dead wood carbon

The total amounts of biomass found in dead wood were measured based on the types of dead wood. For standing dead wood which have branches were measured similar to the biomass estimation allometric equation of above ground biomass. When standing dead wood have not leaves, there was a subtraction of 5-6 percent for softwood/ conifer species and 2-3 percent of hardwood/broadleaved species for their branches [12]. Most of the existing species were conifer species, and hence 5-6 percent reduction from the total above ground biomass of each tree provided estimates of the dead wood carbon stock. The allometric equation confirmed in REDD methodology (2009) were used to estimate the amount of biomass in standing dead wood.

$$\mathrm{BSDW} = \Sigma\,\Sigma\prod_i^n =_0 \frac{1}{3}\left(\frac{D}{200}\right)^2 h \times s \qquad \text{(equ.5)}$$

Where, biomass is expressed in kg, h = length (m), D = tree diameter (cm) and s = specific gravity (g cm^{-3}) of wood. According to[14], specific gravity (g cm^{-3}) of woodis estimated as 0.5 g cm^{-3} as default value.

Then the total biomass of dead wood were obtained by adding the biomass of standing dead woodwhich have branch and standing biomass of wood which have no branch.

$$\mathrm{TBDW} = \mathrm{SBDWB} + \mathrm{SBDWN} \qquad \text{(equ.6)}$$

Where: TBDW= Total Biomass Dead Wood

SBDWB = standing biomass dead wood branched

SBDWN = standing biomass dead wood non branched

Then the carbon content in dead wood is calculated by multiplying total biomass of dead wood with the [15] default carbon fraction of 0.47.

Soil Organic Carbon

According to[12], soil organic carbon estimated by the formula:

$$\mathrm{SOC} = \mathrm{BD} * d * \%\,C \qquad \text{(equ.7)}$$

Where, SOC= soil organic carbon stock per unit area (t ha-1),BD = soil bulk density (g cm-3),

D = the total depth at which the sample was taken (30 cm), and

%C = Carbon concentration (%).

Total Carbon Stock Density

The carbon stock density is calculated by summing the carbon stock densities of the individual carbon pools using the [12] formula;

Carbon stock density of a study area:

$$\text{C density} = \mathrm{CAGB} + \mathrm{CBGB} + \mathrm{C\ Lit} + \mathrm{CDWD} + \mathrm{SOC} \qquad \text{(equ.8)}$$

Where:

C density = Carbon stock density for all pools [t C ha^{-1}]

C_{AGB}= Carbon in above -ground tree biomass [t C ha^{-1}]

C_{BGB} = Carbon in below-ground biomass [t C ha^{-1}]

C_{Lit} = Carbon in dead litter [t C ha^{-1}]

C_{DWD} = Carbon in dead wood biomass [t C ha^{-1}]

SOC = Soil organic carbon [t C ha^{-1}]

The total carbon stock is then converted to tons of CO_2 equivalent by multiplying it by 44/12 or 3.67 [12].

Data analysis

After data collection was completed, data analysis of various carbon pools measured in theforest was accomplished. The data obtained from DBH, fresh weight and dry weight of litter, dead wood and soil were analyzed by using Statistical Package for Social Science (SPSS) software version 20.The significant of measured parameter were tested by one way ANOVA and post-hoc test at α of 0.05 to perform pairwise comparison of means and yield significant results. Pearson correlation was used to test the relationship between forest carbon stocks with altitude.

3. Results and Discussion

3.1. Results

Carbon stock at each pool

As shown in table 1, the average carbon stock per plot for aboveground carbon pool was 281±23.34 ton ha^{-1} with CO_2 equivalent of 1031.2 ± 85.68 ton ha^{-1}. The average belowground carbon stock was calculated as 56.19±4.66 ton ha^{-1} with CO_2 equivalent of 206.24± 17.13 ton ha^{-1}. The mean carbon stock in leaf litter biomass calculated as 0.41 ± 0.008 ton ha^{-1}. In this forest, a few deadwoods were observed in few plots as a result very small standing deadwood biomass and carbon storage was calculated. The dead wood carbon pool accounts only standing dead wood which were branched trees, non-branched trees and logged trees.The total carbon stock mean in deadwood was calculated as 2.37±1.33 ton ha^{-1}. Likewise the total mean soil carbon storage calculated as 183.69 ± 6.17 ton ha^{-1}. The total mean carbon storage of this forest was 523.64±29 ton ha^{-1}.

The aboveground biomass comprised 53%, belowground biomass comprised 11%, Leaf litter biomass comprised 0.07%, and deadwood biomass holds 0.45% and Soil comprised 35.5% carbon within their biomass. The largest stocks was observed on aboveground biomass followed by soil carbon pool, belowground biomass, deadwood biomass carbon storage lastly leaf litter biomass carbon. In general, the aboveground part comprises the total of 53.52% and belowground part takes 46.48% (Table 1).

Table 1. *Summary mean of carbon density (ton ha^{-1}) and distribution of each pool (%).*

	Carbon pools					
	AGBC	**BGBC**	**LBC**	**DWDBC**	**SOC**	**Total.C**
Mean(ton ha1)	281±23.34	56.1 ± 4.66	0.41±0.008	2.37±1.33	183.69 ± 6.17	523.64±29
Percentage	53	11	0.07	0.45	35.5	100

AGBC and BGBC (Above ground and belowground biomass carbon respectively; LBC (Letter Biomass Carbon); DWDBC (Dead wood Biomass Carbon); SOC (Soil Organic Carbon) and T.C (Total Carbon)

Carbon stock at each pool with Altitudinal gradient

In this study, it was observed that carbon stock amount at each pool show variation with altitudinal variation. As shown in table 2 below, it was observed that the middle altitudinal range takes the larger portion of carbon account for both AGB and BGB which have mean total of 324.05 ± 36.68 ton ha^{-1} and 64.81 ± 7.33 ton ha^{-1}, respectively. The higher altitudinal range takes lowest portion of carbon stock with 127.78 ± 30.81 ton ha^{-1} mean total carbon account in AGB and 25.55 ± 6.1 ton ha^{-1}in BGB. Similarly, the higher carbon stock in litter biomass carbon and SOC was estimated at lower altitudinal range of mean (0.55 ± 0.011 ton ha^{-1} and 185 ± 13.6 ton ha^{-1}) and lower amount was found in higher altitude (0.35 ± 0.04 ton ha^{-1} and 182 ± 8.74 ton ha^{-1}). Dead wood biomass carbon mean computed as 7.12 ± 3.89 ton ha^{-1} and only found in lower altitude. Based on the total carbon stocks the rank was ordered as middle altitude > lower altitude > higher altitude.

Table 2. *Estimated carbon stock (ton ha^{-1}) along altitudinal variation at each carbon pools.*

Elevation	C.AGB	C.BGB	C.LB	C.DWDB	C.OS	Total Carbon
Lower	296.89 ± 34.27	59.37±6.85	0.55 ± 0.011	7.12 ± 3.89	185± 13.6	545.80±42.09
Middle	324.05± 36.68	64.81±7.33	0.41 ± 0.04	0	184.37± 10.14	573±45.94
Higher	127.78± 30.81	25.55± 6.1	0.35± 0.04	0	182 ± 8.74	338.69±41.21

Based on one way ANOVA analysis, altitude has significant effect in aboveground biomass carbon (F-Value =5.172, P-value =. 008), belowground biomass carbon (F-Value =5.173, P-value =. 008), deadwood biomass carbon (F-Value =3.354, P-value =.040) and total carbon (F-Value =4.779, P-value =.011) and post hoc test was indicate that the mean difference was significant at α of 0.05 (table 3).

Table 3. *Summary of significant value of altitude on different carbon pool.*

Parameter	Carbon pools	F-Value	P-Value
Altitude gradient	C.AGB	5.172	.008*
	C.BGB	5.173	.008*
	C.LB	2.851	.064
	C.DWDB	3.354	.040*
	SOC	.019	.981
	Total C.	4.779	.011*

*the effect is significant at the $p < 0.05$ level

Correlation of carbon stock with altitude

In order to know their relation, correlation can be mostly applied statistical analysis method. It shows significant, extent and direction of the relationship. AGB and BGB carbon shows weak inverse relation with elevation (R= -.287; P=.012) at α of 0.05. Deadwood biomass and total carbon also shows negative weak correlation with altitude (R = -.264; P =. 022 and R = -259; P = 0.025) at α of 0.05. On the other hand, Litter biomass and SOC shows no correlation with altitude (table 4).

Table 4. *Summary of correlation of carbon pools and altitude.*

Parameter	Carbon pools	R-Value	P- Value
Elevation gradient	C.AGB	-.287*	.012
	C.BGB	-.287*	.012
	C.LB	-.106	.364
	C.DWDB	-.264*	.022
	SOC	-.140	.230
	Total C.	-.259*	.025

* Correlation is significant at the 0.05 level (2-tailed).

3.2. Discussion

The average AGB carbon in Gedo forest (281 t C ha^{-1}) was much more than that of previous estimates about 45.45 t C ha^{-1} for forests of Ethiopia [16], and 193.64 t C ha^{-1} for evergreen forest [23]. 143 ton ha^{-1} carbon for all types of forest for all sub-Saharan Africa [16, 17]; 73t C ha^{-1} for tropical dry forest [15] and 58.9 t C ha^{-1} for eastern and southern Africa forests [18].Gedo forest was also compared with other Dry Evergreen Afromontane forests of Ethiopia (Egdu forest, Chilimo forest, Menagesha Suba state forest and Danaba community forest) based on similarities in agro ecological condition and species composition for their carbon storage capacity (table 5). The AGB and BGB carbon mean of the present study was more than all compared forests and close to Danaba community forest mean result. The large AGB carbon density can be related with the presence of higher density of trees which are more productive and species diversity. This clearly indicated that Gedo forest has better stem size as well as carbon stock because carbon content is directly related to biomass, i.e. the higher the biomass the greater the carbon.Despite this forest has human interference till it has good potential. This suggestion supported by [8].

The mean carbon stock in present study for leaflitter biomass (0.41 ton C ha^{-1}) was less as compared to other forests except Chilimo forest and also very low from previous estimated (2.52-3.69 t C ha^{-1}) for dry tropical forests [19],(1.58 t ha^{-1}) for evergreen forest [20], (1.4-4.8 t C ha^{-1}) for tropical and sub-tropical forest [21]. Litterfall biomass accumulation may associated with forest stand condition such as stand properties (kind of species and diversity of stand, basal area, canopy cover, density of trees) and climate [22]. Similarly, according to [23], tropical forests show higher rate of decomposition because of climatic condition. Therefore, the present study of low amount of litter fall might be due to mainly decomposition rate and stand properties like species type and canopy cover.

SOC amount of the present study (183.69 ton ha^{-1}) was greater than Chilimo and Menagesha Suba state forest but lower than Danaba community forest and Egdu forest (Table 5), This value also high as compared to (55 t C ha^{-1}) for tropical forest soil [24] but lower (209.54 t C ha^{-1}) for evergreen forest [20].The present study area SOC stock amount may be due to climatic condition, amount of soil organic matter and canopy cover.

Table 5. *Comparison of carbon stock (ton ha^{-1}) of the present study with other similar studies.*

Study site	Carbon pools					Total C. mean
	AGBC	BGBC	LBC	DWDC	SOC	
Gedo forest[1]	281	56.1	0.41	2.37	183.69	523.6
Egdu forest[2]	278.08	55.62	3.47	-	277.56	614.72
Chilimo Forest[3]	90.25	17.32	0.386	-	109.40	218.01
Menagasha Suba State Forest[4]	133	26.99	5.26	6.34	121.28	292.87
Danaba community forest[5]	270.03	41.76	1.06	-	186.40	507.29

Source: [1] Present study (2015), [2][25], [3][26], [4] [27], [5][28].

Carbon stocks variation along altitudinal gradient

In present study, all carbon pool result was varied within altitudinal gradient. Altitude has significant difference and inverse correlation with all carbon pools except litter biomass and soil organic carbon. This result consistent with [29] found that negative correlation of tree biomass with altitude. Similarly, in other studies it has been reported that biomass carbon storage decreases as altitude increases [30, 31]. On the other hand, it has been reported by many studies as live biomass carbon increase with altitude increases [25, 32, 33]. In the present study, the middle altitude takes the most AGB and BGB carbon storage and much more depleted carbon storage observed on higher altitude, this result consistent with [34]. Based on the mean result, relatively it is possible to say as altitude decreases biomass carbon increases. According to [19, 35], large individual trees account more carbon proportion. Similarly for the present study, the reason might be related with the presence of more productive stem density in the middle altitude than in higher altitude.

Though [34] study found that deadwood biomass carbon in all altitude classes with more carbon stock in middle altitude, the present study, reported that deadwood biomass carbon found only in lower altitude. Deadwood biomass carbon found only in lower altitude. This is related with more of logged and standing dead wood was foundin the lower altitude. And in the higher altitude, this carbon pool was not found. This might be related to in study area; it seems difficult to collect wood in the higher altitude with steep slope.

Although in many studies it was reported that as altitude increase SOC and LB carbon increases [21, 36, 37]. In present study, it was observed that SOC and Litter biomass means decreases as altitude increases. This result was consistent with [31, 35, 38]. According to [38], Forest stand with dense canopy and higher input of litter can results in maximum storage of carbon stock in the pool. Consequently, the present study area of decreasing trend in SOC stock along elevation might be due to canopy cover, litter biomass accumulation and species diversity. The overall trend of total means carbon stock of the forest show similar pattern with AGB and BGB carbon. This might be due to the fact that total carbon density mostly depending on aboveground biomass carbon pool.

4. Conclusion

Carbon stock study of forests is crucial to show forest potential and role to mitigate climate change risk. Forests have a capability to store substantial amount of carbon within their biomass and soil. Forest carbon stock can influence by different factors such as natural factors, environmental factors, physical factors and human disturbances. This study has shown that forest carbon stock is affected by one of environmental variables i.e. altitude. Each carbon pool shows variation along this environmental gradient. And altitude plays a key role in both aboveground and belowground carbon pool account. This is related with the distribution of productive stem density within the forest.

References

[1] IPCC. (2007). Synthesis Report. 52 p. Fourth Assessment Report. Both of these IPCC publications synthesize information from all Working Groups: WG I--The Physical Science Basis; WG II -- Impacts, Adaptation, and Vulnerability; WG III--Mitigation of Climate Change. Download from: http://www.ipcc.ch/pdf/assessment-report/ar4/syr/ar4_syr.pdf.

[2] Samalca, K. I., Gier, D. L. and Ali, H. (2009). Estimation of tropical forest biomass for assessment of carbon sequestration using regression models and remote sensing. Berua, East Calimantan: Indonesia.

[3] McMahon, S. M., Parker, G. G. and Miller, D. R. (2010). Evidence for a recent increase in forest growth. Proceedings of the National Academy of Sciences.J. Science107: 3611–3615.

[4] Ravindranath, N. H., Chaturvedi, R. K. and Murthy, I. K. (2008). Forest conservation, Afforestation and reforestation in India: Implications for forest carbon stocks. Curr. Sci., 95 (2): 216–222.

[5] Federal Democratic Republic of Ethiopia. (1998). National Action Programme to Combat Desertification.Vol. I: The State of Natural Resources in Arid, Semi-Arid and Dry Sub-HumidAreas. EPA 1998 report, Addis Ababa, Ethiopia, Pp. 28-35.

[6] Yitebitu Moges, Zewdu Eshetu and Sisay Nune. (2010). A review on Ethiopian Forest Resources: current status and future management options in view of access to carbon finances. Prepared for the Ethiopian climate research and networking and the United Nations development programme (UNDP). Addis Ababa, Ethiopia.

[7] Bonan, G. (2008). Forests and climate change: Forcing, feedbacks, and the climate benefits of forests. J. Science320: 1444-1449.

[8] Birhanu Kebede, Teshome Soromessa and Ensermu Kelbessa. 2014. Structure and Regeneration Status of Gedo Dry Evergreen Montane Forest, West Shewa Zone of Oromia National Regional State, Central Ethiopia. Sci. Technol. Arts Res. J., April-June 2014, 3(2): 119-131.

[9] Endalew Amenu. (2007). Use and management of medicinal plants by indigenous people of Ejaji area (CheliyaWoreda) West Shoa, Ethiopia: an ethno botanical approach. Unpublished M.Sc. Thesis, Addis Ababa University, Addis Ababa.

[10] Young, J. (2012). Ethiopian Protected Areas a 'Snapshot. A reference guide for future strategic planning and project funding: Ethiopia.

[11] Bhishma, P. S., Shiva, S. P., Ajay, P., Eak, B. R., Sanjeeb, B., Tibendra, R. B., Shambhu, C., and Rijan, T. (2010). Forest Carbon Stock Measurement: Guidelines for measuring carbon stocks incommunity-managed forests. Funded by Norwegian Agency for Development Cooperation (NORAD).Asia Network for Sustainable Agriculture and Bioresources (ANSAB) publishing, Kathmandu, Nepal, pp.17-43.

[12] Pearson, T. R., Walker, S. and Brown, S. 2005. *Sourcebook for land-use, land-use change and forestry projects.* Winrock International and the Bio-carbon fund of the World Bank. Arlington, USA, pp. 19-35.

[13] MacDicken, K. G. (1997). *A* Guide to Monitoring Carbon Storage in Forestry and Agro-forestry Projects. In Forest Carbon Monitoring Program. Winrock International Institute for Agricultural Development, Arlington, Virginia.

[14] Hairiah, K., Sitompul, S. M., Noordwijk, M. and Palm, C. (2001). Methods for sampling carbon stocks above and below ground. International Centre for Research in Agroforestry. Southeast AsianRegional Research Programme, Bogor, Indonesia, pp. 10-15.

[15] IPCC. (2006). Good practice guidelines for National Greenhouse gas inventories. Switzerland: Intergovernmental panel on climate change. Unpublished document.

[16] Brown, S. 1997. Estimating biomass and biomass change of tropical forests: A primer Rome, Italy, Food and Agriculture Organization of the United Nations (FAO).

[17] Gibbs, H, K. and Brown, S. (2007).Geographical distribution of woody biomass carbon stocks in tropical Africa: an updated database for 2000.

[18] IPCC. (2010). Global Forest Resources Assessment 2010 Main report. FAO. Forestry Paper. 163. Food and Agriculture Organization of the United Nations: Rome.

[19] Brown, S. and Lugo, A. (1982). The storage and production of organic matter in tropical forests and their role in the global carbon cycle. Biotropica14: 161-187.

[20] Mohanraj, R., Saravanan, J., Dhanakumar, S. (2011). Carbon stock in Kolli forests, Eastern Ghats (India) with emphasis on aboveground biomass, litter, woody debris and soils. *J. Forest.* 4: 61-65.

[21] Chang, C., Wang, C., Chou, C., Duh C. (2010). The Importance of Litter Biomass in Estimating Soil Organic Carbon Pools in Natural Forests of Taiwan. Taiwan *Journal of Science.* 25(2): 171-80.

[22] De Castilho, C. V., Magnusson, W. E., de Araújo, R. N. O., Luizão, R. C. C., Luizão, F. J., Albertina, P., Higuchi, N., 2006. Variation in aboveground tree live biomass in a central Amazonian Forest: effects of soil and topography. *J. Forest Ecology Management*. 234: 85–96.

[23] Salinas, N. Malhi1, Y. Meir, P. Silman, M. Roman Cuesta, R. Huaman, J. Salinas, D. Huaman, V. Gibaja, A. Mamani, M. and Farfan, F. (2010). The sensitivity of tropical leaf litter decomposition to temperature: results from a large-scale leaf translocation experiment along anelevation gradient in Peruvian forests. New Phytologist. 10 (1111): 1469-8137 [Available at www.newphytologist.com].

[24] Gorte, R. W. (2009). Carbon Sequestration in Forests. Congressional Research Service. CRS Report for Congress, (Eggleston, H. S, Buendia, L. and Miwa, K. eds). Prepared for Members and Committees of Congress. Greenhouse Gas Inventories Programme. CRS Publishing, USA, Pp. 1-23.

[25] Adugna Feyissa, Teshome Soromessa, Mekuria Argaw. (2013). Forest Carbon Stocks and Variations along Altitudinal Gradients in Egdu Forest: Implications of Managing Forests for Climate Change Mitigation. *J. STAR* 2(4): 40-46.

[26] Getu Shiferaw. (2012). Carbon stocks in different pools in natural and plantation forests of Chilimo, central highland of Ethiopia. Unpublished M.Sc thesis, Addis Ababa University. Addis Ababa.

[27] Mesfin Sahile. (2011). Estimating and Mapping of Carbon Stocks based on Remote Sensing, GIS and Ground Survey in the MenageshaSuba State Forest.M.Sc. Thesis, Addis Ababa University, Addis Ababa.

[28] Muluken Nega Bazezew, Teshome Soromessa, Eyale Bayable. Above- and Below-Ground Reserved Carbon in Danaba Community Forest of Oromia Region, Ethiopia: Implications for CO_2Emission Balance. American Journal of Environmental Protection. Vol. 4, No.2, 2015, pp. 75-82. doi: 10.11648/j.ajep.20150402.11.

[29] Sharma, C. M., Suyal, S., Gairolia, S., Ghildiyal, S. K., 2009. Species richness and diversity along an altitudinal gradient in moist temperate forest of GarhwalHimalaya. *J.American science.* 5(5). 119-128.

[30] Moser, G., Hertel, D. and Leuschner, C. (2007). Altitudinal change in LAI and stand leaf biomass in tropical montane forests: a transect study in Ecuador and a pan-tropical meta analysis. *J. Ecosystems.*10: 924–935.

[31] Sheikh, M. A., Kumar, S. and Kumar, M. (2012). Above and below ground organic carbon stocks in a sub-tropical Pinusroxburghii Sargent forest of the Garhwal Himalayas, Beijing Forestry University and Springer-Verlag Berlin Heidelberg.

[32] Zhu, B., Wang, X., Fang, J., Piao, S., Shen, H., Zhao, S., and Peng, C. (2010). Altitudinal Changes in Carbon Storage of Temperate Forests on Mt. Changbai, Northeast China, *J. Plant Resource.* 123(4): 439-450.

[33] Gairola, S., Sharma, C. M., Ghildiyal, S. K. and Suyal, S. (2011). Live tree biomass and carbon variation along an altitudinal gradient in moist temperate valley slopes of the Garhwal Himalaya (India). *J. Current science.* 100(12): 1862-1870.

[34] Tibebu Yelemfrhat Simegn, Teshome Soromessa. Carbon Stock Variations Along Altitudinal and Slope Gradient in the Forest Belt of Simen Mountains National Park, Ethiopia. American Journal of Environmental Protection. Vol. 4, No. 4, 2015, pp. 199-201. doi: 10.11648/j.ajep.20150404.15.

[35] Mwakisunga, B. and Majule, A. E. (2012). The influence of altitude and management on carbon stock quantities in rungwe forest, southern highland of Tanzania. *Open Journal of Ecology*. 2(4): 214- 221.

[36] Tsui, C., Chena, Z. and Hsieh, C. (2004). Relationships between soil properties and slope position in a lowland rain forest of southern Taiwan. *J. Geoderma*. 123: 131–142. [Available online at www.Science direct.com].

[37] Griffiths, R. P. Madritch, M. D. Swanson, A. K. (2009). The effects of topography on forest soil characteristics in the Oregon Cascade Mountains (USA): Implications for the effects of climate change on soil properties. J. *Forest Ecology and Management*. 257: 1–7.

[38] Sheikh, M. A., Kumar, M. and Bussmann, R. W. (2009). Altitudinal variation in soil organic carbon stock in coniferous subtropical and broadleaf temperate forests in Garhwal Himalaya. *J. Carbon Balance and Management*, Biomed central 4(6): 1-6.

11

Accuracy Assessment of Land Use Land Cover Classification using Google Earth

Abineh Tilahun[1, *], **Bogale Teferie**[2]

[1]Department of Geography and Environmental Studies, Adigrat University, Adigrat, Ethiopia
[2]Department of Geography and Environmental Studies, Dilla University, Adigrat, Ethiopia

Email address:
ababineh@gmail.com (A. Tilahun)

Abstract: This study examines the accuracy assessment of land use land cover classification using Google Earth in the case of Kilite Awulalo, Tigray State, Ethiopia for the year 2014. For this study, Landsat-8 OLI_TIRS image of 2014 was used and analyzed using Arc GIS 10.1. Supervised classification scheme was used to classify the images. Under land use and land cover categories Agriculture land, Settlement land, Grazing land, Forest land, Bush land, Water bodies and Bare/stony land were studied. After classification of land use land cover types, 100 Random Points were generated in Arc GIS and converting random points to KML in order to open in Google Earth. Each random point's value verified from Google Earth for accuracy assessment. Google Earth model was used to measure of how many ground truth pixels are correctly classified. For this study, Free Google Earth which was Build in Date 10/7/2013 was used. The result shows that total (overall) accuracy of land use and land cover for 2014 is 82.00% and Kappa (K) is 77.02% which is acceptable in both accuracy total (overall) and Kappa accuracy.

Keywords: Accuracy Assessment, Google Earth, Kappa, Land Use Land Cover

1. Introduction

The LULC change distribution varies in space and time. This is because physical and social characteristics of communities vary in space and time, so do land-use choices, resulting in a spatial pattern of land-use types (Canute *et al*, 2015). The study of land use land cover pattern is essential for the selection, planning and implementation of the land use schemes to meet the increasing human needs and welfare. This also provides the information for managing dynamics of land use and meeting the demands of increasing human population (Yadav *et al*. 2010). Therefore, showing the results of land use land cover in the form of maps and statistical data is very important for concerned bodies like planning, management and utilization of land for different purpose (Roy and Giriraj, 2008).

Land use land cover analysis can be done from processed aerial photographs, Satellite images (Landsat image, Quick bird image) and Google Earth (Dash, 2005). Since remote sensed data from the earth orbit can be obtained repeatedly over the same area, they have been very useful to monitor and analyze LUCC in various regions of the earth and greatly contribute to planning and management of available resources, especially in the developing countries where other kinds of background data are often lacking (Dash, 2005; Fakeye *et al*, 2015).

After doing of land use land cover classification, the accuracy of special data should be defined. Accuracy assessment or validation is an important step in the processing of remote sensing data which determines the information value of the resulting data to a user (Abubaker *et al*. 2013).

For doing of accuracy assessment we use

1. Ground verification using Global Position System (Observing the area)
2. Compares of the classified image to an image which is assumed to be correct (such as an aerial photograph, Google Earth Image)
3. Asking questions for concerned bodies those they have previous information

In current, researchers tend to use high spatial resolution data in order to obtain more accurate and precise result. In this regard, images with high spatial resolution from Google earth that are free to the public are a good source of imagery including satellite images and air photos. Google earth (http://earth.google.com) provided by Google Inc., is a virtual globe programming that maps the earth by superimposition of high resolution satellite images (Shirkou and Aliakbar, 2013). Since it was released in June 2005, Google Earth has aims to provide viewers with "a more realistic view of the world". Beside Google Earth, map data and positional measurement can be obtained using different methods such as conventional or modern land survey methods, Global positional System (GPS) and remote sensing satellite imagery. Each of these methods is of a known positional accuracy (Nagi *et al*, 2013).

Google earth high-resolution imagery is important for accuracy assessment by comparing of point by point basis. A random set of points is generated for the area and then using the Google Earth the value for each point is identified. Therefore, this study was intended to examine the accuracy of Land Use Land Cover Classification using Google Earth in the case of in Kilite Awulalo, Tigray State, Ethiopia

2. Site Description (Location)

The study area is located in Tigray region; north part of Ethiopia within the geographical grid coordinates of 13°33'37.618"N to 13°57'29.447"N latitude and 39°18'8.606"E - 39°41'44.647"E longitude. The District is bounded with Saesie Tsaedaemba District in the north, Hawzen District in the North West, Enderta District in the south, Degua Temben District in the south west and Atsbi wenberta District in the east.

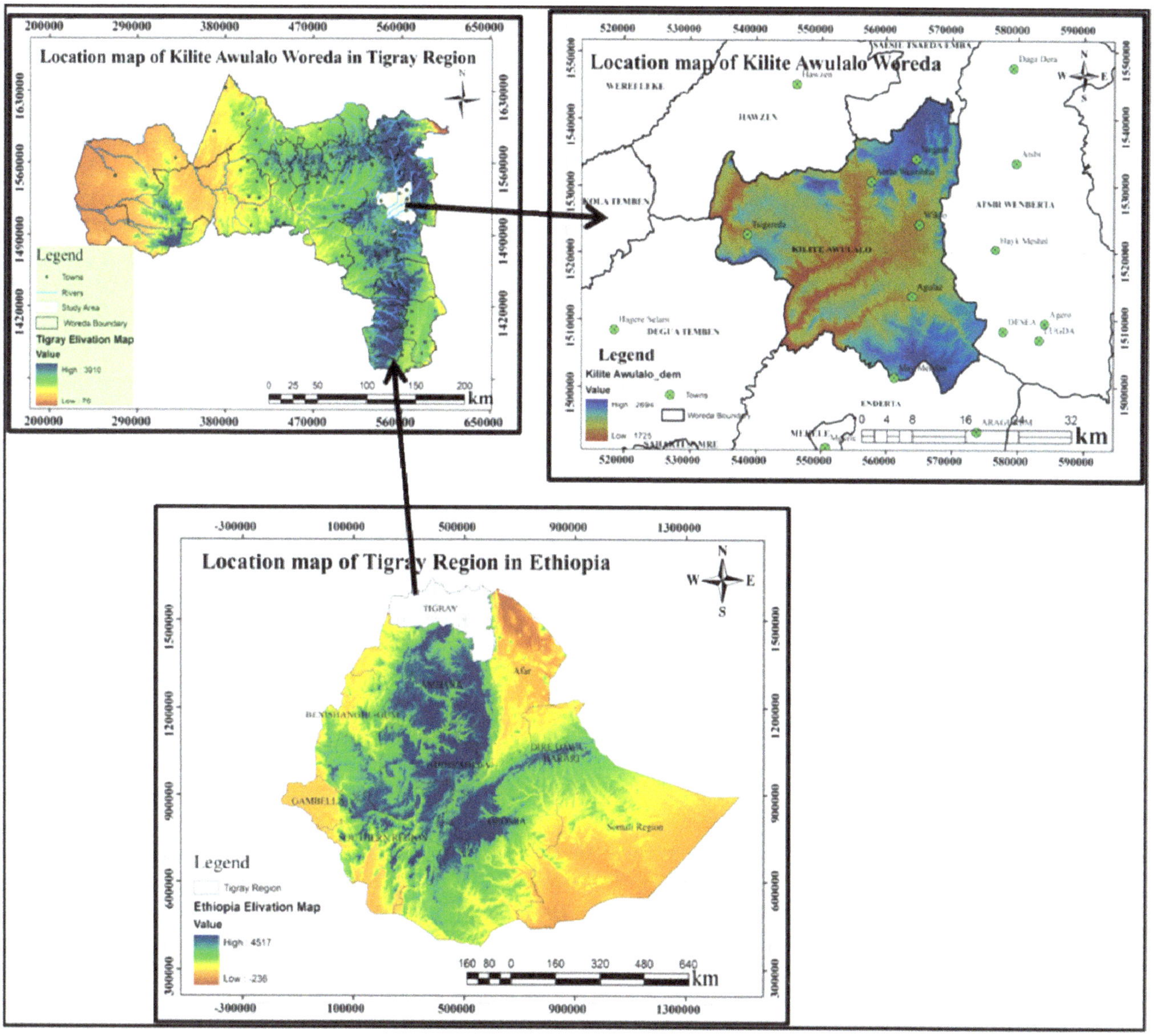

Figure 1. *Location map of the study area.*

3. Materials and Methods

3.1. Data Used

Satellite imagery covering the period 2014 (Landsat-8 OLI_TIRS) were used for land use land cover analysis. The detail of the satellite images is given below.

Table 1. *Source and Satellite Images data collection processing.*

Satellite/ Spacecraft_ID	Sensor ID	Path/row	Date of acquisition	Spatial resolution/ Grid Cell Size (m)	Sun Elevation	Cloud Cover
		168/50	2014-03-07	30m	56.46464264	0.60
Landsat-8	OLI_TIRS	168/51	2014-03-23	30m	61.10639256	3.90
		169/50	2014-03-30	30m	62.39484780	0.48

3.2. Software Used

The following software was used for the processing and/or analysis of data/images.

a) Arc GIS 10.1: Preparation of Location of the project area, Data base generation
b) Erdas Imagine 13: For Image classification
c) Google Earth for creating KML files and verifying of randomly generated points

3.3. Methods of Data Analysis

Landsat image 8, Bands 4, 3 and 2 are used to combine to make true-colour composite images for land use land cover analysis and supervised image classification was done. Which means, the analysts trains the computer to recognize patterns in the data by selecting pixels that represents patterns or land cover features that she recognizes, The signature files thus created are then used in the classification process where each pixel is categorized into the land cover class it mostly resembles. After image classification was done, free Google Earth which was Build in Date 5/17/2013 was used for accuracy assessment.

Accuracy assessment was measured though matrix using user classification and reference image Individual accuracy was measured using equation 1

$$\text{Individual accuracy} = \frac{\text{Reference value}}{\text{row total}} \quad (1)$$

Overall accuracy was measured using equation 2

$$\text{Total (overall) accuracy} = \frac{\text{Number of correct plots (Value)}}{\text{Total number of plots (Value)}} \times 100 \quad (2)$$

Kappa can be used as a measure of agreement between model predictions and reality (Congalton 1991) or to determine if the values contained in an error matrix represent a result significantly better than random (Jensen 1996). Kappa was computed using Equation 3.

$$K = \frac{N\sum_{i=1}^{r} x_{ii} - \sum_{i=1}^{r}(x_{i+} \times x_{+i})}{N^2 - \sum_{i=1}^{r}(x_{ii} \times x_{+i})} \quad (3)$$

4. Results and Discussion

Based on satellite image analysis and observation of the current situation six major land uses and land cover types were identified in the study area. These include Settlement, Agriculture, Forest, Rocky, Bush Land and Water bodies

4.1. Land Use Land Cover Classification for 2014

The land use land cover classification of the area for 2014 from OLI_IRS satellite image (table 2) showed that the majority of the study area is covered by bush land 40573.53 hectares (ha), contributes 40.03% of the total area. Forest land and agriculture land cover an aerial size of 11916.4 ha (11.76 %) and 30402.27 ha (30 %) respectively, whereas the aerial coverage of Rocky/Bare and Settlement land is 13974.39 ha (13.78%) and 4461.48 ha (4.41%) from the total area of the District. There are also artificial waters which covers 1.79 %.

Table 2. *LU/LC classes, their corresponding areas for 2014.*

LULC Categories	2014	
	Area (Hectare)	Area (%0
Forest	11916.4	11.76
Bush land	40573.53	40.06
Agriculture	30402.27	30.00
Settlement	4461.48	4.41
Rock/Bare	13974.39	13.78
Water Bodies	18.09	1.79
Total	101346.12	100.00

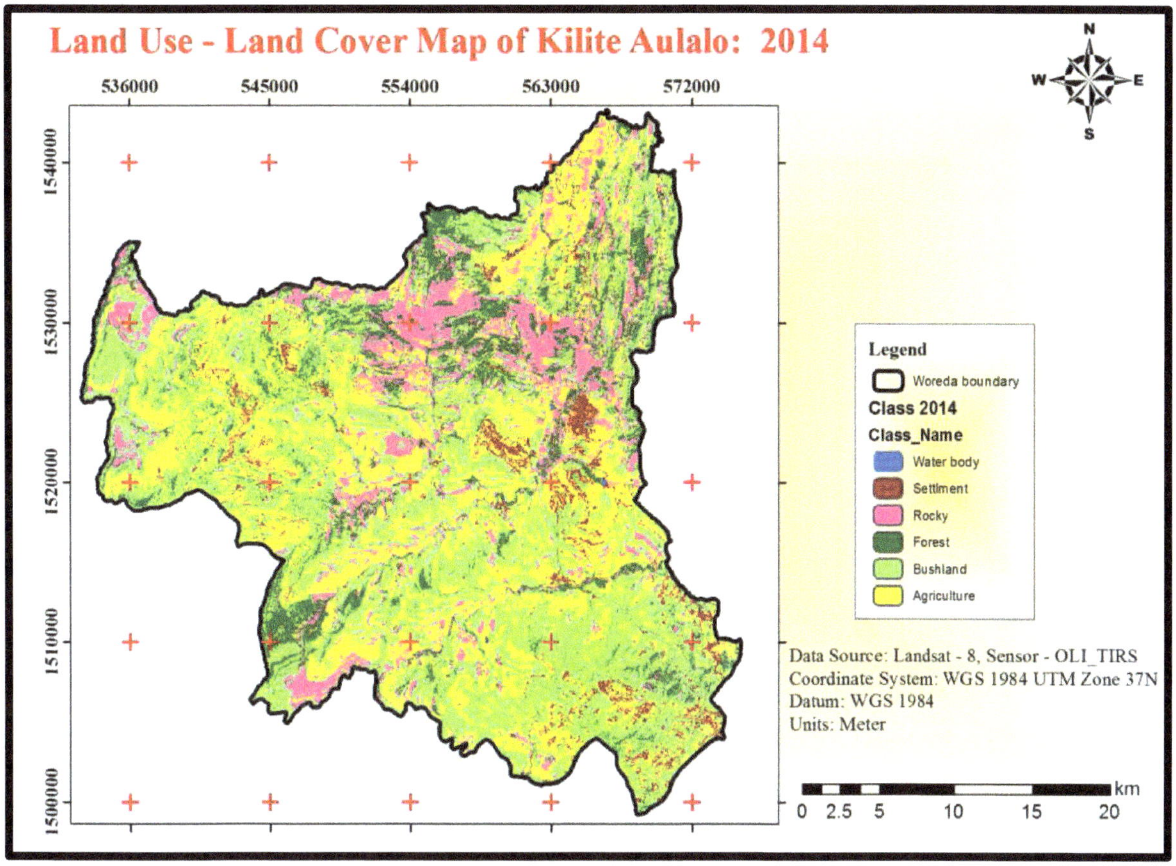

Figure 2. Land use and land cover map of the Kilite Awulalo in 2014.

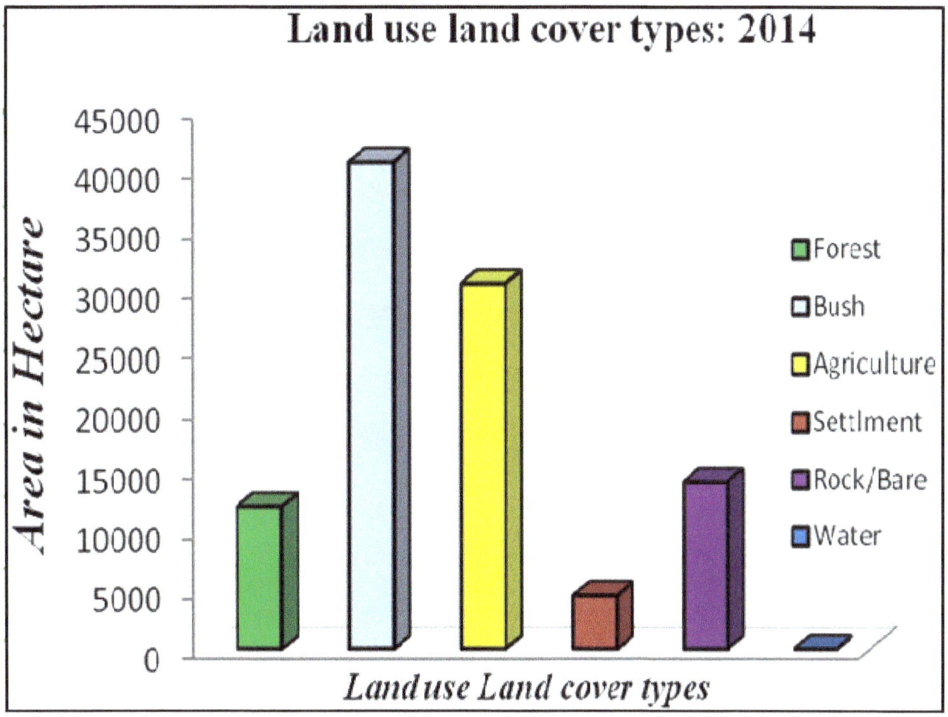

Figure 3. LU/LC classes, their corresponding areas for 2014.

4.2. Accuracy Assessment of Classifications for 2014

Google Earth represents a powerful and attractive source of positional data that can be used for investigation and preliminary studies with suitable accuracy and low cost. Since Images from Google Earth with high spatial resolution are free for public and can be used directly in land use land cover mapping in small geographical extend. A study which was conducted by Abineh and Zubairul in 2015, and the result of accuracy assessment of land use land cover with the help of Google Earth was more than 75% which is acceptable.

After image is classified, generating a set of random 100 points was done in ArcMap (Toolbox >>> Data Management Tools >>> Feature Class >>> Create Random Points >>> create extract values to points. Then the value of each random points were identifying from Google Earth image.

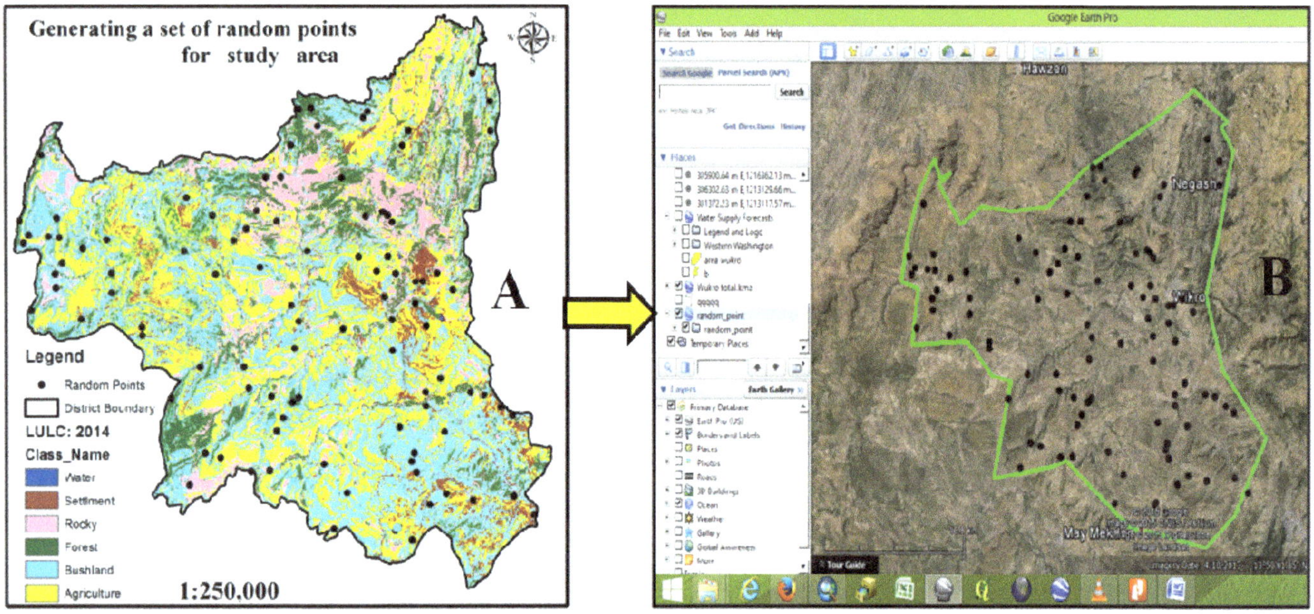

Figure 4. *Generating random points in ArcMap (A) and opening the points in Google Earth (B).*

Table 3. *Accuracy assessment of land use land cover: 2014.*

		Reference from Google earth 2013							
		Settlement (1)	Agriculture (5)	Forest (19)	Rocky (27)	Bush (34)	Water (54)	Row Total	Ind. ACC
User Image (2014 Classified)	Settlement (1)	6	2	0	1	1	0	10	0.6
	Agriculture (5)	0	21	0	0	3	0	24	0.875
	Forest (19)	0	1	16	1	3	0	21	0.79
	Rocky (27)	0	1	0	12	1	0	14	0.857
	Bush Land (34)	0	2	1	0	24	0	27	0.88
	Water (54)	0	0	0	0	1	3	4	0.75
	Column Total	6	27	17	14	33	3	100	0.792

Total (overall) accuracy = ((6+21+16+12+24+3)/100)*100 = 81/100 = 0.85 = 82.00%

$$K = \frac{100\ (6+21+16+12+24+3) - [(10x6) + (24x27) + (21x17) + (14x14) + (27x33) + (4x3)]}{(100)^2 - [(10x6) + (24x27) + (21x17) + (14x14) + (27x33) + (4x3)]}$$

$$K = \frac{100\ (82) - [(60) + (648) + (357) + (196) + (891) + (12)]}{10000 - [(60) + (648) + (357) + (196) + (891) + (12)]}$$

$$K = \frac{8200 - 2164}{10000 - 2164} = \frac{6036}{7836} = 0.7702$$

So Kappa of 0.7716 means there is 77.02% better agreement than by chance alone

5. Conclusion

Google Earth represents a powerful and attractive source of positional data that can be used for investigation and preliminary studies with suitable accuracy and low cost. So Google Earth is very important for mapping of different types of land use/land cover and for accuracy assessment. The result of accuracy shows that total (overall) accuracy of land use and land cover is 82.00**%** and Kappa (K) is 77.02% which is acceptable in both accuracy total (overall) and Kappa accuracy.

References

[1] Abineh Tilahun and I. Zubairul .2015. Use of Google Earth for Land Use mapping in the Case of Gish Abbay Sekela, West Gojjam, Amhara State, Ethiopia. *International Journal of Society and Humanities (ISSN-2319-2070/VOL 6:1-6.*

[2] Abubaker, H. M, Elhag A.M.H. and Salih, A.M. (2013). Accuracy Assessment of Land Use and Land Cover Classification (LU/LC) Case study of Shomadi area-Renk County-Upper Nile State, South Sudan. *International Journal of Scientific and Research Publications*, Volume 3, Issue 5.

[3] Canute Hyandye, Christina Geoffrey Mandara, John Safari. GIS and Logit Regression Model Applications in Land Use/Land Cover Change and Distribution in Usangu Catchment. *American Journal of Remote Sensing*.Vol. 3, No. 1, 2015, pp. 6-16. doi: 10.11648/j.ajrs.20150301.12.

[4] Congalton, R. G. 1991. A review of assessing the accuracy of classifications of remotely sensed data. *Remote Sensing of Environment* 37:35-46.

[5] Dash, P. (2005). Land Surface Temperature and Emissivity Retrieval from Satellite Measurements, Institut fur Meteorologie und Klimaforshung.

[6] David, P. (2008). Horizontal Positional Accuracy of Google Earth's High Resolution Imagery Archive. *Sensors* 2008, *8*, 7973-7981; DOI: 10.3390/s8127973.

[7] Fakeye, Attah Motunrayo, Aitsebaomo, Francis Omokekhai, Osadebe, Charles Chuka, Lamidi, Risikat Bukola, Okonufua, Endurance Omamoke. Digital Modeling of Land Use Changes in Some Parts of Eastern Nigeria. American Journal of Remote Sensing. Vol. 3, No. 3, 2015, pp. 37-42. doi: 10.11648/j.ajrs.20150303.11

[8] Jensen, J. R. 1996. *Introductory Digital Image Processing: A Remote Sensing Perspective* (Second edition). Prentice Hall, Inc., Upper Saddle River, New Jersey, USA.

[9] Nagi, Z. M, Ahmed G. and Hussam E. (2013). Positional Accuracy Testing of Google Earth. *International Journal Of Multidisciplinary Sciences And Engineering, VOL. 4, NO. 6, JULY 2013.*

[10] Roy, P.S. and Giriraj, A. 2008. Land Use and Land Cover Analysis in Indian Context. Journal of applied science, Vol. 8(8): 1346-1353.

[11] Shirkou, J. and Aliakbar, N. (2013). Comparison between Land Use/Land Cover Mapping Through Landsat and Google Earth Imagery. *American-Eurasian J. Agric. & Environ. Sci., 13 (6): 763-768.*

[12] Yadav, P., Kapoor, M. and Sarma, K. 2010. Land Use Land Cover Mapping, Change detection and conflict Analysis of Nagzira-Navegaon Corridor, Central India Using Geospatial Technology. International Journal of Remote Sensing and GIS, Vol. 1(2): 90-98.

12

Impact of Coastal Erosion on Alpha Beach Community in Lagos, Nigeria

Maureen Nkemdilim Chukwu

Department of Biology, School of Science and Technology, National Open University of Nigeria, Lagos, Nigeria

Email address:
mchukwu@noun.edu.ng

Abstract: Coastal erosion is a dominant seasonal climatic factor in Alpha beach community as a result of rise in sea level. The Impact of Coastal erosion on Alpha beach community in Lagos, Nigeria was investigated. The research was motivated by the condition in Alpha beach area as affected by coastal erosion. A total of 60 people living in Alpha beach community in Lagos, Nigeria were interviewed using structured questionnaires. The instrument was validated through peer review and pilot testing. Data collected were summarized and computed using descriptive statistics and Likert scaling test. Results showed that ninety-two percent of the respondents affirmed that coastal erosion destroyed their infrastructure and properties, distorted their business and increased environmental pollution. It impacted negatively on their social lives leading to relocation of some members of the community and retardation of community development. It also reduced the reliability of ocean shipping, impacted negatively on marine lives and activities, thus increasing economic risks. It was recommended that the government and other stakeholders should come to the aid of the community by providing them with affordable health services and armoring the shoreline with seawalls, revetments and jetties to remedy the adverse effects of the coastal erosion.

Keywords: Environmental Pollution, Social Lives, Community Development, Marine Lives

1. Introduction

Coastal erosion is the wearing off of land and the removal of beach or dune sediments by wave action, tidal currents, wave currents, high winds and drainage. Causes of coastal erosion include waves generated by storms, wind or fast moving motor craft, hydraulic action, abrasion impact and corrosion. It may take the form of long-term losses of sediment and rocks, or merely a temporary redistribution of coastal sediments. The key impacts of rising sea levels and increased oceanic activity in coastal communities and waterways connected to the ocean include erosion and inundation. It was reported that the global average sea level had risen by about 1.0 to 2.5 millimeters per year during the last century along the Atlantic and Gulf coasts [1]. However, some parts of Maine and the Pacific Northwest have experienced relative sea level decline because of post-glacial rebound [1 and 2]. According to [3], the Intergovernmental Panel on Climate Change projected a global rise in sea level of between 15 cm and 95 cm, with an estimate of roughly 50 cm by 2100. This implies that the rates of sea level rise will accelerate in future.

Coastal erosion is influenced by both natural factors and human activities. Natural factors contributing to coastal erosion include sand supply, changes in sea level, geologic characteristics; sand-sharing systems of beaches and dunes and the effects of waves, currents, tides and wind. Human activities contributing to coastal erosion include dredging of tidal entrances, construction of jetties and groins, hardening of shorelines with seawalls, beach nourishment and construction of harbors and sediment-trapping dams.

Climate change such as sea level rise, changes in the frequency and intensity of storms, increases in precipitation and warmer ocean temperatures could affect coastal areas in a variety of ways. Rising atmospheric concentrations of carbon IV oxide are causing the oceans to absorb more of the gas and become more acidic. These could have significant impacts on coastal and marine ecosystems.

Coastal habitats such as wetlands, estuaries, dunes and beaches contain a wealth of biological diversity. Beaches are temporal features where sand is always being removed and added. Sandy beaches normally retreat landward in response to sea level rise. According to [4], long term shoreline retreat on the Atlantic Coast averaged about 150 times that of sea level rise. It then follows that an additional rise of 10 cm in sea level could result in 15 meters of beach erosion. Alpha community is presently under threat probably because of global warming, dredging activities on the Eko Atlantic Project; a new city being built at the off shore of Lagos.

Important social, cultural, and environmental activities attract people to the coastal shore. People visit and live near the beach for various reasons ranging from tourism, fishing, surfing, and other recreational activities. Coastal erosion, flooding and water pollution affect man-made infrastructure and coastal ecosystems. Several problems tend to arise in eroding areas which include loss of beach, property loss, damage to infrastructure such as roads transport and water treatment systems.

Several human activities intended to protect coastal development from waves, flooding and erosion have most times alter the behaviour of the coasts considerably [5, 6 and7]. There is evidence that the clearing of coastal forests and trees has increased the vulnerability of coasts to erosion such as in Viet Nam [8], Sri Lanka [9], India [10 and 11] and Thailand [12]. Findings of [13 and 14] showed that the costs of coastal disasters are rising as more people and structures are exposed to hazards. According to [15], many hidden costs relating to coastal hazards are imposed on the business community, individuals, families and neighborhoods, public and private institutions, and natural resources and the environment.

Regulatory measures for shoreline management include specific rules and conditions for land use, construction, and development. In some areas, groins, seawalls, bulkheads, and other measures are used to prevent erosion of the coastline. However, the seriousness of coastal erosion calls for urgent action irrespective of possible acceleration in sea level rise.

2. Objective

The objective of this study is to investigate the impact of coastal erosion on Alpha beach community and environ with a view of finding preventive measures to combat the effects.

In order to find this objective, the following research questions were asked.

i. How does coastal erosion affect the residents of Alpha beach community?

ii. What measures can be taken to combat the effect of the coastal erosion?

3. Methodology

3.1. Study Area

The study area was Alpha Beach on Lekki peninsula, an aspect of Lagos coastal region in South-West Nigeria. It is bounded by the Atlantic Ocean on the South. The beach is located at 6° 25' 19" N x 3° 31' 23" E (Latitude: 6.42212, Longitude: 3.52314) and used to be a popular tourist attraction for many fun seekers. Alpha beach study covers a backshore distance of 1.55km and a surf zone distance of 1.29 km.

Figure 1. *Map showing location of Alpha beach.*

3.2. Research Design

A sample of 60 respondents living in Alpha beach was randomly selected from Alpha beach community in Lagos comprising coastal property and businesses owners as well as other members of the community affected by coastal erosion. Data were collected through observations and personal interviews using structured questionnaire. Section A of the questionnaire focused on the socio-economic characteristics of the respondents while Section B had 18 Likert type items and was designed to assess the impact of coastal erosion on Alpha beach community and their environment. The questionnaire was validated in order to improve the quality and content before administration. The test-reset reliability scale yielded reliability coefficient of 0.80.

Data collected were summarized and computed using frequency counts and percentages while information on the influence of coastal erosion on various activities of the Alpha beach community were computed on a four point Likert Rating Scale (LRS); strongly agreed = 4, agreed =3, disagreed =2, strongly disagreed =1. The mean score will be computed as 4+3+2+1 = 10/4=2.50. Using the interval score of 0.05, the upper limit cut-off was determined as 2.50 ± 0.05 = 0.55 and the lower limit as 2.55 ± 0.05=2.45. On the basis of this, mean scores (MS) below 2.45 (i.e. <2.45) were ranked 'low', those between 2.45 and 2.54 were considered 'medium' (i.e. $2.45 \geq MS \leq 2.54$), while the mean scores greater than or equal to 2.55 (i.e. $MS \geq 2.55$) were considered 'high'.

4. Results and Discussion

4.1. Socio-Economic Characteristics of the Respondents

The summary of selected socio-economic characteristics of members of Alpha beach community is shown in Table 1. Their average age is 40 years; majority of them (79%) were below 50 years old. This is an indication that most of the Alpha beach community members are still young, physically able and energetic with the potential to control the oceanic erosion effectively. Majority of the community members (78%) are married with only 22% singles. Only 11% had attained tertiary education while 39% had secondary education. This shows that if control measures are introduced the community members will not face challenges in implementing it. Such measures can focus on preventing the adverse effects of the oceanic erosion. On the average, 56% of the community members are property owners; an indication that most of them will probably be victims of the coastal erosion hazard. Some of the community members (30%) reside close to the beach while 26% of them have their workplace and business close to the beach.

Table 1. *Socio-economic characteristics of the community members.*

Characteristics		Frequency	Percentage (%)
Gender	Male	30	50
	Female	30	50
Age	21 - 30 years	11	19
	31 – 40 years	19	31
	41 - 50 years	17	29
	51 years and above	13	21
	Average 40 years		
Marital status	Single	13	22
	Married	47	78
Educational attainment	No formal Education	13	22
	Primary	17	28
	Secondary	23	39
	Tertiary	7	11
Ownership of property	Yes	34	56
	No	26	44
Nearness of the beach to residence	Close (1 – 5 kilometers)	18	30
	Far (beyond 5 kilometers)	42	70
Nearness of the beach to place of work/ business	Close (300 meters – 1 kilometer)	16	26
	Far (beyond 1kilometer)	44	74

Source: Field data November, 2014

4.2. Impact of Coastal Erosion on Community Members' Livelihoods

Table 2. *Extent of loss experienced by community members due to Coastal Erosion (August – November, 2014).*

Variable		Frequency	Percentage (%)
Erosion Experience	Experience loss	55	92
	Did not experience loss	05	08
Proportion of loss experienced	80%	12	20
	60%	09	15
	40%	15	25
	20%	24	40

Source: Field data November, 2014

Majority (92%) of the community members encountered loss due to coastal erosion. The erosion washed away a tarred road which was the lifeline of economic activities there. It also destroyed many shops and market stalls being the community members' means of livelihood. It further destroyed a church located on the beach (Figure 2). A community health centre in the area was also destroyed (Figure 3) while a mosque on the beach was reduced to rubbles. In addition, the coastal erosion causes high incidence of water borne diseases, extinction of plants and animals and higher risk of environmental pollution (Figure 4). It had also rendered many of the community members homeless.

Figure 2. *A church building destroyed by coastal erosion at Alpha beach.*

Figure 3. *A community health centre destroyed by coastal erosion at Alpha beach.*

Figure 4. *Shore covered with deposits from coastal erosion at Alpha beach (pollution).*

Houses close to the eroding shore now worth less than similar houses close to relatively stable shorelines as reflected in their selling prices. Coastal property owners expressed strong emotional attachment to their properties; they prefer to redevelop the damaged structures rather than demolish them.

4.3. Influence of Coastal Erosion on the Various Activities of the Community Members

Table 3 shows the impact of coastal erosion on the various activities of the community members. Mean scores greater 2.55 were rated as high. Based on this, coastal erosion was found to have a significant effect on all the activities except community road and government's awareness of the problem. The community members are indeed suffering seriously from the ill effects of the coastal erosion. The government and other stakeholders should as a matter of urgency come to their aid by devising appropriate protective measures to combat the negative effects of coastal erosion on the community.

Table 3. *The impact of coastal erosion on the various activities of the community members.*

Variables	Strongly disagreed	Disagreed	Agreed	Strongly agreed	Total	Mean Score
Influence of coastal erosion on water volume.	0 (0.00)	0 (0.00)	39 (1.95)	21 (1.40)	60	3.35
Influence of coastal erosion on houses in the community.	0 (0.00)	0 (0.00)	7 (0.35)	53 (3.53)	60	3.88
Influence of coastal erosion on landmass.	0 (0.00)	0 (0.00)	5 (0.25)	55 (3.67)	60	3.92
Influence of coastal erosion on community members' abode.	0 (0.00)	0 (0.00)	8 (0.40)	52 (3.47)	60	3.87
Influence of coastal erosion on community development.	0 (0.00)	1 (0.03)	16(0.80)	43 (2.87)	60	3.70
Influence of coastal erosion on farmlands.	1 (0.02)	0 (0.00)	48 (2.40)	11 (0.73)	60	3.15
Influence of coastal erosion on community roads	0 (0.00)	0 (0.00)	17 (0.85)	43 (0.87)	60	1.72
Influence of coastal erosion on the community's social life.	0 (0.00)	0 (0.00)	20 (1.00)	40 (2.67)	60	3.67
Seasonal influence of coastal erosion.	0 (0.00)	5 (0.17)	53 (2.65)	2 (0.13)	60	2.95
Influence of coastal erosion on community members' job.	0 (0.00)	1 (0.03)	52 (2.60)	7 (0.47)	60	3.10
Influence of coastal erosion on plant species.	0 (0.00)	0 (0.00)	55 (2.75)	5 (0.33)	60	3.08
Influence of coastal erosion on animal species.	0 (0.00)	0 (0.00)	55 (2.75)	5 (0.33)	60	3.08
Influence of coastal erosion on the community's population.	0 (0.00)	0 (0.00)	19 (0.11)	41 (2.73)	60	2.84
Loss to the community for the 10 years (quantified in hundreds of Millions of Naira).	0 (0.00)	0 (0.00)	31 (1.55)	29 (1.93)	60	3.48
Loss to individual community members for the 10 years (quantified in tens of Millions of Naira).	0 (0.00)	0 (0.00)	26 (1.30)	34(2.27)	60	3.57
In the next few years more harm will be encured if the situation is left unchecked.	0 (0.00)	0 (0.00)	40 (2.00)	20 (1.33)	60	3.33
You are taking some control measures to salvage this situation	2 (0.03)	25 (0.83)	6 (0.30)	27(1.80)	60	2.96
Government's awareness of the problem.	1 (0.02)	24 (0.80)	25 (1.25)	10(0.67)	60	2.02
It can lead to the disappearance of the community from the map of Lagos state.	0 (0.00)	0 (0.00)	34 (1.70)	26 (1.73)	60	3.43

4.4. Coping Mechanisms Against Coastal Erosion

In the past, Alpha beach community members responded to the oceanic erosion by relocating inland. However, such relocations are no longer possible because most of the land had been sold off. They have devised various coping mechanisms at home, business and workplace. These include setting structures far away from the shore, sand fencing and use of sand mounds, sand dunes and sand bags. The aforementioned mechanisms are not very effective; hence the community members are calling on the government to urgently come to their aid by putting some measures in place to control the coastal erosion.

5. Conclusion and Recommendations

It has been established that the Alpha beach community is at a high risk of damages from coastal erosion. The community members should as a matter of urgency be educated on the coping and control mechanisms to combat the adverse effects of the coastal erosion. The government and other stakeholders should provide affordable health services for the community members. The Federal government should also construct an embankment to save guard the future occurrence of the coastal erosion at Alpha beach and at the same time provide assistance to the coastal disaster victims.

References

[1] Douglas, B.C. (1995). Global sea level change: Determination and interpretation. *Reviews of Geophysics* (supp.) July: 1425–1432.

[2] Leatherman, S.P. (1993). Modes of shoreline behavior: Erosion rate analysis using Geomorphic principles. Pp. 218–223 in Proceedings of International Coastal Symposium, Hilton Head Island, S.C.

[3] Houghton, J.T., MeiraFilho, L.G., Callander, B.A., Harris, N., Kattenberg, A. and Maskell, K. eds. (1996). Climate change 1995: The science of climate change. Report by the Intergovernmental Panel on Climate Change. New York, N.Y.: Cambridge University Press.

[4] Leatherman, S.P., Zhang., K. and Douglas. B. (1999). Sea level rise shown to drive beach erosion. EOS (81): 55-57.

[5] Nicholls, R.J. (2010). Impacts and responses to sea-level rise. In: Church, J.A.; Woodworth, P.L.; Aarup, T., and Wilson, W.S. (eds.), Sea-level Rise and Variability.Chichester, United Kingdom: Wiley-Blackwell, pp. 17-43.

[6] Nicholls, R.J.; Marinova, N.; Lowe, J.A.; Brown, S.; Vellinga, R; De Gusmao, D.; Hinkel, J., and Toi, R.S.J., 2011. Sea-level rise and its possible impacts given a 'beyond 4 C world' in the twenty-first century. Philosophical Transactions A, 369, 1-21.

[7] Nordstrom, K.R, 2000. Beaches and Dunes of Developed Coasts. New York, New York: Cambridge University Press.

[8] Cat, N.N., Tien, P.H., Sam, D.D. & N.N. Bien. (2006). *Status of coastal erosion of Viet Nam and proposed measures for protection.* This volume (abstract).

[9] Samarayanke, R.A.D.B. (2003). Review of national fisheries situation in Sri Lanka. *In:* G. Silvestre, L. Garces, I. Stobutzki, M. Ahed, R.A. Valmonte-Santos, C. Luna, L. Lachica- Alino.

[10] Malini and Rao (2004) reported coastal erosion and habitat loss along the Godavari Delta front owing to the combination of the dam construction across the Godavari and its tributaries that diminishes sediment supply to the coast and continued coastal land subsidence.

[11] Gopinath, G. & Seralathan, P. (2005). Rapid erosion of the coast of Sagar island, West Bengal – India. *Environment Geology,* 48: 1058–1067.

[12] Thampanya, U., Vermaat, J.E., Sinsakul, S. & Panapitukkul, N. (2006). Coastal erosion and mangrove progradation of Southern Thailand. *Estuarine, Coastal and Shelf Science,* 68:75– 85.

[13] Kunreuther, H. (1998). Introduction. Chapter 1 in Paying the price: The Status and Role of Insurance against Natural Disasters in the United States, H. Kunreuther and R. Roth, Sr., eds. Washington, D.C. Joseph Henry Press.

[14] Godschalk, D., Beatley, T., Berke, P., Brower, D.J. and E.J. Kaiser, E.J.(1998). National Hazard Mitigation: Recasting Disaster Policy and Planning. Covelo, Calif.: Island Press.

[15] Heinz Center.(1999). The Hidden Costs of Coastal Hazards—Implications for risk assessment and mitigation. Covelo, Calif.: Island Press.

13

Infectious and Radioactive Waste Management in a Diagnostic and Nuclear Medicine Centre

Tauqir Nasir[1], Shafaqat Ali[1, *], Mujahid Farid[2], Muhammad Shahbaz[3], Muhammad Rizwan[1], Fakhir Hannan[1], Rehan Ahmad[1]

[1]Department of Environmental Sciences and Engineering, Government College University, Faisalabad, Pakistan
[2]Department of Environmental Sciences, University of Gujrat, Hafiz Hayat Campus, Gujrat, Pakistan
[3]Punjab Institute of Nuclear Medicine (PINUM) Hospital, Faisalabad, Pakistan

Email address:
shafaqataligill@yahoo.com (S. Ali), mujahid726@yahoo.com (M. Farid)

Abstract: Hospital and medical waste management practices are very unsatisfactory in developing countries especially in Pakistan. Existing techniques and practices are not in accordance with the guidelines and levels prescribed by the WHO (World Health Organization) and other international bodies concerned with environment and human health. The objective of this study was to design and implement a proper waste management plan for the PINUM (Punjab institute of Nuclear Medicine) hospital Faisalabad, Pakistan. At PINUM Hospital infectious waste is produced as a result of diagnostic and radioactive waste as a result of diagnostic and therapeutic activities in nuclear medicine. This study was carried out in a period of 10 months in two phases. In first phase a proper infectious waste management plan was designed and incorporated at hospital and in 2nd phase proper management plan for radioactive waste was designed and implemented. Results showed a 43% reduction in volume and treatment cost of infectious waste resulting from proper management. There was significant decrease in prescribed` storage time of radioactive waste which is done to bring its activity to normal level. It was evident from the results if radioactive waste properly segregated, lose radioactive waste (gloves, swabs, absorbing sheets, tissue papers etc.) can be disposed off much earlier than the compact radioactive waste (radioactive vials, needles and sharps).

Keywords: Hospital, Infectious, Radioactive, Nuclear, Medicine

1. Introduction

Throughout the world the problem of health and environment is a debatable issue. One of the major environmental problems is the poor hospital waste management, which may significantly increase the exposure of infectious pollutants [1]. Hospital produce a tremendous amount of medical waste, has increased greatly because of the use of more disposable items, such as syringes, gloves, glass and plastic vials, blood bags etc [2].

Medical waste is defined as any solid or liquid waste that is generated from treatment of human beings in a hospital or clinic, from clinical diagnosis, treatment and pathological testing and from medical research and have ability to cause infectious disease [3, 4]. Improper management of health care waste results in transmission of infectious diseases like neurological disorders; typhoid, cholera, HIV and Hepatitis (B and C), cancer, asthma and mutagenicity [5, 6, 7, 8, 9]. It also pollutes water, air and soil, produce foul odor and enhance production and growth of insects, worms, and rodents [10, 11]. 75-90%of the waste produced by the health care facilities is general or ordinary waste in nature [10].

Present study was carried out at PINUM (Punjab institute of nuclear medicine) cancer hospital Faisalabad. It is providing diagnostic and therapeutic services in nuclear medicine and general medical procedures in the town. Different radioisotopes are being used for diagnosis and treatment of patients. Numbers of unsealed radioisotopes are being used at PINUM e.g.99mTc, I-131.Proper management plan for the waste produced by PINUM hospital was developed which includes waste segregation, collection, transportation, storage, disposal and waste minimization and reuse [12]. Protocols and SOPs (Standard Operating Procedures) were designed for infectious and radioactive waste management. Infectious waste is mainly produced by the Clinical, PCR and RIA laboratories and their sample collection rooms. It includes syringes; sample collection tubes, swabs, used tissue papers, blood and serum samples of Hepatitis C (HCV), Hepatitis B (HBV), and TB and other infectious diseases.

Annually approximately 40 thousand investigations are done by these labs collectively and a large amount of them includes infectious diseases like HBV, HCV and TB. Medical waste is considered as subcategory of hospital waste that contains both infectious and ordinary waste. Infectious waste contains pathogens of infectious diseases and can cause infectious diseases to humans [13, 14].The main objective of safe disposal of radioactive waste is to ensure that the radiation exposure does not exceed the prescribed safe limits for public, radiation workers and environment [15].

2. Materials and Methods

2.1. Phase 1

Development and implementation of protocols for proper and effective management of infectious waste produced at PINUM hospital. Infectious waste is generated as the result of diagnostic facilities provided for diseases such as PCR for HCV, HBV, Tuberculosis, and Fungal keratatits, other Clinical procedures and radioimmunology assay. Infectious waste contains syringes, swabs, gloves, and tissue papers etc. which are used to collect blood samples. Infectious waste also contains blood, Blood serum and other body tissues of humans. Infectious waste generated was segregated properly at the point of generation. Peddle baskets (purchased from local market) for infectious and non-infectious waste and were properly labeled with tags, covered from inside with polythene bags to avoid any kind of spillage of liquid waste materials. Waste was segregated at the point of generation by the staff in respective baskets. Stored according to the guidelines and finally transported and treated at disposal site. Staff and waste handlers were educated and awareness was created about the proper and improper waste management and ultimately its consequences on human and environmental health.

In this study to check the presence of Hepatitis B and C virus in the residual ash from incinerator conventional PCR was performed. For this purpose serum of positive samples for HCV and HBV (as Controls) and Ultra-pure water was mixed with ash and PCR was done.

2.1.1. Infectious Waste Incineration

Infectious waste was stored at an isolated area in closed containers and then incinerated on weekly basis in controlled and according to WHO guidelines designed incinerator (INCINER8 Company)

2.1.2. Nucleic Acids Extraction

DNA and RNA were extracted by using GF-1 kit obtained from Vivantis, chemicals and reagents were provided with kit and extraction was done according to the protocol provided with extraction kit.

2.1.3. Nested PCR

Nested PCR was performed (Eppendorf thermal cycler gradient machine) for both HCV and HBV positive blood serum mixed ash and Ultra-pure water mixed ash samples. Primers used for HCV Round 1 Forward primer (S1 5' GCCATGGCGTTAGTATGAGT-3'), Reverse primer (AS1 5-'GTGCACGGTCTACGAGACCT-3') Round 2 Forward primer (S2 5'-GTGCAGCCTCCAGGACCC-3) Reverse primer (AS 25'-CCGTGAGCGTTCGTGGGATA-3') and primers used for HBV were Forward primer (CGGCATATGGAGAACATCACATCAG) Reverse primer (CACCTCGAGTTAAATGTATACCCAAAG). 50 bp marker used for HCV while 100 bp was used for HBV provided by Fermentas.

2.2. Phase 2

Development and implementation of protocols and practices for proper and effective management of radioactive waste under the guidelines of PNRA (Pakistan nuclear regulatory authority).

Radioactive waste at PINUM hospital is comprised of used vials, syringes, swabs, gloves contaminated with 99mTc, I-131. Waste was segregated at the point of generation in polythene bags into two categories one containing sharps such as syringes, vials and second containing contaminated gloves, swabs, and tissue papers. Radioactive waste was stored in closed containers in isolated waste store room according to their half-lives for the period of 3 to 6 months to bring the activity to background level. Containers and baskets were labeled as "RADIOACTIVE WASTE". On the disposal date activity of waste was again measured with the help of digital survey meter. Check sheets and house keeping records were maintained and waste was sent for incineration and then final disposal in closed containers, before final disposal activity from radioactive waste was again measured. Radioactive waste was disposed off according to the guidelines of PNRA (Pakistan Nuclear Regulatory Authority) and IAEA.

2.3. Materials Used

2.3.1. Radiation Survey Meter

Survey Meter for Radiation Detection (DSM-506 by G-M and scintillation) was used to detect and quantify the radiations at the place of radioactive waste storage and at the time of disposal.

2.3.2. Others

1. Lead made peddle baskets were used for segregated radioactive waste at the point of generation before sending it to storage room.
2. Absorbing sheets used during preparation of radioactive element dose for patients to absorb the liquid dose in case of spillage.
3. Radioactive waste trolley (locally manufactured) made up of lead was used to transport the radioactive waste easily and safely.
4. Dose calibrator (Atomlab™ 500Plus) used to calibrate the dose of radioactive substance before administering to patients.

3. Results

3.1. Infectious Waste

Approximately 50 kg of waste was being produced daily from diagnostic labs (Clinical, PCR and RIA), general medical procedures, and chemotherapy for cancer patients, ordinary waste from cafeteria and office work etc. As a result of this research project, waste management plan was implemented at hospital which includes proper segregation of waste, storage, transport, treatment and final disposal. Consequently volume of infectious waste recorded to be 28 kg and ordinary waste was 21 kg approximately. Ordinary waste can be disposed off together with municipal waste without any prior treatment which means that there was 43% reduction in infectious waste treatment cost and its volume due to proper segregation of waste at its point of generation. Cost on treatment of waste (incineration cost) and emissions, residual ash from incinerator was also decreased remarkably.

3.2. PCR Results for Hepatitis C

Ash of incinerated infectious waste was mixed with blood serum of 5 positive (for HCV) samples to check the inhibitory effects (if any) fotabler PCR. All the samples showed positive results after agarose gel electrophoresis (Fig. 1).

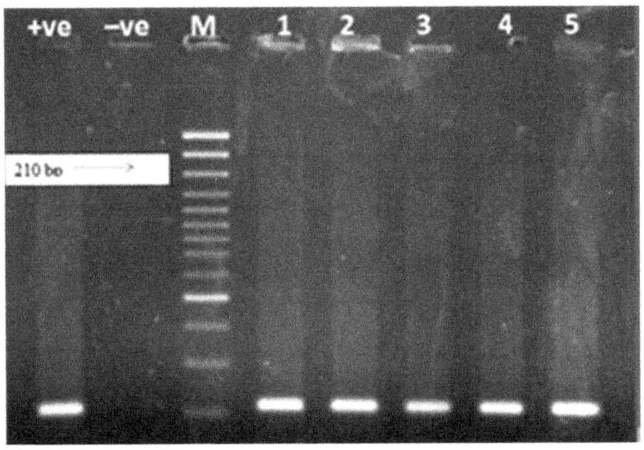

Fig. 1. Conventional PCR agarose gel picture of HCV positive samples mixed with ash of incinerated infectious waste, +ve control, -ve control, M (50bp marker), 1-5 samples.

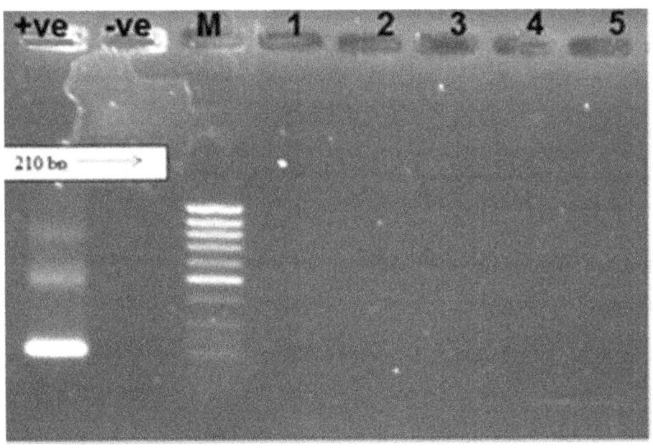

Fig. 2. Conventional PCR agarose gel picture of incinerated infectious waste ash mixed with Ultra-pure water for HCV conformation in ash, +ve control, -ve control, M (50bp marker), 1-5 samples.

Incinerated infectious waste ash was mixed with ultra-pure water to check the presence of HCV in ash. Results were negative (Fig. 2) which showed incineration killed the entire hepatitis C virus present in infectious waste.

3.3. PCR Results for Hepatitis B

Ash of incinerated infectious waste was mixed with blood serum of 5 positive (for HBV) samples. All the samples showed positive results after agarose gel electrophoresis (Fig. 3).

Incinerated infectious waste ash was mixed with ultra-pure water to check the presence of HBV in ash. Results were negative (Fig. 4) which showed incineration killed the entire hepatitis B virus present in infectious waste.

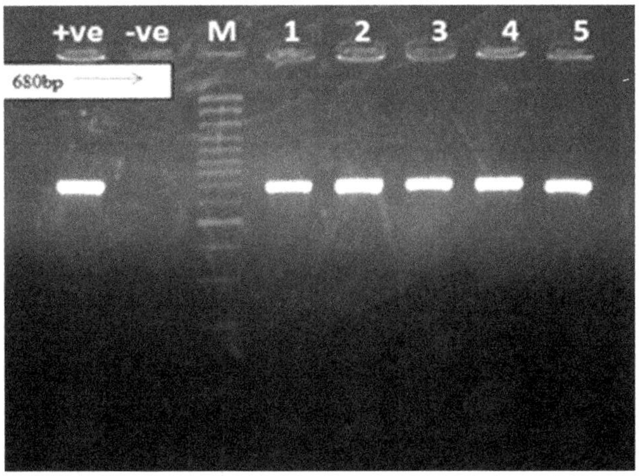

Fig. 3. Conventional PCR agarose gel picture of HBV positive samples mixed with ash of incinerated infectious waste, +ve control, -ve control, M (100bp marker), 1-5 samples.

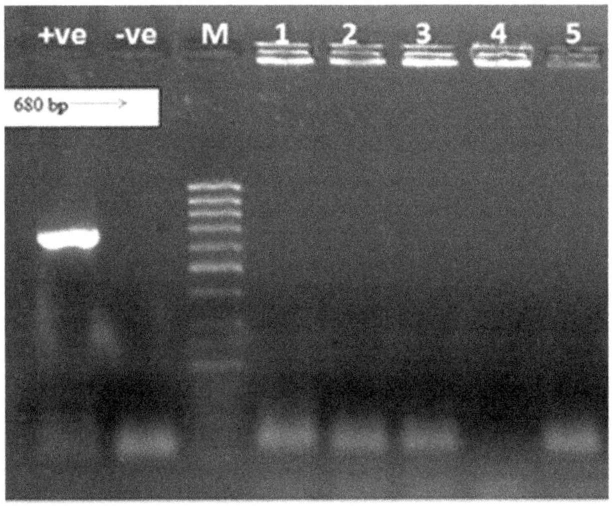

Fig. 4. Conventional PCR agarose gel picture of incinerated infectious waste ash mixed with Ultra pure water for HBV conformation in ash, +ve control, -ve control, M (50bp marker), 1-5 samples.

3.4. Radioactive Waste

Radioactive waste at PINUM includes waste of 99mTc and I131. Waste of I125 results from diagnostic activities by RIA lab.I125 waste from RIA labs is exempted from any prior treatment before disposal by PNRA, so it was disposed off with infectious waste for incineration. 131I is used for thyroid and whole body scan and thyroid therapy. 99mTcis used for different scans e.g. heart, bone, and other scans of the organs of body. PNRA guidelines are to store waste of 131I for 6 months and 99mTC for 3 months to complete its decay process and bring its activity to background level which is not harmful for humans and environment. PINUM is the only center in town providing diagnostic and therapeutic facilities in nuclear medicine. As a result large numbers of patients visit PINUM for different kind of scans and radionuclide therapy. There was limited space for storage of radioactive waste in the hospital. To cope up with storage problem in this study proper segregation of radioactive was implemented. Radioactive waste was divided into two categories.

- Lose radioactive waste: gloves, swabs, tissue papers, syringes etc.
- Compact radioactive waste: glass vials of radio nuclides and needles.

Compact radioactive waste was stored for the exact period of time according to the guidelines by PNRA, as the activity in glass vials and needles remain for longer period of time and also they are not incinerated into ash.

Lose radioactive waste was isolated at the point of generation and was disposed after different time periods by incineration. Activity in waste was measured by survey mater before and after incineration and also in fumes during incineration.

Results showed that lose radioactive waste of 131I can be disposed after 4 months and of 99mTc after 2 months instead of 3. Hence it will produce more space for radioactive waste storage and waste can be managed properly by applying better segregation practice. It also greatly reduces cost on management of waste and risk to human health and environment.

3.5. Radiation Count on Infectious Waste Incineration with Time Limits

3.5.1. Radiation Count from Radioactive Waste of I131 on Weekly and Monthly Basis

Radiation count rate from I131 waste was calculated by using radiation survey meter on weekly basis which included radiation counts before, during and after incineration of waste in the ash (Table 1). Only lose radioactive waste (gloves, swabs, tissue papers, syringes) was incinerated at weekly basis and not the compact radioactive waste (glass vials of radio nuclides and needles) as it contained higher activity and its disposal or incineration on weekly basis is not recommended. Radiation count scale was adjusted on µSv/hr. weight of the radioactive waste was also done before incineration.

Table 1. *Radiation count from radioactive waste of I131.*

Sr #	**Time duration	Weight of radioactive waste	Counts before incineration	Counts in needles and glass vials	Counts in Fumes during incineration	Counts in ash after incineration
	Daily basis					
1	1st day	250 grams	35.8 µSv/hr	-	10.4 µSv/hr	20.2 µSv/hr
2	2ndday	430 grams	33 µSv/hr	-	9.6 µSv/hr	18.3 µSv/hr
3	3rdday	640 grams	32 µSv/hr	-	8 µSv/hr	17.3 µSv/hr
4	4thday	850 grams	31.7 µSv/hr	-	7.6 µSv/hr	15.2 µSv/hr
5	5thday	1.1 kg	30.9 µSv/hr	-	7 µSv/hr	14.9 µSv/hr
6	6th day	1.4 kg	29.8 µSv/hr	-	6.5 µSv/hr	14.1 µSv/hr
	Weekly basis					
1	1st week	1 kg	45 µSv/hr	32.7 µSv/hr	15.3 µSv/hr	25.3 µSv/hr
2	2nd week	2.2 kg	42.2 µSv/hr	30.2 µSv/hr	10.4 µSv/hr	20.2 µSv/hr
3	3rd week	3.5 kg	37.4 µSv/hr	27.2 µSv/hr	7 µSv/hr	18.9 µSv/hr
	Monthly Basis					
1	1 month	4.2 kg	27.1 µSv/hr	23.6 µSv/hr	6.6 µSv/hr	11.7 µSv/hr
2	2 months	8 kg	9.3 µSv/hr	14.2 µSv/hr	Back ground level	Back ground level
3	3 months	12.1 kg	Back groundlevel	Background level	Back ground level	Back ground level

Table 2. *Radiation count from radioactive waste of I131.*

Sr #	**Time duration	Weight of radioactive waste	Counts before incineration	Counts in needles and glass vials	Counts in Fumes during incineration	Counts in ash after incineration
	Weekly Basis					
1	1 week	210 grams	75 µSv /hr	-	10.3 µSv /hr	55.6 µSv /hr
2	2 weeks	390 grams	73.1 µSv /hr	-	9.4 µSv /hr	53.8 µSv /hr
3	3 weeks	550 grams	72.4 µSv /hr	-	9.5 µSv/hr	52.2 µSv /hr
4	4 weeks	800 grams	70.3 µSv /hr	-	9 µSv /hr	50 µSv /hr
	Monthly Basis					
1	2 months	2 kg	63.2 µSv/hr	50.3 µSv/hr	7.3 µSv/hr	47.1 µSv/hr
2	3 months	3.4 kg	40.2µSv/hr	39.5µSv/hr	6µSv/hr	30.1µSv/hr
3	4 months	4.4 kg	18.7µSv/hr	27 µSv/hr	Back ground level	10.2µSv/hr
4	5 months	5.2 kg	Back ground level	9.3 µSv/hr	Back ground level	Back ground level
5	6 months	06 kg	Back ground level	Back ground level	Back ground level	Back ground level

Compact radioactive waste (glass vials of radio nuclides and needles) was also included in this study (Table 1). It was evident that if properly segregated lose radioactive waste (gloves, swabs, tissue papers, syringes) of I131 can be disposed after 4 months storage instead of prescribed 6 months and hence creating space for the storage of more compact radioactive waste.

3.5.2. Radiation Count from Radioactive Waste of 99mTc on Daily / Weekly / Monthly Basis

Radiation count of technetium 99mTc was calculated on daily, weekly and monthly basis (Table 2). Only lose radioactive waste of 99mTc was included as compact waste had high rate of radiation emission when treated on daily basis. Weight of the waste was done before treatment.

It was evident from results that if radioactive waste of 99mTc is properly segregated and stored it could be treated and disposed off after 2 months of storage instead of 3 months.

4. Discussion

Infectious waste management plan was prepared according to better waste management guidelines including proper segregation, storage, transport, treatment and final disposal. Results showed that incineration can be employed effectively in countries like Pakistan with economic issues who cannot afford expensive waste treatment costs of modern technology based practices like autoclaving etc.

The most prevalent cause of infections from health care waste is by needle prick injuries and by sharps that are contaminated with blood and other infectious body fluids. So it is recommended that used sharps, syringes and needles should be handled properly and not placed in polythene or plastic bags which can be easily punctured. To avoid airborne transmission of pathogens open and uncovered waste containers should not be used [16]. According to the WHO code of practices incineration of health care waste is the most commonly used and effective treatment option. During many incineration processes most of the waste is partially burned which is a serious health risk [17]. Improperly operated incinerators emit heavy metals and organic substances which are very harmful for human health and environment [17]. The dumping and disposal of health care waste to general waste dumping sites, present serious hazards to environment and human health by ground water and soil contamination as health care waste contain toxic and infectious elements [18]. Health care waste is 5-15 times more toxic in nature than the general and domestic waste [18, 19]. Most crucial element in hospital waste management is proper and safe packaging of hospital waste prior to final disposal or treatment. Proper packaging reduces the chances of contaminations to waste handlers and also minimize the harmful effects of waste on environment [12]. Substantial literature is present about the input of radioactive waste in the environment and ways to process these radioactive wastes [14]. However, little attention is paid towards the proper management of radioactive waste and its minimization which is produced from health care centre. Amount of radioactive waste is increasing as a result of increased diagnostic and therapeutic activities shifted towards nuclear medicine these days.

It should be ensured that the amount and volume of radioactive waste is kept to minimum achievable level and the waste is managed i.e. collected, stored, transported and disposed off according to the guidelines of regulatory body [18].

Recommendations

- Hospital waste either infectious or radioactive if segregated properly at the point of its generation, it greatly reduces the chances of infection to the staff and community and also less harmful to the environment.
- Frequent training programs for staff involved in handling and managing hospital waste and Public awareness symposiums and seminars should be conducted to minimize the risks of infection and for proper and effective waste management. It will help to create awareness among people about highly infectious diseases such as HBV, HCV and HIV.
- Policies and legislations are very necessary for proper hospital waste management. Implementation of these defined polices and financial assistance for implementing waste management technologies is very important for effective waste management

Acknowledgement

The authors are highly thankful to the PINUM (Punjab Institute of Nuclear Medicine) hospital Faisalabad for all experimental assistance and lab analysis.

Abbreviations

PINUM (Punjab institute of Nuclear Medicine), HCV (Hepatitis C virus), HBV (Hepatitis B virus), PCR (polymerase chain reaction), TB (Tuberculosis), I131 (Iodine-131), 99mTc (Technetium-99m), μSv /hr (micro-Sieverts per hour, a standard unit of measurement for radioactivity), PNRA (Pakistan nuclear regulatory authority).

References

[1] S. R. Dodhy, "Solid Waste and pollution-Silent killer," J Pak Textile, vol. 55, pp. 42-44, 2006.

[2] J. L. Morrison, "The contagious hospital," On the Horizon, vol. 1, pp.11-12, 1992.

[3] N. Al-Mutair, M. Terro, A. L. Al-Khaleefi, "Effect of recycling hospital ash on the compression properties of concrete: statistical assessment and predicating model," Building Environment, vol. 39, pp. 557-566, 2004.

[4] Sabiha-Javied, M. Tufail, K. Sofia, "Heavy metal pollution from medical waste incineration at Islamabad and Rawalpindi," J Microchemical. Pakistan, vol. 90, pp. 77-81, 2008.

[5] M. L. Johannessen, M. Dijkman, C. Bartone, D. Hanrahan, M. G. Boyer, C. Chandra, Health Care Waste Management Guidance Note: Health Nutrition and Population (HNP) Discussion, 2000.

[6] C. A. Smith, "Managing Pharmaceutical Waste-what pharmacist should know," J. Pharm. Soc. Wisconsin, pp. 17-22, 2002.

[7] National Institute for Occupational Safety and Health (NIOSH).Preventing Occupational Exposure to Antineoplastic and other Hazardous Drugs in Healthcare Settings, 2004.

[8] M. Askarian, M. Vakili, G, Kabir, "A total quality management approach to healthcare waste management in Namazi Hospital," Waste Mangement. Iran, vol. 30(11), pp. 2321-2362, 2010.

[9] J. I. Blenkharn, "Lowering standards of clinical waste management: do the hazardous waste regulations conflict with the CDC's universal/standard precautions?," J. Hosp. Infect, vol. 62(4), pp. 467- 472, 2006a.

[10] A. Pruss, E. Giroult, P. Rushbrook, "Safe Management of Waste from Health Care Acivities," World Health Organization, Geneva, Switzerland, 1999.

[11] A. Abdulla, M. Gomei, D. Hyrenbach, G. Notarbartolo-di-Sciara," Challenges facing a network of representative marine protected areas in the Mediterranean: prioritizing protection of underrepresented habitat," ICES Journal of Marine Science, vol. 66, pp. 22-28, 2008.

[12] Ministry of environment Pakistan, "Hospital waste management rule," The Gazette of Pakistan, Ministry of environment notification, 2005.

[13] P. Klangsin, A. K. Harding," Medical waste treatment and disposal methods used by hospitals in Oregon, Washington and Idaho," J Air Waste Manag Assoc, vol.48, pp.516-26, 1998.

[14] Y. A. Levendis, A. Atal, J. Carlson," PAH and soot emissions from combustion of coal and waste tires in fixed beds," Comb. Sci. Technol, vol. 134, pp. 407-431, 1998.

[15] Operational limits," Training workshop on Radiation Safety in Nuclear Medicine and RSO Certification Examination," Murthy BKS, BARC. Mumbai, pp. 6.1–6.6, 2000.

[16] WHO, 2000.In: Rushbrook, P., Chandra, C., Gayton, S., (Eds.), starting health-care waste management in medical institutions. Health-care waste practical information series, No. 1. WHO regional office for Europe, European centre for environment and health, Copenhagen. Samajh ni lagi.

[17] M. C. Agramunt, A. Domingo, J. L. Domingo, J. Corbella, "Monitoring internal exposure to metals and organic substances in workers at a hazardous waste incinerator after 3 years of operation," Toxicology letters, vol.146, pp.83-91, 2003.

[18] M. Tsakona, E. Anagnostopoulou, E. Gidarakos, "Hospital waste management and toxicity evaluation: A case study," Waste management, vol. 27(7), pp. 912-920, 2007.

[19] R. Mohee, "Medical waste characterization in health care institutions in Mauritius," Waste management, vol. 25, pp. 575-581, 2005.

14

An Assessment on the Economic Benefit of Cactus Crop in the Case of Ganta Afeshum District, Eastern Tigray, Ethiopia

Fikre Belay

Department of Geography and Environmental Studies, College of Social Science and Humanities, Adigrat University, Adigrat, Ethiopia

Email address:
fikrebelay23@yahoo.com

Abstract: Cactus is a plant with very thick and fleshy stems. It is often covered with prickles and it usually grows in desert and semi desert area of the world. Cactus has different benefit such as source of food, medicine, chemical and income. Therefore, the objective of this research was carried out to assess the economic benefit of cactus as potential source of food for both human and animal in Ganta Afeshum District, Eastern Tigray region of Ethiopia. For this study, questionnaires, structured interviews, field observations and documentary analysis were applied to collect the necessary information from farm households and concerned bodies. Probability and non - probability sampling methods were used in this study. The sample is selected using simple random sampling from the entire target of population (1230, which is 135(11%). Both quantitative and qualitative methods of data analysis were used to describe the findings. Tables, charts and percentage were used in summarizing the quantitative data. The survey result shows that in the study area out of the total area of the District, 1607.05 hectare of the land is covered by cactus crop. Even though the area is characterized by a potential on cactus crop but the farmers used cactus crop only as diet for human being and as forage for animals particularly for cattle.

Keywords: Cactus, Economic, Social, Benefit

1. Introduction

Ethiopia is one of the most well endowed countries in terms of natural resources including fauna and flora (Gete et al., 2006). However, the country faces different problems like soil erosion (Million and Kassa, 2004). The cause for this is manly associated with land cover change (Woldeamlak, 2002).

Cactus is a plant with very thick and fleshy stems. It is often covered with prickles and it grows in desert and semi desert area of the world which is high tolerant to drought (Barbera, 1995). Cactus crop contains water (92%), carbohydrates (4-6%), protein (12%), minerals (1%), and moderate amount of vitamins mainly A and C (Cantwell, 1991).

Typical values for nutritional components in these studies were found to be: moisture content, 85-90%; crude protein, 5-12%; phosphorus, 0.08-0.18%; calcium, 4.2%; potassium, 2.3%; magnesium, 1.4%; energy, 2.6 Mcal/kg. The in vitro digestibility values were 72% for protein, 62% for dry matter, 43% for crude fiber and 67% for organic matter (Cantwell, 1991).

Clodadose (cactus leaf) is very important to prepare salad, ice cream and juice. Clodadose (cactus leaf) is used for marmalade and cactus fruit also used for marmalade and candy (Mums for Mums, 2004). Cactus serves as raw material for cosmetic products like shampoo, soap,, cream and body lotion. Adhesive and glues, pectins, fibers for hand crafts and papers can be made from cactus; the vegetative material is used with bovine manure as an anaerobic digestion accelerator to produce biogas (Uribe et al, 1992).

According to popular medicine cactus especially its cladode is remedy for diabet mellitus, diarrhea (Corrales-Gracia et al, 2004).

Cactus introduced via Erob District in the 19th century by catholic missionary called Bishop AbuneYackob planted in Erob (Frew, 2007). Today cactus is grown in Eastern Tigray in Erob, Saesie TsadaEmba, and Ganta-Afeshum District. Out of the total area of Ganta-Afeshum District (53035.30 hectare), 1607.05 hectare covered by cactus crop (Ganta Afeshum District administration yearbook, 2012 and Ganta Afeshum Rural and Agricultural office report, 2012).

More over cactus has different economic benefits. Despite the more than 1607.05hectare is covered with cactus crop in Ganta-Afeshum District, however the farmers have little or no knowledge concerning the different economic benefits of cactus crop, therefore the researcher inspired to explore the different economic benefit of cactus and to recommend possible solution.

Objective of the study

The main objective of this study is to assess the economic benefit of cactus as potential source of food for both human and animal and to create an awareness on feeding of cactus crop (as salad, juice and marmalade and bread etc).

2. Site Description (Location)

One of the 36 woredas in the Tigray Region of Ethiopia and part of the Misraqawi Zone, Ganta Afeshum is bordered on the south by Hawzen, on the west by the Mehakelegnaw (Central) Zone, on the north by Gulomahda, and on the east by Saesi Tsaedaemba. Towns in Ganta Afeshum include Adigrat and Bizet. The 1994 national census reported a total population for this woreda of 122,827 of whom 58,398 were men and 64,429 were women; 39,561 or 32.21% of its population were urban dwellers. The largest ethnic group reported in Ganta Afeshum was the Tigrayan (98.07%). The total area of this *District* is 530.35km^2 and with total population 53,339.(Ganta Afeshum District Administration 2012).

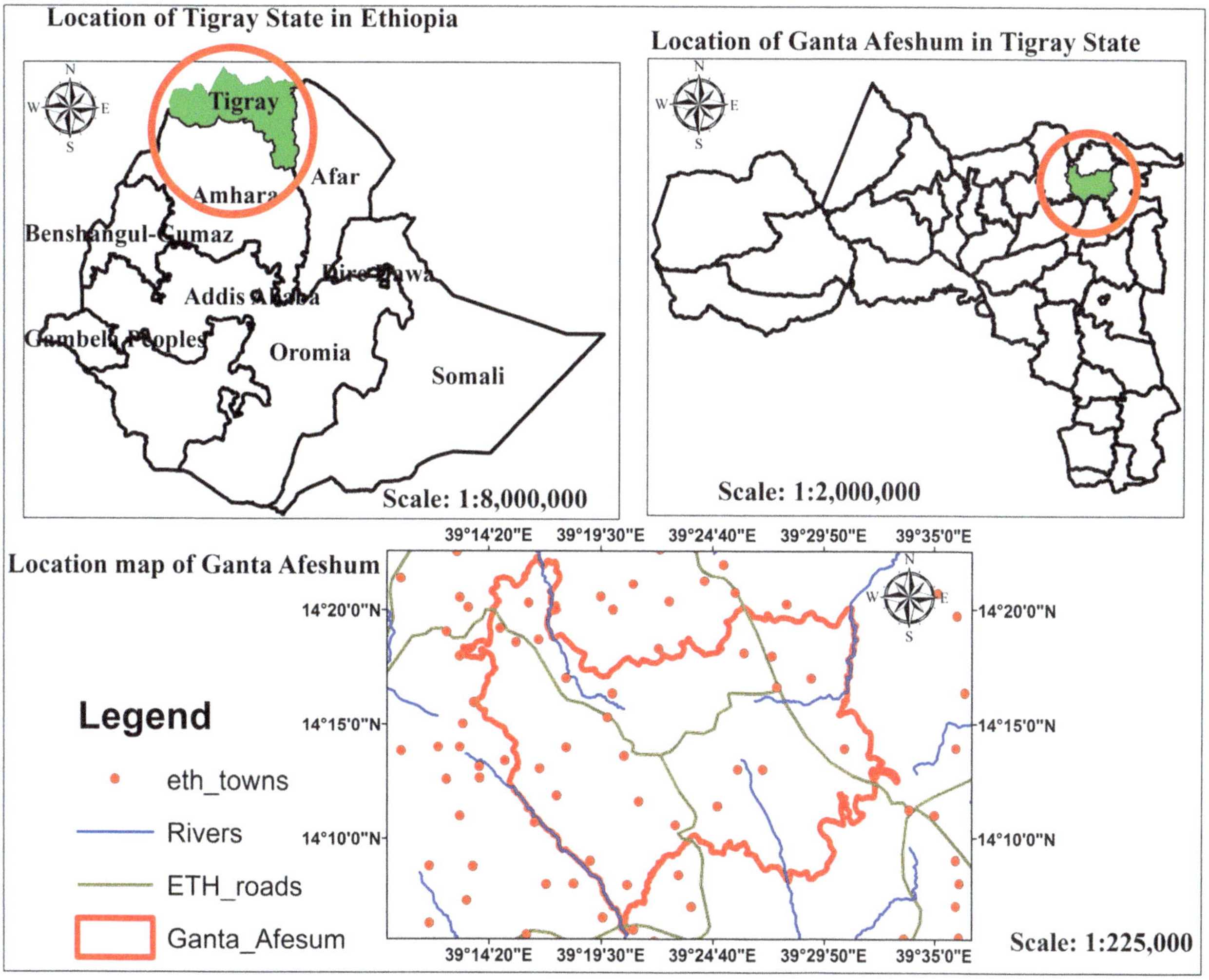

Figure 1. *Location map of the study area.*

3. Methodology

3.1. Sampling Techniques and Designs

The study was conducted in Ganta Afeshum District, Eastern Tigray Zone of Tigray State in 2014. The District was selected due to the fact that there was more Cactus is growing in the area. For this study descriptive type of research was employed to describe the characteristics of variables in the study area.

Probability and non- probability sampling were used in this study. Probability especially systematic sampling was used for households. The households were selected using simple random sampling technique from the entire target of population (1230, which is 135(11%)

3.2. Methods of Data Collection

Both primary and secondary sources of data were used in this study. Secondary sources of data were obtained from various published and unpublished sources of the governmental and non-governmental organizations. Internet sources and research reports were employed for acquiring the necessary information. Primary data were obtained through schedule household questionnaire (structured questionnaires with close-ended and open-ended), Interview (unstructured) and observation.

4. Results and Discussion

4.1. Cactus Crop Classification in the Study Area

From the above table we can understand that (Table 1) 6.66 % of the household answer concerning their house condition that the color of Cactus is white, 80% replied that the color of Cactus is green and 13% of the respondents replayed that the color of Cactus is red.

Table 1. *Types of cactus in Ganta Afeshum.*

Characteristics	Item	Respondents	
		No	Percentage
Color type	A. White	9	6.66
	B. Green	108	80
	C. Red	18	13
	D. No response	0	0
	Total	135	100
The type of cactus grown in the study area	A. Planted by man	14	10
	B. Naturally grown	1	0.7
	C. Both	120	86.9
	D. No response	0	0
	Total	135	100
The type of cactus in terms of spine	A. With spine	100	72.5
	B Without spine	15	10.8
	C. Both	20	14.5
	D. No response	0	0
	Total	135	100

Source: Field data May 2014

Figure 2. *Types of cactus.*

On the other hand concerning the type of cactus grown in the study area, 10% of the households replied that in the area Cactus is grown or planted by man, 0.7% of respondents replayed that Cactus is grown naturally and 86.9% of the respondents replayed that Cactus is grown by man and in naturally. On the other hand, the household's response concerning the spine type, 72.5% replied that cactus with spine in their locality whereas 10.8% answered without spine and 14.5% replied that both.

4.2. Use of Cactus Plant

Identification of medicinal plants with their characteristics is important for botanical drug industry (Esra *et al.* 2015, Francisco *et al.* 2015). Different parts of cactus as a potential source of food for both human and animals. From table 2 we can understand that different parts of cactus like Leaf, Fruit and Cladodes is important for preparing forage, Juice and medicine.

As can be seen in Table 2, among the different parts of Cactus the majority of the respondents (70.4%) they used cactus fruit as a source of diet and 2.96% of the House Holds replied that cactus leaf use as diet. On the other hand the majority of the households (66.66%) used Cactus Cladodes for preparing and uses as forage and 22.2% of the households replayed that Cactus fruit is used as forage.

Table 2. *The benefit of cactus as diet, forage or chemical in Ganta Afeshum.*

Characteristics	Item	Respondents	
		No	Percentage
Different part of cactus uses as diet	A. Leaf of Cactus	4	2.96
	B. Fruit	95	70.4
	C. Both	20	14.8
	D. No response	6	4.44
	Total	135	100
Different part of cactus uses as forage	A. Cladodes	90	66.66
	B. Fruit	30	22.2
	C. Both	15	11.11
	D. No response	0	0
	Total	135	100
Different part of cactus uses for different purpose	A. For Medicine	4	2.96
	B For juice or *Tej*	3	2.22
	C. For marmalades & salads	6	4.44
	D. No advantage for all	120	88.88
	E. No response	2	1.48
	Total		100

Source: Field data May 2014

Figure 3. *Important of cactus for different purpose.*

Table 3. *The benefit of cactus in addition as diet, forage or chemical in Ganta Afeshum.*

Characteristics	Item	Respondents	
		No	Percentage
The benefit of cactus in addition as diet, forage or chemical	A. for fertilizer	2	1.48
	B. for soil conservation	50	70.4
	C. For fence	80	14.8
	D. For timber	3	2.22
	D. No response	0	0
	Total	135	100

Source: Field survey May 2014

Even though different parts of Cactus is used for preparing juice and medicine, but the households didn't use cactus as a source of juice and medicine. Only 2.96% of the households replied that cactus crop used as preparing medicine and 22% of households replied that cactus crop used for preparing juice or *Tej* (local drink).

Cactus has a variety of benefits for the community of study District. As depicted in the above table, more than 70% of the households in the area explained as cactus is important for soil conservation by holding of soils run by erosion. More than 14% of the households replied that cactus is very useful for fence. It is also purposeful for timber production and for fertilizer.

Table 4. *The benefit of cactus as source of income, improving living standard of the households in Ganta Afeshum.*

Use of cactus	Item	Respondents	
		No	Percentage
Commercial value	A. Leaf of Cactus	4	2.96
	B. Fruit	101	74.8
	C. Both	20	14.8
	D. No response	0	0
	Total	135	100
Amount income from cactus fruit per kintal (1 kilo contains average 05-07 pieces of fruits) in terms of Birr	A. 300-500	90	66.66
	B. 600-1000	30	22.2
	C. > 1000	15	11.11
	D. No response	0	0
	Total	130	100
improving living standard of the households	A. High	13	9.62
	B Medium	70	51.8
	C. Low	50	37
	D. No response	2	1.48
	Total	135	100

Source: Field data May 2014

As per the above table, more than 74% of the respondents replied that fruit of cactus has commercial value. Regarding that cactus having income value, 66% of farmers are sure that they gain on average 300 to 500 birr per annum from cactus sales. Hence, it improved living standard of low income households.

5. Conclusion

As obviously known, cactus has different economic benefits and source income in different part of the world but the household should not consider as source income this is duet-to the limited knowledge of the farmers concerning the economic benefit of cactus. In the study area 167.05 hectare of the area was covered by cactus but the housel's conceder its fruit as additional diet for human being and as forage for animals. This is due to the educational level of the society is low and they give less attention to cactus crop than other crops. Therefore, concerned bodies should give awareness and training to the community regarding to the benefits of cactus and establishing cactus association to facilitates the cactus production and provide fertile ground to the farmers to supply cactus for market in order to increase their income.

References

[1] Barbera, G. 1995. History, economic and agro-ecological importance. In: Barbera, G. Inglese, P., Pemineta – Barrior, E. (Eds), Agro-ecology, cultivation and uses of cactus pear, vol.132. FAO Plant Production and Protection Paper, pp. 1-11.

[2] Cantwell, M. 1991. Quality and postharvest physiology of "nopalitos" and "tunas". Proc. 2nd Annual Texas Prickly Pear Conference, Texas Prickly Pear Council, McAllen, Texas. Pp.50-60.

[3] Corrales-Grcia, Jpenavaldvia, CB; Razo Martinez, Y, Sanchez-Hermandaez, M. 2004. Acidity changes and pH-buffering capacity of napalitos (opuntia ssp.).Post harvest Biol.Tech. 32,169-174.

[4] Esra Maltas, Aslı Dageri, Hasibe Cingilli Vural, Salih Yildiz, Investigation on Biochemical and Molecular Analysis of Echinacea Genotypes, *Plant*. Vol. 3, No. 1, 2015, pp. 1-7. doi:10.11648/j.plant.20150301.11.

[5] Francisco das Chagas Dourado de Sousa, Melise Pessôa Araújo, Jesus Rodrigues Lemos, Ethnobotanical Study with Native Species in a Rural Village in Piauí State, Northeast Brazil, *Journal of Plant Sciences*. Vol. 3, No. 2, 2015, pp. 45-53. doi: 10.11648/j.jps.20150302.11.

[6] Frew Tegegn. 2007. Evolution of alternative feed resources for ruminants under arid zone of the tropics and sub tropics: the case of cactus pears in Ethiopia, Humboldt University, Berlin. Tag der Wissenschaftlichen Aussprache. ISBN 978-3-89574-634-5.

[7] Gete Zeleke, Menale Kassie, Pender, J. and Mahmud Yesuf. 2006. Stakeholder Analysis for Sustainable Land Management (SLM) in Ethiopia: Assessment of Opportunities, Strategic Constraints, Information Needs, and Knowledge Gaps. Environmental Economics Policy Forum for Ethiopia (EEPFE). Ethiopia.

[8] Million Tadesse and Kassa Belay. 2004. Factors Influencing Adoption of Soil Conservation Measures in Southern Ethiopia: The Case of Gununo Area. Journal of Agriculture and Rural Development in the Tropics and Subtropics. Vol. 105(1): 49-62.

[9] Mums for Mums. 2004. Preparation of cactus (*Beles*) recipes, Mekelle Institute of Technology, Mekelle.

[10] Uribe, J.M., Varnero, M., I, Benavides, C., 1992. Biomas of prickly pear (Opuntia ficus-indica L. Mill) as bovine manure anaerobic digestion accelerator.Simiente.62, 14-18 (Abstract).

[11] Woldeamlak Bewket. 2002. Land Cover Dynamics since the 1950s in Chemoga Watershed, Blue Nile Basin, Ethiopia. Mountain Research and Development. Vol. 22 (3): 263-269.

15

Validation of the Electromagnetic Physical Processes with Software SPENVIS

Filomena Loffredo[1, 2, *], Mariagabriella Pugliese[1, 2], Maria Quarto[1, 2], Vincenzo Roca[1, 2], Valerio Pisacane[3], Renato Aurigemma[4]

[1]Dipartimento di Fisica, Università di Napoli Federico II, Naples, Italy
[2]Istituto Nazionale di Fisica Nucleare (INFN), Sez. Napoli, Naples, Italy
[3]Euro.Soft srl, Naples, Italy
[4]Consorzio Società Aerospaziale Mediterranea (SAM), Naples, Italy

Email address:
floffredo@na.infn.it (F. Loffredo)

Abstract: The Space Radiation represents a serious risk for astronauts during space missions. The risk related to the space radiation exposure could involve acute and/or late effects. The Solar Cosmic Radiation that consists of protons (≈98%) with a very wide spectrum in energy (up to several GeV), is the major source of exposure for the crew. In this paper we present the results of the validation of the electromagnetic physical processes with the aim to contribute to the study of radiation protection for astronauts, in particular against the radiation due to the Solar Particle Events (SPE). The simulation was performed using MULASSIS, a module to the software SPENVIS, with protons as source, in the energy range from 800 MeV to 1.2 GeV, on a slab of aluminum of mass thickness of 20 g/cm^2. The results obtained by the simulation were compared with PSTAR database of the NIST. Finally, a comparison between SPENVIS and Geant4-9.6p2 was performed.

Keywords: Space Radiation, SPENVIS-MULASSIS, Geant4, Protons, Al Target

1. Introduction

During space missions it is important to consider that the astronauts are subjected to risks from exposure to radiation. On the Earth, three parameters can be considered to limit the exposure to the radiation, exposure time, distance between source and target and shielding. Instead, the radiation protection in the space is a very complex problem because the only possibility to provide it is the shielding. It should have important features: first of all, it has to be light, because very heavy shields are impractical on spaceships and especially must take into account the interaction of radiation with the materials, which produce secondary radiation that can be very harmful to the health of astronauts. The space radiation consist of a mixed field of radiations where, it is possible to distinguish three categories: i) Van Allen radiation belts, ii) Galactic Cosmic Radiation (GCR) and iii) Solar Particle Events (SPE) [4]. In particular the SPE, composed for 98% of protons with energies up to several GeV and high fluences (≈10^{11} particles/cm^2), represent one of the main health risk for long duration manned interplanetary missions [8]. The study of the behavior of protons at energies of about 1 GeV seems to be very interesting because the main contribution to the equivalent dose is provided by them and the remaining is attributed to heavy ions (10%). The Solar Particles Events show a maximum intensity during maximum solar intensity, they are seldom, but their occurrence is not negligible since represent the greatest danger to the unshielded crew. Intense SPE can be life threatening or induce acute radiation syndromes (such as nausea and possibly vomiting). Previous studies showed that for the SPE, the level of shielding for the most dangerous events, needed is established to be about 20 g/cm^2, typical mass thickness of the storm shelters [4].

For the health risk assessment transport codes are used, but these have a high degree of uncertainty, due to the fact that the experimental data available for the validation of different calculation models are still few. Since the limits of career recommended for activities in Low Earth Orbit (LEO) by the National Council on Radiation Protection and Measurements

(NCRP) are estimated by combining data from personal dosimeters of astronauts with the results obtained using the calculation models, it is clear the importance of implementing these using all the information from research conducted both in flight, aboard the International Space Station (ISS), at accelerators dedicated (at the NASA Space Radiation Laboratory at Brookhaven and the National Institute of Radiological Sciences in Chiba).

In this paper, by using MULASSIS, the validation of the electromagnetic physical processes was performed in a range of energy characteristic of SPE. There are other codes, but are not widely available, such as HZETRN which is an analytic code, have shown a good performance [10].

The simulation was performed assuming a slab of aluminum of 20 g/cm^2 as shielding, whose is established to be the level of shielding for the most dangerous events. The primary beam consists of protons of energy varying between 800 to 1200 MeV. The results of the electromagnetic Stopping Power obtained with MULASSIS and Geant4-9.6p2, for different values of energy, and the comparison with the NIST data are reported.

2. Materials and Methods

2.1. SPENVIS – MULASSIS

SPENVIS is an interactive tool developed by the European Space Agency (ESA) and it provides information on some space parameters, such as orbit, total mission duration, solar radiation pressure, perigee and apogee altitude etc. [5]. It has also been made available through the SPENVIS web-site (http://www.spenvis.oma.be/spenvis/). SPENVIS consists of an integrated set of models that facilitates its use in the space environment. The concept of project is the core of SPENVIS. A project is a set of input and output data obtained from a series of related runs.

MULASSIS is one of the radiation analysis tools derived from ESA's investment in the Geant4 Collaboration, in fact it is a module of SPENVIS based on the use of the toolkit Geant4 (Geant4-9.2p1), for study and simulation of the effects of radiation on the shielding in space environment [7, 9]. MULASSIS tool is a significant improvement over conventional engineering tools for simulating shielding effects such as when interact with the radiation. Following, its main capabilities are shown:

- One-dimensional simulation of shielding effects: Multiple layers of predefined or user-defined materials e.g. geometry definition of the target, the number of layers that compose of each thickness, the type and the size of materials in the NIST database (however if the material is not in this database, the software allows to "create" it specifying the name, chemical formula and density).
- General source particle distribution: Standard spectral distribution or pointwise fit; Unidirectional / isotropic / cosine-law angular distribution; Spectra output by other SPENVIS utilities [7].

The physical models that can be implemented in SPENVIS for protons are reported in Table 1:

Tab. 1. *The physical models, for protons, that can be implemented in SPENVIS.*

Energy Range	Physical models
< 10 GeV	Standard EM or Low-Energy EM - G4Binary
> 10 GeV	Above + G4QGSM

In the output of SPENVIS it is possible to choose among four types of the analysis: 1) Fluence analysis, 2) Non-ionizing dose, 3) Energy deposition and total ionizing dose and 4) Pulse-height spectrum analysis. To use MULASSIS, no programming in C++ is required, thus, removing one of the main obstacles for Monte-Carlo simulation becoming a mainstream engineering tool in space radiation analysis [7].

2.2. GEANT4

Geant4 is a versatile and powerful toolkit to simulate the passage of particles through matter. It includes a large variety of physics functionality for each particle type in the energy range from a few eV to several TeV [6]. It is possible to construct a detailed geometry that reproduces the irradiation conditions. Geant4 allows users to choose among a wide range of models, driven by theory or data, or based on parameterisations to describe the different physical phenomena depending on particle type and energy range.

With respect to other general-purpose Monte Carlo systems for particle transport, Geant4 is characterized by the peculiarity of encompassing a very large variety of physics modeling options. Users are responsible for selecting the physics configuration of their experimental applications in fact the nature of Geant4 as a toolkit prevents the definition of any "default" physics configuration. The task of optimizing the physics configuration for a given experimental scenario is guided by the body of knowledge of Geant4 physics validation available in the literature [1], which encompasses both validation tests of the fundamental components of Geant4 models (e.g. cross sections, stopping powers, secondary particle production etc.) and comparisons of complex simulated observables (e.g. energy deposition patterns) with experimental measurements.

2.3. Experimental Setup

The experimental setup adopted for the validation of the electromagnetic physical processes, is reported in Fig. 1. The simulated geometry is constituted by a point source of protons with energies from 800 MeV to 1200 MeV, in the vacuum. This energy range is typical of the protons that constitute the solar particle events. The target is a slab of aluminum ($\rho \approx 2.7$ g/cm^3), in the shape of a parallelepiped, of a mass thickness 20 g/cm^2 in the direction of incidence of the primary beam (z = 7.4 cm) and with surfaces 30x30 cm. To compare the values of the electronic stopping power it is necessary that the energy of the primary beam does not vary appreciably in the target, condition assured by a small thickness. In this configuration, the thickness used aluminum is 5% of the projected-range of the protons, which varies from 300 g/cm^2 to 530 g/cm^2 in the energy range from 800 MeV to 1.2 GeV in referring to the tables in the database PSTAR [3].

The physical model considered for the study of the electromagnetic physical processes is Low-Energy EM (Multiple scattering, Ionization and production of δ rays) , as reported in Table 1. The latter take into account the different features that are relevant to several experimental domains, such as space science. The validation of the electromagnetic physical process was essential to assess the energy loss by ionization and excitation of the protons in aluminum.

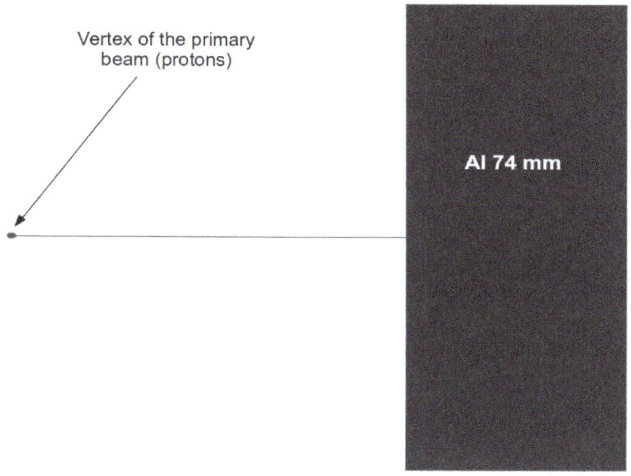

Fig. 1. Experimental Setup: target is a slab of aluminum (ρ ≈ 2.7 g/cm^3) of a thickness 74 mm, in the direction of incidence of the primary beam.

3. Results

The values of the electronic stopping power obtained using the codes, with relative uncertainty, and those calculated by PSTAR reported in the table 2.

The results reported in tab.2 show the expected trend of stopping power, which decreases with increasing energy. With the use of the software SPENVIS-MULASSIS, the stopping power and its relative uncertainty, were obtained from the output 'Energy deposition' that was normalized respect to the mass thickness. The values shown in the PSTAR database of the NIST are supplied without error because the tables of stopping power and range of protons are calculated in accordance with the Bethe Bloch formula. The Stopping Power of the protons in the energy range of interest, in the aluminum slab are reported in fig. 2.

Tab. 2. The comparison, of the stopping power of protons in aluminum, SPENVIS-MULASSIS, Geant4-9.6p2 and PSTAR of the NIST.

Energy (MeV)	Stopping Power SPENVIS-MULASSIS MeV cm^2/g	Stopping Power Geant4-9.6p2 MeV cm^2/g	Stopping Power PSTAR- NIST MeV cm^2/g
800	1.89±0.05	1.87±0.20	1.85
850	1.85±0.05	1.86±0.16	1.82
900	1.83±0.05	1.79±0.14	1.79
950	1.78±0.05	1.79±0.19	1.77
1000	1.78±0.05	1.74±0.16	1.75
1050	1.77±0.05	1.75±0.29	1.73
1100	1.74±0.05	1.70±0.26	1.72
1150	1.69±0.05	1.72±0.16	1.71
1200	1.65±0.05	1.70±0.12	1.69

Fig. 2. Comparison of the values of the electronic stopping power between SPENVIS-MULASSIS software, Geant4-9.6p2 and database PSTAR, of protons in aluminum in the energy range from 800 MeV to 1.2 GeV.

The blue circles represent the Stopping Power values obtained with the MULASSIS module, the red squares represent the values shown in the PSTAR database and the black triangles represent the values obtained with the Geant4 toolkit. The table and graph show that there is a good agreement between simulated data, with SPENVIS-

MULASSIS and Geant4, and the tabulated data PSTAR-NIST because the tabulated value is in the range determined by the standard deviation.

4. Conclusions

The agreement of SPENVIS-MULASSIS and Geant4 stopping powers is not surprising since the first code relies on the second for describing particle interactions with matter. According to the Geant4 release notes, the relevant Geant4 model implementations did not change between the Geant4 version used in MULASSIS (9.2 p1) compared to the stand-alone Geant4 version adopted in the paper (9.6 p2). Hence, the comparison between SPENVIS-MULASSIS and Geant4 reduces to simple verification if Geant4 is correctly used within SPENVIS-MULASSIS. From a theoretical point of view it can be well argued why the difference between Geant4 and NIST-PSTAR stopping powers should be small for protons in Aluminium at these energies. NIST-PSTAR reproduces tables from the ICRU49 report, which, like Geant4, relies on the relativistic Bethe formula at higher energies (i.e. above a few MeV). The only "free" parameter in the Bethe formula is the mean ionization potential, for which Geant4 uses by default NIST values, i.e. the same values as adopted in NIST-PSTAR (166 eV for Aluminium).

In addition to the Bethe formula, ICRU49 and Geant4 both apply higher order correction terms, which include the density effect correction, the Bloch correction, the Barkas correction, as well as shell corrections. For the proton energies discussed in the paper, the largest correction term comes from the density effect, which however contributes still less than 1% to the Bethe term; all other terms contribute less than 0.1% [11]. This means that the expected difference between Geant4 and ICRU49 due to higher order corrections can be expected to be very small, even if Geant4 and ICRU49 would rely on different models for describing the density effect.

Apart from theoretical grounds, one could argue that differences may arise from code details. For example, Geant4 pre-computes many physics quantities in form of tables which are then interpolated during the simulation run. However, stopping powers in this energy regime exhibit only a moderate variation with energy, hence the error due to interpolations is likely small and is probably smaller than the statistical error of the results reported in the paper.

Last but not least, the stopping power models in Geant4 have been verified previously against NIST-PSTAR data [2], in this reference shows a direct Geant4-PSTAR stopping power comparison for Aluminium over a wide energy range, demonstrating a good agreement between Geant4 and the NIST tabulation.

References

[1] S. Agostinelli, J. Allison, K. Amako, J. Apostolakis, et al., Geant4 - A simulation toolkit. Nucl. Instrum, Meth. Phys. Res. A506 (3), 250-303, 2003.

[2] K. Amako, S. Guatelli, V. N. Ivanchenko, M. Maire, B. Mascialino, K. Murakami, P. Nieminen, L. Pandola, S. Parlati, M. G. Pia, M. Piergentili, T. Sasaki, and L. Urban, Comparison of Geant4 Electromagnetic Physics Models Against the NIST Reference Data, IEEE Trans. Nuc. Sci, Vol 52 (2005).

[3] M. J. Berger, ESTAR, PSTAR, and ASTAR: Computer Programs for Calculating Stopping-Power and Range Tables for Electrons, Protons, and Helium Ions (version 1.2.3), National Institute of Standards and Technology, Gaithersburg, MD., [Online] 2005. Available: http://physics.nist.gov/Star.

[4] M. Durante, F.A. Cucinotta, Physical basis of radiation protection in space travel. Rev. Mod. Phy. 83, 1245, 2011.

[5] D. Heynderickx, B. Quaghebeur, J. Wera, E. J. Daly, H. D. R. Evans, New radiation environment and effects models in ESA's space environment information system (SPENVIS) . Proceedings of the 7th European Conference on Radiation and Its Effects on Components and Systems, RADECS 2003.

[6] A. Ivantchenko, P. Nieminen, S. Incerti, G. Santin, V. Ivantchenko, V. Grichine, J. Allison, 40th COSPAR Scientific Assembly. Held 2-10 August 2014, in Moscow, Russia, Abstract F2.2-38-14, 2014.

[7] F. Lei, P. R. Truscott, C. S. Dyer, B. Quaghebeur, D. Heynderickx, P. Nieminen, H. Evans, E. Daly, MULASSIS: A Geant4-Based Multilayered Shielding Simulation Tool, IEEE Trans. Nucl. Sci. 49 (6), 2788–2793, 2003.

[8] R. E. McGuire, T. T. von Rosenvinge, F. B. McDonald, The composition of solar energetic particles, Astrophysical Journal. 301, 938-961, 1986.

[9] P. R. Truscott, F. Lei, C. Dyer, B. Quaghebeur, D. Heyndericks, P. Nieminen, H. Evans, E. Daly, A. Mohammadzadeh, G. Hopkinson, MULASSIS - Monte Carlo Radiation Shielding Simulation for Space Applications Made Easy, European Space Components Information Exchange System, ESCIES, 2003.

[10] J. W. Wilson, F. F. Badavi, F. A. Cucinotta, J. L. Shinn, G. D. Badhhwar, R. Silberberg, C.H. Tsao, L. W. Townsend, R. K. Tripathi, HZETRN: Description of a Free-Space Ion and Nucleon Transport and Shielding Computer Program, NASA TP-3495, 1995.

[11] J. F. Ziegler, The Stopping of Energetic Light Ions in Elemental Matter HZETRN: Description of a Free-Space Ion and Nucleon Transport and Shielding Computer Program, NASA TP-3495, 1995.

16

Removal of Congo Red Dye from Industrial Wastewater by Untreated Sawdust

Mohammad Shafiqul Alam, Rexona Khanom, Mohammad Arifur Rahman

Department of Chemistry, University of Dhaka, Dhaka, Bangladesh

Email address:
rexonakhanom@yahoo.com (R. Khanom)

Abstract: Dyes have widespread applications and exist in the effluents of various manufacturing and processing units; therefore, its discharge has a severe environmental impact. However, the complex aromatic molecular structures of dyes make them very stable, and as a result, more difficult to biodegrade. The objective of the present study is to explore the feasibility of using untreated sawdust in the removal of Congo Red dye from industrial wastewater. The adsorption study was carried out using different particle sizes of adsorbent (90 μm, 144 μm and 355 μm), different pH values ranging from 5.0 to 11.0 along with different adsorbent amounts from 2.0 to 8.0 g. Flow rate was also varied in the range of 0.3 to 0.7 mL/min. Optimization of volume required for saturation of untreated sawdust was investigated. Adsorption method showed the optimum removal of Congo Red dye under the following conditions: constant initial concentration of 10 mg/L dye solution, volume of 200.0 mL, flow rate 0.3 mL/min, adsorbent amount 8.0 g and particle size of 90 μm. This effective removal method may provide a promising solution of removal of Congo Red dyes from textile wastewater in Bangladesh as well as other countries of the world.

Keywords: Removal of Congo Red (Dye), Industrial Waste Water, Untreated Sawdust

1. Introduction

Many manufacturing industries like dye, textile, paper and plastics industries use dyes in order to color their products and also consume substantial volumes of process water. As a result it generates a considerable amount of colored wastewater. Concern has been increased regarding the long term toxic effect of water containing these dissolved pollutants. The dye containing wastewater is usually released directly into the nearby drains, rivers, stagnant ponds or lagoons. Such wastewater disposal may cause damage to the quality of the receiving water bodies, the aquatic eco-system and the biodiversity of aquatic environment. The effluents produced by the industry containing high BOD and COD, suspended solids, toxic compounds and the color that is perceived by human eyes at very low concentration. The presence of small amount of dyes (less than 1 ppm) is highly visible and undesirable and needs to be removed before the wastewater can be discharged into the environment [1]. It is difficult to remove the dyes from the industrial waste, because dyes are not easily degradable [2].

Research to date includes numerous methods to remove dye that were extensively studied such as adsorption, coagulation, flocculation, ultra-filtration, nano-filtration, photo oxidation, adsorption onto activated carbon etc. But these methods are very expensive. At present physical, chemical and biological methods are generally used as a treatment process for removing dyes from textile wastewater. But all these methods are not cost-effective and environmental-friendly. As a result, an alternative environmental-friendly technology has become a necessity for the dye industry. Among these methods, adsorption is a widely used for dye removal from textile wastewater [3]. Granulated activated carbon (GAC) or powdered activated carbon (PAC) is commonly used for dye removal [4, 5]. However, they are expensive and the regeneration or disposal of it has several problems. Thus, the use of several low cost adsorbents has been studied by many researches. They have studied the feasibility of using low cost materials, such as waste orange peel [6]; lemon peel [7], banana pith, cotton waste, rice husk; bentonite clay; neem leaf powder; powdered activated sludge; perlite; bamboo dust, coconut shell, groundnut shell, rice husk, straw; duck weed and sewage sludge; algae, sawdust etc.

However, there are still challenges such as the maximum saturation of untreated sawdust. Over the last few years, adsorption has gained favor of importance in environment protection. Adsorbents are normally used for its environmental-friendly behavior, availability in nature and are very much cost effective. Untreated sawdust is largely composed of cellulose pectin; hemi-cellulose, lignin, etc. It can be used as an efficient and cost effective bio-adsorbent for removing dyes from industrial wastewater. In addition, untreated sawdust is alternative as a adsorbent for its abundance in nature, non-toxicity and bio-degradability. Generally, the most commonly used adsorbent for dye removal is activated carbon, because of its high surface area and thus capability for efficiently adsorbing a broad range of different types of adsorbates. However, its use is limited because of its high cost and the regeneration cost is also very high. Therefore, their use in wastewater treatment may be economically not feasible. There is a need to identify and study the adsorptive characteristics of low cost adsorbents [8]. The use of non-conventional adsorbents, particularly those that can be easily regenerated, to replace activated carbon in the removal of dye from wastewater has been addressed here. The removal of dyes from wastewater using adsorption process by untreated sawdust provides an alternative treatment. Adsorption of Congo Red dye (acid dye) was investigated by untreated sawdust.

Congo Red is one type of acid dyes. Acid dyes are brightest class of dyes [9] and are applied widely in industries. Congo Red is a heterocyclic aromatic chemical compound with molecular formula, $C_{32}H_{22}N_6Na_2O_6S_2$. Figure 1 shows the structure of Congo Red. Its chemical name is sodium salt of benzidinediazo-bis-1-naphthylamine-4-sulfonic acid.

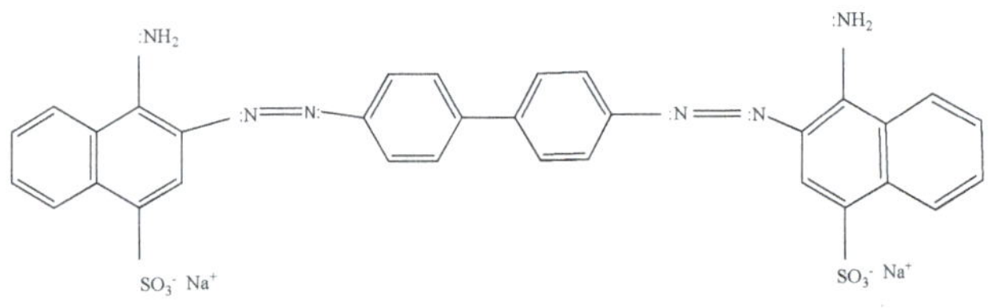

Figure 1. Structure of Congo red diazo dye.

Its formula is illustrated in Figure 1, and molecular weight is 696.66 g/mol. It is a secondary diazo dye having the color index number of 11313.

Due to a color change from blue to red at pH 3.0-5.0, Congo Red can be used as a pH indicator. It has a strong affinity to cellulose fibers. However, the use of Congo Red in the cellulose industries (cotton textiles, wood pulp, and paper industry) has long been abandoned, primarily because of its tendency to change color when touched by sweaty fingers, to run, and because of its toxicity. It can be also used for diagnosis purpose [9].

The literature survey mentioned above suggests that many interesting questions can still be raised regarding average percentage removal of Congo Red and the correlations with the particle size of adsorbent, amount of adsorbent, flow rate, maximum volume of the Congo Red and pH. Under fixed particle size, how does amount of adsorbent influences the percentage removal of the Congo Red dye? Under constant particle size and amount of adsorbents, how do flow rates govern the percentage removal of Congo Red dye? Under fixed particle size, amount of adsorbents, flow rate, how does maximum volume of the adsorbent affects the percentage removal of Congo Red dye? Does the percentage Congo Red dye depend on pH of the Congo Red solution? These are some of the issues that are addressed in this paper. Moreover, in the present investigation, approaches have been made to explore the removal of Congo Red with one waste material, untreated sawdust. The objective of the present study is to explore the feasibility of using untreated sawdust in the removal of Congo Red dye. The attractive features of the study: 1. Collection of adsorbent; 2. Preparation of adsorbent; 3. Optimization of different parameters such as amount of adsorbent amount, particle size of the adsorbent, flow rate, volume of Congo Red solution required for saturation of untreated sawdust, and pH of the Congo Red solution.

2. Materials and Methods

2.1. Materials

The following chemicals were used without further purification: Congo Red (CV) (Merck, Germany); Sodium hydroxide (Merck, Germany); 3. Hydrochloric acid (Analytical Reagent, Bangladesh).

2.1.1. Collection and Preparation of Adsorbent

Sawdust was collected from a Saw-mill of Kawran Bazar, Dhaka, Bangladesh. It was then soaked in de-ionized water for 12 hours. Then floating materials were discarded and dried in oven at 110°C temperature. It was then weighed and further dried until constant weight. After that it was ground into fine particles by grinding machine and sieved into different particle sizes. Particle sizes used for this study were 90 µm, 140 µm, 355 µm.

Then the adsorbent materials (for example 140 µm) were inserted into the column and de-ionized water was passed through the column until removal of dissolved coloring substances of adsorbent.

Figure 2. 140 µm-size Sawdust.

2.1.2. Preparation of Congo Red Dye Stock Solution

A dye stock solution of 1000 mg/L was prepared by dissolving 0.1 g of Congo Red dye in a 100 mL volumetric flask and diluted up to the mark by addition of de-ionized water. An aliquot of 1.0 mL of the stock solution was pipetted in a 100.0 mL volumetric flask and the volume was adjusted by addition of de-ionized water to prepare a solution of 10.0 mg/L.

2.1.3. Preparation of Standard Solutions Regarding Calibration Curve

2.0 mg/L, 4.0 mg/L, 6.0 mg/L and 8.0 mg/L Congo Red azo dye standard Solutions were prepared for the purpose of making the dye calibration curve.

2.1.4. Preparation of Adsorption Column

For the treatment of single adsorbent, glass wool was inserted firstly into the column, which acted as a support for the adsorbent. The soaked adsorbent with de-ionized water was poured into the column and water was then allowed to pass through the column. In this way, dissolved materials and colored substances were washed out from the adsorbent. Water was then drained out. Congo Red dye solution of 10.0 mg/L concentration was then taken into the column and effluent was collected in a clean beaker.

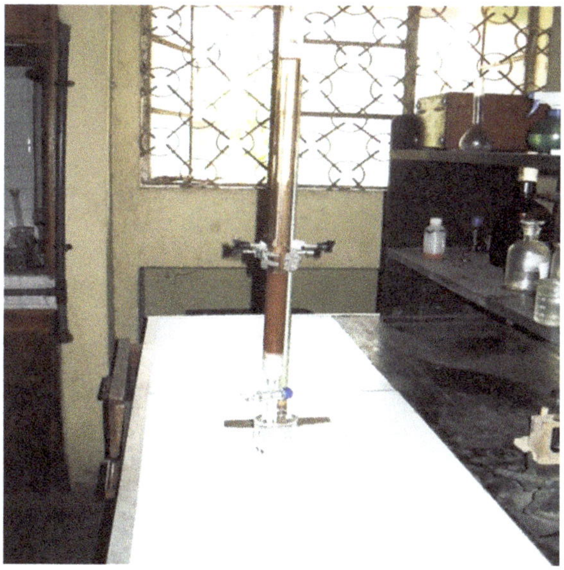

Figure 3. *Experimental set up of column system.*

2.2. Methods

2.2.1. Determination of Maximum Absorbance (λ_{max})

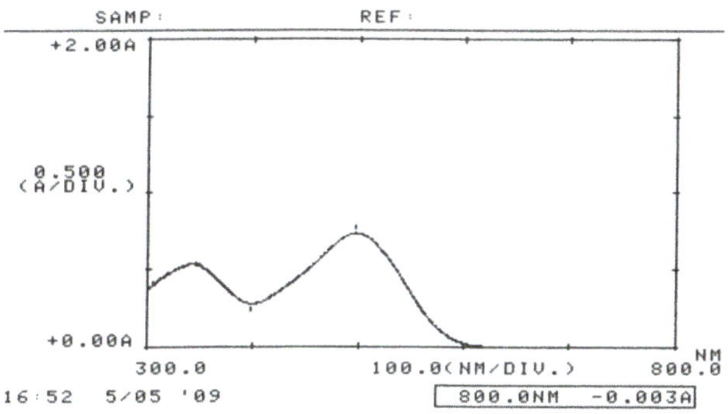

Figure 4. *Determination of λ_{max} of Congo Red dye solution.*

The maximum absorbance of Congo Red dye solution was measured by using UV-visible spectrophotometer (UV-160A, Shimadzu, Japan) is given in Figure 4. Maximum absorbance for the solution of Congo Red was found at 497.0 nm. All the absorbance measurements were carried out as the respective λ_{max} values to attain maximum sensitivity of the instrument.

2.2.2. Measurement of Calibration Curve

A calibration curve in UV-visible spectrometer was obtained by taking absorbance of the standard solution containing known concentration of the dye to be determined and by measuring the absorbance of each solution, there is a linear relationship between the absorbance and concentration is shown below in Figure 5.

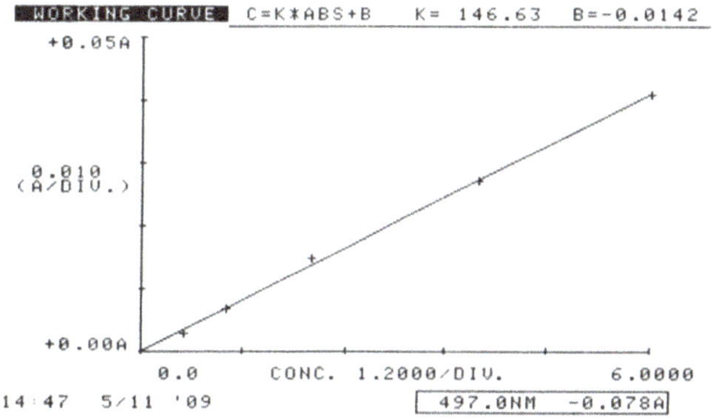

Figure 5. *Calibration curve for different concentration of standard Congo Red dye solution.*

The absorption of a monochromatic beam of light by a homogeneous absorbing system is described by Beer-Lambert law.

Modern absorption measuring instruments can usually display data as either transmittance, %-transmittance, or absorbance. An unknown concentration of an analyte can be determined by measuring the amount of light that a sample absorbs and applying Beer-Lambert's law. If the absorptivity coefficient is known, the unknown concentration can be determined.

2.2.3. Adsorption

A molecule inside the bulk experiences cohesive forces of equal magnetic magnitude in all directions. The resultant cohesive they posses unbalanced or residual attractive force and hold adsorbed particles. The molecules on the surface are pulled downwards and sideways but no forces act on them above as shown in Figure 7. Therefore, they render them reactive towards liquid and gases. At the first available opportunity these surface atoms or molecules will try to satisfy the unstauration by 'capturing' other molecules atoms due to the tendency for the force energy of any surface to decrease [10-11]. If the valency requirement of the atoms on the surface satisfied, adsorption tends to take simply through the Van der Waals force attraction, resulting the physisorption. On the other hand, when the surface is unsatisfied, the adsorption takes place the chemical force for bond formation and termed as chemisorptions.

Monolayer adsorption occurs due to single adsorption with adsorbent in chemisorptions and physisorption. But in some case of physisorption, multiplayer adsorption occurs due to lateral Van der Waals forces interaction.

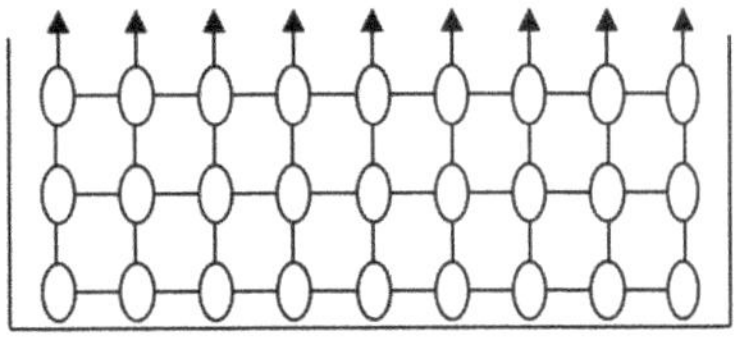

Figure 6. Unsaturation state of adsorbent.

2.2.4. Adsorption Parameters Measurement

(i). Optimization of Particle Size, Amount of Adsorbent and Flow Rate

The effect of three different particle sizes such as 90 μm, 140 μm and 355 μm were investigated. Firstly, initial absorbance of 10 ppm Congo Red dye solution was measured 0.109. Then, 100.0 mL of Congo Red dye solution of 10 ppm was poured to the column of particle size 90.0 μm. Five fractions were collected; volume of each fraction was 20.0 mL. The absorbances were measured and consequently percentage removal of Congo Red dye was subjected to removal efficiency was calculated by using the following equation:

Percentage removal = $[(A_i - A_f) / A_i] \times 100$, [%]

Where, A_i = initial absorbance

A_f = final absorbance

Percentage removal of Congo Red dye was varied from 95.41% to 88.23% with (e.g., particle size 90 μm) and average percentage removal of Congo Red was 91.33% was given in Table 1 and similarly with particle 140 μm and 355 μm and average percentage removal of Congo Red dye were 87.53% and 84.43% respectively.

Table 1. Percentage removal of Congo Red dye for 90 μm-size sawdust adsorbent.

Volume of eluent (mL)	Percentage removal of Congo Red (%)	Average percentage removal of Congo Red (%)
20	95.41	
20	93.24	
20	90.65	91.33
20	89.10	
20	88.23	

After selecting the optimized particle size, the different amounts of adsorbent were subjected to passage of 10 mg/L dye solution in the column with 0.3 mL/min. At this step, amount of adsorbent varied was within 2.0-8.0 g. The experimental column was packed with 8.0 g of sawdust of 90 μm-size. 100.0 mL Congo Red dye solution of 10 ppm was poured to the column. Five fractions were collected; volume of each fraction was 20.0 mL. It was observed that the percentage removal of Congo Red dye was varied from 95.17% to 92.11% and average percentage removal of Congo Red was 94.43% was shown in Table 2. Similar manner observed with 2.0 g, 4.0 g, and 6.0 g sawdust and thus average percentage removal of Congo Red were 86.1%, 91.3% and 93.5% respectively.

Table 2. Percentage removal of Congo Red dye.

Volume of Eluent (mL)	Percentage removal of Congo Red (%)	Average percentage removal of Congo Red (%)
20	95.17	
20	95.17	
20	95.17	94.43
20	94.55	
20	92.11	

Table 3. Percentage removal of Congo Red azo dye for 0.5 mL/min.

Volume of Eluent (mL)	Percentage removal of Congo Red (mL)	Average percentage removal Congo Red (mL)
20	96.33	
20	96.33	
20	90.82	89.27
20	82.57	
20	80.73	

Then the influence of flow rate within the range of 0.3 – 0.7 mL/min on the adsorption process was studied using the optimized particle size and amount of adsorbent. It was observed that absorbances were varied and percentage removal of Congo Red dye was varied from 95.17% to 92.11% with flow rate 0.3 mg/L and average percentage removal of Congo Red was 94.43%. Secondly, it was observed that the percentage removal of Congo Red dye was varied from 96.33% to 82.57% with flow rate 0.5 mg/L and average percentage removal of Congo Red was 89.27% depicted in Table 3. Similarly with 0.7 mL/min flow rate and average percentage removal of Congo Red was 79.12%.

(ii). Optimization of Volume Required for Maximum Saturation of Untreated Sawdust and pH

The different volume of dye solution in the range of 20.0 – 300.0 mL was then injected into the column to study its influence on adsorption. The experimental column was packed with 8.0 g of sawdust of 90 μm-size. 300.0 mL

Congo Red dye solution of 10 ppm was poured to the column. Twelve fractions were collected with flow rate 0.3 mL/min; volume of each fraction was 20.0 mL. It was observed that sawdust was saturated after eluting 200.0 mL, displayed in Table 4.

Table 4. *Required volume until saturated removal efficiency of Congo Red dye.*

Volume of Eluent (mL)	Percentage removal of Congo Red (%)	required volume for saturation of sawdust (mL)
20	95.57	
20	92.03	
20	89.38	
20	83.18	
20	80.53	
20	76.10	
20	69.91	200.0
20	61.94	
20	48.67	
20	44.24	
20	44.24	
20	44.24	

And initial pH of solution was adjusted to 5, 7, 9 and 11 at optimum condition of particle size, adsorbent amount, flow rate, and volume for saturation of sawdust. It was observed that absorbances were varied and the percentage removal of Congo Red dye was varied from 97.92% to 95.50% with pH 5.0 and average percentage removal of Congo Red was 96.0%, inserted below in Table 5. Similarly, for the pH value 7.0, 9.0, 11.0, average percentage removal of Congo Red was 92.50%, 80.0% and 63.52% respectively.

Table 5. *Determination of percentage removal of Congo Red for pH 5.0.*

Volume of eluent (mL)	Percentage removal of Congo Red (%)	Average percentage removal of Congo Red (%)
20	97.92	
20	96.89	
20	96.10	96.00
20	95.70	
20	95.50	

3. Results and Discussion

3.1. Optimization of Particle Size, Amount of Adsorbent and Flow Rate

The plot of percentage removal of Congo Red against particle size is depicted in Figure 8.

Figure refers that removal efficiency of adsorbent for Congo Red decreased with increasing particle size. Highest removal was observed with 90μm-size untreated sawdust. Therefore, particle size 90 μm-size was chosen for next experiments. This phenomenon was probably due to the fact that, with the decrease of particle size, the surface areas of the adsorbent was increased, which provided greater number of adsorption sites for Congo Red to adsorb onto adsorbent.

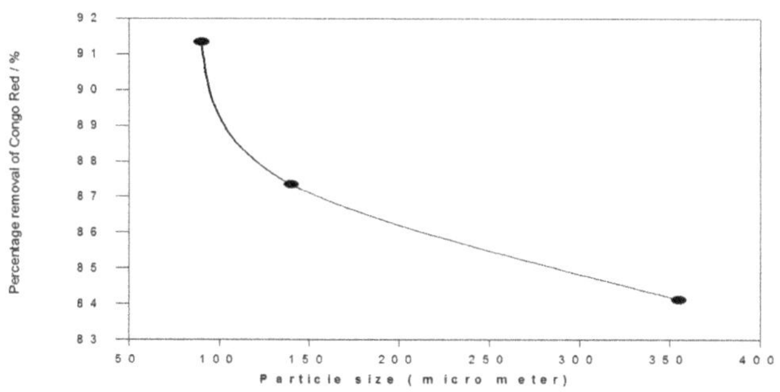

Figure 7. *Optimization of particle size for the removal of Congo Red dye by adsorption.*

Figure 9 which shows that removal efficiency increased with increasing amount of adsorbent. The rise in removal efficiency is due to increase in adsorption sites with increase in amount of adsorbent. It also shows that highest removal efficiency was exhibited by 8.0g untreated sawdust (94.43%). Consequently, 8.0g untreated sawdust of 9μm-size was selected for further study.

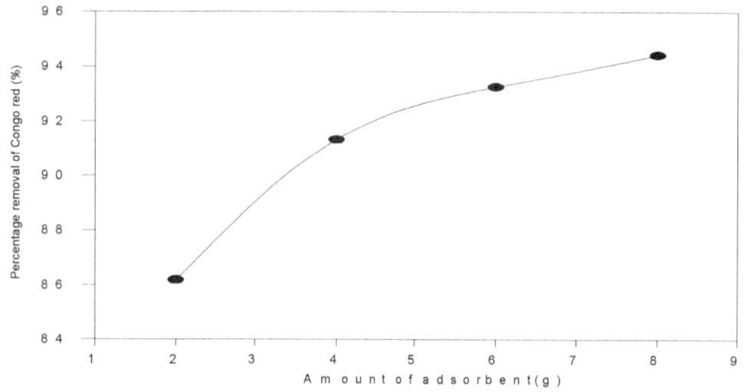

Figure 8. *Optimization of amount of adsorbent (in 'g') on the removal of Congo red by adsorption.*

The influence of the treatment flow rate on the removal of Congo Red is addressed in Figure 10. According to the figure 10, removal efficiency decreased with increasing flow rate. In general, 94.43% of Congo Red was removed at flow rate 0.3 mL/min. Higher removal of dye at slower flow rate was due to the fact that, when the flow rate was slow, Congo Red in the sample solution got more contact time with the active surface of the adsorbent. As a result, flow rate 0.3 mL/min was chosen for the next experiments.

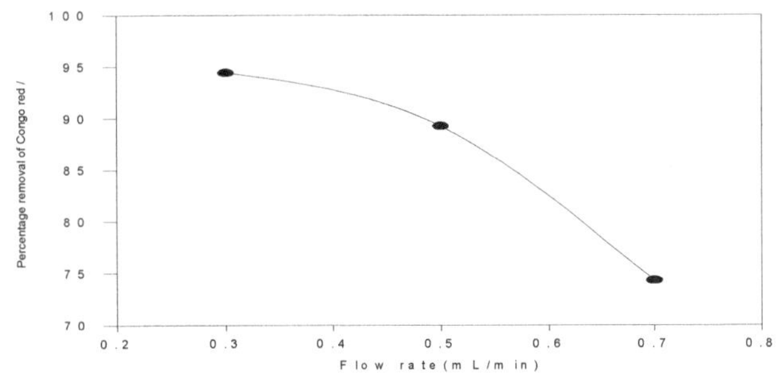

Figure 9. *Optimization of flow rate on the removal Congo Red dye by adsorption.*

3.2. Optimization of Maximum Volume for Saturation of Adsorbent and pH

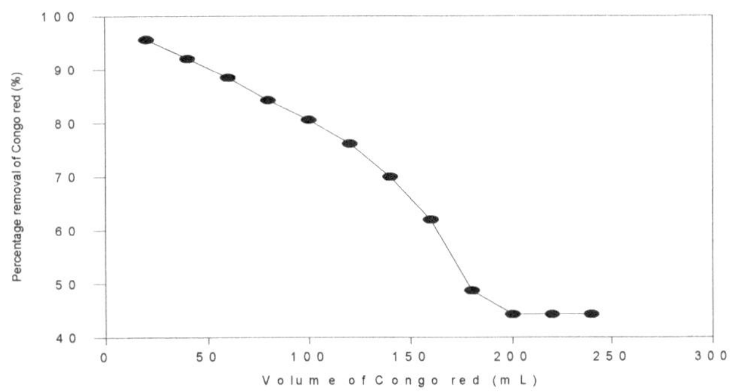

Figure 10. *Optimization of maximum volume for saturation of adsorbent on the removal of Congo Red.*

The effect of the volume on the removal of Congo Red dye is given in Figure 11. The figure shows that removal efficiency decreased with increasing of volume. At a stage of increasing volume, the removal efficiency attained saturation because there was no surface active site to adsorb dye molecule. When volume is 200.0 mL, saturation value was obtained. The volume fixed for saturation was 200.0 mL.

The consequence of pH on the removal of Congo Red in the presence of adsorbent is shown in Figure 12. Figure shows that as the pH increases, the removal of dye from the solution decreased. The decreasing trend of removal of the Congo Red with increasing pH was dependent of the nature of the adsorbent. Adsorption was the highest for pH 5.0. So, this pH was selected further study.

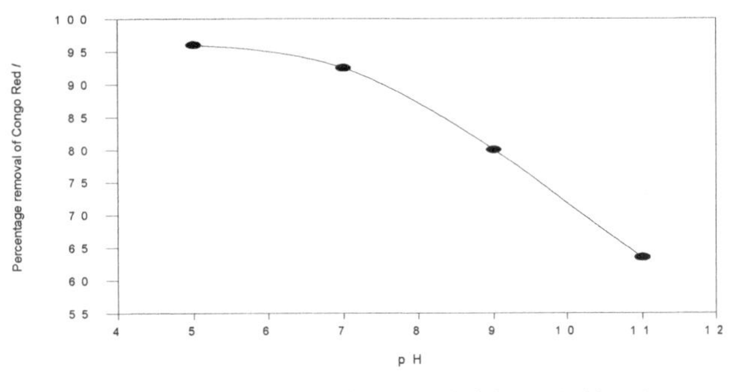

Figure 11. *Optimization of pH on the removal of Congo Red by adsorption.*

3.3. Effect of pH on the Adsorption

Surface charge of adsorbent is dependent on the pH and in some cases structure of the adsorbates species also depends on the pH of the solution. Hence the amount of adsorption and even the mechanism of the adsorption may be affected by the pH of the dye solution. So, pH dependency also gives information about the mechanism of adsorption. The effect of pH on adsorption by sawdust can be explained on the basis of point of zero charge pH_{pzc}, at which adsorbent is neutral, while beyond this the material becomes either positively or negatively charged. The pH_{zpc} value for sawdust $=5.2\pm0.2$ [12]. Below the zero point charge that is at low pH with an increasing of the concentration of the H^+ ion in dye solution, the OH groups of the surface of sawdust would get neutralize by protonation. That is at lower pH, the presence of more H^+ ions increases the number of positively charged adsorbent sites and the increase in the number of positively charged surface sites probably favor the adsorption of negatively charged dye (Congo Red dye). Because, there is an electrostatic attraction increase with decreasing of pH, thus resulting an increase of dye adsorption.

Above point of zero charge that is at higher pH with an increasing concentration of the OH^- ion in dye solution, the surface OH^- groups increases. Consequently the negative charge density would be located more on the dye molecule and this accounts for the lower dye uptake on increasing negatively charged surface sites, because electrostatic repulsion increases with increasing pH.

4. Conclusions

Dyes are generally present in the effluents of textile industries. These are distributed in soils, aquatic environment and are known to be a source of environmental contamination. So, the necessity demands to focus the attention on techniques leading to complete removal of dye molecules. The removal of Congo Red from industrial wastewater using untreated sawdust has been investigated under different experimental conditions. The adsorption of Congo Red was found to be dependent on the adsorbent particle size, amount of adsorbent, flow rate of effluent and volume of dye solution. The uptake of Congo Red was greatly affected by the dye solution pH. The optimized conditions can be applied for the design and fabrication of an economically favorable treatment process.

Acknowledgments

The author thanks Department of Chemistry, Dhaka University, Bangladesh for the financial support to carry out this work and also thank Dr. A. M. Shafiqul Alam, M.Sc. (Dhaka), M.S. in Anal chem. (Lyon, France), D.Sc. (Lyon, France), C. Chem., FRSC (London), Professor, Department of chemistry, University of Dhaka and Dr. M. Arifur Rahman, M.Sc. (Dhaka), Ph.D (Mie University, Japan), lecturer, Department of chemistry, University of Dhaka for supervision, instrumental support and encouragement. The author gratefully acknowledges "Analytical Laboratory" for the use of UV-Visible Spectrophotometer.

References

[1] Vandevivere P. C., Bianchi R, Verstreate, W. (1998), Treatment and reuse of wastewater from the textile Wet-Processing Industry: review of emerging technologies. J. Chem. Technol. Biotechnol, 72, 289-302.

[2] Isik M. and Sponza D. T. (2005), A batch study for assessing the inhibition effects of Direct Yellow 12 in a Mixed methanogenic culture. Process Bio-Chemistry, 40 (3-4-40), 1053-1062.

[3] Alzaydien A. S., Adsorption of Methylene Blue from Aqueous Solution on a Low-cost Natural Jordanian Tripoli, American Journal of Environmental Science, 2009, vol. 5, No. 1, 197-208.

[4] Walker G. M. and Weatherly L. R. (1998), Fixed bed adsorption of acid dyes onto activated carbon, Environmental Pollution, vol. 99, No.1, 133-136.

[5] Chem Jia-Ming and Wu Chia-Yuan (2001), Desorption of dye from activated carbon beds: effects of temperature, pH and alcohol, Water Research, 2001, vol. 35, No.17, 4159-4165.

[6] Kumar K. V. (2007), Optimum sorption isotherm by linear and non-linear methods for malachite green onto lemon peel, *Dyes and Pigments*, 2007, vol. 74, No.3, 595-597.

[7] Ajay Meena. Chitra Rajagopal, Comparative Studies on Adsorptive Removal of Chromium from Contaminated water Using Different Adsorbents, Indian Journal of Chemical Technology, ISSN: 0971-457X, Jan. 2003, Vol. 10, No.1, pp.72-78.

[8] Pappic S., Koprivanac N., Metes A. (2000), Optimizing polymer induced flocculation process to remove the active dyes from wastewater, *Environmental Technology*, 2000, 21, 97-105.

[9] Steensma D. P., Congo red: out of Africa?, *Archives of Pathology and Laboratory Medicine,2001* 125(2), 250–252.

[10] Haque M. and Nawab A. (1974), *Principles of Physical chemistry*, 3rd Edition, p 371, Dhaka, Student Publications, Bangladesh.

[11] Glastone S. (1974), Test Book of Physical Chemistry, 3rd Edition, Pub. Maxmillian Student Editions p 1194.

[12] F. A. Batzias, D. K. Sidiras, Simulation of dye adsorption by beech sawdust as affected by pH, Journal of Hazardous Materials, 2007, 141(3) 668-679.

17

Land Use-Land Cover Change and Drivers of Deforestation in the Patako Protected Area (Center-West of Senegal)

Simon Sambou[1], Anne Mette Lykke[2], Hyacinthe Sambou[1, 3], Idrissa Guiro[1], Bienvenu Sambou[1], Cheikh Mbow[1, 3, 4]

[1]Institute of Environmental Sciences, Faculty of Sciences and Technology, Cheikh Anta Diop University of Dakar. Dakar Fann, Senegal
[2]Department of Biological sciences, Aarhus University, Vejlsovej, Silkeborg, Denmark
[3]Laboratory of Teaching and Research in Geomatics, Polytechnic High School, Cheikh Anta Diop University of Dakar, Dakar-Fann, Sénégal
[4]World Agroforestry Centre (ICRAF), Nairobi, Kenya

Email address:
sambousimon@gmail.com (S. Sambou)

Abstract: This study focus on land use-land cover changes relation to the multidimensional characteristics of spatial and temporal patterns of land dynamics and social practices. The objectives are to characterize the dynamics of land-use and land-cover changes from 1992 to 2015, and to identify the drivers of deforestation and land degradation. Nine (09) types of classes were identified: humid vegetation (HV), savanna/wooded savanna (WS), savanna/shrub savanna (SSS), shrub savanna (SS), mangrove (MA), grass savanna (GS), cropland (CP), water (WA), and bare soil (BS). The process of change was analysed within a 5614.6 ha which includes the Patako protected area. The results have shown significant changes in land use-land cover for more than two decades (1992 to 2015). A high variability of land cover transfer was recorded between periods and class categories. At least, 55% of the HV changed to WS that was manifested by a loss of 16% during the first decade. The most important decline in area was recorded annually within the HV (4.8%), and the WS (2.6%), while the SSS experienced a substantial annual increase of 0.5%. The appearance of the SS (10.3%), during the last fifteen years, is a witness to degradation process as a result of anthropogenic pressure mainly for subsistence raisons. This pressure resulted to a conversion of small forested land to agricultural land on the boundary of the protected area. Tree cover represented 99.4% of the entire forest area in 1992, 97.4% in 2010, and slightly more (+1.2%) in 2015. According to this evolution, the annual rate of deforestation was estimated to 0.09% with an average of 0.11±0.08% for the three defined periods. The drivers leading to changes in land use-land cover stresses the complexity that is related to sustainable management of protected areas. Urgent action is necessary to reduce loss of biodiversity due to deforestation and land degradation. For this purpose, a particular attention must be paid to the implementation of the Patako management plan under a Reducing Emissions from Deforestation and Forest Degradation (REDD+) project.

Keywords: Land Use Change, Savanna, Saloum Delta, West Africa

1. Introduction

Livelihoods and productive systems in African countries are highly related to vegetation changes [1, 2-3]. Gathering landscape scale information to assess land use types, their evolution and deforestation rate requires earth observation methods that allow for large spatial coverage and historical tracking of spatial dynamics [4, 5]. The use of remote sensing enables better design of management plans and thereby improved management of forest resources [6]. Various applied remote sensing methods exist, but despite real progress in applied remote sensing, big challenges are faced in connecting maps with drivers of land use change. The link between image outputs and casual factors of change is sometimes fuzzy, nonlinear or incidental and requires steady disaggregated information on local human practices. Additionally, there is many intertwined connections between drivers with cascade effects that make it difficult to single-out individual factors. In many reports, the most described drivers are forest fires [7, 1-8, 9-10], pasture, farming land extension, fuelwood and timber extraction.

In this context, reversal of land degradation in semi-arid zones requires a clear understanding of the synergetic effect of natural and socio-economic factors [11]. Several studies described these complex dynamics of land use-land cover change in savanna ecosystems of Senegal, e.g. in the groundnut cultivation zone [12]. The rapid development of groundnut production and the subsequent economic crisis and migration are significant causes of change [13]. Natural vegetation has been replaced by farm lands, where trees have been eliminated because of farming techniques and agronomist approaches that aim at modern intensification.

The conversion of forests to agricultural land and/or pasture induce in short and medium terms significant decrease in carbon stock, thereby promoting an increase in CO_2 emissions in the atmosphere. Under UNDESERT (understanding and combating desertification to mitigate its impact on ecosystem services) program, a particular emphasis was given to the restoration of degraded lands, restoration of biodiversity and increasing carbon sequestration. A buffer zone has been created to mitigate population impact on Patako protected forest and several activities such as forestry and agroforestry have been implemented. Therefore understanding of drivers that lead to deforestation at local scale is essential for designing these interventions and develop best policies and measures. Addressing the drivers is also important as strengthening the conservation of forests and its resources are an important part of the development objectives of Senegal. This study focus on land use-land cover changes relation to the multidimensional characteristics of spatial and temporal patterns of land dynamics and social practices. The objectives are to characterize the dynamics of land-use and land-cover changes from 1992 to 2015 in a protected area, and to identify the drivers of deforestation and land degradation.

2. Material and Methods

2.1. Description of the Study Area

The study was conducted in Patako protected area and its surrounding agricultural areas in the center-west of Senegal (13°40’59.59”N and 16°15’59.88”W). Patako covers about 5614.6 hectares and was classified as a protected area in 1934. Patako was originally a tropical dry forest, but has changed into savanna mainly because of human impact such as fire and cutting. The vegetation has changed within the last 80 years.

The precipitation (600-800 mm) is unimodal with a short rainy season from June to October (4-5 months) and a dry season from October to May (8-9 months) [13]. The interannual variability is high ranging from 562 mm to 813 mm within the last 10 years. Mean annual temperature is 26.58°C.

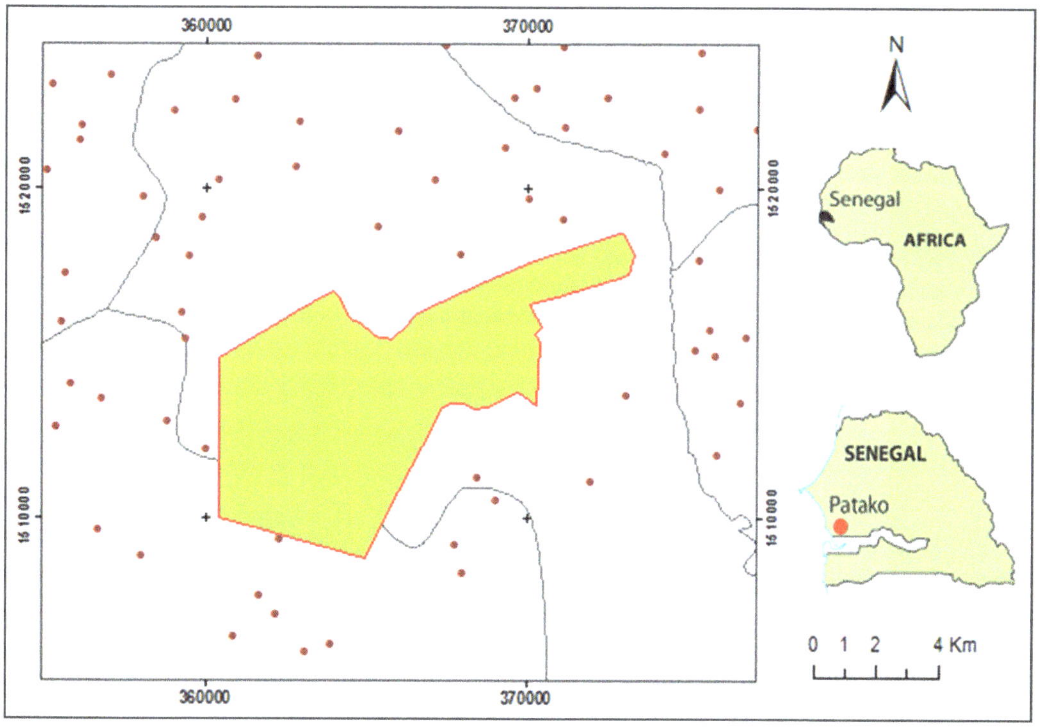

Figure 1. *Location of the study area. Red dots represent villages and red line is the border of the protected area in green.*

The area is dominated by ferruginous tropical soils, poorly evolved soils, hydromorphic soils on alluvial plains and gley-salted soils along valleys. Croplands with groundnuts, millet and maize occupy vast spaces, with mosaics of other land use-land cover types, mainly savannas, plantations of *Anacardium occidentale*, humid vegetation and vegetable gardens. There is a sharp border between the agrosystem with open cultivated land and the protected area. The agricultural lands have an average density of less than 2 trees/ha (range from 0 to 4 trees/ha) and a deforestation rate of 3% per year [14].

2.2. Satellite Data Used

The whole area is covered by the Landsat scene (Path 205, Rows 50) for each of relevant dates. The selected images are acquired in the same period of the year during the dry season to reduce problems related to phrenology change in vegetation (exaggeration of the herbaceous plant cover) and soil humidity difference. The advantage of this period is to have a low cloud cover. Thus, the images used are: an old cover TM of 5 March 1992, ETM + of January 9, 2001 and December 28, 2010 and a recent Landsat 8 OLI (Operational Land Imager) of January 9, 2015. All images were taken from the database of the United State Geological Survey (USGS).

2.2.1. Images Processing and Classification

The first step of image processing was the visual analysis. It corresponds to a visual interpretation, *ie* the application of the conventional method of photo-interpretation to satellite image. This method involves identifying the different homogenous units [15]. The interpretation was often facilitated by knowledge of the study area and field surveys. The images used for this study were already georeferenced (UTM WGS 84), with radiometric corrections and did not need geometric corrections. In order to avoid mismatches due to difference in the Landsat sensors and optimize the superposition of images for diachronic analysis, the data from 1992, 2001 and 2010 were orthorectified in reference to 2015 on the basis of seamarks points. The images were mosaicked and a smoothing of the dynamic performed. In order to harmonize and facilitate the superposition of images for diachronic analysis, the image of 1992 and 2001 have been readjusted compared to that of 2015. The wedging error (RMS = 0.124) is less than a Landsat pixel.

After wedging images, we created colored compositions by combining channels 5 for infrared [0.75-0.90 μm], 4 for red [0.63 to 0.69 μm] and 3 for green [0.52 to 0.60 μm]. In the colorimetric order red, green, blue were chosen for the images 1992, 2001 and 2010. As for the image 2015, the channels 6 for infrared [1.560 to 1.660 μm], 5 for red [0.845 -0.885 μm] and 4 for green [0.630 to 0.680 μm] were chosen. On this basis the study area was extracted in order to quantify the land use and its evolution over time.

The supervised classification method by maximum likelihood was used for all images. The choice of this method can be explained by the heterogeneity of land use in the study area and the probability of confusion for some components, namely bare soil and tans. The hierarchical classification starts with the determination of the number of classes and the choice of training plots to achieve a cartographic synthesis through successive selections of the best classified elements. The determination of the number of classes is based on the clustering performed upstream and information gathered in the field.

Furthermore, to eliminate isolated pixels or small unclassified areas and to homogenize the results of the thematic classification a filter of 3x3 pixels was applied from majority analysis. To check the validity and performance of the classifications, witness areas as well as reference data were collected in the field. The selected classes were also confronted with exogenous documents that provided information on the type of land use. These documents are thematic and synthesis maps. The analysis of the colored compositions allowed to identify at least nine (09) classes interpreted using the following abbreviation codes: humid vegetation (HV), savanna/wooded savanna (WS), savanna/shrub savanna (SSS), shrub savanna (SS), mangrove (MA), grass savanna (GS), cropland (CP), water (WA), and bare soil (BS). The type of vegetation was defined in the basis of the nomenclature of Yangambi [16] modified by [17].

2.2.2. Land Use-Land Cover Change Analysis

The result obtained from the supervised classification was exported to vector format in ARC GIS software. This transfer of map files helped to complete the processing and analysis of images and produce land use land cover maps. Changes detection were perform based on these maps. The crossing of the 1992-2001, 2001-2010 and 2010-2015 land use-land cover maps allowed to establish the maps of change. This cross was made based on classes' codification and a code comparison between two maps. Statistics of detected changes were computed by comparing image class shifts between the four images 1992, 2001, 2010 and 2015.

The trends of changes were calculated based on typical rate of land cover transition for each category. Transition intensities were estimated relatively to the size of a land cover class in the initial year to its size in the subsequent year. These changes can be positive (increase or perspective of gains) or negative (decrease or perspective of losses). Three situations have been observed and taken into account in the analysis (e.g. land cover stability, conversion and modification). Land cover stability refers to all classes that have not been affected by changes during the studied period. The changes was expressed as conversion and/or modification. By conversion, we mean the transition from one category to another (*e.g.* from woodland to cropland), while modification represents a change within one land use or land cover category [18, 19] due to changes in its physical or functional attributes (e.g. from savana to shrub savana to degraded savana). Matrices of land use-land cover transfer was generated using changes from one layer to another.

2.2.3. Annual Rate of Deforestation

Deforestation is explained as a conversion of forested areas to nonforest land use. This type of degradation notify a reduction of forest quality and a decrease of its overall cover due to native forest land clearing for different purposes [20]. The annual rate of deforestation was calculated by comparing the area under forest cover at two different times. The period covered in estimating the rate of deforestation is from 1992 to 2015. Different methods have been used to estimate the annual rate of change. But in order to standardize the methods and avoid confusion, [21] devised a new formula which is used in the context of this study:

$$Rate = \left(\frac{1}{t_2 - t_1}\right) \ln\left(\frac{S_2}{S_1}\right) \quad (1)$$

where S_1 is the class area at the initial time (t_1) and S_2 is the class area at the final time (t_2) according to the land use-land cover maps.

2.3. Drivers of Deforestation

Characterizing the drivers of deforestation first involve their identification and grouping in different categories. We identified two main drivers (direct and indirect) which were divided into two categories. The direct drivers concerned natural effects that includes climate variability and change, flood and parasitic attack. The direct anthropogenic drivers are actions or factors that directly impact forest cover. They are considered as subsistence drivers. The indirect or underlying drivers of deforestation are complex interactions of demographic, economic, cultural and institutional that in combination intensify direct human causes. Semi-structured interviews were conducted with resource people for three target groups of local stakeholders (local population, local authorities and local technical staff of forest service). The interviewees were selected from 6 villages located in a 0–3 km radius around the protected area to help better understand the complexity of the drivers. Furthermore, on the basis of field knowledge observations were made into and outside Patako protected area for triangulating the obtained information.

3. Results

3.1. Land Use-Land Cover from 1992 to 2015

The land use-land cover maps show that three characteristic classes were dominant during the studied period (1992-2015). These classes related to the CP in the periphery, the SSS and the WS in the Patako protected area. The land cover is highly dynamic (Table 1). In 1992, the periphery of the protected area was covered by a mixture of CP and SSS. Within the protected area the SSS represented (56.6%) while the WS and the HV occupied respectively 28.1% and 14.5%. A significant difference was observed between the latter and the rest of classes (WA, GS, CP and BS) as totaling less than 1%. These classes have increased slightly (1.2%) nine years after (in 2001). As for the WS a substantial increase of its size was observed (51.1%) at the expense of the SSS (44.3%) and the HV (3.4%). In opposition to the protected area the periphery was highly deforested and substituted to agricultural lands.

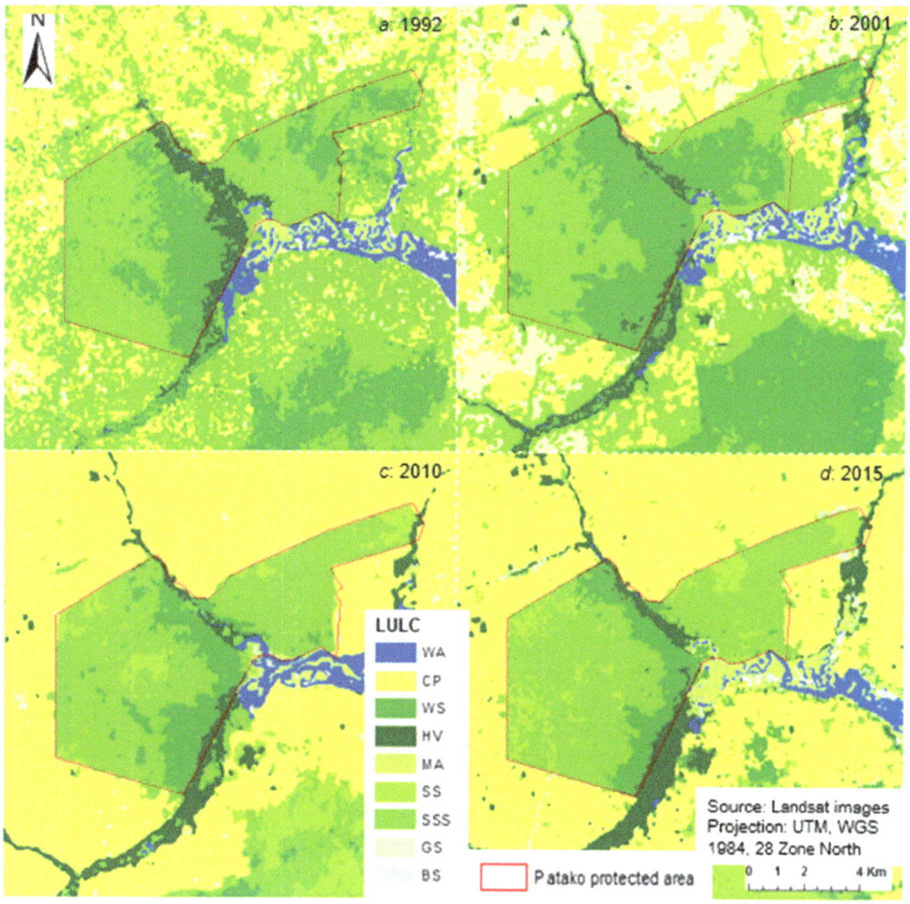

Figure 2. Land use-land cover in Patako protected area and its periphery.

Table 1. Dynamics of land use-land cover in Patako Forest.

Code	LULC classes	1992		2001		2010		2015	
		ha	%	Ha	%	ha	%	Ha	%
HV	Humid vegetation	812.3	14.5	191.4	3.4	195.3	3.5	269.5	4.8
WS	Woodland to wooded Savanna	1578.3	28.1	2869.5	51.1	1091.9	19.4	844.1	15.0
SSS	Savanna to shrub savanna	3177.9	56.6	2489.5	44.3	3688.4	65.7	3603.1	64.2
SS	Shrub savanna	0	0.0	0	0.0	473.3	8.4	796.2	14.2
MA	Mangrove	9.9	0.2	14.0	0.3	19.6	0.3	21.0	0.4
WA	Water	29.9	0.5	21.7	0.4	41.0	0.7	29.4	0.5
GS	Grass savanna	1.0	0.0	4.6	0.1	10.4	0.2	2.9	0.1
CP	Cropland	5.3	0.1	23.9	0.4	94.6	1.7	47.9	0.9
BS	Bare soil	0	0.0	0	0.0	0	0.0	0.5	0.0
Total		5614.6	100	5614.6	100	5614.6	100	5614.6	100

In 2010 there is almost an elimination of the SSS in the periphery. Within the protected area the latter class cover remained the most significant type (65.7%). Therefore the WS and the HV covered respectively 19.4% and 3.5%. Marginal classes covered a total of 2.9% of the area of the protected area, an increase of more than 41% compared to 2001. Between 2001 and 2010, we noted the emergence of a new class, the SS representing (8.4%). It results from the fragmentation of the SSS and generally located in the center and north-eastern parts (Figure 2c).

Concerning 2015 there are some increase in land cover for many classes compared to 2001 and 2010 (Figure 2d). The SSS remained the most significant class (64.2%) despite the fragmentation it undergoes and which causes the increase of the SS (14.1%). This relative extension came at the expense of the WS (15.3%). The MA was characterized by a mean increase of 0.3% ranging from 0.2% to 0.4% between 1992 and 2015. This is also the case of the CP which substantially increased for 5 to 95 ha between 1992 in 2010 before decreasing to 48 ha in 2015. Other land cover (WA, GS, etc.) were poorly represented inside the protected area.

3.2. Land Use-Land Cover Change Analysis

Results from LULCC showed significant changes in land use and land cover. Between 1992 and 2001, many land use were converted to other types in the agrosystem. These changes affected mostly the SSS and GS (Figure 3a, b) and processes of change continued to occur during the period 2001-2010. Between 2010 and 2015 (Figure 3c) the agrosystem became more stable due to the almost total use of farmland resulting in a scarcity of fallows. As the changes in the agricultural lands were unidirectional we focused the analysis within the protected area.

Regarding the land use-land cover transfer between class categories it was found that generally modification was more important in term of frequency within the Patako protected area. This type of change affected mostly the HV from 1992 to 2001. At least 83% of its area moved to WS. The latter moved for 31.6% to the SSS and inversely for 36.9%. For this, a stability of 64.6% and 62.3% was respectively observed for these classes. Concerning the conversion the most important land use-land cover transfer was found in the CP and GS. More than 56.6% moved from the CP to GS and in the opposite direction 91.4%.

From 2001 to 2010 the HV became more stable (67.3%) compared to the first period. An average modification (19.3%) was observed towards the WS. This latter was the class that decreased in favor of the SSS (58.5%). This transfer caused a progressive shrinking of the class cover. As for the conversion it was characterized mainly by the GS which stability had become *quasi* null. 25.8% of the CP was abandoned and replaced by the SSS. Changes from vegetation cover to small classes (WA, GS, GS and CP) was generally weak (less than 10 %) but the transfer in the opposite direction was more significant despite the good stability (60.4%) of the CP.

From 2010 to 2015 the stability of some land use-land cover classes is rather good (> 50%) compared to others like the GS (3.4%). The vegetation cover modification affected more the WS and the SS which areas were respectively transfered towards the SSS for more than 36% and the 54%. The results from the matrix of land use-land cover transfer reveal that conversions were mainly observed in the GS and the CP. Despite its small area 54.5% of the GS was converted to SS while 18%, 12% change respectively towards the CP and the SSS. These changes explain the low stability of that class. As for the cultuvated land more than 35% were converted to the SSS and 15% to the SS.

When we consider the matrix of land use-land cover transfer from the overall period of study (1992 to 2015) we note that the HV and the WS experienced a low stability. For the first class at least 55% was modified to WS and 18% to SSS. For the second one the major modification took place towards the SSS as representing 72.2%. The transfer of area from the SSS towards the SS was done gradually going from 12.6% in 2010 to 14.9% in 2015 with an average of 20.2% between 1992 and 2015. It is also important to note that the intrusion of the arable lands generally took place in the border between the protected area and the agricultural zone especially where the farmers do not have a precise reference mark of delimitation.

Several changes occurred within the protected area (Figure 4). The most important decline in vegetation cover were held from 1992 to 2001 in the HV (16.0%). We also noted a substantial decline in the same period for the SSS (2.7%) while the WS experienced a gain of 6.6%. By cons, it recorded a loss of 10% of its area between 2001 and 2010 and 4.8% between 2010 and 2015. During the last period, substantial losses (0.4%) have been also demonstrated in the SSS. This decline was offset by a gain in area in the SS (10.3%). Globally the rate of change corresponding to these types showed a deline of 4.8%.yr^{-1} in HV and 2.6%.yr^{-1} in the WS, while the SSS recorded an extension of its area during the study period (1992-2015). Regarding the dynamics, MA stand out with significative changes. However, one of the most disturbing change was the influence of the agricultural zone of the protected area. Between 1992 and 2010, the CP experienced an increase of 16.3% before declining to 13.8% for the last five years (2010 and 2015).

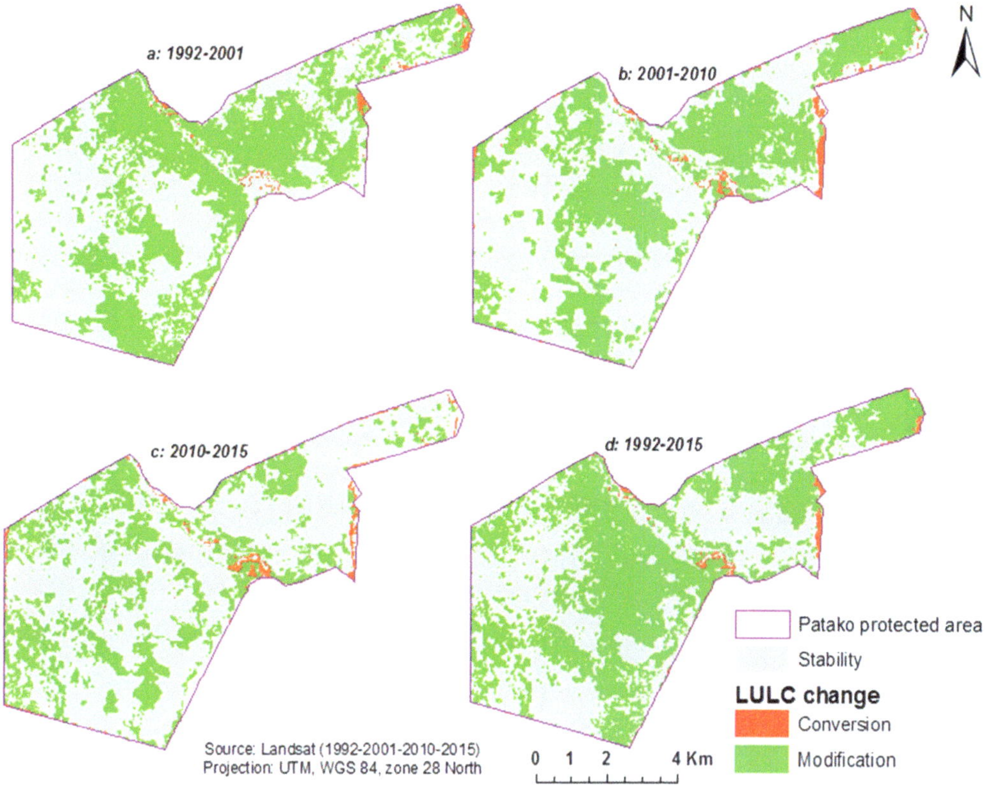

Figure 3. *LULCC in Patako area from 1992 to 2015.*

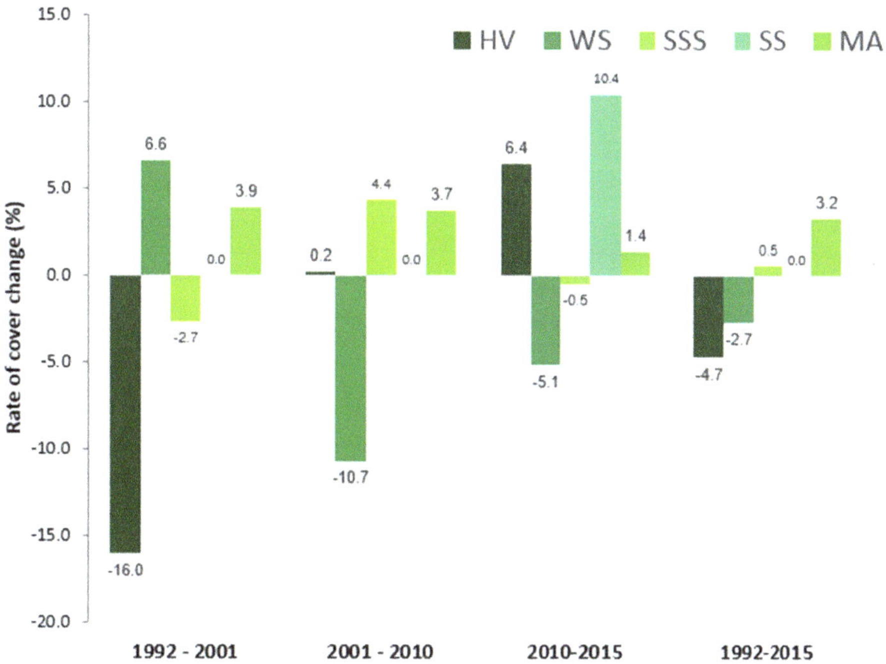

Figure 4. *Rate of cover change in Patako protected area.*

3.3. Rate of Deforestation

Deforestation is the process by which forested land is replaced by another type of non-forested land, for example agricultural land. The total area of the Patako protected forest is estimated to 5614.6 ha. The conversion of forested land in this area is mainly connected with small-scale shifting agriculture or land clearing by sedentary landholders who want to expand their crop.

Taking into account the vegetation cover an annual rate of deforestation was estimated. The forested area decreased slightly during the study period. It covered 5578.4 ha (99.4%) in 1992, 5564.5 ha (99.1%) in 2001 and 5468.5 ha (97.4%) in 2010. In 2015 a slight increase (1.2%) was recorded compared to the coverage of 2010.

The results showed a differential annual rates of deforestation between the defined intervals of the studied period which are 0.03% from 1992 to 2001. During the following 9 years (2001-2010) the rate has increased to 0.19% before decreasing to 0.13% from 2010 to 2015. The average deforestation rate of the three periods is estimated to 0.11±0.08% /year, while the overall annual rate of the studied period (1992-2015) is 0.09%, which are globally a slow rate.

3.4. Drivers of Deforestation

Deforestation implies a long-term or permanent loss of forest cover and its transformation into another land use. There are a multitude of drivers that can be classified into two categories: direct and indirect. At the local level the drivers of deforestation are subjected to complex interactions with short or long-term impacts. Figure 5 summarizes the main drivers and their probable impacts in the study area.

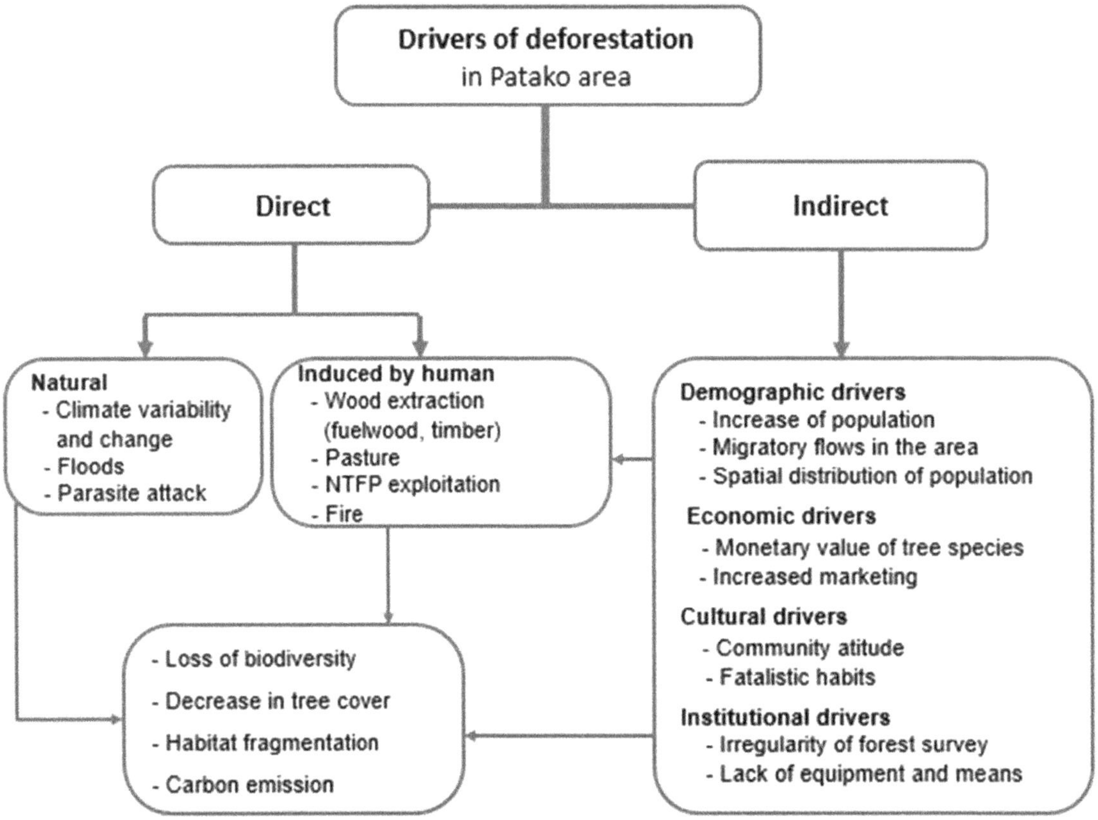

Figure 5. *Drivers of deforestation in and around Patako protected area.*

The localities around the protected area did not have the same impact on forest degradation. Villages of both sides within a radius of 3 km were grouped into four locations. Those located in the western, northern and southern parts of Patako protected area had more facilities for access than the eastern part. This has resulted by an increase of the deforestation front line. Djikoye river is a natural barrier that reduced the flow of forest operators from the eastern side. The nearest protected area of Baria played an important role in meeting fuelwood and pasture for local communities.

4. Discussion

4.1. Land Use-Land Cover from 1992 to 2015

The supervised classification approach resulted in good classification. The HV category presented both dense trees, thickets and aquatic invasive species (*Typha domingensis* and ferns). Pixels falling into this mixed category of vegetation had high similarity in spectral signature and were difficult to discriminate. It have been reported that degraded forest may be difficult to distinguish from certain types of grassland [22, 23-24]. Such difficulties arise particularly in wetlands and riparian areas [25, 26]. The 30 m resolution of images used may be a limit in spatial analysis and can influence the accuracy of the classification of small areas. Therefore field visits are invaluable to avoid lousy visual interpretation of images.

The LULC maps showed a progressive increase in land under cultivation over the years of study. The results have shown a rapid evolution of WS between 1992 and 2010. In 2001, the observed extend of this category might be explained by different phonological process of some species. In the other hand, this may be as a result of increased precipitation in 1999 and 2000 in the sudanian zone compared to earlier years. An increase of grass cover was also observed at the expense of the SSS. This category included areas currently under crop and land under preparation. In 2010 and 2015, it was highly difficult to discriminate GS in the agrosystem area because of the agricultural practices. Famers used to apply one time of weed control to avoid competition with grass until a certain height of millet crops. As for peanut cultivation, they usually weed two times and export straw after harvest (food for livestock). Such practices show a character of bare soil during the dry season.

4.2. Dynamics of Land Use-Land Cover Change

Land use and land cover changes (LULCC) are complex processes involving multiple driving forces that are location specific and context dependent. This complexity becomes even more important that the interactions between direct, indirect and underlying drivers are dominated by anthropogenic activities or influences.

Significant changes in the LULC have been observed in the agrosystem and within the protected area during the period 1992-2015. Some categories have been systematically transformed in varied proportions. These transformations concerned mainly land conversion. This type of change, from SSS or GS to CP, was quite intense and common in the studied landscape. Indeed, since the 1970s, many protected areas in the sudanian zone of Senegal have known an important spatial and temporal dynamics [12]. However, despite the human pressure agricultural incursion within the protected area was negligible. In other hand land modification affected HV, WS, SSS and SS, but these changes can be subtle [18]. Some authors supported that a detailed analysis of spatial dynamics allows a better knowledge of human activities impact on habitats [27, 28]. Changes resulting from land use are generally complex and contributes to alter natural habitats. [29] described this alteration as leading to the establishment of a mosaic of savanna in Patako protected area.

Changes affected mostly the periphery of the protected area. It has been shown that almost all lands in the agrosystem were cultivated in 2010. An important increase of land converted to agricultural land was observed, which supports earlier studies from 1972 to 2002 [30]. These changes resulted from an intensification of the use of animals for cultivation during many years ago in the Groundnut Basin. We argued that such increasing trend of agricultural production was to fulfill growing food demand and commercial opportunities. The cultivation techniques used promote land clearing. Likewise, canopy fragmentation had an impact on diameter structure its opening led to tree density decreasing and habitat fragmentation [31, 32]. Combining field observations of tree cover, height and diameter structure helped to characterize subtle changes in land cover.

4.3. Rate of Land Cover Change

In the study area, the combination of drivers led to a fast rate of land cover change. These drivers constituted a wide variety of human activities including climate variability and change. The most affected vegetation types in Patako protected area were HV and WS. This could be related to the types of species exploited in these categories, mainly large-diameter trees for timber (*Khaya senegalensis, Daniellia oliveri, Pterocarpus erinaceus* and *Cordyla pinnata*). However, the SSS have been increasing after two decades despite of its serious modification at the middle and the northeast. Knowing how and why forest area changes over time is important for managing forests sustainably because such changes may result in long-term forest conversion to agriculture. From 1992 to 2015 the current study found an annual rate of deforestation which is globally very low compared to that estimated in the whole of Senegal by [1], approximately for the same period (0.47% annually). Compared to the protected area the agrosystem has a slightly higher rate of deforestation as estimated by [14]. Nowadays, this area is dominated by a low tree density of *Cordyla pinnata* with aging population. Indeed, such aged stands of trees increased the vulnerability to fire and an absence of natural regeneration recruitment.

4.4. Drivers of Deforestation

This section reviews some of the most important drivers of deforestation that are either direct or indirect. Some direct drivers are natural. It is the case of climate variability and change and its repercussions on wetlands. Since the 70s, as other Sahelian countries in West Africa, Senegal has experienced a long period of rainfall deficits that affected natural ecosystems. Mean rainfall decreased by 20–49% between the periods 1931 – 1960 and 1968 –1997 [33]. The declining rainfall has modified the hydrological regime of the rivers around Patako protected area. This change has resulted in the development of vegetation in the riverbeds. However, since the early 2000s there has been an increase in rainfall despite a sharp intermonth and interannual variability.

Floods became a cause of tree mortality in the valleys. It was perceived that the number of flood events has been increased on average from less than 2 per year before 1990 to more than 10 during the 2000s [34]. The deposits of organic matter and silt drained by the water favored the establishment of an invasive perennial herbaceous plant (*Typha domingensis*). This invasive species has an important asphyxiating power that inhibits the development of other species. Such ecological conditions increased tree mortality along the river banks and created small openings on the canopy.

Parasite attacks, although weakly expressed are not negligible for describing natural drivers of deforestation. The genus Tapinanthus is the most common and often lead to tree mortality in the study area. As observed by [35], the susceptibility for trees to be infested was aided by senescence, fissure of bark and decrease of ability for resistance. This phenomenon is common in the local area where the *Cordyla pinnata* parkland consists of aging population.

An important part of direct drivers of deforestation was determined by human activities such as wood extraction, pasture, NTFP exploitation and the use of fire. Agricultural expansion was the main driving force of deforestation in the periphery. Contrary to the agrosystem, this problem is mostly dominated, in the core protected area, by wood harvesting which is driven by illegal logging.

Wood extraction has been demonstrated as one of the strong drivers of deforestation in the study area. The surrounding communities use wood as a primary source of energy for cooking. Unlike fuelwood, charcoal usually comes from trunks or large limbs and requires cutting trees. Even if charcoal may not always be the primary driver of deforestation, it can contribute significantly to degradation and destruction of the vegetation strata that have already been disturbed. As for the timber woods, there is a selective logging both among species and individuals into a cohort of one species. It was supported that this practice constituted a major agent of landscape modification due to the intensity use of machines for illegal timber extraction [36]. The most threatened species in the savanna formations in relation to wood timber extraction are *Pterocarpus erinaceus, Cordyla pinnata, Bombax costatum, etc.* By cons, in the HV, *Khaya senegalensis* is abusively exploited due to the economic value of its wood. Even if proximity to Patako protected area is determinant in considering the problem of deforestation, it has been perceived that people engaged in illegal activities are mostly non-residents. Indeed, beyond the satisfaction of the local market in furniture, the attraction is also due to supply some fisheries areas for making boats.

These types of forest exploitation has been documented as a problem in many rural areas in Africa [37, 38-39, 40]. [41] considered wood extraction for fuelwood and timber as contributing to forest degradation even without significant changes especially when the initial area was well bordered. As stated by [42] continuous and/or widespread human impacts can lead to vegetative changes. The damages can be an irreversible loss of species and carbon stock. Indeed, repeated actions of tree cutting and overgrazing resulted in open savanna types (SSS and DSS) with a significant frequency of stumps and cut tree stems. These elements have been considered as a key indicator of forest degradation [43, 44-45]. The presence of several tracks on any direction in Patako protected area illustrated the pressure level on natural vegetation, as it has been shown in other protected areas in the region [46].

Human pressure was also manifested by the incursion of the Fulani transhumant pastoralists during the dry season. This incursion was considered a driving force behind degradation because requiring a significant animal load. The environmental problems caused by overgrazing in Patako area are the destruction of vegetation and soil erosion.

The predominant motivation of actors in NTFP harvesting is the quest for income generation. Exudates from plants mainly such as resins are processed to give local communities added value in the marketplace. As an example, human pressure on the perfumes products of *Daniellia oliveri* was manifested by unsustainable practices of harvesting. Other activities such as the use of plants (leaves, barks, roots) in traditional medicines can be a source of deforestation or degradation because they become more vulnerable to fire.

Unsustainable practices of forest exploitation are often a key variable of annual fire regimes in Patako area. Here, fires are caused accidentally or simply used by forest operators to burn grasses in order to facilitate access during the dry season. Even if its impact was considered negligible on adult trees [47], fire is one of the significant driving force in the dynamics of savannas [48, 49].

Indirect drivers contributed mostly to increase deforestation in Patako area. Four categories of indirect drivers were described: demographic, economic, cultural and institutional.

Rapid growth in population and increase of socio-economic needs implied a cumulative pressure on forest resources. The study area recorded also an important flow of migration due to space availability and social context. This context explained the spatial distribution of land use [50]. National agricultural policies in the 1960s and 1970s encouraged farmers to clear their fields of most trees to facilitate crop planting and harvesting. Then, trees density and diversity has declined as well because of the use of animal traction technologies for cultivation.

The economic value of some species used for timber and fuelwood has dramatically increased in recent years because of the local and external strong demand. In fact, the creation of weekly markets in areas around a radius of 8 km from the border of the protected area opened increasingly a way to fraudulent and opportunistic exploitation of forest resources. Some villages have a major advantage due to the presence of fertile valleys with fresh water, a favourable framework for market gardening. Other ones, where this opportunity did not exist people developed logging activities to supply economic needs in dry season. In this context, [51] considered the existence of the communities in many surrounded villages as closely linked to the use of forest resources.

Cultural drivers can be catalysts that influence flows entering into the protected areas. A significant part of the informants considered the natural resources in the protected area as belonging to local people. This representation has led to attitudes hardly compatible with a sustainable management of forest resources [51].

The institutional drivers of deforestation are multiform and complex. Two technical agents were mandated by the Forest Service for managing the Patako protected area. However, they are frequently faced with a problem of logistics and the fact their service is 22 km far from the site of intervention, justify the irregularities on forest survey. The lack of close monitoring paved the way for fraudulent exploitation. The consequence is that savannas support two weights (anthropogenic pressure and devastation caused by fires) causing a constant spatial dynamics. Underlying drivers must be treated with cautious because they have a cumulative effect on forest resource.

5. Conclusion

In this study, the combination of direct and indirect anthropogenic drivers has demonstrated the complexity of land cover change analysis. The results have shown significant changes in land use-land cover for more than two decades (1992 to 2015). These changes are characterized by two main situations which are either an increase or a decrease in the area of the specified classes. Within the Patako protected area, the rate of change varied from one type of land use to another. Globally, the area of HV and WS declined due probably to the selective logging of some species exploited for their wood timber. These species are subjected to a covetousness and constitute one of the main causes of the loss in vegetable cover. The SSS has experienced a small extension of its area despite some fragmentations in stands. The persistence of this fragmentation has contributed to the emergence of the SS which is a key indicator of human pressure. Some limitations have been shown using Landsat images of medium spatial resolution for discriminating riparian tree vegetation to invasive aquatic grasses. Detecting all stages of degradation was difficult because the scale at which the degradation takes place is often at sub-pixel resolution. This means that the nature of the degradation affects areas smaller than the detection capability of the remotely sensed pixel. So mapping land cover change can introduce substantial errors in data processing and interpretation. To avoid inappropriate interpretation a generic name HV (humid vegetation) was given to this land cover instead of gallery forest. In the agrosystem the rate of land cover change was high because of the agricultural practices. To solve the problem of deforestation, a reforestation project has been implemented in a participatory way with rural communities. The main goal was to create a buffer zone by restoring plant biodiversity in order to reduce the pressure on the protected area for which in spite of a low rate of deforestation experienced many modifications.

Acknowledgements

The authors are grateful to UNDESERT Program (www.undesert.neri.dk) for the financial support. UNDESERT (EU FP7 243906), "Understanding and combating desertification to mitigate its impact on ecosystem services" is funded by the European Commission, Directorate General for Research and Innovation, Environment Program. We acknowledge the Department of Bioscience of Aarhus University (Denmark) for the collaboration.

References

[1] FAO, 2010. Global Forest Resource Assessment. *Food and Agriculture Organization of the United Nations* (Ed.), Rome, pp. 378.

[2] Houghton, R., 2012. Carbon emissions and the drivers of deforestation and forest degradation in the tropics. *Current Opinion in Environmental Sustainability* 4, 1–7.

[3] Hansen, M.C., Potapov, P.V., Moore, R., Hancher, M., Turubanova, S.A., Tyukavina, A., Thau, D., Stehman, S.V., Goetz, S.J., Loveland, T.R., Kommareddy, A., Egorov, A., Chini, L., Justice, C.O., Townshend, J.R., 2013. High-resolution global maps of 21st-century forest cover change. *Science* 342, 850–853.

[4] Huang, B., Babcock, H., Zhuang, X., 2010. Breaking the diffraction barrier: super-resolution imaging of cells. *Cell* 143, 1047–1058.

[5] Hansen, J., Ruedy, R., Sato, M., Lo, K., 2010. Global surface temperature change. *Reviews of Geophysics* 48, 1–15.

[6] Kadioğullari, A.İ., Sayin, M.A., Çelįk, D.A., Borucu, S., Çįl, B., Bulut, S., 2014. Analysing land cover changes for understanding of forest dynamics using temporal forest management plans. *Environmental Monitoring and Assessment 186*, 2089–2110.

[7] FAO, 2005. Global Forest Resource Assessment 2005. *Food and Agriculture Organization of the United Nations* (Ed.), *Forestry Papers. Rome,* pp. 147.

[8] Mbow, C., Nielsen, T.T. Rasmussen, K., 2000. Savanna fires in East-Central Senegal: distribution patterns, resource management and perceptions. *Human Ecology* 28, 561–583.

[9] Mbow, C., Nielsen, T.T., 2009. Senegal case study. SAFNet, Africa environment outlook case studies: impacts of fires on the environment (UNEP and SARDC). Nairobi-Kenya, *UNEP-AEO: 48.*

[10] Brink, A.B., Eva, H.D., 2009. Monitoring 25 years of land cover change dynamics in Africa: a sample based remote sensing approach. *Applied Geography 29,* 501–512.

[11] Clevers, J., Bartholomeus, H., Müchers, S., de Witt, A., 2004. Land cover classification with the Medium Resolution Imaging Spectrometer (MERIS). *EARSeL eProceedings* 3.

[12] Tappan, G.G., Sall, M., Wood, E.C., Cushing, M., 2004. Ecoregions and land cover trends in Senegal. *Journal Arid Environment* 59, 427-462.

[13] Mbow, C., Mertz, O., Diouf, A., Rasmussen, K., Reenberg, A., 2008. The history of environmental change and adaptation in Eastern Saloum, Senegal: driving forces and perceptions. *Global and Planetary Change* 64, 210–221.

[14] Christensen, S.N., 2010. Socio-economic and ecological determinants of local scale tree distribution, diversity and dynamics in agro ecosystems in West-central Senegal. *Master thesis, Aarhus University*, 19 p.

[15] Hammi S, Simonneaux V, Alifriqui M, Auclair L, Montes N, 2007. Evolution des recouvrements forestiers et de l'occupation des sols entre 1964 et 2002 dans la haute vallée des Ait Bouguemez (Haut Atlas central, Maroc). *Sécheresse* 18: 271-7.

[16] Aubreville A, 1957. Accord de Yangambi sur la nomenclature des types africains de vegetation. Bois et Forets des Tropiques 51: 23-7.

[17] De Wispelaere G, 2002. Cartographie de la végétation du complexe du « W ». Montpellier: Cirad editions.

[18] Lambin, E.F., Geist, J.H., Lepers, E., 2003. Dynamics of land-use and land-cover in tropical regions. *Annual Review of Environmental Resources* 28, 205-241.

[19] Sambou, H., Sambou, B., Diaw, A.T., Mbow, C., Traore, V., 2014. Remote Sensing Mapping of the Rice Field and Vegetal Cover and the Impacts of Soil Factors on Herbaceous in the Sub-Watershed of Boutolate (Lower Casamance, Senegal). *American Journal of Environmental Protection*, 3, 73-82.

[20] FAO, Food and Agriculture Organization (2007). Manual On Deforestation, Degradation, And Fragmentation Using Remote Sensing And GIS. Strengthening Monitoring, Assessment and Reporting on Sustainable Forest Management In Asia. Food and Agriculture Organization of the United Nations. *Working Paper No. 5*. Italy Rome.

[21] Puyravaud, J.P., 2003. Standardizing the calculation of the annual rate of deforestation. *Forest Ecology and Management* 177, 593–596.

[22] Jezching, T., Sticklen, J., Jain, A.K., 1991. Knowledge-based segmentation of Landsat images. *IEEE Transactions on Geoscience and Remote Sensing*: 222–232.

[23] Foody, G.M., 2002. Status of land cover classification accuracy assessment. *Remote Sensing of Environment* 80, 185–201.

[24] Hansen, M.C., Roy, D.P., Lindquist E., Adusei, B., Justice, C.O., Altstatt, A., 2008. A method for integrating MODIS and Landsat data for systematic monitoring of forest cover and change in the Congo Basin. *Remote Sensing of Environment* 112, 2495–2513.

[25] Corey, B., Lawrence, R., Montagne, C., Patten, D., 2006. Mapping wetlands and riparian areas using Landsat ETM+ imagery and decision-tree-based models. *Wetlands* 26, 465–474.

[26] Gong, P., Wang, J., Le, Y., Zhao, Y., Zhao, Y., 2013. Finer resolution observation and monitoring of global land cover: first mapping results with Landsat TM and ETM+ data. *International Journal of Remote Sensing* 34, 2607–2654.

[27] Hepinstall, J.A., Sader, S.A., 1997. Using Bayesian statistics, thematic mapper satellite imagery, and breeding bird survey data to model bird species probability of occurrence in Maine. *Photogrammetric Engineering and Remote Sensing* 63, 1231-1237.

[28] Sperduto, M.B., Congalton, R.G., 1996. Predicting rare orchid (small whorled Pogonia) habitat using GIS. *Photogrammetric Enginnering and Remote Sensing* 62, 1269-1279.

[29] Mbow, C., Chhin, S., Sambou, B., Skole, D., 2013. Potential of dendrochronology to assess annual rates of biomass productivity in savanna trees of West Africa. *Dendrochronologia* 31, 41–51.

[30] Guiro, I., Mbow, C., Baret, F., Diaw, A.T., 2012. Dynamique de l'occupation du sol de la forêt classée de Patako et de sa périphérie de 1972 à 2002. *Revue de Géographie du Laboratoire Leïdi* 10, 257.

[31] Curran, L.M., Trigg, S.N., McDonald, A.K., Astiani, D., Hardiono, Y.M., Siregar, P., Kasischke, E., 2004. Lowland forest loss in protected areas of Indonesian Borneo. *Science* 303, 1000–1003.

[32] Fischer, J., Lindenmayer, D.B., 2007. Landscape modification and habitat fragmentation: a synthesis. *Global Ecology and Biogeography* 16, 265–280.

[33] Dore, M.H.I., 2005. Climate change and changes in global precipitation patterns: What do we know? *Environment International 31*, 1167–1181.

[34] Sarr, B., 2011. Return of heavy downpours and floods in a context of changing climate. Climate change in the Sahel. A challenge for sustainable development. *AGRHYMET Monthly Bulletin (Special issue)*, pp. 9-11.

[35] Edagbo, D.E., Ajiboye, T.O., Borokini, T.I., Ighere, D.A., Alowonle, A., Michael, C., 2013. A study of the conservation status of *Citrus sinensis* as affected by the African Mistketoe, *Tapinanthus bangwensis* in Moor plantation, Ibadan, South-West, Nigeria. *International Journal Current Agricultural Sciences 3*, 5–9.

[36] Mon, M.S., Mizoue, N., Htun, N.Z., Kajisa, T., Yoshida, S., 2012. Factors affecting deforestation and forest degradation in selectively logged production forest: a case study in Myanmar. *Forest Ecology and Management* 267, 190-198.

[37] Madubansi, M., Shackleton, C.M., 2006. Changing energy profiles and consumption patterns following electrification in five rural villages, South Africa. *Energy Policy* 34, 4081–4092.

[38] Van der Plas R.J., Ali Abdel-Hamid, M. 2005. Can the woodfuel supply in sub-Saharan Africa be sustainable? The case of N'Djamena, Chad. *Energy Policy* 33, 297–306.

[39] Ndayambaje, J.D., Mohren, G.M.J., 2011. Fuelwood demand and supply in Rwanda and the role of Agroforestry. *Agroforestry Systems* 83, 303–320.

[40] Hiemstra-van der Horst, G., Hovorka A.J., 2009. Fuelwood: the other renewable energy source for Africa? *Biomass and Bioenergy* 33, 1605 –1616.

[41] Shi, M., 2008. Literature Review : Changes and feedbacks of land-use and land-cover under global change 15 pp.

[42] Lykke, A.M., 1998. Assessment of species composition change in savanna vegetation by means of woody plants size class distributions and local information. *Biodiversity and Conservation 7*, 1261–1275.

[43] Lambin, E.F., 1999. Monitoring forest degradation in tropical regions by remote sensing: some methodological issues. *Global Ecology and Biogeography* 8, 191–198.

[44] Frasera, E.D.G., Dougilla, A.J., Mabeeb, W.E., Reeda, M., McAlpine, P., 2006. Bottom up and Top down: Analysis of participatory processes for sustainability indicator identification as a pathway to community empowerment and sustainable environmental management. *Journal of Environmental Management* 78, 114–127.

[45] Kotwal, P.C., Kandari, L.S., Dugaya, D., 2008. Bioindicators in sustainable management of tropical forests in India. *African Journal of Plant Science* 2, 99–104.

[46] Tappan, G.G., Hadj, A., Wood, E.C., Lietzow, R.W., 2000. Use of argon, corona, and landsat imagery to assess 30 years of land resource changes in West-central Senegal. *Photogrammetric Engineering and Remote Sensing* 66, 727–736.

[47] Lloret, F., Estevan, H., Vayreda, J., Terradas, J., 2005. Fire regen- erative syndromes of forest woody species across fire and climatic gradients. *Oecologia 146*, 461–468.

[48] Sow, M., Mbow, C., Hély, C., Fensholt, R., Sambou, B., 2013. Estimation of herbaceous fuel moisture content using vegetation indices and land surface temperature from MODIS data. *Remote Sensing 5*, 2617–2638.

[49] Mbow, C., Goïta, K., Bénié, G.B., 2004. Spectral indices and fire behavior simulation for fire risk assessment in savanna ecosystems. *Remote Sensing of Environment* 91, 1–13.

[50] Uneke, C., 2009. Deforestation and malaria in sub-Saharan Africa: an overview. *The Internet Journal of Tropical Medicine* 6, 1.

[51] Diop, M., Sambou, B., Goudiaby, A., Guiro, I., Niang-Diop, F., 2011. Ressources végétales et préférences sociales en milieu rural sénégalais. *Bois et Forêts des Tropiques* 4, 57–68.

18

Importance of Integrated Water Resources Management in Flood and Flash Flood Management

Nino Tkhilava

Department of Environmental Policy and International Relations, Ministry of Environment and Natural Resources Protection, Tbilisi, Georgia

Email address:
Nino.tkhilava@moe.gov.ge

Abstract: Water resources management setup and practices in Georgia are relied on administrative approaches. Gradual introduction and implementation of the river basin based Integrated Water Resources Management (IWRM) will require corresponding legal and institutional adjustment as well as significant capacity building. The IWRM model for Georgia is intended to address prioritized concerns regarding water resource management, based on extensive consultations with stakeholders from a wide range of agencies and institutions related to water issues. Managing water efficiently, balancing conflicting uses of water resources, eliminating the results of catastrophic floods causing loss of human lives and damages to infrastructure and settlements and ensuring environmental stability of the river basins are critical issues which should be solved through ecosystem-based Integrated Water Resource Management approaches. The manuscript briefly reviews why the issue of shifting to IWRM is a priority for Georgia, what actions are needed to be implemented, and what are the sequences of these actions.

Keywords: Water, Resources, Management, Floods

1. Introduction

Water resources management setup and practices in Georgia are relied on administrative approaches. Gradual introduction and implementation of the river basin based Integrated Water Resources Management (IWRM) will require corresponding legal and institutional adjustment as well as significant capacity building. The IWRM model for Georgia is intended to address prioritized concerns regarding water resource management, based on extensive consultations with stakeholders from a wide range of agencies and institutions related to water resources management.

Managing water efficiently, balancing conflicting uses of water resources, eliminating the results of catastrophic floods causing loss of human lives and damages to infrastructure and settlements and ensuring environmental stability of the river basins are critical issues which should be solved through ecosystem-based Integrated Water Resource Management approaches.

The manuscript briefly reviews why the issue of shifting to IWRM is a priority and what actions are needed to be implemented and what are the sequences of these actions.

2. Prepositions for Improved Water Resources Management in Georgia

2.1. Freshwater Resources in Georgia

2.1.1. Freshwater Supply

Georgia has an abundance of fresh water resources - rivers, lakes, reservoirs, glaciers and springs. There are over 26,000 rivers with total channel length of some 60,000 km, 99.5% of which are rivers with a length less than 25 km. Due to the Georgia's mountainous nature, there are very few rivers with a long channel length or large basin size - only 273 rivers are more than 25 km long.

Rivers in Georgia belong to two hydrological basins: the Black Sea Basin in the west and the Caspian Sea Basin in the east of the country. The Black Sea Basin is significantly richer in water resources compared to the Caspian Basin.

Rivers in Georgia have a good potential for all types of

water uses, especially hydropower development and fisheries because of high channel slopes and high flow velocity. In terms of trans-boundary waters, Georgia is both an upstream and a downstream riparian, as water flows into Georgia from the Republic of Turkey and the Republic of Armenia, and flows out of Georgia – directly into the Black Sea, or indirectly into the Caspian Sea via Russian Federation or the Republic of Azerbaijan. Georgia is rich in ground water resources. The total volume of fresh ground water is estimated to be about 24 million cubic meters. Georgia also has more than 800 fresh water lakes.

Information on groundwater resources in Georgia is sparse, due to very limited hydro-geological monitoring data. Groundwater monitoring in Georgia stopped in the early 1990s, therefore estimates on groundwater reserves largely date back to the 1970-1980s. It is assumed that only about one third of the "known" groundwater resources have been surveyed in detail, and that total groundwater resources are estimated to be 18 BCM with 67% in Western Georgia and 33% in Eastern Georgia [1].

Figure 1. *Georgia's trans-boundary surface waters*

The mean annual precipitation value within Georgia is 1,338 mm, and the mean annual precipitation volume is 93.3 km3. The mean annual total flow of rivers in Georgia is about 61.45 km3, generated in upstream the Republic of Turkey and the Republic of Armenia (8.68 km3) as well as within the country (52.77 km3) (table 1).

Table 1. *National Water Balance in Georgia, km3*

Territory	Local Flow	Transit Flow	Total Water Resources	Outflow		
				Into the Black Sea	Into the neighboring countries	Total
West Georgia	41.52	6.48	48.0	48.0		48.0
East Georgia	11.25	2.20	13.45		13.45	13.45
Total	52.77	8.68	61.45	48.0	13.45	61.45

2.1.2. Water Abstraction & Consumption

While overall Georgia is rich in water resources, accounting for that wealth is critical as multiple uses expand and overall demand increases. In addition, water is redistributed unevenly throughout the country. While the Western Georgia is abundant in water, the Eastern regions frequently suffer from water shortages.

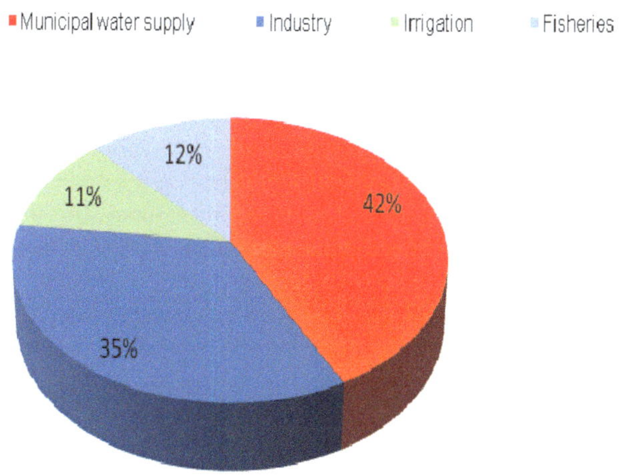

Figure 2. *Water consumption by sectors in Georgia, excluding hydropower generation*

Total water abstraction in Georgia in 2011 reached 22,767.4 MCM, of which 381.21 MCM from groundwater sources, or 1.67%. Direct water consumption in the country, excluding hydropower generation totaled 1044.72 MCM, divided over the sectors of irrigation – 114.91 MCM (11%), fisheries – 126.04 MCM (12%), industry – 357.92 MCM (35%), from which 291.95 MCM falls on thermal power generation, and municipal & drinking purposes – 439.15 MCM (42%), while an additional volume of 20557.9 MCM was non-consumptively used by the hydropower sector [1].

2.2. Changes in River Flows

2.2.1. Inter-Annual Variation

Rivers in Georgia show significant inter-annual and seasonal variations in river flow, typically depending on variations in climatic conditions from year to year and season to season. Additionally, the expansion of human activities since the mid-20th century have impacted also on water use and river discharge in Georgia.

2.2.2. Seasonal Variation

The total amount of water being seasonally discharged through the country's rivers depends on the amount of snow in winter and rain in other seasons. Typically, the largest portion of the total annual discharge volume passes the rivers in spring, during the snow melting months of March through June/July. With envisioned climate change, typically an increase in the seasonal variation of river flow is expected, related to an increase/decrease of overall precipitation or an increase in the frequency and intensity of large storms, which may be associated with specific seasons.

2.2.3. Climate Change

There are two main climatic zones in Georgia, roughly separating the eastern and western parts of the country along the Likhi Ridge. The Black Sea coastal zone and much of Western Georgia are located within the humid subtropical zone with mild winters, hot summers and abundant precipitation. The prevalent climate in Eastern Georgia is drier, ranging from arid sub-tropical in the lowlands to alpine in the mountainous regions. Within these main zones climate conditions still are very diverse, largely conditioned by the heterogeneous topography. The Greater Caucasus Mountain Range plays an important role in moderating Georgia's climate, protecting the country against the penetration of colder air masses from the north. The Lesser Caucasus Mountains partially protect the country from the influence of dry and hot air masses from the south.

According to the climate forecast models climate change is expected to have significant impacts on the water resources of Georgia in terms of increased temperatures, reduced rainfall and reduced river flow in the coming few decades. In addition, climate change is expected to cause more intense and frequent extreme weather events including flooding/flashfloods, heat waves and droughts. To respond to these challenges, it is recommended to develop flood risk maps and flood hazard maps and develop flood management plans on their basis as it is outlined in the EU Directive on Assessment and Management of Flood Risks. It is also essential to develop early warning systems and conduct sufficient public awareness-raising to increase public's preparedness for emergency situations.

A complex mountainous topography makes the country more prone to the hydro-geomorphological processes and climatic hazards. As such, Georgia is vulnerable to natural hazards including floods, flash floods, earthquakes, droughts, landslides, avalanches, and mud flows. Catastrophic events that have an annual probability of occurrence of 0.5 percent threaten an economic loss for Georgia that exceeds 20 percent of the country's GDP. Floods, including flash floods are the catastrophic events of such category of high probability [2, 3].

2.3. Extrime Events

From extreme climate related events floods and flash floods are frequently observed in Georgia. The 1987 flood on the Lower Rioni River caused inundation of a vast territory including human settlements for a long period of time, resulting in some casualties and huge damage to the local agricultural sector. In 1997, the flood events in the Tbilisi-Gori-Kvemo Kartli region killed 7 people, affected 500 others and incurred a reported economic loss of $US 29.5 million. In June 2005, the flood in the Mtskheta-Tianeti region killed 2 people, affected 51 others and caused an economic loss of $US 2 million. On 13 May 2012, the flood in Tbilisi killed 5 people and caused several million dollars in damages. In the period of 1995-2010 in total 164 flood events occurred throughout Georgia with the total impact of $US 272.3 million [4].

While Georgia emerges as the most vulnerable in the broader region of Eastern Europe and Central Asia for the period of 1980–2000, measured by the mortality rate among those exposed to floods, since 2004 considerable improvements in the institutional landscape to manage flood risks and support timely recovery have occurred. However, much remains to be done and levels of vulnerability and exposure are still high [5].

Some traditional flood risk management measures have a negative impact on the quality and quantity of waters. Examples can be building of dams or dikes which change the river flow by reducing water for related ecosystems in the area of which accentuate problems in dry seasons by alerting the natural flow of the river and may even increase flooding downstream [6].

Increased population pressure and enhanced economic activities in floodplains, such as the construction of buildings and infrastructure, further increase the risk of flooding. Floodplains provide excellent, technically easy livelihood opportunities in many cases. In developing countries with primarily agricultural economies, food security is synonymous with livelihood security. Floodplains contribute substantially to the food production that provides nutrition for the people of these countries [7].

The urgent need for increasing flood security is also illustrated by the fact that the majority of the members of Georgian society that are vulnerable to floods are women, children and the poor. Internally Displaced People (IDPs), a large portion of who inhabit western regions of the country are among the most socially disadvantaged groups in the country.

2.4. Improvement of Water Resources Management

2.4.1. EU – Georgia Association Agreement

The European Union (EU) - Georgia relations started in 1992 after Georgia regained its sovereignty in the wake of the break-up of the Soviet Union following the recognition of Georgia's independence by the European Union. Relationships with EU upgraded after Georgia became a part of the Eastern Partnership initiative since its launch in 2009. Under this initiative, closer political association and economic integration with the EU is being promoted. The EU and Georgia completed the negotiation of the Association Agreement between the European Union and the European Atomic Energy Community and their Member States, of the one part, and Georgia, of the other part (AA), including the Deep and Comprehensive Free Trade Area (DCFTA) element, in July 2013 and initialed the Association Agreement at the Eastern Partnership Vilnius Summit in November 2013. Ultimately, both sides signed the Agreement on 27th June 2014.

The Association Agreement creates new legal framework for co-operation with the EU. It is a comprehensive and politically stronger document with the potential to lead co-operation between Georgia and the European Union towards qualitatively new and higher levels in many important sectors. Besides political and economic association the AA covers a broad range of topics, including regulatory approximation in the field of environment, climate change and natural resources protection to some extent.

The environmental parts of the Association Agreement envisage a progressive opportunity for gradual harmonization with EU environmental rules and standards with the objective to protect environment and ecosystem and encourage effective environmental governance including water resources management. The Association Agreement includes specific provisions of six EU Directives concerning water quality and resource management, including marine environmental policy [8].

2.4.2. Decentralizing Water Management to a Basin Level

In order to comply with the requirements of the Association Agreement Georgia faces a number of structural environmental challenges such as improving environmental legislation, increasing awareness of stakeholders, improving monitoring, inspection and enforcement systems, and strengthening the knowledge for adequate policy-making.

Although some progress has been made in Georgia in water sector development, there are still significant difficulties in terms of legal and institutional frameworks and law enforcement. The Government of Georgia is currently working towards harmonization of its institutional settings and legislation with those of the European Union, including water resources management in line with the requirements of the United Nations Economic Comission for Europe (UNECE) Convention on the Protection and Use of Transboundary Watercourses and International Lakes, which also includes improvement of water resources management.

The purpose of adopting IWRM and its principles is to improve the way water resources are managed in a situation where improved management is necessary, conditioned by increasing pressures on the resources from human activities and climate change. According to the Global Water Partnership, "Integrated Water Resources Management is a process which promotes the coordinated development and management of water, land and related resources, in order to maximize the resultant economic and social welfare in an equitable manner without compromising the sustainability of vital ecosystems"[9].

Currently, water resources management is strongly centralized in Georgia with the Ministry of Environment and Natural Resources Protection being a key entity and the government is striving to shift to IWRM according to the guidelines of the UNECE Convention on the Protection and Use of Transboundary Watercourses and International Lakes and the obligations under the Association Agreement.

Main concerns relate to the inefficient use of water and to its environmental and socio-economic consequences: low river flows, water shortages, salinisation of freshwater bodies in coastal areas, human health problems, loss of wetlands, desertification and reduced food production. The main challenge of modern management is to ensure a sustainable management of water resources, avoiding overexploitation and degradation, so as to maintain an adequate supply of freshwater of suitable quality for human use and to support aquatic and other ecosystems. This implies reducing losses, using more efficient technologies and increase recycling, and applying an integrated approach to the management of freshwater resources by river basin. It further requires applying the user pays principle to all types of uses [10].

2.4.3. Transition Process

The transition period from one management system to another in the proper resources management is measured in multiples of years, and requires many legal and institutional changes. As the first step it is necessary to create a platform for establishing water management at a basin level. River basin is a unit for implementing IWRM and river basin authorities should be responsible for the development and implementation of river basin management plans. The Ministry of Environment and Natural Resources Protection of Georgia can have a lead role in guiding and supporting river basin authorities through this process, especially at the initial stages. It is also necessary to create a river basin councils - a platform for involvement and participation of all water user stakeholders in the river basin management.

Climate change also expected to aggravate extreme natural events such as floods. To respond to these challenges, it is recommended to develop flood risk maps and flood hazard maps and develop flood management plans on their basis as it is outlined in the EU Directive on Assessment and Management of Flood Risks. It is also essential to develop early warning systems and conduct sufficient public awareness-raising in terms of preparedness for emergency situations.

In the first phase will be initiated data collection and other activities for preparation to development of flood risk maps and flood hazard maps; climate parameters will be collected and analyzed; early warning systems will be developed; and flood risk maps and flood hazard maps will be developed based on the previous preparation activities. In the second phase modeling (application of selected models) for making a prognosis on impacts of climate change on water resources will be done and climate change adaptation plans will be developed based on data obtained in the first phase and on the basis of the forecasts made based on the models. In addition, flood risk management plans will be developed based on the flood risk maps and flood hazard maps. Through the both phases awareness raising activities will be implemented to increase public's awareness on climate change adaptation as well as overall preparedness for emergency situations.

2.4.4. Development of Flood Risk Maps and Flood Hazard Maps

Development of flood risk maps and flood hazard maps requires the undertaking of preliminary flood risk assessment in each river basin management unit. This involves gathering of information on the floods that have occurred in the past, the likelihood of future floods and estimated consequences of these floods. Based on this information river basins are categorized according to flood risks. The next step will be development of flood risk maps and flood hazard maps.

Flood risk maps and flood hazard maps identify all areas at risk of flooding and indicate the probability of flooding (high, medium or low) for each of those areas. Flood hazard maps also show the extent of possible flooding, while flood risk maps indicate the potential damages that could occur under each flood scenario [11].

2.4.5. Development of Flood Risk Management Plans

Based on flood risk maps and flood hazard maps, flood risk management plans should be developed. Floor risk management plans will include measure focusing, reducing the probability of flooding and the potential consequences of flooding, covering all aspects for floods management: prevention, protection and preparedness.

2.4.6. Obtain Climate Parameters for the Climate Change Forecast Model

The use of most updated climate change models will enable increased accuracy in predictions. Predicting climate change impacts on water resources and water availability will be critical for long-term planning and sustainable development. The main activities for this issue are:

- Digitalize historical hydro-meteorological data
- Select climate change model(s);
- Select model(s) for climate change impact prognosis;
- Obtain climate parameters for the climate change forecast model;
- Apply the climate change impact forecast model(s) to different scenarios;
- Assess the results of modeling for different climate change scenarios in terms of impacts on water resources.

2.4.7. Increase the Public Awareness and Overall Preparedness for Emergency Situations

Extensive public awareness activities will be needed in order to increase overall awareness of the public on climate change and adaptation as well as on preparedness for emergency situations. This will help to successful implementation of the climate change adaptation measures and also will ensure timely and effective implementation of emergency measures when needed.

3. Results

River basin management should be implemented based on river basin management plans. River basin management plans are developed by river basin authorities according to the requirements of the EU Water Framework Directive (WFD). Development of river basin management plans requires extensive preparation activities including studying physical, chemical and hydro-morphological characteristics of river basins, identifying impacts, conducting economic analysis of water use, developing systems for water quality status assessment, developing program of measures etc. Detailed guideline for the development of river basin management plans is provided by the EU WFD.

As for water policy and law making, Georgia is required to identify river basin districts and analyze their characteristics, establish programmes for monitoring water quality and prepare river basin management plans in consultation with public:

- "Identification of River basin districts and establishment of administrative arrangements for international rivers, lakes and coastal waters";
- Preparation of river basin management plans, consultations with the public & publication of these plans" [12].

For assessment and management of flood risks, Georgia is required to undertake flood assessments, prepare flood hazard and risk maps and establish flood risk management plans.

The Directive on the Assessment and Management of Flood Risks complements the Water Framework Directive. Its objectives are to reduce the adverse consequences for human health, the environment, cultural heritage and economic activity associated with all types of floods and assessment and management of flood risks and it requires that prevention and management measures are organized on a basin level.

The Directive regulates: carrying out an assessment of potential risks, undertaking preliminary flood assessment; having a basis for priority setting and decision-makings regarding flood risk management; maps are supposed to show the potential adverse consequences associated with different flood scenarios, preparation of flood hazards maps and flood risks maps; implementation of measures for management of flood risks including costs and benefits; establishment of flood risk management plans:

- "Undertaking preliminary flood assessment";
- "Preparation of floods Hazards map and flood risk maps";
- "Establishment of Flood Risk management Plans" [13].

An Integrated Flood Management (IFM) approach is also an essential component of Integrated Water Resources Management and such an approach can help to balance flood risk management and development needs. Integrated Flood Management is a process, promoting an integrated – rather than fragmented – approach to flood management. It integrates land and water resources development in a river basin, within the context of IWRM, and aims at maximizing the net benefits from the use of floodplains and minimizing loss of life from flooding [6].

At different times floods caused inundation of vast territories, including human settlements resulting in huge damage to the local economy and agriculture and causing even casualties. It is predicted, that climate change will cause more intense and frequent extreme weather conditions, including flooding. Preparation for these events will enable flood risks to be assessed and flood damages to be minimized/prevented. EU Directive on Assessment and Management of Flood Risks provides a common framework and detailed guidelines for assessing and reducing the risk that floods pose to human health, economy and the environment. The EU Floods Directive is closely linked with the EU Water Framework Directive, so as it requires that prevention and management measures are organized on a basin level.

4. Conclusions

As an integral part of Integrated Water Resources Management (IWRM), Integrated Flood Management (IFM) faces similar challenges. The effective implementation of both IWRM and IFM requires an enabling environment in terms of policy and legislation; clear institutional roles and functions; and management instruments for effective regulation, monitoring and enforcement.

References

[1] IWRM plan for Georgia, GEF/UNDP project "Reducing transboundary degradation in the Kura-Aras river baisin", 2014, unpublished.

[2] Adapting to climate change in Europe and Central Asia, World Bank, 2009, p.89.

[3] D. Kereselidze, L. Matchavariani, V. Trapaidze, L. Lagidze, G. Dokhnadze, N. Tkhilava, N. Gokhelashvili. Evaluation and Management of the Risk of Flooding River Bank. Engineering Geology for Society and Territory. Vol.3: River Basin, Reservoir Sedimentation and Water Recourses. SPRINGER International Publishing Switzerland, 2015, 463-469.

[4] EM-DAT, The International Disaster Database, The Centre for Research on the Epidemiology of Disasters – CRED http://www.emdat.be/result-country-profile.

[5] Developing climate resilient flood and flash flood management practices to protect vulnerable communities of Georgia, Adaptation Fund/UNDP project, 2012, unpublished.

[6] Towords better environmental options for flood risk management, European Comission, unpublished.

[7] Integrated flood management, concept paper, world Meteorological Organization, WMO-No. 1047,

[8] Association Agreement between the European Union and the European Atomic Energy Community and their Member States, of the one part, and Georgia, of the other part.

[9] Integrated water resources management, Global Water Partnership Technical Advisory Committee, 2000, Stockholm. Technical Background Paper No. 4.

[10] Key environmental Indicators, OECD, 2008, 20-23.

[11] J. Schanze, Flood risk management - a basic framework - hazards, vulnerability and mitigation measures. Springer, 2006, pp. 1-20.

[12] Directive 2000/60/EC of the European Parliament and of the Council, WFD, Articles 3 (1) – 3(7), 13, 14.

[13] Directive 2007/60/EC of the European Parliament and of the Council on the assessment and management of flood risks, Articles 4 ,5, 6, 7.

Levels, Distribution and Sources of Polycyclic Aromatic Hydrocarbons in Surface Water in the Lower Reach of Qua Iboe River Estuary, Nigeria

Eno Anietie Moses*, Bassey Anie Etuk, Essien Daniel Udosen

Department of Chemistry, University of Uyo, Uyo, Nigeria

Email address:
enomoses27@yahoo.com (E. A. Moses), basseyetuk2@gmail.com (B. A. Etuk), essienandong@yahoo.com (E. D. Udosen)

Abstract: The lower reach of Qua Iboe River estuary houses one of the largest crude oil production facility in the Niger Delta region of Nigeria. Frequent oil spills has led to severe deterioration of the water system, This study determined 16 polycyclic hydrocarbons (PAHs) in surface waters of sites across the estuary in the wet and dry seasons. Gas chromatography with flame ionization detector (GC / FID) was used to measure PAHs in water samples following extraction. The sum of PAHs ranged from7.827E-04 for the wet season to 1.500E – 02 mg/l for the dry season with significantly higher ($P<0.05$) PAH concentration in the wet season. Total PAHs for the water sample at all sampling sites were above the permissible limit. The composition pattern of PAHs were dominated by high molecular weight PAHs, constituting 4-6 ring compounds. Diagnostic ratios of Fluoranthene/Fluoranthene + Pyrene, Anthracene/Anthracene + Phenanthrene and low molecular weight-PAH to high molecular weight PAH were calculated to evaluate possible sources of PAH compounds These ratios reflected predominantly pyrogenic sources of PAH to the water body originating from gas flaring from a petrochemical facility located close to the sampled stations. Cluster analysis was used to assess similarities between individual PAHs in water from QIRE for the dry and wet seasons. It revealed three primary PAH clusters in both seasons indicating different levels of anthropogenic activities in the cluster areas. There is a need for appropriate regulatory legislation on the control, treatment and discharge of oilfield effluents into QIRE and its adjourning creeks.

Keywords: Polycyclic Aromatic Hydrocarbon, Water Pollution, Qua Iboe River Estuary, Gas Flaring, Oil Pollution

1. Introduction

Water pollution occurs when the quality of water is changed in a way which affects the organisms that depend on it, or its suitability for recreational and domestic purposes [1]. Pollution of water is enhanced by industrialization, urbanization and technological development as these processes result in increased discharge of pollutants to receiving water, causing undesirable effects to the aquatic environment. Some contaminants of water include trace metals, pesticides and polycyclic aromatic hydrocarbons. Polycyclic aromatic hydrocarbons (PAHs), are formed by incomplete combustion of organic matter containing fuels such as oil, wood, coal, diesel, fat, garbage and incense [2]. PAHs occur naturally in coal, crude oil, and gasoline and are present in products made from fossil fuels such as asphalt, creosote, coal tar. Natural sources of PAHs include oil seeps, volcanic eruption, forest fires, decaying organic matter while anthropogenic sources include emission from petroleum refineries, gas flaring, power generation from fossil fuels, atmospheric fall out, inadvertent oil spill [3, 4]. PAHs are classified as low molecular weight (LMW), which constitute the 2 to 3 ringed compounds, and high molecular weight (HMW) of 4 to 6 rings. HMW PAHs are of significant concern because of their carcinogenicity, mutagenic properties and bioaccumulation in organic tissues due to their lipophilic character. Prolonged exposure to low levels of PAHs leads to adverse health effects due to their toxicity, carcinogenicity and mutagenicity [5, 6]. rude oil provides 20% of Nigeria's gross domestic product, 65% of its budgetary revenues and 95 % of the country's export earnings [7]. However, the exploration, refining and exploitation of crude oil have resulted in environmental degradation which affects communities near and far away from offshore and onshore installations of oil companies. Usually oil exploration activities have resulted in the release of large amounts of hydrocarbons into the terrestrial and aquatic environment (8). This has led to disruption of socio-economic activities, water scarcity, poor esthetics and deterioration of water quality in rivers of the Niger Delta [9]. Qua Iboe terminal (QIT) located at the Qua Iboe River estuary, is one of the largest crude oil production facility in the Niger Delta region of Nigeria [10]. Activities at the terminal has facilitated the establishments of smaller oil servicing companies at the estuary, resulting in urban migration and resultant population growth. Consequently, there is an increase in the volume of industrial, domestic and municipal waste generated and discharged into the water system. Studies on of the QIRE and associated creeks were basically on the concentrations of trace metals and PAHs. However no data is available on the modeling of PAHs measured on the water system. This study therefore seeks to determine the contamination status, distribution and source apportionment of sixteen priority PAHs in the QIRE system.

2. Materials and Methods

2.1. The Study Area and Site Description

Qua Iboe river is one of the major rivers in the Niger Delta region of Nigeria. It originates from a waterfall in Umuahia hills and flows southwards across Umudike into Akwa Ibom state. Five sampling sites are located at the lower reach of Qua Iboe River in Ibeno local government area close to a petrochemical effluent treatment and discharge plant while the control site is located at Ekpene Ukpa in Etinan local government area of AkwaIbom State, about 27km from the examined sites and is free from oil exploration and production activities. Ibeno lies on the eastern side of Qua Iboe River about 3km from the river and is one of the largest fishing settlements in the Nigerian coast. The global positioning system (GPS) coordinates of the sampling sites are: Okoroutip (4°55'5"N - 7°54'47"E), Ukpenekang (4°27'2"N - 8° 3'5"E), Iwochang (4°36'50"N – 7°50'03"E), Douglas creek (4°30'55"N – 8°07'E), Stubb creek (4°34'41"N – 7°59'47"E), EkpeneUkpa(4°47'90"N – 7°50'03"E). Figure 1 is a map of the study area indicating the sampling sites.

2.2. Samples and Sampling

Sample containers and glassware for PAH analyses were rinsed with hexane and dichloromethane to remove adhering polar and non-polar compounds [11]. Surface water samples for the analysis of PAHs were collected using an amber-coloured borosilicate 1litre capacity glass bottles with teflon-lined caps to prevent photochemical degradation and 3 drops of 1M HCl was added to prevent bacterial degradation of the sample respectively [12]. Sampling was conducted monthly from November, 2013 to October, 2014. Sub-samples from five points per sampling site were homogenized to form a composite sample. A total of three hundred and sixty subsamples and 72 composite samples were collected.

Extraction and Clean up

The extraction of the samples was carried out by liquid-liquid extraction method protocol described by [13] and the clean-up was carried out in a column chromatography packed with Na_2SO_4 and silica gel.

2.3. Determination of PAHs Concentration

The concentration of PAHs was determined using standard protocol described by [12]. The sample was automatically detected as it emerged from the column by a flame ionization detector (FID) by measuring the retention time. The identification of the PAH compounds was achieved using Chemstation software and was based on matching their retention time with calibrated PAH standards while quantification was obtained from the corresponding areas of the respective chromatograms. Procedural blanks and solvent blanks were analysed and quantified with no PAHs found in these blanks. Prior to use, the GC was calibrated using a five point calibration curve established using dichloromethane-based standards (Accustandard PAH mix, 1000μg/ml in CH_2Cl_2). The coefficient of determination values (R^2) were greater than 0.87. Prior to extraction, four surrogate standards were added to the sample to take care of unusual matrix effect. These internal surrogate standards were acenaphthene-d_{10}, phenanthrene-d10, chrysene-d12 and perylene-d12 from Smart solutions, USA.

Chromatographic Conditions

The gas chromatograph was Agilent Hewlett Packard 5890 series II, coupled with flame ionization detector (FID) powered with HP Chemstation Rev. A. 0901 (10206) software. To identify and quantify PAH components, the following GC operating conditions was utilized:

Detector: Hydrogen at 35 ml/min; air at 250ml/mins and nitrogen at 30 ml/min.

Oven: Initial temperature 65°C; final temperature; 325°C, Run time: 30 minute

Inlet: Splitless injection was adopted using a rubber septum and the volume injected was 1μl. The inlet temperature was 275°C, with a pressure of 14.8 Psi and total flow of 65.4ml / min.

Column: The column was lined with 1,3-dimethyl polysiloxane with capillary HPS type of 30m length, 0.32mm wide bore diameter, 0.25μm film diameter.

2.4. Identification of PAH Sources

Diagnostic ratios were used to identify the possible sources of PAH in the water bodies. The following ratios were used as source indicators: Fluoranthene/Fluoranthene + Pyrene ratio (Flue / Flue + Pyr), Anthracene/Anthracene + Phenanthrene ratio (Ant / Ant + Phe) and LMW-PAH to HMW-PAH ratio.

2.5. Statistical Analysis

All values were expressed as mean of six determinations ± standard deviation. Student's t-test was used to compare between the mean of total PAH values in both the dry and wet seasons and a P<0.05 was considered statistically significant. Pearson correlation coefficient was used to examine the relationship between PAH pairs in water and cluster analysis was used to assess common pollution sources between sampling sites. Statistical analyses were performed using SPSS statistics 17.0 windows.

Figure 1. It is a map of the study area indicating the sampling sites.

3. Results and Discussion

Concentration of the 16 priority PAHs in water from QIRE in both dry and wet seasons are shown in Tables 1, 2, 3, and 4. The mean total PAHs ranged from 7.827E-04 for the wet season to 1.500E – 02 mg/l for the dry season. Results revealed significantly higher (P<0.05) PAH concentration in the wet season, compared to the dry season. In addition mean PAH concentrations were significantly higher (P < 0.05) at the sampled sites than the control. In the dry and wet seasons, the total PAHs for the water sample at all the sampling sites were above the permissible limit except at the control site. The high levels of PAHs in water at the examined sites may be due to atmospheric fall-out from continuous gas flaring and waste water discharge from a petrochemical plant situated close to the two creeks. Studies by [14] showed that the pathways for PAHs to enter surface water include atmospheric fallout, urban-run off, municipal effluents, sewage discharge, industrial effluents and oil spillage while High levels of B(k)F and Ind (1,2,3-cd)P in water from the Niger Delta were linked to run-off from petroleum by-products (4). Higher PAH values recorded in the wet season compared to the dry season may be due to flooding, possible pipeline vandalisation and oil spillage that occurred in the month of July during the sampling period [15]. The level of individual PAHs from the present study were lower than levels reported for Cankiri, Turkey [16], but comparable to levels in other areas of the Niger Delta [4, 17].The occurrence of low molecular weight PAHs (LMW- PAHs) ranged from 0.49% to 22.7% in the dry season and 0.07%to 3.67% in the wet season. The occurrence of High molecular weight-PAHs (HMW-PAHs) ranged from 77.1% to 99.5% in the dry season and 98.6% to 99.9% in the wet season. Low percentage of low molecular weight PAHs (LMW-PAHs) compared with high molecular weight PAHs (HMW-PAH) was observed in this study. The low percentage of LMW PAHs may be due to their possible volatilization to the atmospheredue to their relatively high vapour pressure. Also, 2-3 ring PAHs are more easily biodegraded and volatilized compared with 4-6 ring PAHs. Moreover, the higher levels of HMW-PAHs in this study may be due to their attachment on soot produced from continuous gas flaring from an oil exploration industry which eventually deposits into the water [18]. The Findings from this study are different from the findings of other studies with reports of dominance of 2-3 rings PAHs [19, 20]. The differences observed from other studies were attributed to differences in anthropogenic inputs of PAHs to the water bodies. For instance, PAHs in estuarine water from Wenzou, China were sourced from leather industries.

Table 1. *Concentration of low molecular weight polycyclic aromatic hydrocarbon with fluoranthene and pyrene in water from QIRE in the dry Season.*

PAH mixture		Okoroutip	Ukpenekang	Iwochang	Douglas	Stubb	EkpeneUkpa
Naphthalene	Mean	1.639E-04 ± 4.327E-05	1.025E-04 ± 1.522E-05	2.213E-04 ± 2.828-06	2.055E-04 ± 5.03-05	6.730E-05 ± 2.081-06	8.9135E-05 ± 1.5044-06
	Range	1.333E-04 to 1.945E-04	9.180E-05 to 1.130E-04	1.326E-04 to 3.100E-04	1.287E-04 to 2.824E-04	5.604E-05 to 7.803E-05	8.642E-05 to 9.85E-05
Acenaphthylene	Mean	1.779E-06 ± 3.7260E-07	5.3175E-06 ± 2.560E-07	2.006E-06 ± 2.000-08	4.632E-06 ± 4.163-07	1.868E-05 ± 2.516-06	2.337E-06 ± 1.958-07
	Range	1.373E-06 to 2.105E-06	4.134E-06 to 6.501E-06	1.962E-06 to 2.129E-06	3.325E-06 to 5.939E-06	4.715E-06 to 4.257E-05	4.342E-06 to 3.335E-07
Acenaphthene	Mean	6.3457E-06 ± 4.979E-07	3.2945E-06 ± 5.820E-08	4.6215E-07 ± 3.511-09	2.785E-06 ± 2.00-07	3.2561E-06 ± 2.413-07	2.552E-06 ± 3.964-08
	Range	2.186E-06 to 1.323E-05	2.359E-06 to 4.230E-06	2.253E-07 to 6.990E-07	N.D. to 1.114E-06	6.950E-07 to 7.257E-06	3.250E-06 to 4.779E-06
Fluorene	Mean	6.8507E-06 ± 1.7039E-07	6.372E-05 ± 1.12E-06	5.589E-05 ± 4.041E-06	1.2507E-05 ± 2.0816-07	1.5902E-04 ± 3.1-06	6.823E-05 ± 2.431-06
	Range	1.350E-06 to 1.260E-05	1.350E-06 to 1.836E-05	3.356E-06 to 1.636E-04	4.487E-07 to 3.375E-05	4.406E-08 to 3.187E-04	4.474E-05 to 1.323E-04
Phenanthrene	Mean	3.774E-05 ± 6.185-06	3.006E-05 ± 5.781-06	5.872E-05 ± 1.634-06	6.076E-05 ± 4.505-07	2.608E-05 ± 2.00E-07	9.3975E-06 ± 2.081E-08
	Range	1.066E-06 to 1.388E-04	2.741E-06 to 6.452E-05	1.548E-06 to 1.159E-04	1.002E-05 to 1.115E-04	1.391E-06 to 7.361E-05	7.2225E-06 to 1.157E-05
Anthracene	Mean	3.714E-05 ± 8.976E-07	2.618E-05 ± 7.919E-08	8.5325E-06 ± 2.641E-07	9.042E-06 ± 2.645E-08	3.0550E-08 ± 3.0550E-08	3.655E-06 ± 4.321E-07
	Range	3.618E-06 to 4.176E-06	2.562E-06 to 2.674E-06	4.145E-06 to 1.292E-05	3.325E-06 to 1.476E-05	5.981E-06 to 4.500E-06	6.035E-08 to 1.057E-05
Fluoranthene	Mean	2.250E-04 ± 8485E-05	3.902E-04 ± 1.760E-05	1.084E-05 ± 1.527E-06	1.969E-04 ± 4.932E-06	9.939E-05 ± 2.00E-06	1.443E-04 ± 3.911E-06
	Range	2.904E-06 to 6.260E-04	6.219E-06 to 6.290E-04	3.456E-06 to 3.200E-04	7.325E-05 to 3.206E-04	2.821E-06 to 2.911E-04	6.325E-07 to 2.88E-04
Pyrene	Mean	1.525E-05 ± 7.767E-07	1.904E-06 ± 7.735E-08	7.571E-04 ± 5.131E-05	8.758E-04 ± 3.963E-05	8.784E-04 ± 2.532E-05	8.221E-05 ± 2.516E-07
	Range	6.474E-06 to 2.676E-05	1.119E-06 to 3.265E-06	3.224E-06 to 2.216E-03	1.199E-04 to 2.155E-03	3.855E-06 to 2.578E-03	7.325E-07 to 1.637E-04

Results are presented as Mean±SD

Table 2. *Concentration of high molecular weight polycyclic aromatic hydrocarbon in water from QIRE in the dry Season.*

PAH mixture		Okoroutip	Ukpenekang	Iwochang	Douglas	Stubb	EkpeneUkpa
B(a)A	Mean	6.454E-06 ± 6.807E-07	8.377E-05± 2.347E-06	1.333E-05± 3.055E-07	8.580E-06± 3.577E-07	5.603E-06± 2.594E-07	1.8205E-06± 3.2145E-08
	Range	5.340E-07 to 1.291E-05	8.211E-05 to 8.543E-05	6.336E-06 to 2.033E-05	7.325E-06 to 9.836E-06	1.286E-06 to 1.320E-05	2.281E-07 to 3.413E-06
Chysene	Mean	3.445E-05± 4.582E-06	1.8293E-05± 1.7785E-07	5.790E-05± 2.902E-06	3.260E-05± 2.503E-06	1.824E-05± 3.214E-07	7.8973E-06± 2.523E-07
	Range	2.322E-06 to 1.253E-04	1.722E-05 to 2.041E-05	1.008E-05 to 1.091E-04	6.335E-06 to 8.199E-05	3.336E-06 to 4.767E-05	4.346E-07 to 1.536E-05

PAH mixture		Okoroutip	Ukpenekang	Iwochang	Douglas	Stubb	EkpeneUkpa
B(b)F	Mean	4.589E-03± 3.732E-04	1.368E-03± 4.46E-04	1.975E-02± 2.081E-03	3.267E-02± 1.552E-03	1.111E-02± 2.289E-04	3.0118E-04± 4.509E-06
	Range	2.187E-04 to 1.565E-02	1.699E-04 to 1.882E-03	3.652E-04 to 5.462E-02	1.429E-04 to 3.542E-02	4.347E-05 to 2.614E-02	7.001E-05 to 4.236E-04
B(k)F	Mean	2.822E-05± 3.214E-07	4.303E-05± 8.695E-07	3.011E-04± 1.518E-05	2.503E-05± 2.383E-06	4.552E-05± 052E-04	3.5312E-06± 4.875E-08
	Range	3.461E-07 to 1.009E-04	3.962E-06 to 1.199E-04	2.362E-06 to 3652E-04	2.932E-07 to 7.146E-05	3.652E-05 to 1.247E-04	3.645E-07 to 6.698E-06
B(a)P	Mean	1.9488E-05± 4.061E-06	2.945E-05± 1.603E-06	1.376E-05± 2.013E-07	2.354E-05± 2.511E-07	2.342E-05± 5.312E-06	4.1585E-06± 2.865E-07
	Range	4.532E-06 to 3.250E-04	3.365E-06 to 5.261E-05	8.669E-06 to 2.001E-05	4.733E-07 to 6.914E-05	7.965E-06 to 4.263E-05	1.890E-05 to 6.421E-06
Lnd (1,2,3,cd) P	Mean	3.931E-03± 4.061E-05	2.949E-05± 1.603E-06	1.1634E-02± 7.29E-03	2.380E-03± 3.04E-05	1.9921E-03± 1.111E-04	4.427E-05± 8.821E-07
	Range	2.077E-03 to 7.256E-03	2.350E-04 to 1.967E-02	2.256E-05 to 1.237E-03	3.117E-04 to 3.506E-03	8.555E-05 to 2.567E-03	8.321E-07 to 8.772E-05
D(ah)A	Mean	4.188E-06± 5.519E-07	2.830E-03± 3.785E-05	4.3416E-05± 2.081E-07	3.922E-05± 1.0094E-06	5.6102E-05± 1.6258E-07	5.4545E-06± 8.5040E-07
	Range	3.267E-06 to 9.735E-06	2.826E-05 to 8.401E-03	1.901E-05 to 7.161E-05	5.245E-07 to 4.352E-05	6.474E-07 to 1.143E-04	3.583E-06 to 7.326E-06
B(ghl)P	Mean	2.238E-05± 2.081E-07	1.376E-05± 6.021E-04	4.687E-05± 1.907E-06	1.471E-03± 4.006E-05	5.327E-05± 3.399E-06	1.724E-04± 8.934E-05
	Range	1.862E-06 to 4.690E-05	7.677E-05 to 1.307E-04	4.068E-06 to 9.630E-05	3.815E-07 to 4.076E-03	6.6638E-07 to 8.603E-05	1.724E-04 to 8.934E-05

Results are presented as Mean±SD

Note: B(a)A = Benzo(a)anthracene, B(b)F = Benzo(b)Fluoranthene, B(k)F = Benzo(k)Fluoranthene, B(a)P = Benzo (a) Pyrene, Ind (1,2,3-cd)P = Indeno(1,2,3-cd)Pyrene, D(ah)A = Dibenzo(a,h) Anthracene, B(ghi)P = Benzo(ghi)Perylene

Table 3. *Concentration of low molecular weight polycyclic aromatic hydrocarbon with fluoranthene and pyrene in water from QIRE in the wet Season.*

PAH mixture		Okoroutip	Ukpenekang	Iwochang	Douglas	Stubb	EkpeneUkpa
Naphthalene	Mean	6.116E-06 ± 3.0E-07	6.5417E-06 ± 1.123E-07	2.211E-05 ± 3.324E-06	1.849E-04 ± 3.6055E-06	9.987E-06 ± 6.245E-08	1.581E-07 ± 2.2E-09
	Range	2.252E-06 to 1.236E-05	2.532E-07 to 1.806E-05	1.6204E-06 to 0.681E-04	2.416E-06 to 5.216E-04	6.352E-05 to 1.230E-04	2.963E-05 to 3.392E-05
Acenaphthylene	Mean	1.6544E-06± 3.511E-08	2.236E-06± 2.0816E-08	2.041E-04± 1.00E-06	1.261E-06± 3.6055E-08	1.4618E-05± 3.605E-07	5.274E-08± 2.972E-09
	Range	1.600E-06 to 1.600E-06	1.312E-07 to 3.087E-06	2.341E-06 to 6.097E-04	3.562E-07 to 1.528E-06	3.733E-06 to 3.54E-05	N.D to 2.111E-07
Acenaphthene	Mean	4.340E-06± 3E-08	7.048E-06± 2.0816E-08	1.5781E-04± 4.582E-06	6.7925E-06± 2E-08	6.228E-05± 2.6457E-08	3.613E-07± 5.9197E-08
	Range	1.256E-07 to 1.219E-05	7.0483E-06 to 7.090E-06	3.336E-07 to 4.726E-04	1.008E-06 to 2.215E-04	3.369E-06 to 1.207E-05	N.D to 1.445E-06
Fluorene	Mean	4.287E-06± 4.5092E-08	1.677E-05± 5.859E-07	2.9237E-05± 3.869E-06	2.135E-06± 5.033E-07	2.246E-05± 5.131E-07	1.134E-05± 3.131E-07
	Range	8.772E-07 to 1.103E-05	9.25E-07 to 5.407E-05	7.331E-07 to 1.079E-04	2.500E-07 to 4.822E-05	7.056E-06 to 1.207E-05	N.D to 4.537E-05
Phananthene	Mean	1.240E-06± 3.055E-08	1.740E-6± 2E-08	2.348E-05± 1.1135E-06	2.1169E-05± 4.358E-07	9.4216E-05± 1.9139E-06	9.236E-06± 2.668E-08
	Range	1.342E-07 to 2.951E-06	1.136E-06 to 2.916E-06	1.119E-06 to 6.329E-05	2.066E-07 to 1.116E-05	7.251E-07 to 3.261E-04	7.112E-06 to 2.983E-05
Anthracene	Mean	2.289E-06± 9.6090E-08	3.876E-06± 7.5605E-08	1.988E-05± 3.5118E-07	1.390E-06± 2.345E-05	1.8714E-05± 7.767E-07	2.403E-06± 3.5118E-05
	Range	1.441E-06 to 3.212E-06	4.073E-07 to 7.345E-06	6.041E-07 to 1.455E-05	1.242E-05 to 4.1633E-07	1.012E-07 to 7.251E-05	1.321E-07 to 5.231E-06
Fluoranthene	Mean	1.9135E-05± 7.2111E-07	8.211E-06± 5.957E-07	6.2706E-06± 2.565E-07	1.5300E-05± 8.1247E-06	4.417E-05± 5.514E-07	4.778E-06± 2.3985E-07
	Range	6.055E-06 to 2.823E-05	1.532E-07 to 1.200E-06	2.307E-06 to 1.218E-05	3.346E-07 to 2.605E-05	1.112E-06 to 4.125E-05	2.761E-06 to 8.675E-06
Pyrene	Mean	6.820E-06± 3.417E-07	3.929E-07± 1.5557E-08	1.4587E-05± 4.428E-07	3.175E-06± 2.083E-07	1.592E-06± 3.031E-07	1.938E-06± 2.328E-07
	Range	7.659E-08 to 2.246E-05	3.265E-08 to 1.280E-06	6.523E-08 to 3.593E-05	6.398E-07 to 6.334E-06	6.445E-07 to 3.897E-06	6.034E-07 to 5.0344E-06

Results are presented as Mean±SD

Table 4. *Concentration of high molecular weight polycyclic aromatic hydrocarbon in water from QIRE in the wet Season.*

PAH mixture		Okoroutip	Ukpenekang	Iwochang	Douglas	Stubb	EkpeneUkpa
B(a)A	Mean	1.793E-05± 7.150E-07	1.715E-05± 2.396E-06	8.888E-06± 3.428E-07	1.0769E-05± 2.4542E-07	6.1503E-06± 1.0400E-07	1.411E-07± 1.6623E-09
	Range	3.063E-06 to 5.617E-05	2.115E-07 to 4.312E-05	6.590E-07 to 2.903E-05	8.973E-07 to 3.251E-05	6.450E-06 to 7.216E-06	N.D to 5.642E-07
Chysene	Mean	1.392E-05± 5.686E-07	4.355E-06± 2.318E-07	1.312E-05± 4.956E-07	1.220E-06± 2.645E-08	6.198E-06± 4.0722E-07	5.450E-07± 3.3350E-08
	Range	3.250E-06 to 4.201E-05	2.456E-07 to 1.047E-05	4.223E-07 to 1.540E-06	2.063E-07 to 1.48E-06	6.630E-06 to 4.0722E-07	
B(b)F	Mean	1.703E-02± 3.723E-03	9.327E-03± 3.400E-04	2.928E-02± 1.250E-03	1.272E-02± 1.527E-03	2.073E-02± 1.760E-03	1.715E-06± 4.450E-08
	Range	7.56E-03 to 1.123E-02	6.712E-04 to 3.429E-03	1.128E-02 to 6.868E-03	6.546E-03 to 3.251E-02	1.857E-02 to 2.145E-02	N.D to 6.859E-06
B(k)F	Mean	3.260E-06± 4.1186E-08	7.774E-06± 1.6522E-07	2.781E-06± 3.605E-08	2.354E-06± 4.725E-08	9.975E-05± 2.516E-07	1.715E-06± 4.450E-08
	Range	2.16E-07 to 7.11E-06	6.342E-06 to 1.185E-05	2.135E-07 to 8.717E-06	1.367E-07 to 5.326E-06	3.162E-07 to 3.542E-04	N.D to 6.859E-06
B(a)P	Mean	1.4950E-05± 4.387E-07	8.398E-06± 6.5574E-08	2.391E-05± 5.8966E-07	1.2938E-05± 4.725E-07	2.826E-05± 4.041E-07	9.943E-07± 9.068E-06
	Range	3.542E-06 to 3.165E-05	6.321E-06 to 1.221E-05	5.513E-06 to 2.826E-05	2.063E-07 to 7.561E-06	1.126E-06 to 1.985E-05	2.516E-06 to 3.214E-07
Lnd (1,2,3,cd) P	Mean	1.029E-02± 1.527E-03	2.594E-02± 7.00E-03	4.942E-03± 3.785E-04	4.685E-02± 3.605E-03	3.338E-02± 4.358E-03	5.872E-04± 7.571E-05
	Range	3.245E-03 to 1.213E-02	7.55E-03 to 6.372E-02	2.531E-04 to 1.17E-02	2.456E-03 to 7.694E-02	1.746E-03 to 6.523E-02	3.323E-05 to 1.332E-03
D(ah)A	Mean	2.062E-05± 7.0116E-07	2.466E-05± 4.9328E-07	7.285E-06± 2.023E-07	3.1473E-05± 1.619E-06	2.598E-06± 6.00E-08	2.4217E-06± 5.2915E-08
	Range	2.145E-06 to 7.467E-05	1.461E-06 to 1.325E-05	4.763E-08 to 1.078E-05	3.112E-07 to 9.217E-05	4.451E-07 to 5.259E-06	3.115E-08 to 4.714E-06
B(ghl)P	Mean	1.842E-05± 3.6055E-07	3.451E-05± 7.088E-06	2.3753E-05± 7.023E-07	1.759E-06± 7.211E-08	1.054E-06± 7.505E-07	1.596E-05± 6.557E-07
	Range	4.256E-06 to 4.421E-05	1.086E-05 to 6.142E-05	2.116E-07 to 3.342E-05	9.321E-07 to 2.263E-06	8.321E-06 to 1.172E-05	3.402E-06 to 2.852E-05

PAH diagnostic ratio is used for identifying pollution sources [21-23]. In this study, Ant / Ant + Phe, Flue / Flue + Pyr and ∑LMW/∑HMW ratio were used for the identification of pyrogenic and petrogenic origins of PAHs. For molar mass of 178, ratio of Ant / Ant + Phe less than 0.1 indicate petrogenic origin while ratios greater than 0.1 indicates pyrogenic origin. For molar mass 202, Flue / Flue + Pyr> 0.5 indicates pyrogenic PAH origin and the ratio < 0.4 indicate petrogenic origin. ∑LMW/∑HMW ratio less than 1 indicates pyrogenic origin while ∑LMW/∑HMW ratio greater than 1 indicates petrogenic source [22, 24]. Table 5 reveals the result of diagnostic ratio for PAHs in water from QIRE during the dry and wet season. In the dry season, the diagnostic ratio used in this study reveals pyrogenic PAH sources except at Iwochang where Flue/Flue + Pyre ratio is less than 0.5 signifying petrogenic PAH source. In the wet season, pyrogenic PAH sources was also observed for all the stations except for Iwochang for Flue/Flue + Pyr and Douglas creek for Ant/Ant + Phe.

The predominance of combustion (pyrogenic source) showed that PAHs in water from the examined sites are from combustion sources. The combustion sources of PAH in the water were likely to predominantly originate from constant gas flaring from a petrochemical facility located close to the sampled stations. Other combustion sources may have been emissions from combustion of fuel from automobiles and engine boats. The PAHs detected were predominantly HMW-PAHs and therefore may have entered the water bodied through atmospheric deposition. Petrogenic PAHs are emitted directly to a water body while pyrogenic PAHs are first emitted into the air and enters the water through atmospheric fall-out [24]. In addition, mixed sources of PAH observed in this study compares with the result obtained by [25, 26].

The result of the correlation analyses conducted among the concentration of individual PAHs in water samples to assess PAH interrelationship and their origin during the dry and wet seasons are presented in Tables 6 and 7. Strong positive significant correlation was observed between some individual PAHs in water during the dry season. B(a)A and D(ah)A showed the highest PAH association with correlation coefficient of 0.994, followed by Ind(1,2,3-cd)P and chrysene with correlation coefficient of 0.942 and Ind(1,2,3-cd)P and B(b)F with correlation coefficient of 0.938. They were significant at 0.01 levels. B(k)F and chrysene interrelated strongly at 0.05 levels.

During the wet season, significant positive correlation was observed between Ind(1,2,3-cd)P and B(a)P (r = 0.719) at 0.01 levels while some PAH pairs recorded significant positive correlation at 0.05 levels. They are : B(b)F/ chrysene (r = 0.789), D(ah)A / B (a)P (r = 0.657), D(ah)A / B(a)A (r = 0.516). The result for correlation analysis between individual PAHs in water from QIRE revealed that the compounds were derived from a common anthropogenic origin [27]. The strong positive correlation between B(k)F and chrysene observed in this study may be due to close relatedness in molecular weight [28].

Table 5. Diagnostic ratios of PAHs in water during the dry and wet seasons.

	DRYSEASON			WET SEASON		
Station	ΣLMW/ΣHM	FL/ FL + PY	AN/ AN + PH	ΣLMW/ΣHMW	FL/ FL + PY	AN/ AN + PH
Okoroutip	0.0049	0.9365	0.8960	0.0001	0.7371	0.6484
Ukpenekang	0.0480	0.9951	0.4665	0.0011	0.9545	0.6896
Iwochaang	0.0106	0.0125	0.1268	0.0133	0.3007	0.4449
Douglas creek	0.0078	0.6923	0.1295	0.0036	0.8279	0.0616
Stubb creek	0.0195	0.5300	0.1459	0.0041	0.9654	0.1656
EkpeneUkpa	0.2863	0.6370	0.2880	0.0381	0.7110	0.2064

Fl / FL + PY =Fluoranthene/Fluoranthene + Pyrene , AN/ AN + PH = Anthracene/Anthracene + Phenanthrene

Table 6. Correlation matrix for high molecular weight PAHs in water from QIRE during the dry season.

	B(a)A	Chrysene	B(b)F	B(k)F	B(a)P	Ind(1,2,3-cd)P	D(ah)A
B(a)A	1						
Chrysene	-0.163	1					
B(b)F	-0.315	0.575	1				
B(k)F	-0.035	0.837*	0.325	1			
B(a)P	0.640	-0.012	0.241	-0.119	1		
Ind(1,2,3-cd)P	-0.268	0.942**	0.426	0.938**	-0.216	1	
D(ah)A	0.994**	-0.272	-0.382	-0.130	0.621	-0.367	1

** Correlation is significant at 0.01 level (2-tailed), * correlation is significant at 0.05 level (2-tailed)

Cluster analysis was used to assess similarities or dissimilarities among sampling sites based on the level of PAHs in water for both the dry and wet season and the results presented in Figures 2 and 3 revealed three principal clusters for both seasons. In the dry season, group one was made up of Stubb creek and Okoroutip with the highest concentration of total PAHs and HMW-PAHs, while the second group was made up of Douglas creek and Iwochang with intermediate concentration and the third group was Ukpenekang and EkpeneUkpa with the lowest concentration. These sites especially the control site was free from industrial activity and the level of PAHs here is low and may be due to domestic waste incineration and sewage only. During the wet season, three clusters were identified belonging to a group of three similar PAHs. The highest total PAHs was recorded by group one (Douglas, Ukpenekang and Stubb creek), while Iwochang and Okoroutip belonged to group two and the control site with the lowest total PAH concentration was classified as group three.

Table 7. Correlation matrix for high molecular weight PAHs in water from QIRE during the wet season.

	B(a)A	Chrysene	B(b)F	B(k)F	B(a)P	Ind(1,2,3-cd)P	D(ah)A
B(a)A	1						
Chrysene	-0.037	1					
B(b)F	-0.164	0.786*	1				
B(k)F	-0.189	-0.031	0.271	1			
B(a)P	-0.158	-0.454	-0.110	-0.213	1		
Ind(1,2,3-cd)P	0.188	-0.411	0.033	0.368	0.719**	1	
D(ah)A	0.516*	-0.111	-0.111	-0.461	0.657*	0.572	1

** correlation is significant at 0.01 level (2-tailed), * correlation is significant at 0.05 level (2-tailed)

Hierarchical dendrogram was used to assess similarities between individual PAHs in water from QIRE for the dry and wet seasons are presented in Figures 4 and5. Three primary clusters were identified in both seasons. The first group and second group comprised of B(b)F and Ind (1,2,3-cd)P respectively while the third group comprised of the remaining carcinogenic PAHs. The observed difference in the cluster for the two seasons may be due to the effect of flooding and possible dilution. Water volume and flow rate of the river is usually lower in the dry season which suggests higher organic pollution [29]. However, the effect of dilution expected to result in total PAH decrease is compensated for by wet deposition of soot as a result of gas flaring. The grouping in Figures 3 and 4 may be due to the effect of concentration and similar anthropogenic source. Strong positive correlation between PAHs in the third group corroborates with the correlation matrix for water during the wet season. Several authors have reported the high levels of B(b)F and Ind (1,2,3-cd) P in environmental samples from the Niger delta, which may be anthropogenic [17, 30].

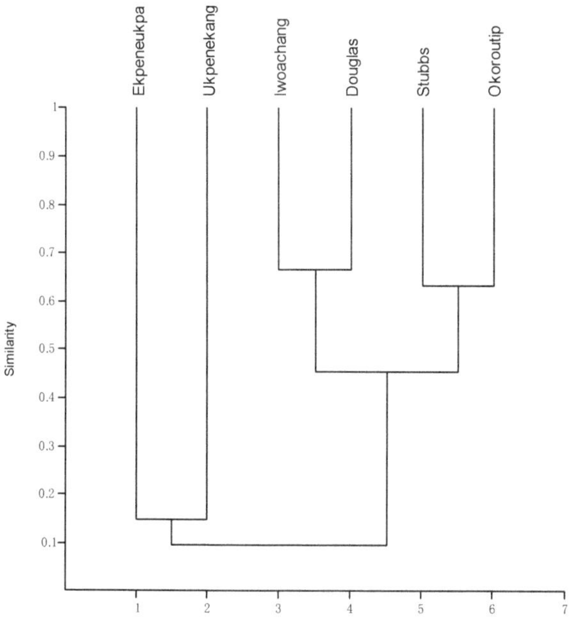

Figure 2. Cluster analysis showing the spatial distribution of PAHs among different sampling sites in water from QIRE during the dry season.

4. Conclusion

The level of PAHs in water from QIRE was evaluated in this study. Data analysis using diagnostic ratio, correlation studies and hierarchical cluster analysis was used to identify the sources of PAHs in this study. The total PAHs and total C-PAHs at all the sites except the control site were higher than the standard of European Union and WHO respectively. The values of B(b)F and Ind(1, 2, 3-cd)P at all the sampling sites during both seasons, were above the WHO standard of 5 x 10^{-5}mg/l for individual PAHs in coastal and surface water. The PAHs detected were predominantly HMW-PAHs predominantly from pyrogenic sources such as gas flaring from and vehicular emissions. Atmospheric deposition may have been the pathway for entry of PAHs into the water bodies. Three primary PAH clusters were identified in both seasons indicating different levels of anthropogenic activities in the cluster areas. The result revealed high levels of B(b)F and(1,2,3-cd)P, two carcinogenic PAHs in lower reach of the Qua Iboe River estuary. Carcinogenic risk assessment of PAH from the water of QIRE and strict legislation on gas flaring in the area is suggested.

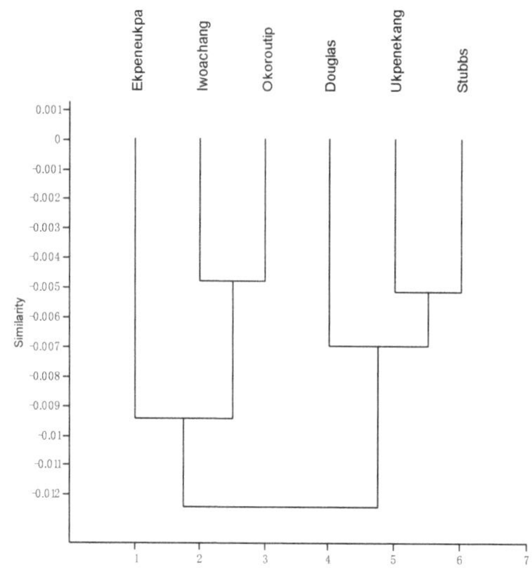

Figure 3. Cluster analysis showing the spatial distribution of PAHs among different sampling sites in water from QIRE during the wet season.

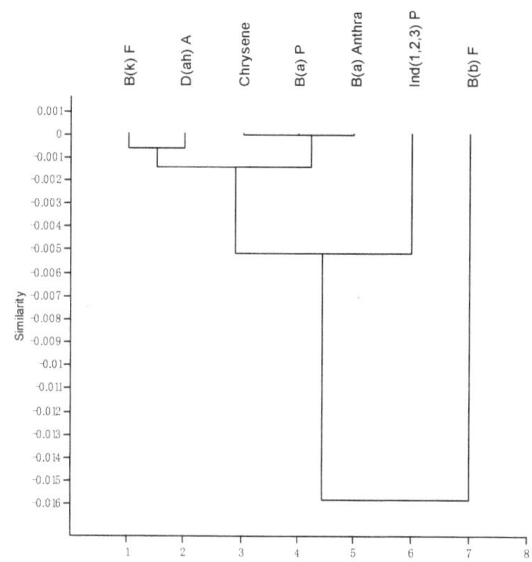

Figure 4. Hierarchical dendrogram for carcinogenic PAHs in water from QIRE during the dry season.

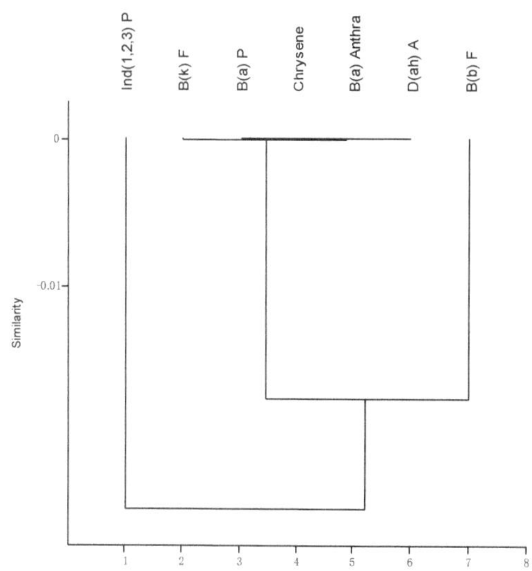

Figure 5. Hierarchical dendrogram for carcinogenic PAHs in water from QIRE during the wet season.

References

[1] G. Oze, R. Oze, C. Anunuso, H. Ogukwe, and Okorie, K, "Heavy metal pollution of fish of Qua Iboe river estuary: Possible implication for neurotoxicity," Int. J. Toxicol, vol. 3(1) pp 12-15, 2005.

[2] K. Srogi, "Monitoring of environmental exposure to polycyclic aromatic hydrocarbon: A review. Environ. Chem. Letters," vol. 5, pp 169 – 195, 2007.

[3] Agency for Toxic Substances and Disease Registry. (ASTDR). "Toxicological profile for polycyclic aromatic hydrocarbons (PAHs)" PB / 95 / 264370. Atlanta, USA. 1995.

[4] G. R. Ana, K. C. Sridher, and O. G. Emerole, "Contamination of surface water by polycyclic aromatic hydrocarbons in two Nigerian coastal communities," J. Environ. Hlth. Res., vol. 11(2) pp 77-86, 2010.

[5] W. Wilcke, "Polycyclic aromatic hydrocarbons (PAHs) in soil-a review. J. Plant Nut. Soil Sci., vol. 163, pp 229-248, 2000.

[6] Y. Z. Zhang, S. Tao, H. Z. Shen, J. M. Ma, "Inhalation exposure to ambient polycyclic aromatic hydrocarbons and lung cancer risk of Chinese population" Proceedings of the National Academy of Sciences, vol 106, 21063-21067, 2009.

[7] T. O'Neil, Curse of the Black man's gold: Hope and betrayal in the Niger Delta. National Geographic Magazine, p. 12. February, 2007.

[8] A. Sarma, and H. Sarma, "Enhanced biodegradation of oil products by some microbial isolates supplemented with heavy metals," *Int. J. Bot.*, vol. 6: pp 441 - 445, 2011.

[9] R. Nwilo, and O. Badejo, "Oil spill problems and management in the Niger Delta". International oil spill conference, Fluorida, USA, pp 15-19 May, 2005.

[10] Odugwu, "NOSCP and oil spill management," The pioneer, p.10, April 15, 2013.

[11] E. Manoli, and C. Samara, "Polycyclic aromatic hydrocarbon in natural waters: Source, occurrence and analysis," Trends Anal. Chem. vol. 18(6): pp 417 – 420, 1999.

[12] M. A. Gorleku, D. Carboo, L. Palm, W. Quasie, and A. Armah, "Polycyclic aromatic hydrocarbon pollution in marine waters and sediments at Temaharbour, Ghana," Acad. J. of Environ. Sci., vol. 2(7) pp 108-115, 2014.

[13] C. Anyakora, and H. Coker, "Determination of Polycyclic aromatic hydrocarbon in selected water bodies in the Niger Delta," Afri. J. Biotech., vol. 5(21) pp 2024- 2031, 2006.

[14] D. K. Essumang, C. K. Adokoh, J. Afriyie, and E. Mensah, "Source assessment and analysis of PAHs in the Oblogo waste disposal sites and some water bodies in and around the Accra metropolis of Ghana," J. Water Res. and Prot. vol 1 pp 456 – 468, 2009.

[15] I. Isine "Akwa Ibom community locks down Exxon Mobil," premium times, July 9, 2014. Assessed on September 17 2015 at http//www.premiumtimesng.com/regional/sso.

[16] E. Caylak, "Health risk assessment for trace metals, polycyclic aromatic hydrocarbons and trihalomethanes in drinking water of Cankiri, Turkey," E-Journal of chemistry, vol 9(4): pp 1916-1991, 2012.

[17] E. Inam, E. Owhoeke, and J. Essien, "Human carcinogenic risk assessment of polycyclic aromatic hydrocarbon in freshwater samples from Ogba/ Egbema / Andoni communities in River state, Nigeria," J. Chem. Soc. Nig. vol 39 (2) pp 15 -22. 2014.

[18] C. Anyakora and H. Coker, "Assessment of Polycyclic aromatic hydrocarbon contamination threat on groundwater: A case study of the Niger Delta region of Nigeria," Int. J. Risk Assess. Mgt., vol 13(2) pp 150- 161, 2009.

[19] J. Li, X. Shang Z. Zhao, R. L. Tanguay, Q. Dong and C. Huang, "Polycyclic aromatic hydrocarbons in water, sediment, soil, and plants of the Aojiang River waterway in Wenzhou, China," J. Hazard Mater. vol 173(1-3) pp75-81, January, 2010.

[20] R. J. Law, V. I. Dawes, R. J. Woodhead, and P. Matthissen, "Polycyclic aromatic hydrocarbons (PAHS) in seawater around England and Wales," Mar. Poll. Bull., vol 34(5) pp 306-322, 1997.

[21] H. H. Soclo, P. Grrigue, and M. Ewald, Origin of Polycyclic aromatic hydrocarbons (PAHs) in coastal marine sediments: Case studies in Cotonou (Benin) and Aquitaine (France) areas," Mar. Pollut. Bull., vol 40(5) pp 387-396, 2000.

[22] M. B. Yunker, R. W. Macdonald, R. Vingerzan, R H. Mitchell, D. Goyette, and S. Sylvestre, "PAHs in the Fraser River basin: A critical appraisal of PAH ratio as indicators of PAH source and composition," Organic Geochemistry, vol. 33(4) pp 489-515, 2002.

[23] Z. Zhang, J. Huang, G. Yu, and H. Hong, "Occurrence of PAHs, PCBs and organochloride pesticides in the Tonghui River of Beijing, China," *Environmental Pollution*, vol. 130(2) pp 249-261, 2004.

[24] M. Tobiszewski, and J. Namiesnik," PAH diagnostic ratios for the identification of pollution emission sources," Env. Poll. vol. 162 pp 110-119, 2012.

[25] K. Opuene, I. E. Agbozu, O. Adegboro, "A critical appraisal of PAHs indices as indicators of PAHs source and composition of Elelenwo creek, southern Nigeria," Environmentalist, vol. 29 pp 47-55, 2009.

[26] M. Sinae, and A. Mashinchian, "Polycyclic aromatic hydrocarbons in the coastal sea water, the surface sediment and mudskipper (*Boleophthalmusdussumer*) from coastal areas of Persian Gulf: Source investigation, composition pattern and spatial distribution," J. Environ. Health Sci. Engr. vol. 12 pp 59 -63, 2014.

[27] E. Gilbert, D. Dadoo, R. Okai-Sam, D. Essumang, and K. Quagraine, "Characterization and source assessment of heavy metals and PAHs in sediment of Fosu Lagoon, Ghana," J. Environ. Sci. Hlth. vol. 41 pp 2447- 2755, 2006.

[28] D. H. Ogbuagu, C. G. Okoli, C. L. Gilbert, and S. L. Madu, "Determination of the concentration of groundwater sources in Okrika mainland with polycyclic aromatic hydrocarbons," British J. Environ. Clim. Change, vol. 4(3) pp 90-102, 2011.

[29] S. Farooq, S. Equani, R. Malik, Y. Zhang, J. Li, L. Xang, K. Jones, and Z. Shinwari, "Occurrence, fingerprinting and ecological risk assessment of PAHs in Chenab river," Parkistan. J. Environ. Monit. vol. 13 pp 3207- 3215, 2011.

[30] K. W. Nkpa, M. O. Wegwu, and E. B. Essien, "Assessment of polycyclic hydrocarbons (PAHs) levels in two commercially important fish species from crude oil polluted waters of Ogoni-land and their carcinogenic health risks," J. Env. And Earth Sci. vol. 3(8) pp 128- 137, 2013.

Land Use and Land Cover Change in a Community-Managed Forest in South-Eastern Senegal Under a Formal Forest Management Regime

Laurice Codou Faye[1], Hyacinthe Sambou[2], Boateng Kyereh[3], Bienvenu Sambou[2]

[1]Department of Civil Engineering, College of Engineering, Kwame Nkrumah University of Science and Technology, Kumasi, Ghana
[2]Institute of Environmental Sciences, Cheikh Anta Diop University of Dakar, Dakar, Senegal
[3]Department of Silviculture and Forest Management, Faculty of Renewable Natural Resources, Kwame Nkrumah University of Science & Technology, Kumasi, Ghana

Email address:
fayelaurice@gmail.com (L. C. Faye)

Abstract: While most studies of community forests in Senegal address issues in institutional and political arrangements for managing forests, this study was carried out to find out how these political and institutional changes embodied in the new approach to forest management impact on land use and land cover change in the country. Using the Missirah Forest in south-eastern Senegal, as a case study, the objective was to quantify the land use and land cover changes that have happened over a twenty four- year period, from 1990 to 2014 using remote sensing. Six land use and land cover types were identified and mapped, namely, gallery forest, tree savanna, shrub savanna, degraded shrub savanna, croplands and settlements. The area of croplands and settlements expanded between 1990 and 2014. The conversion from natural vegetation to croplands (14.45%) was higher than the conversion from cropland to natural vegetation (3%). Between 1990 and 2003, the expansion in croplands was higher than between 1990 and 2003 but the reverse was the case for settlements. Regarding vegetation types, they decreased in cover between the two periods with the exception of shrub savanna that experienced an increase of 1.46% from 1990 to 2003. Transition to less wooded vegetation (31.58%) was higher than transition to more wooded vegetation (13.91%). This study shows that deforestation and forest degradation are still in progress despite the implementation of a management plan for a full rotation.

Keywords: Land Cover Changes, Vegetation Transition, Community-Forest Management, Senegal

1. Introduction

In sub-Saharan Africa, land use and land cover change trends are extremely fast and the direction and rate of change are unstable [1]. Land use and land cover changes are often related to both natural and anthropogenic causes. But the most important driver of land cover and land use change is the intensive use of natural resources by local communities to satisfy their daily needs [2, 3] particularly in sub-Saharan countries where communities' livelihoods depend mainly on natural resources. Changes in land use and land cover contribute significantly to alter the environment and ecosystem services that support human needs [4, 5]. To natural resources dependent-societies, land use and land cover change constitute a major challenge to sustainable livelihoods aspirations. To implement remedial strategies to cope with the issue, a good understanding of the direction of change and their extent is needed.

Land use is defined as human' intervention on land [6] and involves both the manner in which the land is manipulated and the intent that motivated that manipulation [7]. Land cover refers to the biophysical attributes of the earth's surface [8]. The causes of land use land cover change can be summarized in two major categories namely the proximate or direct causes and the underlying causes [9, 10]. The underlying causes are factors that trigger the proximate causes and refer to economic, demographic, institutional political or technological factors that mostly occur at regional or global levels [11]. The proximate causes refer to immediate actions that affect directly the land cover [12, 14]. The proximate causes of land use and land cover changes are manifold but the most cited in Africa are wood extraction, agricultural expansion, and infrastructure extension [9, 15-18].

After forest clearing for agriculture, wood extraction mostly for household energy consumption is the major driving force of vegetation dynamics in sub-Saharan countries [19, 20] where woodfuel either used directly as firewood or converted into charcoal is the primary source of domestic energy [21]. Indeed, despite the capacity of tropical forest species to regenerate after cutting for charcoal that allows forest recovery [22, 24], the pressure exerted by charcoal production sometimes results in devastating ecological and environmental effects [25] particularly deforestation and forest degradation [26, 27]. The environmental effects of deforestation often prompt countries with high dependence on fuel wood to develop strategies for coping with the problem. In Senegal a major charcoal producing country, the concern for unsustainable production of charcoal and the need to halt the process of deforestation and forest degradation induced by charcoal production led to the evolution of forest management in which formal forest regulation backed by management plans was introduced in local communities [28].

In different studies conducted in Senegal, the evolution of decentralization of the forest management process [29], the effect of the institutional pluralism on the decentralization and the management of forest resources [30], the function of the forest management plan in the new Senegalese Forest code [31], the evolution of forest management in Senegal [32] were analysed. With the exception of a study by [33] that examined the effects of charcoal production on woodland regeneration, all these studies focused on the political and institutional framework of forest management. However, the ultimate impact of these new forest management arrangements on deforestation is yet to be determined. Therefore, this study was carried out to find out how these political and institutional changes embodied in the new approach to forest management impact on land use and land cover change in the country. Using the Missirah Forest in south-eastern Senegal, as a case study, the research sought to quantify the land use and land cover changes that have happened over a twenty four- year period, from 1990 to 2014 using remote sensing.

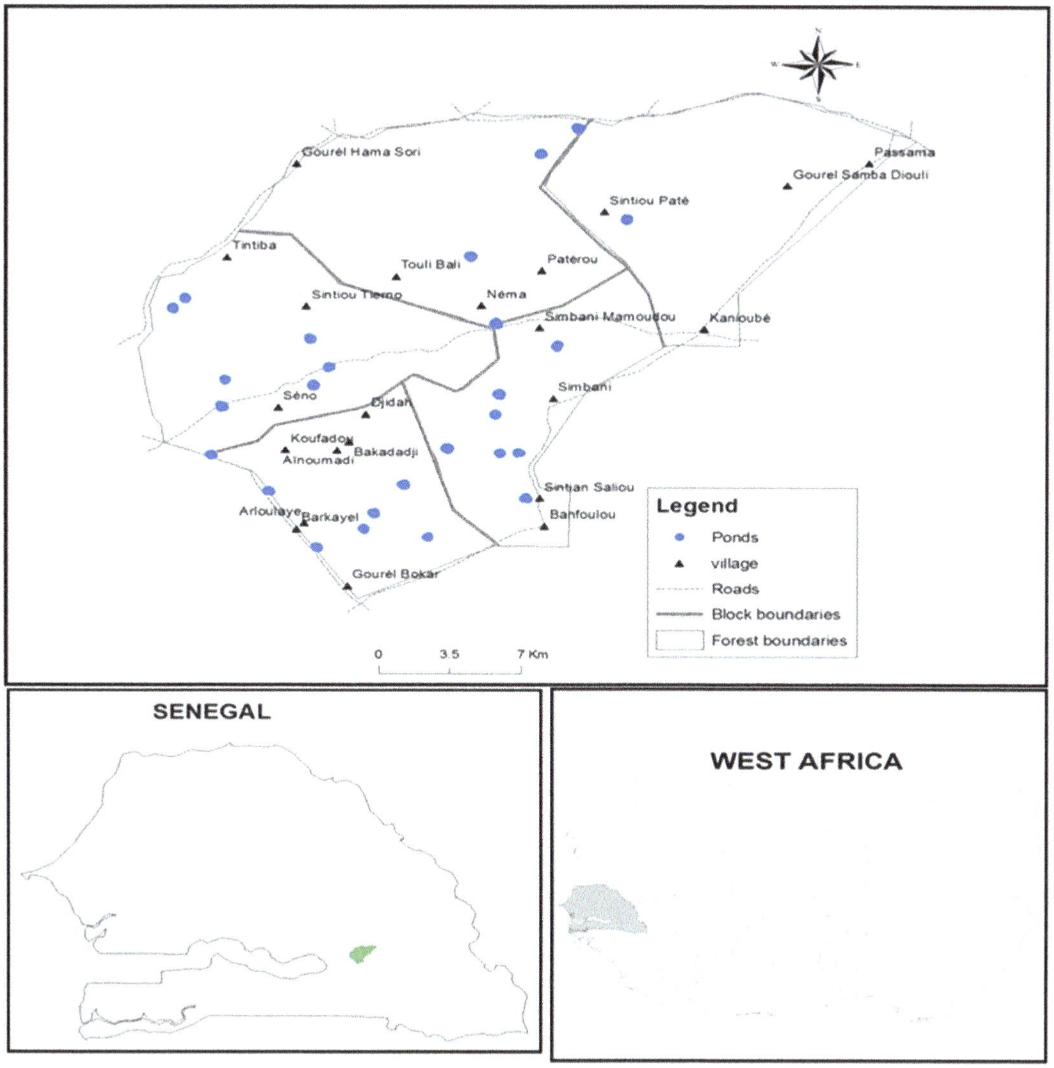

Figure 1. Location of Missirah Forest.

2. Materials and Methods

2.1. Study Area

This study was conducted in Missirah Forest which is located in the Sudanian zone in south-eastern Senegal. The total area of the forest is 63123.54 ha and the vegetation is dominated by species such as *Combretum glutinosum, Cordyla pinnata, Pterocarpus erinaceus, Bombax costatum, Lannea acida,* and *Acacia macrostachya.* The study area is characterized by distinct dry and wet seasons. The mean annual rainall from 1960 to 2014 is estimated at 754 mm. Temperatures range from 42 to 16°C with a mean of 29°C. The management of Missirah Forest dates back to 2004 with the support of a World Bank project. The management is based on a plan that specifies when, where and the volume of wood to be cut, methods to be used and reforestation activities for replacing the losses [28]. The forest is divided into five (5) blocks split up each into eight (8) plots of exploitation. This corresponds to eight years rotation period. Gallery forest and croplands were classified as protected areas in the management plan and consequently proscribed to charcoal production activities. Resident communities were allowed to maintain their croplands but forbidden to extend or open new farms.

2.2. Estimation of Land Use and Land Cover Change

2.2.1. Remote Sensing and Field Data

The dynamics of the land use and land cover were studied using time series of Landsat images obtained from Landsat TM (Thematic Mapper), of 11/29/1990, ETM+ (Enhanced Thematic Mapper) of 01/01/2003 and Landsat OLI (Operational Land Imager) of 02/24/2014 corresponding to scene 203/51. The images were acquired approximatively for the same period, at the beginning of the dry season to ensure that the phenological stages of plant cover were not too different between dates. Moreover the beginning of the dry season is a suitable period to distinguish the various Sudano-Sahelian land-cover types because the contrast between the croplands and the natural environment is more marked [34]. Also, images captured in the dry season have the advantage of low cloud cover. Images used were downloaded from the United States Geological Survey (USGS) Landsat Earth Resources Observation Systems (EROS) Data Center. A ground survey was also carried out in the beginning of the dry season in conformity to the period of image acquisition. A set of 111 GPS points were collected in the different vegetation types and land use types. More GPS points were collected in classes that were more difficult to separate. The vegetation types were identified based on the classification of Yangambi [35]. Species in the vegetation types visited were also recorded.

2.2.2. Image Processing

As a first step we did a visual analysis that corresponds to the traditional method of photo-interpretation of the satellite images. This method consists of identifying the different homogenous units [36]. The interpretation was often made easy by our knowledge of the vegetation and field surveys.

The images used were already all geo-rectified to UTM WGS 84 Zone 28 North coordinates with radiometric corrections. To avoid geographical deviation between images due to differences in sensor when superimposing them for change detection analysis, the images of 1990 and 2003 were georectified to the image of 2014 already corrected using ground survey. The images were georectified using an image to image adjustment with an error estimated to less than the value of one pixel.

After adjusting the images, coloured compositions were created by combining channels 5 for the infrared [0.75-0.90 μm], 4 for the red [0.63 to 0.69 μm] and 3 for the green [0.52 to 0.60 μm] that display respectively the red, green, and blue colours for images 1990 and 2003. Regarding 2014 image, that was done associating the channels 6 for the infrared [1.55-1.66 μm], 5 for the red [0.84-0.88 μm] and 4 to the green [0.63-0.68 μm]. Then, the study area was extracted from the scene to determine the land cover land use types by classifying images.

A supervised classification was performed using the maximum likelihood algorithm. The choice of the classification type was motivated by heterogeneity of the study area that may favour high probability of confusion between some classes. The classification was parameterized by digitizing training areas. Prior to the determination of training areas, the number of classes was defined. The determination of the number of classes was based on the unsupervised classification performed up stream and the information collected in the field. To homogenize the classification, the images were subjected to 3x3 pixels filter. To check the accuracy of the classifications, a confusion matrice was created for each year to determine the overall accuracy and the kappa coefficient. The accuracy assessment showed good classification with kappa coefficient of more than 80% for 2014 image taken as reference. The different classes derived from the images were also compared with existing documents (thematic and synthetic maps of the study area). All the processing was performed using Envi 4.7.

The classified images were vectorized and processed using Arc Gis to produce land cover maps for the different periods. The post-classification comparison which is the most appropriate way to compare multi-source data [34] was used for change detection analysis. The analysis of the land use and land cover change was done taking into consideration the modifications and conversion processes observed between the different periods. Modifications refer to changes that affect the character without changing the category while conversions concern the replacement of one category by another [37]. A second temporal analysis of transition among vegetation types was carried out to determine the proportion of transition to less and more wooded vegetation.

3. Results

3.1. State of Land Use Land Cover in 1990, 2003 and 2014

The land use and land cover mapping identified six classes namely gallery forest, tree savanna, shrub savanna, degraded shrub savanna, croplands, and settlements. Land cover statistics (Table 2) showed the respective dominance of shrub savanna, tree savanna, croplands, gallery forests, and settlements in 1990 and 2003. In the 2014 image a new land cover type designated as degraded shrub savanna was identified as a distinctive land cover type. It thus became the fourth land cover type with 10.82% of the total area next to shrub savanna (42.35%), tree savanna (25.25%) and croplands (18.15%). Degraded shrub savanna is dominated by grasses with many dead tree trunks and few shrubs. In some spots it is even devoid of shrubs and thus differs from shrub savanna in terms of woody cover.

Table 1. *Land use/land covers statistics of Missirah Forest in 1990, 2003, and 2014.*

Land use land cover types	1990		2003		2014	
	Area (ha)	Percentage	Area (ha)	Percentage	Area (ha)	Percentage
Gallery forest	2799.02	4.43	2428.43	3.85	1655.87	2.62
Tree savanna	26234.68	41.56	19948.09	31.60	15938.27	25.25
Shrub savanna	29626.66	46.94	30547.82	48.40	26730.72	42.35
Croplands	4231.00	6.70	9939.36	15.75	11459.45	18.15
Settlements	230.18	0.36	257.83	0.41	508.89	0.81
Degraded shrub savanna	–	–	–	–	6828.34	10.82
Total area	63121.54		63121.54		63121.54	

For the three different assessment occasions the forest was dominated by shrub savanna and tree savanna that together represented 88.50% of the area in 1990, 80% in 2003, and 67.60% in 2014. Tree savanna and gallery forest gradually cover decreased between 1990 and 2014. The area covered by tree savanna decreased progressively from 41.56% in 1990 to 25.25% in 2014. Gallery forest accounted for 4.43% of the area in 1990 but only 2.62% in 2014. Croplands more than doubled between 1990 and 2003 from 6.70% to 15.75% but expanded more slowly from 2003 to 2014. The settlements presented an inverse scenario. They almost doubled between 2003 (0.41%) and 2014 (0.81%) whereas for the first period a slight increase of 0.05% was noticed. Shrub savanna the most dominant land cover type for the three periods increased by 1.46% between 1990 and 2003 and showed a decrease of 6.05% from 2003 to 2014 due to degradation. The diachronic land cover maps obtained for the three periods are presented in Figure 2.

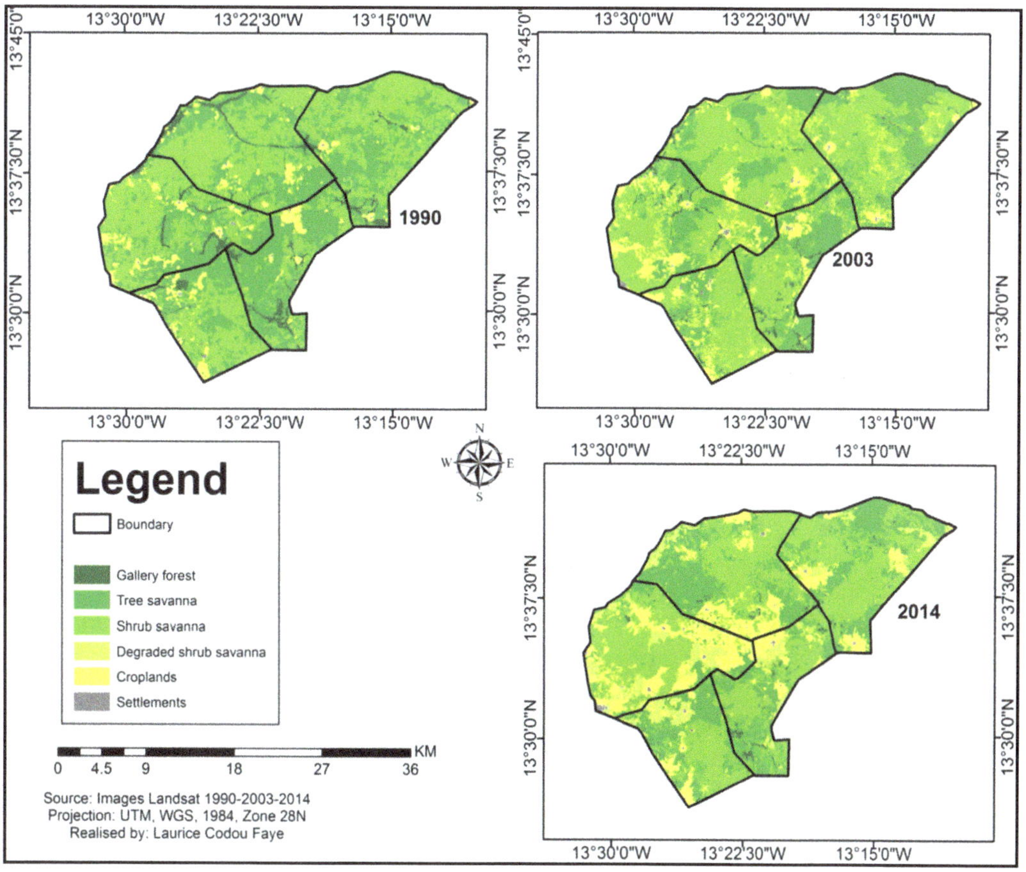

Figure 2. *Missirah Forest in 1990, 2003, and 2014.*

3.2. Change Detection Analysis

3.2.1. Land Use Land Cover Dynamics

Annual rate of land cover change was unidirectional for all land cover types with the exception of shrub savanna (Figure 3). The rate of change was positive in all periods for croplands and settlements contrary to tree savanna and gallery forest. The overall annual rate of change (1990-2014) was highest for croplands with an increase of 11.45% and lowest for settlements (0.44%). Tree savanna recorded the greatest loss between 1990 and 2014 (-16.31%).

The matrices of land cover change presented in Table 2 reveal that from 1990 to 2003, 43.61% of the area remained stable, 16.20% got converted and 40.19% was modified while from 2003 to 2014 conversions reached 23.30% of the total area with only 38.60% remaining stable. The unchanged area of each class is presented in diagonal elements (bold figures) in Table 2. For the first period (1990-2003), shrub savanna with 53.20% of its surface had the highest proportion of unchanged area followed by croplands (48.60%). In the second period, settlements and shrub savanna were the most stable land cover types with 51.20% and 46.40% of their area respectively remaining unchanged.

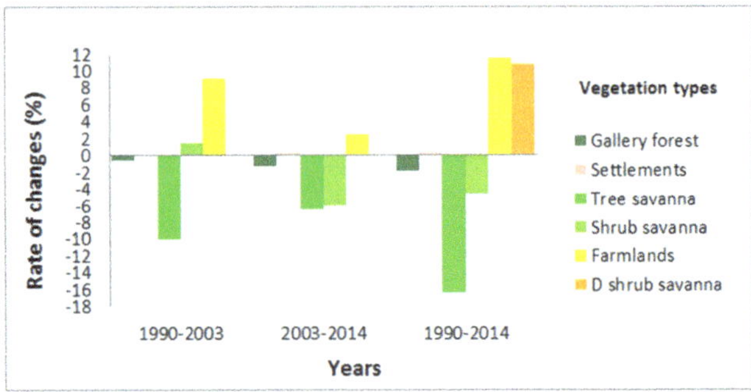

Figure 3. *Annual rates of land cover change in Missirah Forest between 1990 and 2014.*

Table 2. *Matrices of land cover change in Missirah Forest between 1990 and 2014.*

	From 1990 to 2003				
	Gallery forest	**Tree savanna**	**Shrub savanna**	**Croplands**	**Setllements**
Gallery forest	27.06	3.59	2.22	1.61	0.00
Tree savanna	23.45	33.99	31.98	20.35	16.09
Shrub savanna	37.01	48.03	53.2	25.77	25.24
Croplands	12.48	14.19	12.53	48.6	44.28
Setllements	0.00	0.2	0.07	3.66	13.9
	From 2003 to 2014				
	Gallery forest	Tree savanna	Shrub savanna	Croplands	Setllements
Gallery forest	9.73	2.53	2.2	2.42	0.44
Tree savanna	25.56	32.01	22.95	19.25	0.18
Shrub savanna	41	42.2	46.4	31.17	17.64
Degraded shrubsavanna	5.47	10.14	11.99	10.05	3.57
Croplands	17.38	12.59	15.78	35.66	23.74
Setllements	0.85	0.54	0.67	1.43	51.2
		From 1990 to 2014			
	Gallery forest	Tree savanna	Shrub savanna	Croplands	Setllements
Gallery forest	10.15	2.28	2.33	1.88	2.21
Tree savanna	23.40	28.14	25.33	8.82	9.71
Shrub savanna	44.00	45.01	42.88	20.63	27.89
Degraded shrubsavanna	5.16	7.87	13.67	12.94	9.79
Croplands	16.63	16.25	14.83	52.94	42.16
Setllements	0.65	0.46	9.73	2.78	29.96

The dynamics concerned mostly changes from tree savanna and gallery forest to shrub savanna in both periods 1990-2003 and 2003-2014. In the first period, 48.03% of tree savanna and 37.01% of gallery forest were changed into shrub savanna. In the second period, tree savanna was still most affected with 42.2% of its area but the proportion of gallery forest that changed into shrub savanna (41%) was higher than the previous change. From gallery forest to tree savanna, the change was also quite high affecting more than 20% of its area. With regard to shrub savanna, the change was mostly toward tree savanna with 31.98% and 22.95% of its area respectively in the first and the second period. The transition of tree and shrub savanna into gallery forest was not important in both periods. The transition into degraded shrub savanna newly appeared in 2014 images, and was more important for shrub savanna followed by tree savanna and degraded gallery forest respectively with 11.99%, 10.14%, and 5.47% of their area.

With regard to conversions from natural vegetation to cropland between 1990 and 2003, it occurred as follows: tree savannah (14.19%), shrub savannah (12.53%), and gallery forest (12.48%); and between 2003 and 2014, conversions appeared in the following order: gallery forest (17.38%) shrub savannah (15.78%) and tree savannah (12.59%). Conversions from cropland to natural vegetation in the first period took place in the following order shrub savannah

(25.77%), tree savannah (20.35%), and gallery forest (1.61%); in the second period, the same order was observed except that the conversion into degraded shrub savanna (10.05%) was higher than conversion into gallery forest (2.42%).

Conversions from natural vegetation to croplands were more important in terms of magnitude than those from cropland to natural vegetation. Area of natural vegetation converted into cropland in the first period and second period were estimated at 12.33% and 14.28% respectively. However, from croplands to natural vegetation conversions were estimated at 3.20% in the first period and 8.32% in the second period. Generally from 1990 to 2014, the conversions into croplands reached 14.45% of the total area whilst from croplands to natural vegetation, they represented only 3%. Figure 4 presented the spatial distribution of these transitions. The transition from natural vegetation to cropland is considered as "new croplands", resulting from forest clearing, from cropland to natural vegetation "abandoned croplands", that indicates also a recovery of the vegetation cover and the unchanged area of cropland as "permanent croplands".

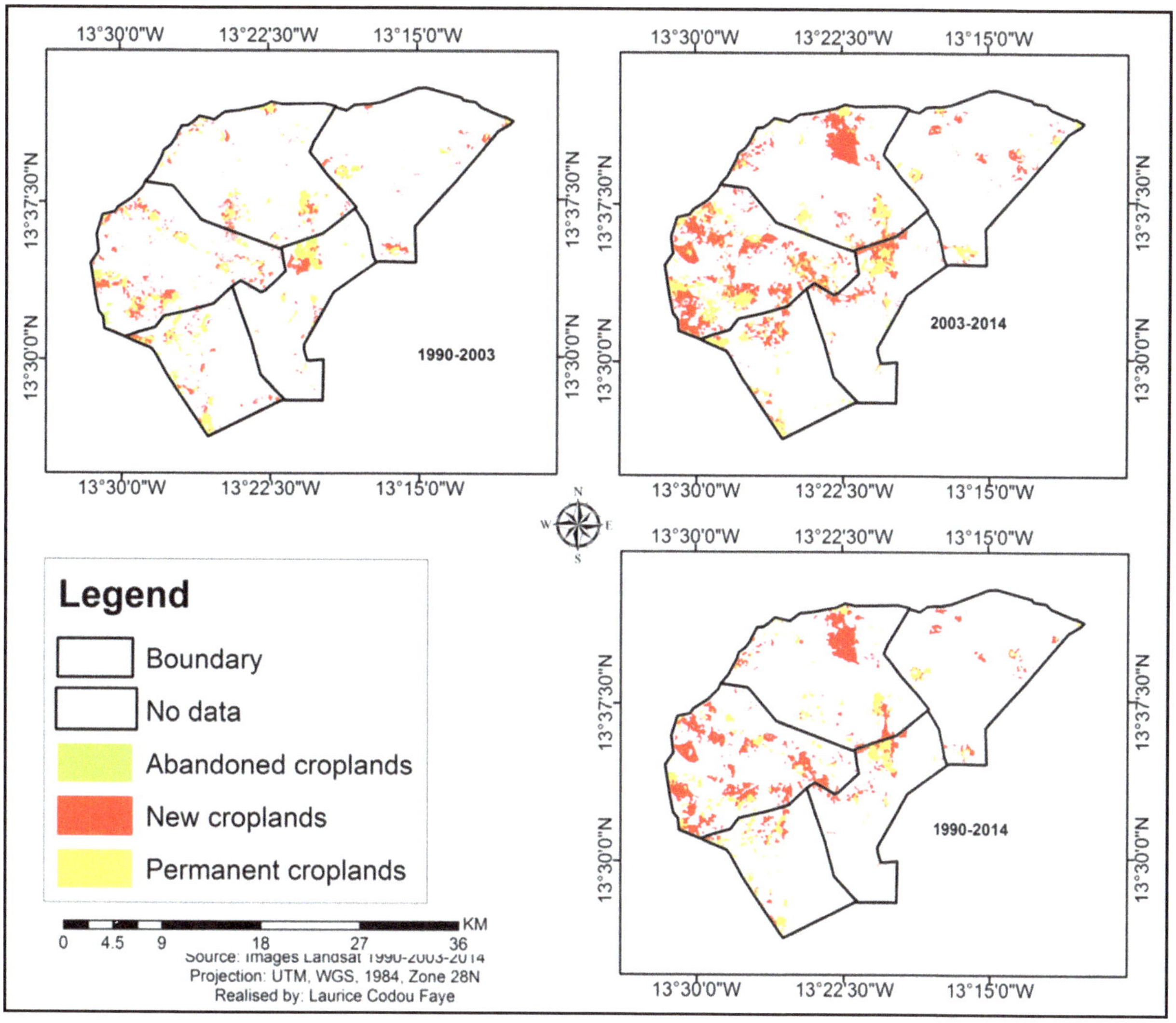

Figure 4. *Transition between natural vegetation and croplands in Missirah Forest for the three periods (1990-2003; 2003-2014; 1990-2014).*

3.2.2. Transition Among Vegetation Types

In Table 3 are found the statistics of the transition among natural vegetation cover types. The transition to less wooded vegetation was of a higher magnitude than transition to more wooded vegetation in both periods. However, the difference was more marked in the second period when a management plan was implemented. While in the first period the transition to less wooded vegetation was 5.08% more than the transition to more wooded vegetation, in the second period, it almost doubled. The transition among vegetation types was mainly characterized by transition between tree savanna and shrub savanna. The transition from shrub savanna to tree savanna balanced each other in the two periods respectively with 11.11 % in the first period and 11.88 % during the second period. But regarding transition from tree savanna to shrub savanna the highest value was recorded in the first period with 19.96% of the forest whereas in the second it was estimated at 13.33%. The transition into degraded shrub savanna in the second period was also important affecting 10% of the total forest area.

Table 3. *Temporal analysis of transition among vegetation types in Missirah Forest between 1990 and 2014.*

(a) 1990-2003			
Transition to less wooded vegetation (%)		Transition tomore wooded vegetation (%)	
Gallery forest to tree savanna	1.04	Tree savanna to gallery forest	1.49
Gallery forest to shrub savanna	1.64	Shrub savanna to gallery forest	1.04
Tree savanna to shrub savanna	19.96	Shrub savanna to tree savanna	11.09
Total	22.62	Total	17.54
(b) 2003-2014			
Transition to less wooded vegetation (%)		Transition to more wooded vegetation (%)	
Gallery forest to tree savanna	0.98	Tree savanna to gallery forest	0.79
Gallery forest to shrub savanna	1.57	Shrub savanna to gallery forest	1.06
Gallery forest to deg. shrub savanna	0.21	Shrub savanna to tree savanna	11.11
Tree savanna to shrub savanna	13.33		
Tree savanna to deg. shrub savanna	3.2		
Shrub savanna to deg. shrub savanna	5.8		
Total	25.09	Total	12.96
(c) 1990-2014			
Transition to less wooded vegetation (%)		Transition to more wooded vegetation(%)	
Gallery forest to tree savanna	1.03	Tree savanna to gallery forest	0.94
Gallery forest to shrub savanna	1.95	Shrub savanna to gallery forest	1.09
Gallery forest to deg. shrub savanna	0.22	Shrub savanna to tree savanna	11.88
Tree savanna to shrub savanna	18.7		
Tree savanna to deg. shrub savanna	3.27		
Shrub savanna to deg. shrub savanna	6.41		
Total	31.58	Total	13.91

4. Discussion

The dynamics of the vegetation was characterized by a process of degradation and deforestation that manifested respectively as transition from wooded to less wooded vegetation and the reduction of forest cover following forest clearing. Conversions affected tree savanna more than the other vegetation cover types (14.19%) in the first period and gallery forest in the second period (17.38%). In both periods shrub savanna came in second position but with a higher proportion in the second period (15.78%). The importance of the proportion of shrub savanna converted in the second period may be due to areas harvested for charcoal that were subsequently converted to croplands. Agricultural expansion taking advantage of logging or tree harvesting for charcoal is a common occurrence in developing countries [38]. One important finding of this study was the appearance of a new land cover type that we called degraded shrub savanna in 2014. This land cover type is mainly the result of tree and shrub density decline in tree savanna and the shrub savanna vegetation types. It is an open-vegetation consisting mainly of grass cover with a few shrubby species. A similar observation was made in eastern and southern Africa where charcoal production led to the transition of woodland to bush and bush to scrub [39]. It was observed that a reduction of about 20.9% of the forest cover and an expansion of cropland areas occurred within the period 1990 to 2014. This change is higher than the decrease in forest cover (4.1%) observed at national level by [40] between 1965 and 2000, a period of 35 years. The difference may be explained by two factors. First, they took into account the whole vegetation cover at national level including protected areas where there is no human footprint because they are protected (protected forests reserves and sanctuaries) and are consequently devoid of human pressure. Second, pressure from exponential growth of human population and livestock was probably not so high at that time. Our estimation also exceeded their findings on deforestation for the region where the study was conducted. Here the reduction of the woody cover following cutting for charcoal production was estimated between 15 and 20% in 1965 against 5 to 20% in 2004. The same trend was observed in other parts of the country specifically in the peanut basin located in west-central Senegal [41]. and in the Ferlo, northern part of Senegal where from 1976 to 1995 the woody cover decreased from 10-15% to 1-5% [42].

The croplands were characterized by a continuous expansion between the two periods to the detriment of natural vegetation cover. This result matches with the findings of many authors [2, 34, 43-44] who using remote sensing, documented an increase of croplands following vegetation cover clearance. Similar results have also been reported elsewhere in the central part of Senegal in Patako Forest [45] and in the peanut basin [46]. However, our findings contradict those of [47] in the river valley (northern Senegal) and [48] in south-western part of Senegal who documented a reduction in croplands. This difference could be explained by the fact that both studies concern agriculture in flooded areas where a combination of rainfall decrease and increase in temperature has exacerbated soil salinity leading to the abandonment of many ricefields by farmers. The increase in cropland observed in the study was more important between 1990 and 2003 (150%) than from 2003 to 2014 (30%). This finding is consistent with the scenario observed at the national level where the annual expansion rate of cropland was estimated at 27.72% between 1965 and 1985 and 20.53% from 1985 to 2000 [40]. The high increase of cropland between 1990 and 2003 may be explained by migration of farmers from the west towards the south-eastern part of Senegal under the project "terres neuves" that relocated people and created new villages to increase the production of groundnut [49-50]. Indeed, migration is an important factor driving land use change as argued by [37, 51]. Furthermore, this period coincided with a growing interest in cotton production with the establishment of the "Société de Développement et des Fibres Textiles" (SODEFITEX). It boosted the production of cotton by providing technical support, seeds, fertilizer, and credit for equipment purchase. Then, in this favourable context of technology and marketing, the area cultivated for cotton experienced a significant increase. The experience in Burkina Faso is the same [11, 52]. In the second period (2003-2014), in spite of the population growth and settlement expansion recorded in the area, the expansion rate of croplands was lower. However, contrary to the findings of [53] in the Volta Basin of Ghana who concluded that the slowing in cropland expansion was due to agricultural intensification, in Missirah Forest the crisis in cash crop production (cotton and ground nut), and the introduction of charcoal production in the area by a World Bank project which is more economically rewarding than agriculture, could be the two main explanatory factors.

5. Conclusion

Forest management was introduced to improve forest quality and ensure security of local livelihoods. Our results show that the extent of changes in land use and land cover types does not point to sustainable use of resources. Missirah Forest has been significantly cleared for cropland expansion whilst tree harvesting for charcoal production has led to more open vegetation. The change has been more important in the second period signifying a trend towards more degradation. This means the implementation of the forest management plan ought to be strengthened.

Results from the present investigation provides information on historical and current land use land cover changes that may be useful in improving existing forest management especially spatial processes. Suggestions derived from this study can be summarized into three points. First, there is a necessity to discourage croplands expansion by making effective the intensification of agriculture to cope with the increase in demand for food and/or sale and land degradation. Second, the promotion of other livelihoods other than farming and charcoal production will reduce the pressure exerted on forest which is the main source of incomes. Third, law-enforcement should be strengthened to avoid clearing of protected areas.

Acknowledgments

This study was supported by the West African Science Service Center on Climate Change and Adapted Land use (WASCAL) programme on Climate Change and Land Use which is funded by the German Federal Ministry for Education and Research (BMBF).

References

[1] Yeshaneh, E., Wagner, W., Exner-Kittridge, M., Legesse, D., & Blöschl, G. (2013). Identifying Land Use/Cover Dynamics in the Koga Catchment, Ethiopia, from Multi-Scale Data, and Implications for Environmental Change. *International Journal of Geo-Information, 2*, 302-323.

[2] Nacoulma, B. M. I., Schumann, K., Traoré, S., Marcus, B. R., Karen, H., Rudiger, W., & Thiombiano, A. (2011). Impacts of land-use on West Africa savanna vegetation: a comparison between protected and communal area in Burkina Faso. *Biodiversity and Conservation*. doi: Springer [DOI 10.1007/s 10531-011-0114-0].

[3] Kadeba, A., Nacoulma, B. M. I., Ouedraogo, A., Bachmann, Y., Thiombiano, A., Schmidt, M., & Boussim, J. I. (2015). Land cover change and plants diversity in the Sahel: A case study from northern Burkina Faso. *Annals of Forest Research, 58*(1), 109-123.

[4] Le, Q. B., Parkb, S. J., Vlek, P. L. G., & Cremersc, A. B. (2008). Land-Use Dynamic Simulator (LUDAS): A multi-agent system model for simulating spatio-temporal dynamics of coupled human–landscape system. I. Structure and theoretical specification. *Ecological I nformatics 3*, 135–153.

[5] Mayaux, P., Pekel, J.-F., Desclée, B., Donnay, F., Lupi, A., Achard, F.,... Nasi, R. (2013). State and evolution of the African rainforests between 1990 and 2010. *Philosophical Transactions of the Royal Society B: Biological Sciences, 368*(1625), 20120300.

[6] Meshesha, D. T., Tsunekawa, A., & Tsubo, M. (2010). Continuing land degradation: cause-effects in Ethiopia's central rift valley. *Land Degradation & Development*. doi: 10.1002/ldr.1061.

[7] Turner Ii, B., Skole, D., Sanderson, S., Fischer, G., Fresco, L., & Leemans, R. (1995). Land-use and land-cover change. Science/Research plan.

[8] Lambin, E. F., Turner, B. L., Geist, H. J., Agbola, S. B., Angelsen, A., Bruce, J. W.,... Folke, C. (2001). The causes of land-use and land-cover change: moving beyond the myths. *Global Environmental Change, 11*(4), 261-269.

[9] Geist, H. J., & Lambin, E. F. (2002). Proximate Causes and Underlying Driving Forces of Tropical Deforestation. *BioScience, 52*(2), 143-150.

[10] Ikpa, T. F., Debra, B. A., & Jande, J. A. (2009). Biodiversity conservation: why local inhabitants destroy habit in protected areas. *Science World Journal, 4*, 22–27.

[11] Ouedraogo, I., Tigabu, M., Savadogo, P., Compaore, H., Ode, P. C., & Ouadba, J. M. (2010). land cover change and its relation with population dynamics in burkina faso, west africa. (2010). *Land Degradation and Development..* doi: 10.1002/ldr.981 www.interscience.wiley.com.

[12] Braimoh, A. K. (2004a). Seasonal migration and land-use change in Ghana. *Land Degradation & Development 15*, 37–47.

[13] Fox, J., & Vogler, J. B. (2005). Land-use and land-cover change in montane mainland southeast Asia. *Environmental Management, 36*(3), 394-403.

[14] Hosonuma, N., Herold, M., De Sy, V., De Fries, R. S., Brockhaus, M., Verchot, L.,... Romijn, E. (2012). An assessment of deforestation and forest degradation drivers in developing countries. *Environmental Research Letters, 7*(4), 044009.

[15] Carr, D. L. (2004). Proximate population factors and deforestation in tropical agricultural frontiers. *Population and environment, 25*(6), 585-612.

[16] Hansen, C. P., Lund, J. F., & Reue, T. (2009). Neither fast, nor easy: the prospect of Reducing Emissions from Deforestation and Degradation (REDD) in Ghana. *International Forestry Review, 4*, 439-455.

[17] Norris, K., Asase, A., Collen, B., Gockowksi, J., Mason, J., Phalan, B., & Wade, A. (2010). Biodiversity in a forest-agriculture mosaic–The changing face of West African rainforests. *Biological Conservation, 143*(10), 2341-2350.

[18] Damnyag, L., Saastamoinen, O., Blay, D., Dwomohb, F. K., Anglaaere, L. C. N., & Pappinen, A. (2013). Sustaining protected areas: Identifying and controlling deforestation and forest degradation drivers in the Ankasa Conservation Area, Ghana. *Biological Conservation 165*, 86–94.

[19] Arnold, M., Köhlin, G., & Person, R. (2005). Woodfuels, livelihoods and policy interventions: changing perspectives. *World Development, 34 (n° 3)*, 496-511.

[20] Kouami, K., Yaovi, N., & Honan, A. (2009). Impact of charcoal production on woody plant species in West Africa: A case study in Togo. *Scientific Research and Essay 4*(9), 881-893.

[21] Karekezi, S. (2002). Poverty and energy in Africa – a brief review. *Energy Policy, 30*, 915–919.

[22] Ribot, J. C. (1993). Forestry policy and charcoal production in Senegal. *Energy Policy, 21*, 543–585.

[23] Nygård, R., Sawadogo, L., & Elfving, B. (2004). Wood-fuel yields in short-rotation coppice growth in the north Sudan savanna in Burkina Faso. *For Ecol Manage, 189*, 77–85.

[24] Aguilar, R., Ghilardi, A., Vega, E., Skutsch, M., & Oyama, K. (2012). Sprouting productivity and allometric relationships of two oak species managed for traditional charcoal making in central Mexico.. *Biomass Energy, 36*, 193–207.

[25] Chidumayo, E. N., & Gumbo, D. J. (2013). The environmental impacts of charcoal production in tropical ecosystems of the world: A synthesis. *Energy for Sustainable Development, 17*, 86-94.

[26] Luoga, E. J., Witkowski, E. T. F., & Balkwill, K. (2002). Harvested and standing wood stocks in protected and communal miombo woodlands of eastern Tanzania. *For Ecol Manage, 164*, 15–30.

[27] Mwampamba, T. H. (2007). Has the woodfuel crisis returned? Urban charcoal consumption in Tanzania and its implications to present and future forest availability. *Energy Policy, 35*(8), 4221-4234. doi: http://dx.doi.org/10.1016/j.enpol.2007.02.010.

[28] Agrawal, A., & Ribot, J. (1999). Accountability in decentralization: a framework with South Asian and West African cases. *The Journal of Developing Areas*, 473-502.

[29] Poteete, A. R., & Ribot, J. C. (2011). Repertoires of Domination: Decentralization as Process in Botswana and Senegal. *World Development, 39*(3), 439-449.

[30] Faye, P. (2006). Décentralisation, pluralisme institutionnel et démocratie locale : Étude de cas de la gestion du massif forestier Missirah/Kothiary (région de Tambacounda, Sénégal). Conseil pour le développement de la recherche en sciences sociales en Afrique World Resources Institute Centre de Coopération Internationale en Recherche agronomique pour le Développement.

[31] Ribot, J. C. (2009). Analysis of Senegal's draft forestry code: With special attention to its support for decentralization laws. USAID (pp. 68). Dakar, Senegal.

[32] Ribot, J. C. (2001). *Historique de la gestion forestière en Afrique de l'Ouest: Ou: comment la" science" exclut les paysans*: International Institute for Environment and Development, Programme zones arides.

[33] Wurster, K. (2010). *Management matter? Effects of charcoal productions management on woodland regeneration in Senegal.* (Ph. D dissertation), University of Marylan, USA.

[34] Ruelland, D., Levavasseur, F., & Tribotte, A. (2010). Patterns and dynamics of land-cover changes since the 1960s over three experimental areas in Mali. *International Journal of Applied Earth Observation and Geoinformation, 12*, 11-17.

[35] Aubreville, A. (1957). Accord à Yamgambi sur la nomenclature des types africains de végétation. *Bois et Forêts des Tropiques, 51*, 23-27.

[36] Hammi, S., Simonneaux, V., Alifriqui, M., Auclair, L., & Montes, N. (2007). Evolution des recouvrements forestiers et de l'occupation des sols entre 1964 et 2002 dans la haute vallee des Ait Bouguemez (Haut Atlas central, Maroc). *Secheresse 18*, 271-277.

[37] Lambin, E. F., Geist, H. J., & Lepers, E. (2003). Dynamics of land use and land cover change in tropical regions. *Annual Review of Environment and resources, 28:* 205–241. doi: 10.1146/annurev.energy.28.050302.105459.

[38] Chidumayo, E. N., Masaileti, I., Ntalasha, H., & Kalumiana, O. S. (2001). Charcoal potential in southern Africa (CHAPOSA) — report for Zambia. Stockholm: Stockholm Environment Institute.

[39] Arnold, M., & Persson, R. (2003). Reassessing the fuelwood situation in developing countries. 2003; 5:. *International Forestry Review, 5*, 379–383.

[40] Tappan, G. G., Sall, M., Wood, E. C., & Cushing, M. (2004). Ecoregions and land cover trends in Senegal. *Journal of Arid Environments 59*, 427-462.

[41] Wood, E., Tappan, G., & Jacobs, D. (1995). *Monitoring Senegal's natural resources using airborne videography. Proceedings.* Paper presented at the 15th Biennial Workshop on Color Photography and Videography in Resource Assessment. Terre haute, Indiana. 315-325.

[42] CSE. (1998). La vegetation ligneuse du Sene´gal: distribution, dynamique et fonction (pp. 52). Dakar, Senegal.

[43] Ouédraogo, I. (2006). Land use dynamics in Bieha district, Sissili province; southern Burkina Faso, West Africa. *Umoja: Bulletin of the African and American Studies, 1* (2), 18-34.

[44] Pare, S., Soderberg, U., Sandewall, M., & Ouadba, J. M. (2008). Land use analysis from spatial and field data capture in southern Burkina Faso, West Africa. *Agriculture, Ecosystems & Environment, 127*, 277–285.

[45] Guiro, I., Mbow, C., Baret, F., & Diaw, A. T. (2012). Dynamique de l'occupation du sol de la forêt classée de Patako et de sa périphérie de 1972 à 2002. *Revue de Géographie du Laboratoire Leïdi.*

[46] Tappan, G., Hadj, A., Wood, E., & Lietzow, R. (2000). Use of argon, corona, and landsat imagery to assess 30 years of land resource changes in west–central Senegal. *Photogram metric Engineering and Remote Sensing, 66*(6), 727–735.

[47] Sarr, M. A. (2002). Cartographie des changements de l'occupation du sol entre 1990 et 2002 dans le nord du Sénégal (Ferlo) à partir des images Landsat URL: http://cybergeo.revues.org/index 22707. html. *Cybergeo: European Journal of Geography, Environnement Nature Paysage.*

[48] Sambou, H., Sambou, B., Diaw, A. T., Mbow, C., & Traore, V. B. (2014). Remote sensing mapping of the rice field and vegetal cover and the impacts of soil factors on herbaceous in the sub-watershed of Boutolate (lower Casamance, Senegal. *American Journal of Environmental Protection, 3*(2), 73-82.

[49] Fall, A. S. (1992). Une réponse à la crise de l'agriculture. La migration des Sereer du Siin (Sénégal) *Sociétés-Espaces-Temps, 1*(1), 138–149.

[50] Mbow, C., Mertz, O., Diouf, A., Rasmussen, K., &Reenberg, A. (2008). The history of environmental change and adaptation in eastern Saloum–Senegal—Driving forces and perceptions. *Global and Planetary Change, 64*, 210–221.

[51] Ouedraogo, I., Savadogo, P., Tigabu, M., Cole, R., Oden, P. C., & Ouadba, J. M. (2009). Is rural migration a threat to environmental sustainability in Southern Burkina Faso? *Land Degradation & Development, 20*, 217–230.

[52] Gray, L. C. (2005). What kind of intensification? Agricultural practice, soil fertility and socioeconomic differentiation in rural Burkina Faso. *Geographical Journal 171*, 70–82.

[53] Braimoh, A. K., & Vlek, L. G. (2004b). Land-Cover Change Analyses in the Volta Basin of Ghana. *Earth Interactions, 8*, 17.

The Role of the Public in Environmental Decision-Making

Nino Gokhelashvili

Department of Environmental Policy and International Relations, Ministry of Environment and Natural Resources Protection, Tbilisi, Georgia

Email address:
n.gokhelashvili@moe.gov.ge

Abstract: Many cases prove that involvement of the public in decision-making process on environmental issues has a real influence on proposed or planned activities and the final decisions are acceptable for all main stakeholders: government, the public, and developer of activity. Despite the facts demonstrating the role of the public in environmental decision-making, various experts approach to the public participation process with doubt and some criticism. From their point of view, many aspects of public participation are still questionable, such as who, why, how, at what stage should participate, etc. This article does not aim to answer each existing question. But along with the criticism of various experts, highlighting some important aspects of public participation, the article aims to: demonstrate a rationale of public participation in environmental decision-making; show the trend in this direction based on the analysis of violations of public participation procedures in some countries; demonstrate the real influence of the public on final decisions with regard to the environment, focused on the concrete examples. The arguments and conclusions made on this overview creates additional inspiration for further discussions and exploration of different aspects of public participation in environmental decision-making.

Keywords: Public, Government, Participation, Environment, Decision-Making

1. Introduction

Considering democratic principles of the countries around the world, all stakeholders should be involved in the decision-making process, where the public[1] is one of the main players. Public participation is becoming extremely important when it comes to the matter of common interest. It is not arguable that the environment is the issue of common concern – from the individual citizens to the government officials. Despite the different approaches to the environment and natural resources we have, there is a common consent that all we need is a healthy environment. This is a reality which should consolidate us in acting in an environmentally friendly manner and being careful with our environment.

From the 90's of the last century, many theories have been developed concerning public participation in the environmental decision-making process. The more the work in this direction exists, the more it raises discussion topics and questions. The issue is quite a complex and thus it is difficult to reach a consensus. However, there would be no doubt that the public should participate in the decision-making process when it comes to the environment, but how – it definitely needs discussion.

Besides the basic principles of human rights, granting people both to leave in a healthy environment and to participate in decision-making, existed a prevalent agreement that all interested parties should participate in environmental matters generally caused by the complexity of environmental issues. Therefore everyone has its own duty towards the environment. For example, the government has an obligation to protect the environment and the public should have a responsibility to take care of it. It is well known that public involvement encourages and, at the same time, forces the governments to carry out a policy in compliance with the requirements of society considering their needs. Moreover, such approach often creates a basis for less conflict situations.

Providing some examples of public participation in the activities related to the environment, the presented article demonstrates at what extent the public can influence on final decision, and what the final decision results in.

1 According to the definition of the United Nations Economic Commission for Europe (UNECE) Convention on Access to Information, Public Participation in Decision-making and Access to Justice in Environmental Matters (Aarhus Convention) „'the Public' means one or more natural or legal persons, and, in accordance with national legislation or practice, their associations, organizations or groups." Done at Aarhus, Denmark on 25 June 1998.

This article is organized in six main parts. In the first one it provides the arguments on the importance and necessity of public participation in the environmental decision-making process. The second part, based on the opinions of some experts in the field, the criticism on the issue is presented. In the third section the article explores the facts about violations of public's right – to take part in the environmental decision-making processes in the countries of the UNECE region, who are the parties to the Aarhus Convention. In the fourth part some examples when the public had a real influence on the activities related to the environment are presented. In the fifth section the article provides results and discussion topics. In the final sixth part some conclusions are made and, based on the information provided in this paper, it opens the floor for further research.

2. General Arguments for Public Participation in Environmental Decision-Making

The United Nations Conference on Environment & Development (UNCED) in Rio de Janerio, Brazil in 1992, with participation of 172 government representatives, including 116 heads of states, about 2400 representatives of non-governmental organizations (NGOs) and 17,000 other representatives recognized the public participation as one of the main principles (among other 27 ones) for sustainable development, stating that "environmental issues are best handled with the participation of all concerned citizens, at the relevant level. At the national level, each individual shall have"..."the opportunity to participate in the decision-making processes. States shall facilitate and encourage public awareness and participation by making information widely available..." [1].

It is recognized as well that environmental problems cannot be solved by governments alone [2]. In order to make effective solutions to the environmental problems, participation of only experts and political elite is not enough. Environmental protection is a cooperative process that requires coordinated action of government, NGOs, scientists, the private sector and individual citizens. Therefore, public participation in environmental decision-making is extremely important. Such approach will contribute to each above listed group and the whole society in general [3].

Moreover, public involvement in the environmental discussions, promotes transparency of the process and raises accountability of the governing bodies, which ensures the democratic legitimacy of decision-making. The good environmental governance depends on the participatory approach. Besides that, neglecting information provided from the public leads to legitimacy questions and potential conflicts [4]. Possibly, public participation in environmental decision-making could increase the commitment among stakeholders, which strengthens enforcement of environmental laws. Through joint action and mutual efforts better environmental governance is expected to be achieved by making the most appropriate decision possible [5],[6]. "...in the field of the environment, improved access to information and public participation in decision-making enhance the quality and the implementation of decisions, contribute to public awareness of environmental issues, give the public the opportunity to express its concerns and enable public authorities to take due account of such concerns" [7].

Many experts and public servants have analyzed the importance of public participation in the rational environmental decision-making. Up to now many treaties, theories and mechanisms have been developed about public participation. In this article there is a reference to the two documents, remaining the governments their duty – to serve the public and requiring from them to set up relevant procedures, and develop appropriate mechanisms to support the public to exercise its right of being included in the environmental decision-making process adequately.

One from the above mentioned documents is the European Union Directive on public participation in decision-making and access to justice in environmental matters. According to the directive "effective public participation in the taking of decisions enables the public to express, and the decision-maker to take account of, opinions and concerns which may be relevant to those decisions, thereby increasing the accountability and transparency of the decision-making process and contributing to public awareness of environmental issues and support for the decisions taken" [8]. To support the public participation in the decision-making process in the field of the environment, the presented directive requires from the EU member states implementation of the following activities:

1. adoption of national legislation and designation of competent authority/ies;
2. establishment of a mechanism for providing the public with information;
3. establishment of a mechanism for public consultation;
4. establishment of a mechanism for public comments and opinions to be taken into account in the decision-making process.

This directive, requiring from the member states facilitation of public participation procedures, underlines the importance of the public as one of the main stakeholders in environmental issues.

The second document mentioned in this article is the Aarhus Convention. This treaty recognizes the right of every person "to live in an environment adequate to his or her health and well-being, and the duty, both individually and in association with others, to protect and improve the environment for the benefit of present and future generations"[7]. The Convention requires from the contracting parties to guarantee the public's rights on access to environmental information, participation in the decision-making process and access to justice in environmental matters in order to contribute to the protection of human right – to live in a healthy environment. The Aarhus Convention considers public participation as the integral part of democratic governance. "The Aarhus Convention remains the most ambitious venture in the field of environmental democracy under the auspices of the United Nations. The Convention is the only international legally binding instrument giving the public broad and concrete rights of participation in decision-making and access to information and justice regarding the environment. The Aarhus Convention twin protections for environment and human rights, and its focus on involving the public, provide a mechanism for holding governments to account in their efforts to address the multi-dimensional challenges facing our world today, including climate change, biodiversity loss, poverty reduction, increasing energy demands, rapid urbanization, and air and water pollution" [9].

The Aarhus Convention Implementation Guide states that responsible public authorities take advantage of the interest, knowledge and the energy of the public. The public often has the desire to participate in the process of collecting information and discussing options for decision-making, both out of self-interest and because of their wish to protect the environment. But for proper participation of the public in mentioned processes an open, regular and transparent procedures should be ensured by the government officials in which the public can have confidence [10].

In general, public involvement in the environmental decision-making process has two main reasons, among others. On the one hand, this is an obligation of public authorities as their function is to "serve" the public and to work for the whole society. On the other hand, public participation gives decision-makers a chance to receive a feedback from the public (information on the expectations of the public, compliance of activities or decisions of public authorities with these expectations) [11]. Furthermore, it should not be forgotten that the environment belongs to everyone and each of us must be involved in the activities related to it. Participatory approach gives the opportunity to gather more respective information and everyone's knowledge. All opinions to be expressed which will contribute to a thorough analysis of the subject and this, in its turn, lead to better decisions.

3. Criticism of Public Participation in Environmental Decision-Making

Public participation in decision-making is not an easy process. That is why the issue is the subject of constant criticism. According to the opinions of some experts many administrative problems arise from public interaction and also plenty of problems with participation programs from the public's perspective [12]. Indeed, it is very difficult to rich consensus, when there are stakeholders of different views and values seated at the discussion table. According to the opinions of some experts, participative democracy could not guarantee the substantive environmental benefits if there are competing views of what the environment should be like and what it is valuable for [13].

Many of aspects are questionable for the experts, such as: what are the mechanisms for effective public participation; who has to take part in the decision-making process; when and how the public participation should be for the better final decision; what their purpose is; to what extent public participation improves final decision; whether public participation always contributes to the better decision or on the contrary – prevents it; what the "better decision" means and how to estimate it; etc. Consequently, the participatory approach to environmental decision-making is becoming a controversial topic.

With regard to the public participation, permanent discussions will continue and different views will exist. However, with the analysis of the factual material, it is possible to answer some of the questions.

4. Violations of Public Participation Procedures in Environmental Decision-Making in the Contracting Parties[2] to the Aarhus Convention

The presented section provides an overview of all communications submitted to the Aarhus Convention secretariat, which covers the years from 2003 to 2014. According to the established procedure under the Convention, the secretariat informs the Compliance Committee3 of any submissions that it receives and circulates to it. Compliance Committee is one of the bodies of the Aarhus Convention that provides findings with recommendations on violations of implementation of the provisions of the Aarhus Convention on the basis of communications, submissions or referrals. Information on violations of the provisions of the Aarhus Convention can be submitted to the secretariat of the Convention from the different interesting party through the different procedures, in particular:

1. any member of the public by submitting a communication;
2. any Party to the Aarhus Convention on the compliance of another Party or its own compliance by making a submission;
3. Aarhus Convention secretariat by making a referral.

Below there are data elaborated by the Aarhus Convention Compliance Committee, which could be interesting for further discussion.

4.1. Submitted Communications to the Aarhus Convention Compliance Committee in the Period 2003-2014 [4]

In the above mentioned period (2003-2014) 95 communications to the Compliance Committee out of the 96 ones were submitted from the public.

2As of January 2015, there are 47 Parties to the Aarhus Convention, including European Union.

3 http://www.unece.org/env/pp/cc.html

4 http://www.unece.org/index.php?id=35290

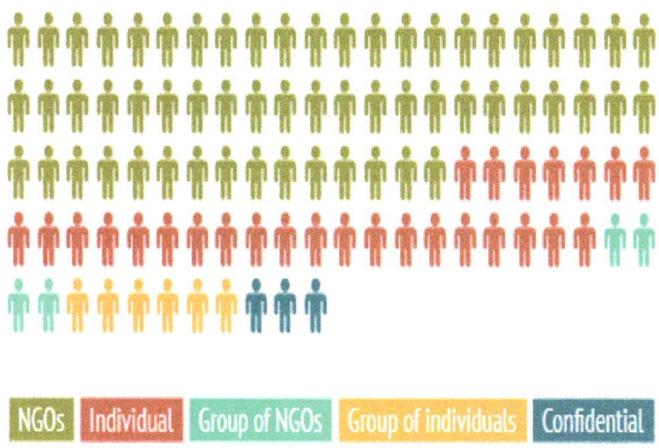

Figure 1. *Communications to the Compliance Committee in the period 2003-2014.*

The Figure 1 shows that the vast majority of communications to the Compliance Committee from the public were submitted by NGOs (59 from 96).

4.2. The Articles of the Aarhus Convention Found in Non-Compliance in the Period 2003-2014[5]

According to the data presented below, 47 findings were adopted by the Committee in the period 2003-2014, from which 3 out of 5 found in non-compliance with the requirements of the Aarhus Convention. Below there are the provisions of the Convention found in non-compliance:

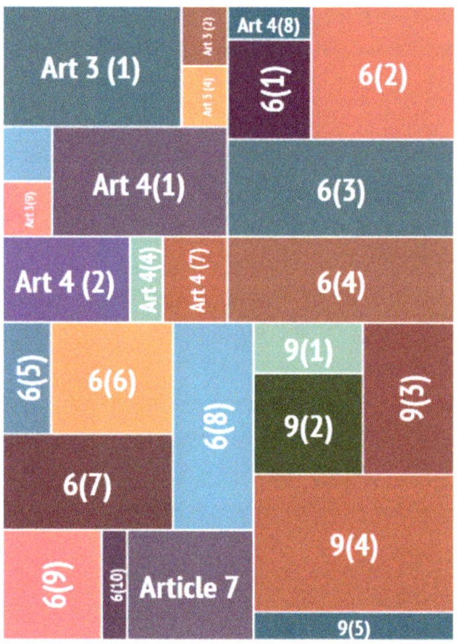

Figure 2. *Aarhus Convention articles found in non-compliance (2003-2014).*

Based on the information provided in the Figure 2, which shows which articles of the Aarhus Convention have been found out in non-compliance by the Compliance Committee, it is possible to make percentage calculation on violations of the three pillars of the Convention, such as:

- General provisions – 12%
- Access to environmental information – 15%
- Public participation in decisions on specific activities, and concerning plans, programmes and policies related to the environment – 50%
- Access to justice – 23 %.

According to data provided above, it is obvious that the main part of the communications refer to the violations of public participation procedures regarding the environmental issues.

4.3. Number of Communications Submitted to the Compliance Committee in the Period 2003-2014[6]*.*

Reviewing communications to the Compliance Committee in each intersessional period[7], we can see a dynamic of submitted communications.

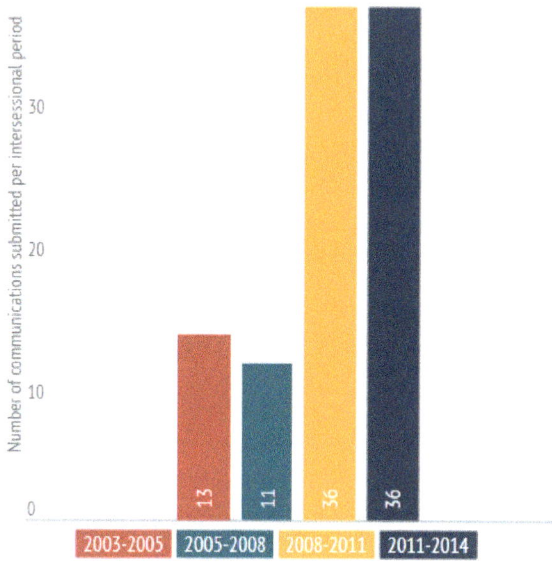

Figure 3. *Communications to the Compliance Committee (2003-2014).*

According to the information provided in the Figure 3, it is obvious that number of communications have been increased considerably since 2008:

- 2003-2005: 13 communications
- 2005-2008: 11 communications
- 2008-2011: 36 communications
- 2011-2014: 36 communications.

5. Public Influence on Environment Related Activities

Nowadays, when environmental issues are becoming increasingly challenging, the public participation, as one of the main parts of dealing with the environmental problems, has become more actual. „...there are countless cases where unexpected or hidden factors became apparent only through a public participation process, with the result that potentially

5 http://www.unece.org/index.php?id=35294

6 http://www.unece.org/index.php?id=35290.

7 Under "intersessional period" time between the Meetings of the Parties to the Aarhus Convention is considered.

costly mistakes were avoided. Furthermore, even where the original proposal is not substantially changed as a result of public participation, the successful implementation of the final decision can be promoted through the active and real participation of the public during the decision-making process. Conversely, public participation that is merely pro forma – i.e., that takes place when options are already closed – can injure the chances for successful implementation of a decision because of the questionable legitimacy of the process" [10].

To confirm the above mentioned proposal, it will be reasonable to present the real facts. Therefore, a brief description of some cases that took place in Georgia, when public participation had a real influence on the process is presented in the presented article. Advocacy campaigns for all the cases provided below were carried out by the NGO Green Alternative [14]:

1. A road rehabilitation project in one of the regions of the country was planned. Before starting of particular activities, it was found out that in case of the implementation of the proposed project, anticipated social and environmental impacts would be very high. Revealing an inconsistency in a timely manner, goal-oriented lobbying and activation of the public resulted in changing of already planned activities. Public's involvement finally brought tangible results in improving the environmental and social aspects of the project in terms of efficiency and effectiveness.
2. From one of the three routes offered for the construction and rehabilitation of the transmission line, the less costly route was selected by the government and investor. It was revealed that selected route could seriously affect the biodiversity. As a result of advocacy campaigns of the NGO Green Alternative, the route with the least impact on the biodiversity has been selected, for which additional funds have been allocated. As a result of the competent and targeted public participation, the most environmentally acceptable route was in place. Despite the fact that this route was more expensive than the one planned before, due to the proper advocacy and active public participation all stakeholders agreed on the proposed environmentally friendly version.
3. Regarding construction of a hydro power cascade in one of the regions of the country, it was found out that its construction and further exploitation would be problematic from the environmental, social and economic point of view. As a result of intensive debates and negotiations, where public participated actively, further development of the construction of this hydro power cascade has been stopped for that stage.

All the aforementioned cases clearly show that as a result of public participation decisions were made, which were acceptable for the local population, public authorities, project developers and, what is more important, for the environment.

6. Results and Discussions

According to data provided by the Aarhus Convention Compliance Committee in section 4 above, we have the following information: most part of violations refers to the public participation procedures (Fig.1); majority of communications are submitted by NGOs (Fig.2); and number of communications were increased in the last period (2008-2014). Based on the information interesting discussion might be held, assuming the following:

a) the public interest towards the environmental issues has been raised and, consequently, the public demand be involved more actively in the decision-making process when activity is addressing the environment has been increased;

b) the public has become more competent in environmental issues and its responsibility – to take care of the environment, and its right – to take part in environmental decision-making. As a result, the public started protesting its ignorance in the environmental decision-making process;

c) NGOs are more active, organized or competent part of the public to trigger the issue;

d) it became more frequent from public authorities to violate the public's right of taking part in the environmental decision-making process;

e) since the communications to the Compliance Committee are submitted only after national legal procedures of particular countries, it gives the reason to think that number of countries do not have properly regulated procedures for public participation, and consequently, cannot resolve the issue within the frames of their national legislation. Such approach, in its turn, leaves the assumption that there is no capacity in respective countries to establish appropriate procedures and develop proper implementation mechanisms for the public involvement in environmental decision-making or there is no political will to regulate the process;

f) the public have acquired more freedom in expressing its opinions, which in its turn indicates the high quality of democracy in the countries; etc.

To explore what is happening in this direction and which of the mentioned proposals would be relevant, a thorough study should be conducted addressing all aspects of public participation.

7. Conclusions

From the article, which mainly reviews the UNECE region and presents Georgia's examples in terms of public participation in the decision-making process related to the environmental issues, as a positive sign of the level of activism of the public in the countries could be seen.

Despite many questions and controversial issues, it is a reality that nowadays there are many examples when the public participation in environmental decision-making contributes to the sound decisions. The main reason not to use the general term "better decision" here is caused by the fact that, as it was mentioned above, it leads to questions and doubts among some experts. But the following explanation would not be questionable or doubtful: in my understanding, "better decision" in this context means a decision that would be acceptable to the public and local communities in terms of improved social conditions, for the government – in terms of economic progress and security of the country, and for the environment – in terms of minimizing the impact on it and its maintenance. Hereby, the significant share of environmental factor in the concept of "better decision" should be highlighted, referring to the consideration that minimization of impact on the environment finally leads to the economic growth of the country and improvement of social conditions of the population.

Based on the content of the communications to the Aarhus Convention Compliance Committee and its findings, it could be assumed that countries are still not ready or don't make sufficient efforts to ensure adequate public participation in the decision-making process of environmental issues. In this direction development of national legislation in compliance with generally recognized principles of respective international environmental law and treaties, and establishment of appropriate mechanisms for its implementation is crucial.

A proper public participation in the environmental decision-making process requires a lot of efforts. On the one hand, the existence of a political will is necessary, and on the other hand, the desire of the public and the sense of civil responsibility to be actively involved in the process is essential. Moreover, the public is obliged to take care of its natural surroundings for the wellbeing of itself and its generation. To that it is very important that the public has a confidence, that the government respects its opinions and takes them into account while making final decisions concerning environmental matters.

Involvement of the public at early stage, when the options are open, is especially important. Public participation at the early stage avoids additional time and finances in case of some changes in the project are requested from the public. Moreover, such approach prevents the conflict that may arise during the implementation of particular activities, when the public does not agree with them.

Hereby, two aspects that are very important for the successful participatory process should be mentioned. One of them is that the confidence should be existed between the actors. The second considerable aspect is that all the parties included in the process have to realize that they are not barriers of each other, they are only partners the efforts of which are strived to reach a consensus on the subject and achieve a better final decision.

During the analysis of the public participation topic importance of environmental education and awareness raising on environmental issues should be emphasized as well. When there is a lack of environmental knowledge among the population, it is possible that the public unintentionally becomes a partner to the player, who is not a great supporter of environmental principles and standards. There is a risk that such population will be "fascinated" by offering temporarily improvement of economic social conditions, which will be implemented at the expense of inadequate approach to the environment, but in the long term perspective, will turn out into worse conditions. Besides, unaware public, that has a lack of knowledge about its environmental rights, will easily become the subject of ignorance or even fraud by dishonest player. Unfortunately, there are examples of it.

Based on the facts, arguments and brief analysis provided in this article and considering the existing reality, that active discussion on public participation in the environmental decision-making process continues around the world, it is obvious that the issue is of high importance. Therefore, public participation in environmental decision-making needs well-considered and wise approach in order not to make the process unintentionally wasted or, even more, harmful for final decision. It is very important to bear in mind that "without a coordinated vision, new and commendable initiatives in public participation may ultimately yield little more than wasted effort and frustration"[12].

Finally, analyzing the information of the Aarhus Convention Compliance Committee on the existing gaps in public participation procedures in the countries of the UNECE region, who are the parties to the Aarhus Convention, it is clear that public reaction on violations of its right – be included in the environmental decision-making process – has been increased nearly three times since 2008. This fact does not allow us to make a conclusion, whether the number of violations has been increased or the public have become more active, but it stimulates research to bring the issue up for further discussion.

References

[1] Principle 10. Public participation. The Rio Declaration on Environment and Development, (1992), United Nations "Conference on Environment and Development" (UNCED).

[2] Pring, G. and Noé, S.Y. (2002) "The Emerging International Law of Public Participation Affecting Global Mining, Energy, and Resource Development" in Zillman, D.M., Lucas, A. and Pring, G.(eds) Human Rights in Natural Resource Development: Public Participation in the Sustainable Development of Mining and Energy Resources, Oxford University Press, Oxford, p. 76.

[3] Public Participation in Making Local Environmental Decisions. The Aarhus Convention Newcastle Workshop. Good Practice Handbook. United Kingdom. 2000.

[4] Coenen, Frans H. J. M. (Ed.), Public Participation and Better Environmental Decisions, The Promise and Limits of Participatory Processes for the Quality of Environmentally Related Decision-making, (2009), VIII, p. 209.

[5] Bulkeley, H. and Mol, A.P.J. (2003), "Participation and Environmental Governance: Consensus, Ambivalence and Debate", Environmental Values 12 (2): pp.143–54.

[6] Fischer, F. (2000) Citizens, Experts and the Environment, Durham, NC: Duke University Press, p. 222.

[7] UNECE Convention on Access to Information, Public Participation in Decision-making and Access to Justice in Environmental Matters (Aarhus Convention. 1998). Preamble.

[8] Directive 2003/35/EC providing for public participation in respect of the drawing up of certain plans and programmes relating to the environment and amending with regard to public participation and access to justice Directives 85/337/EEC and 96/61/EC. Preamble (3).

[9] Ban ki-Moon, Secretary-General of the United Nations. United Nations Economic Commission for Europe. The Aarhus Convention. An Implementation Guide. Second edition, (2014). United Nations. p.3.

[10] United Nations Economic Commission for Europe. The Aarhus Convention: Implementation Guide. Second edition, (2014). Pillar II, Public Participation in Decision-making. Purpose of the Public Participation. p. 119.

[11] European Commission. Implementation of the Aarhus Convention. Guidelines for Public Servants for Georgia. Kiev, Ukraine (2004).p.51.

[12] Nancy Perkins Spyke Public Participation in Environmental Decision-making at the New Millennium: Structuring New Spheres of Public Influence. Boston College Environmental Affairs Law Review. Volume 26 | Issue 2 Article 2. 12-1-1999.

[13] Davies, A. (2001). "What silence knows – planning, public participation and environmental values", Environmental Values, 10: pp. 77–102.

[14] Green Alternative. Advocating for Environmental and Social Justice. 10 Advocacy Campaigns. (2011). pp.3-5; 11-13; 18-20.

22

Resilience of Ecosystems to Climate Change

Mengistu Asmamaw[1, *], Argaw Ambellu[2], Seid Tiku[2]

[1]Department of Biology, Debre Berhan University, Debre Berhan, Ethiopia
[2]Department of Environmental Health Sciences and Technology, Jimma University, Jimma, Ethiopia

Email address:
mengistilash99@gmail.com (M. Asmamaw), aambelu@yahoo.com (A. Ambelu)

Abstract: Human-induced climate change is occurring as an unprecedented rate resulting in ecosystem deterioration and biodiversity loss. Climate change exacerbates environmental hazards, ecosystem modification as well as loss of biological organisms. Anthropogenic releases of greenhouse gases, mainly CO_2 has warmed the Earth's surface. The average global temperature has risen more linearly for the last 100 year (1906-2005) by 0.74°C (.56-0.92°C); than the corresponding period (1901-2000) of 0.6°C (0.4-0.8°C). The objective of this review was to examine the potentials of ecosystems in mitigating climate change and in building resilience. A review of relevant literature was employed as possible approach to compile this document. Ethiopia is an agrarian country where agriculture is both the cause and victim to extreme climate variability like elevated temperature and precipitation fluctuation. These have resulted recurrent drought, flooding and reduction in agricultural productivity as a whole. The occurrence and spatial distribution of drought is projected to continue in the upcoming future. The effects of climate variability significantly reduced farmers' adaptive capacity while exacerbating their vulnerability to further weather events. The current trends of ecosystem degradation could not be controlled only through protection of parks. Instead, it involves a large-scale and integrated approach addressing the whole land and sea-escapes. This large scale and integrated ecosystem management approach conserves biodiversity, builds system resilience as well as to ensures the sustainable production of ecosystem services from which the majorities of the rural poor depend on. Resilient ecosystems have a wide range of biodiversity and ecosystem services, resist and recover from extreme events more quickly and are potential to mitigate and adapt to climate change while sustaining livelihoods. Thus, ecosystem based approach through adaptation and mitigation is instrumental to maintain ecosystem health so as to sustain system resilience in the face of climate change.

Keywords: Adaptation, Climate Change, Ecosystem, Mitigation, Resilience

---

1. Introduction: Climate Change and Consequences

1.1. Current Trends and Future Projections of Climate Change

Human-induced climate change is occurring [33, 35, 39] as an unprecedented rate resulting in considerable consequences on ecosystems and human well-being [32, 39, 43]. Anthropogenic releases of greenhouse gases, mainly carbon dioxide (CO_2) [35] has warmed the Earth's surface. Eleven of the last twelve years (1995-2006) rank among the warmest years in record of global surface temperature since 1850. The average global temperature has risen more linearly for the last 100 year (1906-2005) by 0.74°C; than the corresponding period (1901-2000) of 0.6°C (Fig.3). As a result, this period (20thc) is the largest and fastest warming period of the planet [35] leading to occurrences of extreme weather events, such as rising sea-levels, recurrent droughts, intensive hurricanes and floods, etc. [49].

In line with atmospheric temperature, global sea level was increased at average rate of 1.8 mm/yr and at 3.1 mm/yr since 1961 and 1993, respectively. This has resulted in thermal expansion, melting of glaciers and ice caps, and polar ice. On the other hand, Arctic sea ice has reduced by 2.7 percent per decade since 1978; where a reduction by 7.4 percent per decade was recorded during summer seasons [35] (Fig 1).

1.2. Consequences of Climate Change

Climate change is a universal challenge; threatening

ecosystems, biodiversity and livelihoods. However, it becomes worse particularly affecting the developing poor who dependence on climate-sensitive sectors, such as agriculture, livestock and eco-tourism for their survival [35] and is highly vulnerable and suffer from the effects of environmental changes [2, 57, 35]. Some common climate-induced impacts are presented as follows.

1.2.1. Climate Change Increases the Risks of Climate-Related Disasters

Climate-induced hazards, such as floods, storms, landslides, droughts, etc. are risks to all nations throughout the world damaging ecosystems, humans, and their economies; thereby deteriorating social-ecological resilience [49]. Their occurrence and intensity are expected to continue in the future even causing magnified disaster risks [35], reducing ecosystem's resilience against the effects of climate change [74]; whilst increasing risk of disasters [35, 49]. The number of climate-induced disasters for the last 70 years has been illustrated in figure 2 and shows climate related extreme events are sharply increasing since 1960s and their occurrences dramatically accelerated rapidly since 1990s Fig. 2 [49].

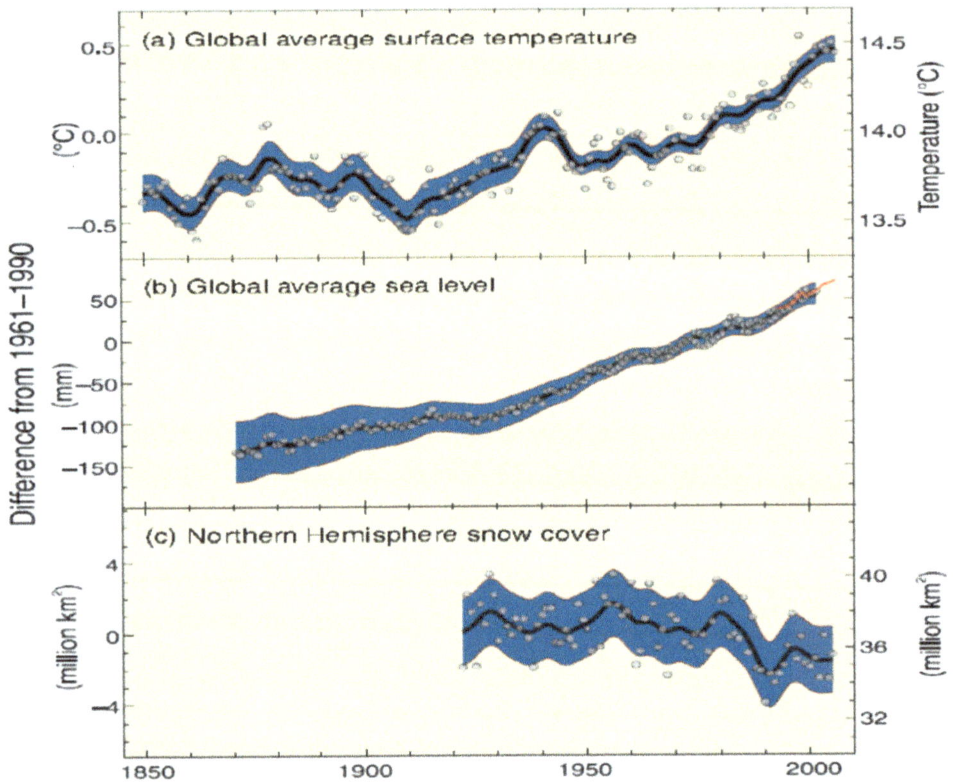

Figure 1. *Changes in Temperature, Sea level and Northern Hemisphere Snow cover [35].*

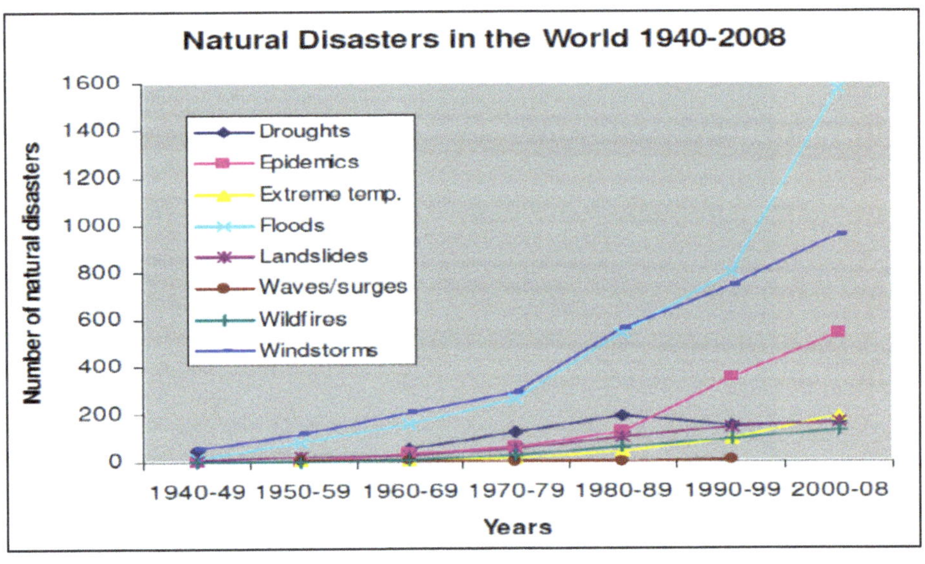

Figure 2. *Number of Climate-induced Disasters in the World during 1940s to 2008 (Source: International Disaster Database cited in [49]).*

1.2.2. Climate Change Exacerbates Ecosystem Degradation and Biodiversity Loss

The socioeconomic and biophysical environment of the Earth is rapidly changing [38]; particularly for the last five decades [47, 14]. It could not be denied that ecosystem changes have contributed to societal developments. However, they have resulted in environmental degradation by deteriorating ecosystem services, aggravating environmental changes and vulnerability [47, 14].

For instance, the Global Biodiversity Outlook-3, published by the Convention of Biological Diversity in 2010 reveals that natural habitats are shrinking significantly during the last century all over the world; where mangroves, forests and wetlands declined by 35%, 40% and 50%, respectively [58]. Consequently, one third of the abundance of vertebrates has been declined and almost a quarter of plant species are likely to be threatened with extinction during the period of 1970 to 2006 [16].

Climate change triggers ecosystem transformation, biodiversity loss [12] and substantial modifications of ecosystem services [47]. The structure and function of terrestrial, freshwater and marine ecosystems are considerably changing [49]. For instance, about 60 percent of ecosystem services and almost 70 percent of regulating and cultural services have been degraded and over-exploited globally [47]. It is predicted that drivers to social-ecological transformation [47] and associated consequences on ecosystems and biodiversity are expected to become stronger and complex in the future [11, 42, 35] (Fig 3).

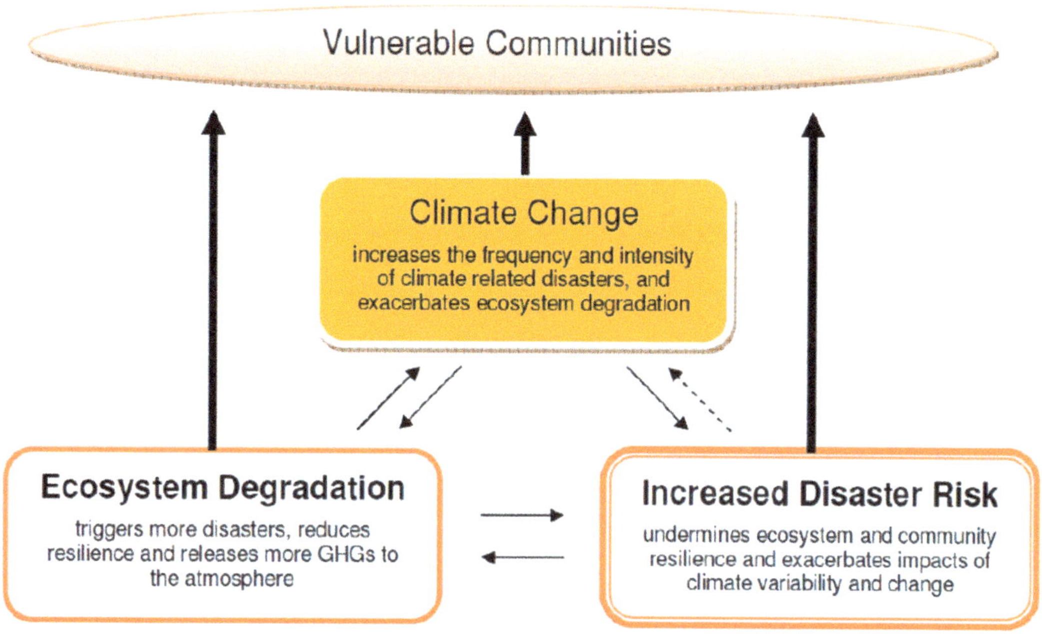

Figure 3. *Simplified Illustration of the Interaction between Climate Change, Ecosystem Degradation and Disaster Risk (source: [64]).*

1.2.3. Climate Change Impacts on Human Communities and Livelihoods

Climate change is threatening the agriculture, tourism, resource management, industry, water resources, health and others [64]. For instance, fluctuation in frequency and severity of extreme climate events has profound impacts on food and forestry production, availability of water and fodder, as well as on the health and productivity of the whole ecosystem [22]. It is predicted that the food production will be in 5 to 25% short of demand and price is expected to rise in 30-50% by 2050 [51].

Scientific evidences disclose that climate change has substantial impacts on crop productivity and associated food security [28, 73]. In Africa for example, around 75-250 millions of people are expected to be vulnerable to climate-induced water stress by 2020; a 50% decline in agricultural yields in 2020; lowland densely populated areas will suffer from sea level rise; and an estimated temperature rise of arid and semi-arid regions in 5-8% by 2080 [35]. Moreover, the Sahel experiences worldwide prominence, where more than 300,000 humans and millions of animals died, during the catastrophic drought of the early 1970s [48].

1.3. Climate Change in Ethiopia

Ethiopia is an agrarian country located in the Horn of Africa where agriculture constitutes 52 percent of the gross domestic product, absorbs 80 percent of labor force and generates more than 85 percent of the foreign exchange earnings [17]. The sector is the principal source of food production and prime contributor to food security [15]. Agriculture is often rain fed, traditional subsistence small-scale mixed crop and livestock production (8 million households) with minimal agricultural productivity; where irrigation accounts only one percent of the arable land area of the country. Trends like traditional farming system, soil degradation, poor complementary services like extension, credit, marketing, infrastructure and climatic factors, such as recurrent drought and flood have been identified as the major limiting factors that hinder agricultural productivity in the country [20, 3, 5, 75]. These factors are responsible to lessen adaptive capacity by increasing the vulnerability of farmers to forthcoming uncertainty such as climate change that affect the efficiency of the agriculture [19]. Due to its limited adaptive capacity, the country has been placed as among the most vulnerable countries in Africa in recent mapping of vulnerability and poverty in Africa [61].

Historically, Ethiopia has experienced five major droughts since 1980s [1].The effects of climate change in Ethiopia are not restricted only to recurrent droughts [40]; instead fluctuations in other critical climatic variables are apparent highlighting the multifarious consequence of climate [61]. For instance, the average minimum and maximum temperature of the country has risen by 0.25°C and 0.1°C, respectively during the last decades. Climate prediction models show that temperature in Ethiopia continues rising in the next years [59].It is projected that the mean temperature will rise by 1.7-2.1°C by 2050 and 2.7-3.4°C by 2080 [60]. Most models are consistent with increasing patterns of temperature. However, depending on the type of models used, both increasing and decreasing patterns are projected in precipitation forecasts [59]. However, [19] disclose that rainfall pattern has showed an extreme variability during the last 50 years. Furthermore, the occurrence and spatial distribution of drought has been intensified during the last decades as well as projected to continue in the upcoming [40].

This study was designed to highlight the current states of climate change and variability with a special emphasis in developing countries including in Ethiopia as well as to assess the potential of ecosystems in mitigating the effects of climate change and in building system resilience. Review approach was used to compile this study and more than 168 literatures were collected where most relevant 75 of them were selected and analyzed in this study.

2. Managing Landscapes for Ecosystem Resilience

The etymology of the word "resilience" comes from Latin "resilire" to mean "to spring back" or "to rebound" and described as springiness or elasticity. The term resilience was introduced into ecological studies in late 1950s and denoted as the amplitude of changes brought about by disturbance and the dynamics of post disturbance recovery [25].

The landmark for the application of resilience to ecosystems is laid by [30]. Nowadays, the term has different definitions and applied in various disciplines, including: ecology, sociology, economics, psychology and political science and engineering science [27, 9]. It is the capacity of an ecosystem to absorb change and re-organize itself, whilst undergoing change, in order to retain its character and ecological functioning [10].The basic notion for resilience thinking is the likely of ecosystem crossing a threshold or tipping point, beyond which it will become fundamentally, and often irreversibly different [70].

Resilience in social-ecological systems refers the capacity of a system to absorb disturbances and identifies the ongoing changes so as to still maintain the same function, structure, identity and feedbacks [70]. It is the tendency to avoid basic long term changes in main characteristics, like level of primary productivity and to maintain fertile soil (ecological system) or the presence of efficient livelihood alternatives and health care (social system) [50].

Ecosystem resilience is also defined as the capacity of an ecosystem to withstand perturbations without losing any of its functional properties. Currently, resilience denotes the capacity of a system (e.g. a community, society or ecosystem) to cope with disturbances (e.g. financial crises, floods, drought, hurricane, tsunami or fire) without shifting into a qualitatively different state [29]. A resilient system has the potential to cope with shocks, surprises and recover from disturbances; as resilience is both the capacity of a system to deal with change and continue functioning [8, 9].

2.1. Characteristics of Resilient Ecosystem

Basic characteristics of resilient ecosystem are summarized as follows [4].

Diversity and redundancy

Diversity is recognized as the most determinant factor to contribute resilience and more frequently cited by several authors [13, 27, 30, 55]. Sustaining diverse functional groups within ecosystem maintains ecosystem health and assists sustainability [30].The existence of different groups of organisms performs different functions and balance major components of an ecosystem which will be supportive in preventing the loss of key ecological resources [4]. [7] explained the concept of response diversity pointing that higher functional diversity enhances an ecosystem to be more resilient to disturbances; hence different organisms respond to disturbances in different mechanisms. Redundancy is the ability of different species to carry out similar activity. This redundancy property of a system is instrumental as the loss of certain system components doesn't jeopardize the whole system; instead it allows sustaining the function of a system to continue even with shocks [24].

Participation and community engagement

The contributions of community engagement, ownership, participation and indigenous/local knowledge to social-ecological resilience have been reported in different scholars [44, 45, 52, 6, 54, 7]. Studies underline that community participation is supportive in building legitimacy, promoting monitoring and enforcement, enhance recognition of system dynamics as well as encourage system's capacity to sense changes [7]. Therefore, participation of local communities, particularly who largely benefited from ecosystem services, influence the success of ecosystem management [7]; promotes belongingness [4] and is essential to maintain ecosystem health [54].

Polycentricism, decentralization and flexibility

Ecosystems exhibiting flexible, decentralized and well integrated management of different local and regional institutions maintain sustainability and resilience [27, 54, 56, 7]. Unlike the command and control approach of ecosystem management, decentralized management allows the incorporation of scale specific knowledge in major decisions to regulate ecosystem health [7]. This strategy is supportive for management flexibility of ecosystems and maintains the system not to tip over dis-functionality [52].

Learning, experimentation and innovation

Ecosystem management and associated resilience of ecosystem is critically determined by learning and experimentation. It is understood that change and uncertainty are an unavoidable in any ecosystems. Consequently, continual learning and knowledge revision is essential to effective ecosystem management [7]. This characteristic is especially important where ecosystems are restored from disturbances, as they are more vulnerable to similar shocks. Learning from experiences ensures that a system capable to bounce back better and resist if similar disturbances exist again [18].

Connectivity, networks and cross-scalar linkages

Connectivity is the extent in which resources, species or social actors disperse, migrate or linked to ecological and social landscapes [7]. In the context of ecosystem management, connections and networks can build resilience, as it enables a system to quickly either restore or prepared for changes [52]. In building community resilience to disasters, great interrelation across scales of governance during early warning systems results in enhanced resilience to disaster [63].

2.2. Resilience, Vulnerability and Adaptive Capacity

Vulnerability is antonym to resilience and refers to the degree of sensitivity that associates the ecosystem services to global changes and also the degree to which sectors, such as agriculture, forestry, water management, energy and nature conservation relying on ecosystem services is unable to adapt to the changes [37, 33]. Anthropogenic disturbances such as rigorous land use change makes species and their habitats vulnerable [46].Vulnerability to climate change is the degree to which a system is subject to/or unable to acclimatize adverse effects of climate change, including climate variability and extremes [33].

Vulnerability is central concept to adaptation and is dependent up on exposure, sensitivity, and adaptive capacity of a system. For instance, high exposure or sensitivity and low adaptive capacity induces high vulnerability of a system to environmental changes [41]. Sensitivity is the degree to which a system is affected, either adversely or beneficially, by climate-related stimuli. Adaptive capacity is the ability of a system to adjust to climate change (including climate variability and extremes), to moderate potential damage, to take advantage of opportunities, or to cope with the consequences [33, 35]. Adaptive management is not only maintains ecological stability but also builds more flexible institutions for resource management [31].

Resource management strategies through ecosystem-based approaches are crucial to lessen vulnerability of ecosystems and promote resilience of the communities who depend on natural resources [31]. The ecosystem concept argues that proper planning along with ecosystem changes whilst considering the role of human behavior on the environment is essential over temporal and spatial scales [62].

3. Working on Ecosystems to Mitigate Environmental Changes

Ecosystem approach is a strategy for the integrated management of land, water and living resources that promotes conservation and sustainable use in an equitable way. It is instrumental in increasing ecological connectivity with a view to increasing resilience, and contributes to address conservation objectives [21]. Ecosystem-based approaches are cost-effective, accessible to rural poor, contribute to relieve poverty and promote sustainable development [26, 49]. Ecosystem based approach is a win-win approach with multifaceted benefits and summarized as follows.

3.1. Ecosystem-Based Approach Enhances Climate Change Adaptation and/or Mitigation and for Disaster Risk Reduction

Disaster risk reduction is a practice of reducing disaster risks through reducing exposure to hazards, reducing vulnerability of social-ecological system, wise management of land and the environment and improves preparedness for adverse events [68]. Adaptation contributes to disaster risk reduction in being prepared for climate related impacts; whereas, disaster risk reduction is crucial to adaptation and it serves as a basis to achieve long-lasting climate impacts. Both climate change adaptation and disaster risk reduction rely on regulating services of ecosystems and are instrumental to build ecosystem health and resilience [53, 69].

The IPCC defines climate change adaptation as an adjustment in natural or human systems in response to actual or expected climatic stimuli or their effects, which moderates harm or exploits beneficial opportunities [33]. It is the use of natural capital by humans to acclimatize to climate change impacts; that in turn possesses manifold co-benefits for mitigation, protection of livelihoods and poverty alleviation [50].

The basic objective of adaptation is to lower humans/ecosystems vulnerability to the harmful effects of climate change (sea level rise, rising temperature, rainfall variability, drought, flood, etc.); instead creates an opportunity on existing weather events (increasing ecosystem productivity) [36]. It is a strategy designed to address the fundamental relations between climate change, biodiversity and sustainable resource management and can be functional in both developing and developed nations [50].

Adaptation options range from technological like in increasing sea defense or flood-proof houses on stilts, to behavioral changes at individual level, for instance reducing water utilization during drought, using mosquito net (insecticide-sprayed) to prevent mosquito bites. Besides, improved water management system, improved risk management, biodiversity conservation, early warning for extreme events and establishing insurance options are also adaptation strategies mostly recommended to the societies [65, 66].

Mitigation is an anthropogenic intervention involving reducing greenhouse gases concentration either by reducing their generation (burning of fossil fuels, heat or transport) or improving the potentials of their sinks (oceans, wetlands, forests and soil) to reduce the impacts of climate change [34]. The main purpose of mitigation is to avoid the unsafe human intervention in the climate systems and to maintain the concentration of greenhouse gases in level where ecosystems able to adapt naturally to climate change; so as to ensure the health and productivity of social-ecological system [36].

Mitigation action is crucial to lessen new emissions and/or enhance sinks. However, it should be integrated with adaptation [35] and implemented in broader development objectives, such as economic development, energy security, public health, air quality and local environmental protection for enhanced response to climatic impacts [67, 65, 64]. Sustainable management of key biological resources and habitats through mitigation and adaption strategies is helpful especially for rural poor, marginalized and indigenous communities by regulating ecosystem services and their access to the community in times of drought and disaster [74] and summarized in Fig. 4.

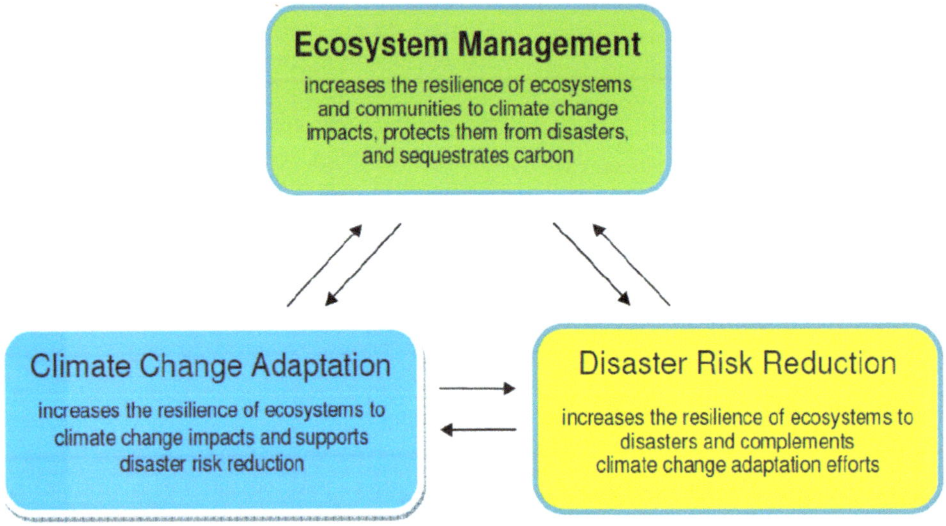

Figure 4. *Ecosystems as a foundation in Disaster risk reduction and Climate Change Adaptation [64].*

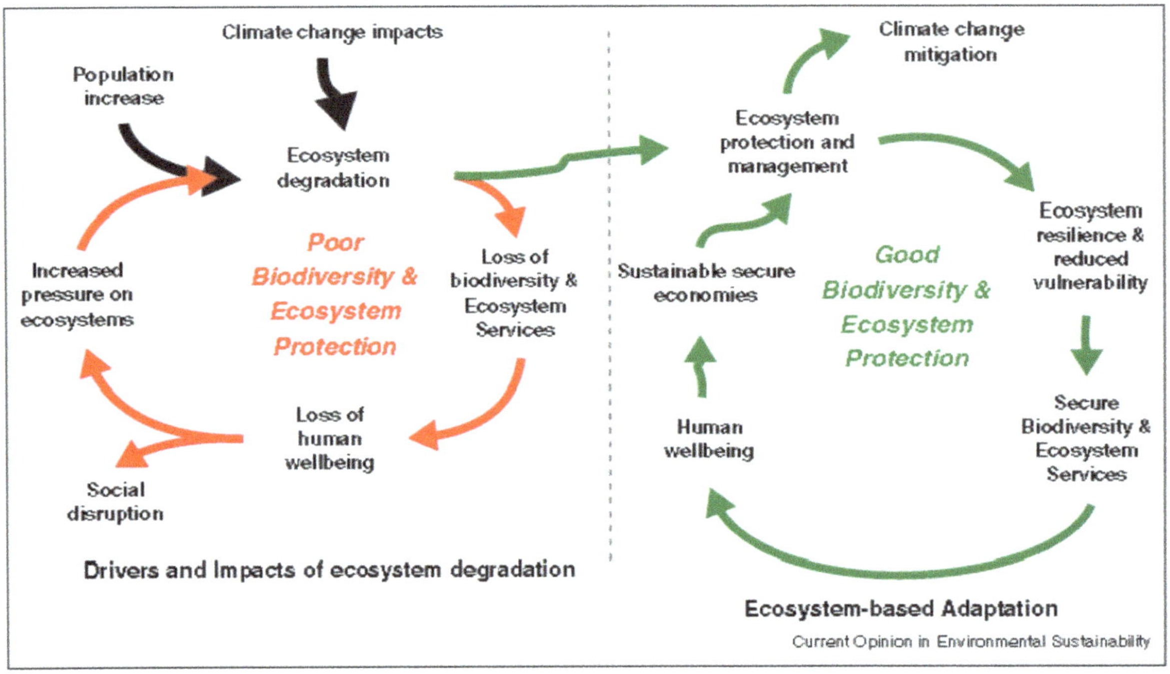

Figure 5. *The Vicious Cycle of Poverty, Ecosystem Degradation and Climate Change [49].*

3.2. Ecosystem-Based Approach Conserves Biodiversity, Maintains Ecosystem Services and Resilience

Maintaining high species diversity and functioning ecosystems across the wider terrestrial, freshwater and marine environment is essential to climate proofing and building resilience [26]; as they are effective in storing significant amount of carbon [64] and sustain unforeseen disturbances [26]. In effect, it is understood that biodiversity rich ecosystems are highly resilient to external shocks, such as extreme climatic events [64].

One of the objectives of ecosystem management is to maintain health and enhance resilience; whilst reducing vulnerability in the face of climate change. It creates substantial social, economic and cultural advantages, support biodiversity conservation and build on traditional knowledge and trends of indigenous people, including women participation [49]. Resilient ecosystems have a greater potential to mitigate and adapt to climate change and to reverse global warming. They resist and recover more easily from extreme weather events and provide a wide range of ecosystem services [26] Fig. 5.

4. Conclusion

It is real that global climate is changing at unprecedented rate resulting in considerable consequences on ecosystems and human well-being. Anthropogenic releases of greenhouse gases, mainly CO2 has warmed the Earth's surface; where its concentration has risen more linearly for the last 100 year (1906-2005) by 0.74°C than the corresponding period (1901-2000) of 0.6°C. This has resulted in occurrences of extreme weather events, such as rising sea-levels, recurrent droughts, intensive hurricanes and floods, etc. It is uncertain to predict the potential of developmental sectors if environmental change continues. Unless ecosystems become resilient enough against environmental disturbances, no life could exist on this planet. Accordingly, ecosystem resilient should be built through adaptation and mitigation strategies so as to reverse current emissions as well as to adapt to live with uncertainty.

References

[1] (ACCCA, 2010) Farm – Level Climate change Perception and Adaptation in Drought Prone Areas of Tigray, Northern Ethiopia. Improving decision-making capacity of smallholder farmers in response to climate risk adaptation in three drought-prone districts of Tigray, northern Ethiopia Vol 3. Advancing capacity to Support Climate Change Adaptation (ACCCA), Mekelle, 2010.

[2] W. Adger, P. Neil, M. Kelly, and N. Huu, *Living with Environmental Change: Social Vulnerability Adaptation, and Resilience in Vietnam*. Rutledge, London, 2001.

[3] A. D. Alene, Improved production technology and efficiency of small holder farmers in Ethiopia: Extended Parametric and non-parametric approaches to production efficiency analysis. PhD Thesis. Department of Agricultural Economics, Extension and Rural Development. University of Pretoria, South Africa, 2003.

[4] A. Bahadur, M. Ibrahim, and T. Tanner,. *The resilience renaissance? Unpacking of resilience for tackling climate change and disasters.* Brighton, UK: Institute of Development Studies, 2010.

[5] K. Belay, Agricultural extension in Ethiopia: the case of participatory demonstration and training extension system. Journal of social development in Africa 18(1), 50-83, 2003.

[6] F. Berkes, Understanding uncertainty and reducing vulnerability: Lessons from resilience thinking. *Natural Hazards* 41: 283-295, 2007.

[7] R. Biggs, M. Schlüter, D. Biggs, E. L. Bohensky, S. BurnSilver, G. Cundill, V. Dakos, T. M. Daw, L. S. Evans, K. Kotschy, A. M. Leitch, C. Meek, A. Quinlan, C. Raudsepp-Hearne, M. D. Robards, M. L. Schoon, L. Schultz, and P. C. West, Toward principles for enhancing the resilience of ecosystem services. Annual *Review of Environment and Resources,* 37: 421-448, 2007.

[8] F. Brand, *Ecological resilience and its relevance within a theory of sustainable development.* UFZ-Report 03/2005. UFZ Centre for Environmental Research, Leipzig, Germany.

[9] F. S. Brand, and K. Jax, Focusing the meaning(s) of resilience: Resilience as a descriptive concept and a boundary object. *Ecology and Society* 12, 1: 23, 2007.

[10] M. Cadman, C. Petersen, A. Driver, N. Sekhran, K. Maze, and S. Munzhedzi, *Biodiversity for Development: South Africa's landscape approach to conserving biodiversity and promoting ecosystem resilience*. South African National Biodiversity Institute, Pretoria, 2010.

[11] F. Capra, The Hidden Connections: A Science for Sustainable Living. New York: Anchor Books, 2002.

[12] S. Carpente, H. Mooney, J. Agard, D. Capistrano, R. DeFries, S. Diaz, T. Dietz, A. , A. Oteng-Yeboah, E. Pereira, C. Perrings, W. Reid, J. Sarukhan, R. Scholes, A. Whyte, Science for managing ecosystem services: Beyond the Millennium Ecosystem Assessment. *Proceedingsof the Natural Academy of Science* (106) 5: 1305-1312, 2009.

[13] S. R. Carpenter,J. J. Cole, J. R. Hodgson, J. F. Kitchell, M. L. Pace, D. Bad, K. L. Cottingham, T. E. Essington, J. N. Houser, and D. E. Schindler, Trophic cascades, nutrients and lake productivity: wholelake experiments. *Ecological Monographs* 71:163-186, 2001.

[14] Carpenter SR., et al. 2012. Program on ecosystem change and society: an international research strategy for integrated social–ecological systems, Curr Opin Environ Sustain, doi:10.1016/j.cosust.2012.01.00.

[15] CEEPA, Climate change and African agriculture. University of Pretoria, South Africa. Policy note No. 25, 2006.

[16] Convention on Biological Diversity (CBD), *Global Biodiversity Outlook-3.* Convention on Biological Diversity, 2010.

[17] CSA (Central Statistics Authority), The Federal Democratic Republic of Ethiopia Statistical abstract. Addis Ababa, Ethiopia, 2005.

[18] Department for International Development (DFID) *Defining Disaster Resilience: A DFID Approach Paper.* Department for International Development: London, UK, 2011.

[19] T. T. Deressa, R. M. Hassan, C. Ringler, Perception of an adaptation to climate change by farmers in the Nile basin of Ethiopia. J Agri Sci 149:23–31, 2011.

[20] S. Devereux, Food security in Ethiopia: A discussion paper for DFID. Institute for Development Studies, Sussex, 2000.

[21] N. Dudley, S. Stolton, A. Belokurov, L. Krueger, N. Lopoukhine, K. MacKinnon, T. Sandwith, N. Sekhran, *Natural Solutions: Protected areas helping people cope with climate change.* IUCN-WCPA, TNC, UNDP, WCS, The World Bank, WWF, Gland, Switzerland, 2010.

[22] W. E. Easterling, P. K. Aggarwal, K .M. Batima, L. Brander, S. M. Erda, A. Howden, J. Kirilenko, J. F. Morton, J. Soussana, and F. N. Tubiello, Food, fibre and forest products. Climate Change 2007: Impacts, Adaptation and Vulnerability. Contribution of Working Group II to the Fourth Assessment Report of the Intergovernmental Panel on Climate Change, M.L. Parry, O.F. Canziani, J.P. Palutikof, P.J. van der Linden and C.E. Hanson, Eds., Cambridge University Press, Cambridge, UK, 273-313, 2007.

[23] [23]J. Eliasch, The Eliasch Review: Climate Change: Financing Global Forests, United Kingdom Office of Climate Change (OCC), London, UK, 2008.

[24] T. Elmqvist, C. Folke, M. Nystrom, G. Peterson, J. Bengtsson, B. Walker and J. Norberg, Response diversity, ecosystem change, and resilience. *Frontiers in Ecology and the Environment* 1, 488–494, 2003.

[25] C. S. Elton, The ecology of invasions by animals and plants. London:_Methuen, 1958.

[26] European Commission, White Paper: Adapting to climate change: Towards a European framework for action. COM(2009) 147 final, 2009.

[27] C. Folke, Resilience: The emergence of a perspective for social-ecological systems analyses. Global Environmental Change 16: 253-267, 2006.

[28] H. C. Godfray, I. R. Beddington, L. Crute, D. N. Haddad, J. F. Lawrence, J. Muir, S. Pretty, S. M. Robinson, and C. Toulmin, Food security: the challenge of feeding 9 billion people. Science 327:812, 2010.

[29] L. H. Gunderson, C. S. Holling, Panarchy: understanding the transformations in human and natural systems. Island Press, Washington, 2002.

[30] C. S. Holling, Resilience and Stability of Ecological Systems. Annual Review of Ecology and Systematics 4: 1-23, 1973.

[31] C. S. Holling, editor. *Adaptive Environmental Assessment and Management.* Wiley, London, 1978.

[32] T. Hughes, A. H. Baird, D. R. Bellwood, M. Card, S. R. Connolly, *et al,* Climate Change, Human Impacts, and the Resilience of Coral Reefs. Science 301: 929-933, 2003.

[33] Intergovernmental Panel on Climate Change (IPCC), "Working Group I Third Assessment Report." *Cambridge University Press.* Cambridge, UK. 881 pp, 2001.

[34] Intergovernmental Panel on Climate Change (IPCC), Climate Change 2001: Impacts, adaptation, and vulnerability. Contribution of Working Group II to the Third Assessment Report of the Intergovernmental Panel on Climate Change. McCarthy, J.J., Canziani, O.F., Leary, N.A., Dokken D.J., and White, K.S. (Eds.), Cambridge University Press, Cambridge, 1032 pp, 2001a.

[35] Intergovernmental Panel on Climate Change (IPCC), Climate Change 2007: Synthesis Report, 52 pp (http://www.ipcc.ch/pdf/assessment-report/ar4/syr/ar4_syr_spm.pdf), 2007) Date accessed: 22 February 2015.

[36] Intergovernmental Panel on Climate Change (IPCC), *Climate Change 2014: Impacts, Adaptation, and Vulnerability. Part A: Global and Sectoral Aspects. Contribution of Working Group II to the Fifth Assessment Report of the Intergovernmental Panel on Climate Change* [Field, C.B., V.R. Barros, D.J. Dokken, K.J. Mach, M.D. Mastrandrea, T.E. Bilir, M. Chatterjee, K.L. Ebi, Y.O. Estrada, R.C. Genova, B. Girma, E.S. Kissel, A.N. Levy, S. MacCracken, P.R. Mastrandrea, and L.L. White (eds)]. Cambridge University Press, Cambridge, United Kingdom and New York, NY, USA, 1132 pp, 2014.

[37] J. X. Kasperson, R. E. Kasperson, and B. L. Turner, *Regions at Risk: Comparisons of Threatened Environments,* 1995. United Nations University Press, NY.

[38] R. W. Kates, T. M. Parris, Long-term trends and a sustainability transition. In Proceedings of the National Academy of Sciences of the United States of America, 100:8062-8067, 2003.

[39] J. Kiehl, Lessons from the Earth's past. Science 2011, 331:158-159, 2011.

[40] S. Lautze, Y. Aklilu, A. Raven-Roberts, H. Young, G. Kebede, L. Leaning, Risk and Vulnerability in Ethiopia: Learning from the Past, Responding to the Present. Preparing for the Future, Addis Ababa, 2003.

[41] B. Locatelli, V. Evans, A. Wardell, A. Andrade, and R. Vignola, Forests and climate change in Latin America: linking adaptation and mitigation. Forests 2: 431-450, 2011.

[42] J. Lovelock, The Revenge of GAIA. New York: Basic Books, 2007.

[43] Y. Lu, R. Wang, Y. Zhang, H. Su, P. Wang, A. Jenkins, R. C. Ferrier, M. Bailey, and G. Squire, Ecosystem health towards sustainability._Ecosystem Health and Sustainability 1(1):2, 2015.

[44] S. B. Manyena, The concept of resilience revisited. *Disasters,* 30(4): 433-450, 2006.

[45] J. S. Mayunga, *Understanding and applying the concept of community disaster resilience: A capital-based approach.* Draft working paper prepared for the summer academy, Megacities as Hotspots of Risk: Social Vulnerability and Resilience Building, Munich, Germany, 22-28 July 2007.

[46] M. J. Metzger, M. A. Rounsevell, L. Acosta-Michlik, R. Leemans, D. Schroter, The vulnerability of Ecosystem services to land use change. Agric.Ecol.Environ.114, 69-85, 2006.

[47] Millennium Ecosystem Assessment (MEA) Ecosystems and Human Well-being: Synthesis. Millennium Ecosystem Assessment Series. Island Press and World Resources Institute. Washington, D.C, 2005.

[48] M. Mortimore, W. M. Adams, Working the Sahel: environment and society in northern Nigeria. Routledge, London, 1999.

[49] R. Munang, I. Thiaw, K. Alverson, M. Mumba, J. Liu, and M. Rivington, Climate change and Ecosystem-based Adaptation: a new pragmatic approach to buffering climate change impacts, Curr Opin Environ Sustain, 2013.

[50] S. Naumann, G. Anzaldua, P. Berry, S. Burch, M. Davis, A. Frelih- Larsen, H. Gerdes, M. Sanders, Assessment of the potential of ecosystem-based approaches to climate change adaptation and mitigation in Europe. Final report to the European Commission, DG Environment. Ecologic Institute and Environmental Change Institute, Oxford University Centre for the Environment, 2011.

[51] C. Nellemann, M. MacDevette, T. Manders, B. Eickhout, B. Svihus, A. G. Prins, and B. P. Kaltenborn eds. The environmental food crisis: The environment's role in averting future food crises. A UNEP Rapid Response Assessment. Arendal, Norway: GRID-Arendal, United Nations Environment Programme, 2009.

[52] D. R. Nelson, W. N. Adger, and K. Brown, Adaptation to environmental change: Contributions of a resilience framework. *Annual Review of Environment and Resources 32*(1): 395–419, 2007.

[53] K. O'Brien, L. Sygna, R. Leichenko, W. N. Adger, J. Barnett, T. Mitchell, Disaster Risk Reduction, Climate Change Adaptation and Human Security. A Commissioned Report for the Norwegian Ministry of Foreign Affairs, 2008.

[54] E. Ostrom, A general framework for analyzing sustainability of socio-ecological systems. *Science,* 325: 419–422, 2009.

[55] Resilience Alliance, *Resilience* Available at: www.resalliance.org/576.php. Accessed on: 21/10/2015, 2002.

[56] Rockefeller Foundation*Building climate change resilience.* Rockefeller Foundation White Paper. Rockefeller Foundation: New York, USA, 2009.

[57] B. Smit, and O. Pilifosova, Adaptation to climate change in the context of sustainable development and equity. In: *Climate Change 2001: Impacts, Adaptation and Vulnerability. Contribution of the Working Group II to the Third Assessment Report of the Intergovernmental Panel on Climate Change* [McCarthy, J., O. Canziani, N. Leary, D. Dokken, and K. White (eds.)]. Cambridge University Press, Cambridge, UK and New York, NY, USA, pp. 879-912, 2001.

[58] South Africa Convention on Biological Diversity (SCBD), Connecting Biodiversity and Climate Change Mitigation and Adaptation: Report of the Second Ad Hoc Technical Expert Group on Biodiversity and Climate Change Montreal: SCBD, 2009.

[59] K. Strzepek, A. Mccluskey, District Level Hydro-climatic Time Series and Scenario Analysis to Assess the Impacts of Climate Change on Regional Water Resources and Agriculture in Africa. CEEPA Discussion Paper No. 13, Pretoria, South Africa, 2006.

[60] A. Tadege, Climate Change National Adaptation Program of Action (NAPA) of Ethiopia. NMS (National Meteorological Agency), Federal Democratic Republic of Ethiopia Addis Ababa, 2007.

[61] Y. Tessema, S. C. Awoke, and G. S. Endris, Understanding the process of adaptation to climate change by small-holder farmers: the case of east Hararghe Zone, Ethiopia. Agricultural and Food Economics, 1: 13, 2013.

[62] E. L. Tompkins, and W. N. Adger, Building resilience to climate change through adaptive management of natural resources, Tyndall Centre Working Paper 27, 2003.

[63] J.Twigg, *Characteristics of Disaster- Resilient Community.* University College London: London, UK, 2009.

[64] United Nations Environment Programme (UNEP), The Role of Ecosystem Management in Climate Change Adaptation and Disaster Risk Reduction. UNEP Copenhagen Discussion Series, 2009.

[65] United Nations Framework Convention for Climate Change (UNFCC), Report of the Conference of the Parties on its thirteenth session, held in Bali from 3 to 15 December 2007.

[66] United Nations Framework Convention on Climate Change (UNFCCC), United Nations Framework Convention on Climate Change: Ecosystem-based approaches to adaptation: compilation of information. Nairobi, Kenya, 2011.

[67] United Nations Framework Convention on Climate Change (UNFCCC), United Nations Framework Convention on Climate Change: Handbook. UNFCCC Secretariat. Bonn, Germany, 2006.

[68] United Nations International strategy for Disaster Reduction (UNISDR), Global Assessment Report on Disaster Risk Reduction. United Nations International Strategy for Disaster Reduction Secretariat, Geneva, Switzerland, 207 pp, 2009.

[69] N. Uy, R. Shaw, The role of ecosystems in climate change adaptation and disaster risk reduction. In Community, Environment and Disaster Risk Management. Edited by Uy N, Shaw R. Emerald Group Publishing Limited, 41-59, 2012.

[70] B. Walker, and D. Salt, Resilience *Thinking: Sustaining Ecosystems and People in a Changing World.* Island press, Washington DC, 2006.

[71] B. Walker, C. S. Holling, S. R. Carpenter, and A. Kinzig, Resilience, adaptability and transformability in social-ecological systems. *Ecology and Society* 9(2), 2004.

[72] Water Resource Institute, United Nations Development Programme, United Nations Environmental Programme and World Bank (WRI/UNDP/UNEP/World Bank). World Resources, Roots of Resilience—Growing the Wealth of the Poor. Washington: World Resources Institute, 262p. http://www.wri.org/publication/world-resources-2008-roots-of-resilience, 2008.

[73] T. Wheeler, and J. Von Braun, Climate change impacts on global food security. Science 341:508–513.–818, 2013.

[74] World Bank, Convenient solutions for an inconvenient truth: Ecosystem-based Approach to Climate Change. The World Bank, Washington D.C, 2009.

[75] C. T. Yirga CT, The dynamics of soil degradation and incentives for optimal management in Central Highlands of Ethiopia. University of Pretoria, Pretoria, South Africa, 2007.

23

Research of Devdoraki Glacier Based on Satellite Remote Sensing Data and Devdoraki Glacier Falls in Historical Context

G. Kordzakhia[1], L. Shengelia[1], G. Tvauri[2], M. Dzadzamia[3]

[1]Georgian Technical University, the Institute of Hydrometeorology, Tbilisi, Georgia
[2]I. Javakhishvili Tbilisi State University, Nodia Institute of Geophysics, Tbilisi, Georgia
[3]National Environmental Agency of the Ministry of Environment and Natural Resources Protection of Georgia, Tbilisi, Georgia

Email address:
giakordzakhia@gmail.com (G. Kordzakhia), larisa.shengelia@gmail.com (L. Shengelia), gena_tvauri@yahoo.com (G. Tvauri), murmandzadzamia@yahoo.com (M. Dzadzamia)

Abstract: The problem of the transportation safety for Mtskheta-Stepantsminda-Larsi road is overviewed. The scientific study of glaciers in the Caucasus started in the first half of the 19th century. From that time the terrestrial observations on glaciers were carried out. The outputs of terrestrial observations resulted in different series of various temporal duration characteristics of glaciers due to the difficulties of field works organization. The data received contain uncertainties and have no sufficient resolution in time and space. With the invention of the Earth's satellites it was determined that satellite remote sensing is the best technology allowing to receive the data with needed regularity both in terms of time and space resolution. The results of various researches show that sometimes the satellite remote sensing data is difficult to interpret and determine needed glacier characteristics. In such cases the data of the terrestrial observations are used for validation of glacier characteristics along with the expert knowledge of corresponding glaciers. This complex approach gives the possibility to adjust glaciers contours and calculate more accurately various characteristics of a glacier. At present the main problem for the safety of the Georgian Military Road in the territory of Georgia is the glacial and hydrological disasters such as ice blocks falls from the Kazbegi glaciations accompanied with the debris, followed by the river bed blockage and natural dam formation, with a consequent break and pass of catastrophic mudflows. The glacial catastrophes of Kazbegi glaciers: Devdoraki, Mna, Gergeti, Abano and Chachi have great impact on Mtskheta-Stepantsminda-Larsi road safety on the territory of Georgia. The biggest catastrophes on the Georgian territory took place due to Devdoraki glacier fell in 1776, 1778, 1785, 1808, 1817, 1832. The Devdoraki glacier listed falls during those years were connected with the pulsation of the glaciers and their increase. In Kazbegi glaciation the environmental disasters occurred in 2002, 2007 and 2014 (twice). It was the result of ice layers movement. Based on the analysis, it was concluded that the creation of the May 17 and August 20, 2014 catastrophic mudflows was not connected with the glacier pulsation.

Keywords: Remote Sensing, Glaciers, GIS, Melting & Accumulation, Climate Change

1. Introduction

A thousand years before people from the highland areas needed to study glaciers for reaching the neighboring valleys. Scientific interest in the study of glaciers was pronounced three centuries ago. It was caused by different factors: a study of the dynamics of glaciers and research of scientific facts about the past glaciation ages. At present, to the listed factors, glaciers research necessity is stipulated by: the water provision, surface water level rise in oceans/seas, indication of climate change at regional and global scales.

The importance of the Glaciology is proved by the fact that the glaciers occupied approximately 11% (16 million square kilometers) of the earth's land area. The glaciers contain a large number of fresh-water (30 million km^3), which is the 2/3 of the earth's fresh water supply.

Modern Glaciology, as a science, was established in 18th century in Switzerland, in 1760. Swiss scientist Benedict de

Saussure began the research of the Alps. The results were published in his book "Travelling in Alps; 1779-1796". At the same time Benedict de Saussure was the first, who climbed the highest peak of the Alps, Mont Blanc. Thus, he is considered as the founder of the Glaciology and mountain climbing.

Scientific study of the glaciers in the Caucasus was started during the second half of 19th century and was mainly concentrated on transportation safety. At that time the biggest transportation artery was the Mtskheta-Stepantsminda-Larsi road (founded in 1861) connecting Georgia with Russia passing over the Caucasus range. The main problem for the safety of transportation route are glacial and hydrological disasters occurred in the Kazbegi glaciation and having an impact on the mentioned road. Several scientists were researching the existing glaciers characteristics participating in field works trying to determine several characteristics of the glaciers [1-3]. The safety of the transportation was important problem as well. Due to glaciers falls Georgia faced heavy economic losses and casualties. At present this problem remains topical.

2. Study Area & Glaciers Survey

Georgia is a transcontinental country, located in the South Caucasus, between the Black Sea to the West and the Caucasus mountains to the North. The border of Georgia with Russian Federation is located in the North part of the country. The Mtskheta-Stepantsminda-Larsi road passes over the Caucasus range connecting Georgia with Russian Federation.

The present research overviews the problem of the transportation safety for Mtskheta-Stepantsminda-Larsi road (Georgian Military Road). The interests of several neighbors (Armenia, Turkey, partially Azerbaijan) are focused on this road, the shortest way from Georgia to Russian Federation.

The study area is mainly concentrated on the glaciers from Kazbegi glaciation located in Georgian territory. Mountain Kazbegi from the paleographical view is the young (several thousand years) volcanic cone, which belongs to extinct volcano. It seems that the snow-ice cover exist a long time and at present it is much degraded. Snow-ice masses are spread on the massive volcanic peaks and slopes. From the snowfields around the cone ice streams, i.e. glaciers: Devdoraki, Mna, Gergeti, Abano, etc. passes nearly to each gorge.

In fig. 1 Aster DEM (a.) and false color Landsat images (b.) of Kazbegi glaciation is presented.

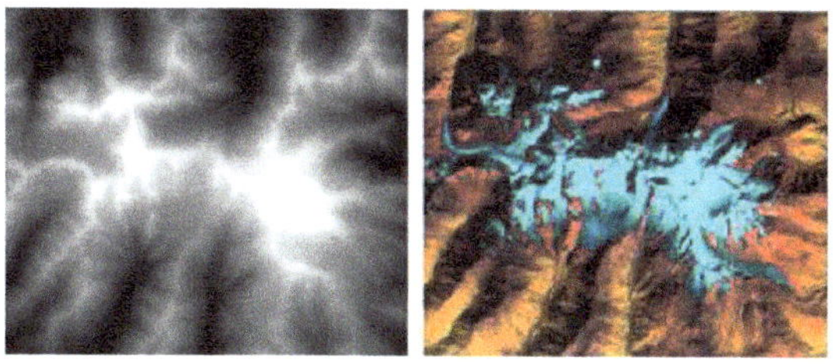

(a) (b)

Figure 1. *Aster DEM (a.) and False color Landsat images (b.) of Kazbegi glaciation.*

At present the main problem for the safety of transportation route are glacial and hydrological disasters such as glacier falls, the river bed blockage and natural dam formation, with a consequent break and passing of catastrophic flash floods and/or mudflows. Kazbegi glaciers: Devdoraki, Mna, Gergeti, Abano and Chachi are the main source of glacial catastrophe threats of road safety. They are located in the river feeds of the rivers Devdoraki, Mnestskali, Chkeri, Amali. These rivers are the left tributaries of the river Terek.

The biggest catastrophes from Georgian territory took place due to Devdoraki glacier fall in 1776, 1778, 1785, 1808, 1817, 1832 and 2014 (twice) [4, 5].

In fig. 2 the 3D visualization of Landsat images and Aster DEM (a.) and Google Earth image (b.) of the Kazbegi glacial mountain array are shown. In left parts of a. and b. of this picture the Devdoraki glacier visualizations are presented.

(a) (b)

Figure 2. *The 3D visualization of Landsat images and Aster DEM (a.) and Google Earth image (b.) of the Kazbegi glacial mountain array are presented.*

It should be noted that the glacier fall took place during the periods (18th - 19th centuries), when the glaciers parameters (area, length, volume) were mainly increased [1, 2, 4]. It is determined that during the second half of the last century the characteristics of the Georgian glaciers are steadily diminishing. The area of the Georgian glaciers decreased by 36% and the volume - by 48% consequently. Some glaciers melted away, but the total number of glaciers increased. The length of glaciers was reduced by 600 m in average. In the lower part of glaciers ice thickness decreased by 50-150 m. At the upper part ice thickness was reduced by 20-30 m [6-9]. The elevation of the lower boundary of the glaciers increased.

In the last century the terrestrial observations on glaciers were carried out. Due to the difficulties of organization and conducting of terrestrial observations the outputs resulted in different series of various temporal duration characteristics of glaciers. The data received contain an uncertainties. From other hand carrying out of such field works are expensive and data gathered have no sufficient spatial and temporal resolution.

With the launch of the Earth's satellites it was determined that satellite remote sensing is the best technology allowing to receive data with needed regularity in terms of both time and space resolution. Some uncertainities remain in the data as the observational tool is too far away from the Earth's surface. So, the necessity for the strong quality assessment/quality control (QA/QC) remains. A lot of studies show that the best method for investigation of glaciers are application of satellite remote sensing combined with terrestrial observations and expert knowledge of separate glaciers. Actually, this is the important steps for the QA/QC.

Usually the user is responsible to acquire the data from airborne sensors or/and application of the such data that have already been collected.

The situation with regards to space borne data is rather diverse. Some operational space borne systems collect all possible data free of charge, whenever others gather the highest resolution imagery data only according to special demand, which is rather expensive.

Whenever disasters early warning system is to be realized the special operational space borne system, i.e. specifically satellite images must be applied.

The satellite remote sensing gives possibility of universal, systematic and complex research of glaciers properties and changes. This is due to the fact that satellite remote sensing allows to cover efficiently the interesting regions and dynamically to study such important parameters, as the glacier area, terminus positions at the end of its location, firn line elevation, accumulation and ablation area, the hypsometry [9,10].

3. Research Methods

Changes in glacier area and terminus positions are relatively easy to determine using multispectral satellite images.

As mentioned above, glacier contours combined with a digital elevation model (DEM) give possibility to define various glacier parameters. Delineation of glacier contours is possible both manually and automatically based on the corresponding software by the band combination techniques. Human interpretation remains the best tool for extracting high level information from the satellite imagery. Tedious, manual digitalization of glacier boundaries by an operator with good knowledge of the region can produce glacier boundary contours of high quality and accuracy. Automatic delineation of glacier relies on different spectral properties of glacial ice and snow in the visible and the IR part of the electromagnetic spectrum. The greatest difficulty in glacier mapping from remote sensing is the presence of debris cover on glaciers. In those cases, higher resolution satellite images and DEM are used.

According to the methodology developed the following actions are performed for specifying the contours of the glaciers:

- The Google software product Google Earth use satellite images and the satellite sensor ASTER data based on the generated Digital Terrain Model (Aster DEM). It would be noted that the spatial resolution of the satellite Aster DEM is 30 m. GIS systems Google Earth, BEAM Visat and Quantum GIS Lisboa are applied for satellite data processing [11];
- For the glaciers identification its various characteristics (glaciers contours, numerical characteristics of glaciers, etc.) stored in the GLIMS database are used. GLIMS-database protected contours are generated based on the Aster sensor data [12].
- For glaciers contours validation Soviet topographical maps (1: 50,000), as well as, the terrestrial observations data are used. It is clear that such data exists only for the certain glaciers and for some years;
- Expert knowledge is applied to research these glaciers.
- Google Earth software package provides high spatial resolution (0.6-1 m) for the Kazbegi glacier array. Satellite image dated by September 1, 2010 allows high accuracy digitalization of the contours of all Kazbegi glaciers.

4. Results & Discussions

The Kazbegi glaciation characteristics: maximum length, area, minimum and maximum elevations, firn line elevation, ablation area determination on the basis of satellite remote sensing are overviewed in [10,13].

In GLIMS- database Devdoraki glacier is presented with three contours. One of them is named by Devdoraki 044517E42715N and in World database is registered as SU4G08011048. The contours of two relatively small glaciers are registered only in GLIMS database, namely by G044532E42709N and G044537E42706N codes. These two contours, after detailed study of high resolution satellite data, as well as based on the analysis of hypsometric curves are assigned to the Devdoraki glacial basin.

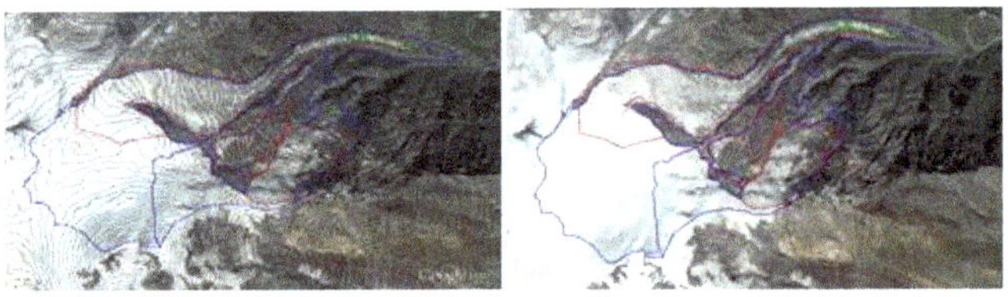

(a) (b)

Figure 3. *The Devdoraki glacier contours together with height isoclines. Part a. Devdoraki glacier adjusted contours are given together with height isoclines and in part b. consequently the corrected precise contours of Devdoraki glacier are shown. Registered contours from the GLIMS database are indicated with red color, the adjusted contours defined from satellite remote sensing data are presented with blue color and the glacier surface areas free of debris are shown by green color.*

In fig.3 the Devdoraki glacier contours are shown. In fig.3 part a. Devdoraki glacier adjusted contours are given together with height isoclines and in part b. consequently the corrected precise contours of Devdoraki glacier are shown. Registered contours from the GLIMS database are indicated with red color, the adjusted contours defined from satellite remote sensing data are presented with blue color and the glacier surface areas free of debris are shown by green color.

As it is shown in fig. 3 the significant part of the Kazbegi plateau refers to Devdoraki glacier. This decision was made as a result of the use of digital terrain model. In particular, the Devdoraki glacier watershed was corrected using the isoclines received from Aster DEM. Thus, the glacier borders were estimated based on corrected watershed. The elevation isoclines are traced by steps 5, 10, and 30 m resolution.

In fig 4 the comparison of Devdoraki contours with the former USSR topographic map is presented.

Fig. 4 includes 30 m resolution drawn isoclines and adjusted contours defined on the basis of precise watershed. Thus, the significant difference between registered contours from GLIMS database and the adjusted contours received from the data of satellite remote sensing is a result of more precise definition of watersheds. Hence, the contours received from the remote sensing are more precise.

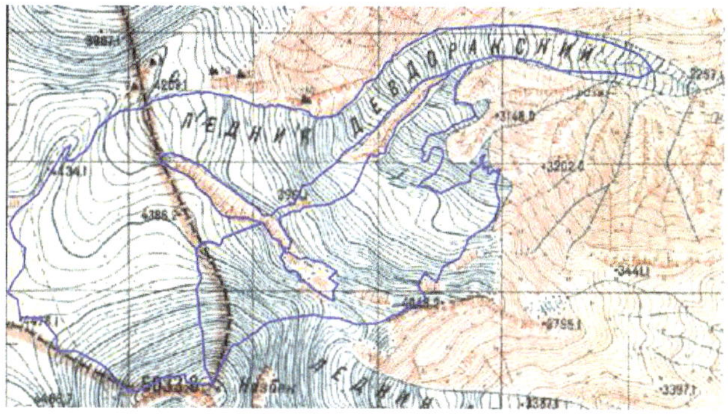

Figure 4. *The comparison of Devdoraki contours with topographic map.*

Table 1. *The characteristics of the Kazbegi glaciations in Georgian territory. Data received from the satellite remote sensing are presented in columns marked by (1); Data from USSR catalogue are shown in the columns marked by (2).*

Glacier parameters			Devdoraki	Mna	Gergeti	Abano	Chachi
Maximum lenght (km)	Whole glacier	(1)	6.87	3.32	7.10	3.83	3.22
		(2)	7.30	4.10	8.50	4.10	3.20
	Free of debries	(1)	6.32	2.43	7.10	2.26	3.10
		(2)	6.50	3.80	8.50	2.50	3.20
Area (km^2)	Whole glacier	(1)	6.24	3.14	5.15	1.68	2.28
		(2)	7.00	4.60	8.30	2.00	2.80
	Free of debries	(1)	6.04	2.95	4.63	1.00	2.07
		(2)	6.60	4.30	7.10	1.40	2.80
Minimum elevation (m)	Whole glacier	(1)	2,392	3,036	3,091	3,020	3,239
		(2)	2,260	2,860	2,870	2,950	3,230
	Free of debries	(1)	3,549	3,330	3,091	3,462	3,299
		(2)	2,540	3,060	2,870	3,410	3,230
Maximum elevation (m)		(1)	5,038	4,612	4,936	5,038	4,447
		(2)	5,030	4,606	5,030	5,030	4,430
Firn line elevation (m)		(1)	3,326	3,568	3,793	3,735	3,665
		(2)	3,260	3,480	3,659	3,700	3,660
Ablation area (km^2)	Whole glacier	(1)	0.74	0.36	1.50	0.87	0.78
		(2)	1.00	0.60	2.30	0.90	0.60
	Free of debries	(1)	0.48	0.17	1.20	0.18	0.53
		(2)	0.60	0.30	1.10	0.30	0.60

Comparison of a topographic map with contours received from the data of satellite remote sensing (fig. 4) shows that these fields formed entire Devdoraki glacier in the past.So, for comparison of Devdoraki glacier characteristics (area, length etc.) created from satellite remote sensing the joint contour is applied. This data is compared with the Devdoraki glacier data from the USSR glaciers catalogue. In reference to the glacier square: the Devdoraki (SU4G08011048 contour) direct area is equal to 4.05 square km, united G044532E42709N and G044537E42706N contours area is 2.19 square km. Their total area is 6.24 square km.

In Table 1 the Kazbegi glaciation (glaciers in Georgian territory: Devdoraki, Mna, Gergeti, Abano, Chachi) parameters, such as: length, area, minimum elevation, firn line elevation, ablation area defined from the satellite remote sensing data (columns marked by (1)) and from the USSR glaciers catalogue (columns marked by (2)) are presented.

Comparison of the received results from satellite remote sensing with the data from the USSR catalogue [14] is presented in Table 1 (the time gap between them is approximately 50 years). It should be noted that complex research of the glaciers using the data from different sources such as: satellite remote sensing, terrestrial observations and expert knowledge are effective for accurate determination of glaciers main characteristics.

Based on the detailed comparison of the output data received using the satellite remote sensing provided with similar data from the glaciers catalogue it can be concluded that during the last 50 years all the characteristics of the Kazbegi glaciers change:

- The minimum elevation and firn line altitude are increasing;
- The length and area as well as the ablation area are decreasing.

3D visualization of Devdoraki glacier contours are presented on fig.5 for the evidence. Registered contours from the GLIMS database are indicated with red color, the adjusted contours received from the data of satellite remote sensing are presented with blue color and the glacier surface areas free of debris are shown by green color. The big differences due to adjusting the Devdoraki glacier contours are clearly shown comparing the red and blue colors on fig.5.

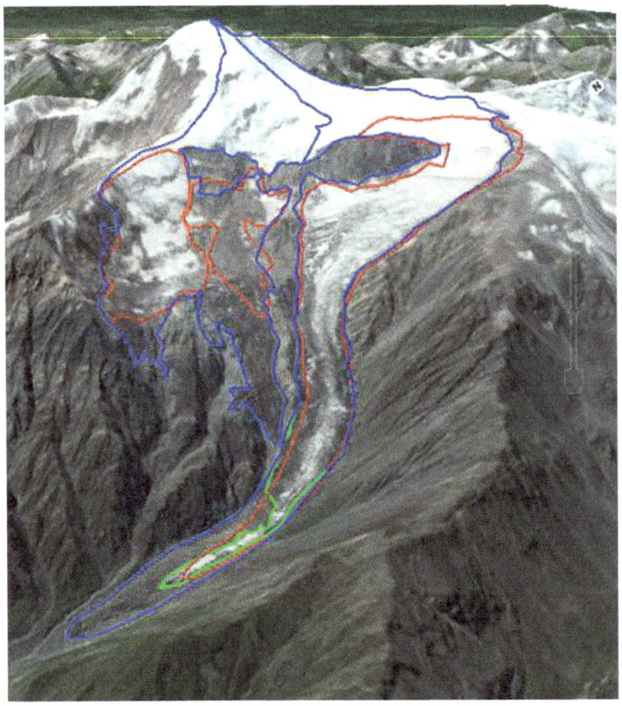

Figure 5. 3D visualization of Devdoraki glacier contours. Registered contours from the GLIMS database are indicated with red color, the adjusted contours received from the data of satellite remote sensing are presented with blue color and free of debris areas are shown by green color.

In fig. 6 Devdoraki glacier satellite images: a. August 31, 1989 from Landsat 4 TM, (generating thematic maps) and b. August 28, 2014 from Landsat 8 OLI are presented. Glacier Devdoraki contours digitalization presented on these images are manually done according to the Google Earth Digital Globe high resolution satellite image, dated January 1, 2010.

(a) (b)

Figure 6. Devdoraki glacier false color satellite images: (a.) August 31, 1989 from Landsat 4 TM and (b.) August 28, 2014 from Landsat 8 OLI.

The latest changes in Devdoraki glacier development in light of glacier fall of the 2014 can be defined by comparison of the images from fig.6 part a. and part b. This allows to define latest modification for the period preceding the August 20, 2014. In fig. 6 part b. is clearly shown the place (the rose circle on the figure) of the collapse of the Devdoraki glacier in 2014.

The biggest scale and the strongest natural catastrophes are observed from the Kazbegi glaciations, namely: Devdoraki glacier from Georgian side and Kolka glacier from the Osetian side. The detailed information about Devdoraki mudflows comes from the time of the Mtskheta-Stepantsminda-Larsi road construction. The biggest catastrophes on the Georgian territory took place due to Devdoraki glacier fell in 1776, 1778, 1785, 1808, 1817, 1832. The largest scale glacier disaster was observed for Devdorki glacier in 1832. The elevation of created dam was 90 m and generated reservoir, fully covered the territory of village Gveleti. Recently (2002) the mudflow from the Kolka glacier killed 2 hundred humans and several mln. cubic meter of stone is still kept in Karmadon gorge. Devdoraki glacier advances in 1842, 1843 and 1855 caused small ice-stone mudflows not reaching r. Terek valley. Devdoraki glacier advances in 1865-1875, 1886-1887 and 1893-1904 at the distances of 241.0, 44.8 and 55.5 m correspondingly did not cause mudflows.

The main reason of the glacial mudflows genesis are consequences of the glaciers subduction. Most of the mountainous glaciers are characterized by rapid pulsing movement. This is a result of the processes going inside the glacier, i.e., fast change of glacier bed conditions and distribution of substances in accumulation and ablation area. There are pulsating glaciers (e.g. Kolka and Devdoraki), which are characterized by sharp advancing of the glacier body the so-called surge accompanied by glacier falls.

When glacier retreats the glacier tongue, due to restricted feeding (result of decreasing of snow precipitation and reduction of its accumulation in firn field) is grinded. Therefore, the crevices are created on the glacier tongue that make prerequisites for moraine material accumulation (rocks and their debris that falls on glacier surface).

In these places due to high temperatures and frequent precipitation on the surface of glacier the certain amount of water is accumulated. It will break the created dams capturing all the remaining ice and moraine material. As a result the destructive mudflows are created.

Obviously, the mudflows that took place in the past were the result of the glaciers such pulsation. At that time periods the characteristics of the Georgian glaciers were mainly increasing.

Last catastrophic events occurred in May and August, 2014 during the period when the characteristics of the Georgian glaciers were steadily diminishing [5].

Based on video, photo material analysis, it can be concluded that the creation of the May 17 and August 20, 2014 catastrophic mudflows are not connected with the pulsation of the Devdoraki glacier. In May 17, 2014 from Kazbegi slope at 4, 700 m elevation was cut ice block, with the volume of 5 million cubic meters. In Fig. 3 part b. the area of the collapse of the glacier is clearly shown and the place of the accident is no longer covered by ice. This mass fell down along 80-85 degrees inclined slope below 2, 500 m. Then it jumped over the Devdoraki glacier tongue and stroke the left side of the narrow gorge. The ice mass from the mountain slope was accompanied with the vulcanic slag. Vise versa can also be considered, i.e., vulcanic slag was off first and then it was accompanied with ice.

***Figure 7.** The red trace of the volcanic slag on the sides of the gorge and river bed.*

The red trace of the volcanic slag is well reflected in the area's of the strike of the gorge sides and debris flow trace from the source to the confluence (fig.7). Mudflow material surface that arrived in the r. wTerek valley also was red due to the vulcanic slag.

The torrential stream generated due to the ice block collapse into pieces. Water/ice masses mixed with moraine material and vulcanic slag (red slag pieces, including andesite & dacite rock material). Generated semi-viscous mass passed through narrow Devdoraki-Amali gorge with the speed of 100 km/h in 4-5 min maximum.

Semi-viscous mass stream blocked r. Terek riverbed and artificial reservoir was established (fig. 8).

***Figure 8.** The Terek river bed blockage.*

The consequences were heavy. Outgoing pipeline was exploded. Casualties totaled to 4 persons. Established artificial reservoir occupied the area about 2 hectares and the pond was 15 m deep. Luckily water was discharged by under a construction derivation tunnel of hydropower station. The Mtskheta-Stepantsminda-Larsi strategic road was closed for a long-term time.

It should be noted that in May 17, 2014, the following extra phenomenon was observed: the white colour gas appeared in 1-1.5 hour from the glacier torrent pass, in the foot of the areal of ice block fall. In May 18, 2014 instead of this gas the black color fume appeared in the same place. From the distance it was seen as heavy, viscous "mushroom-shaped" formation. Despite the wind the smoke (maybe dust) did not change its shape and covered the surface for 3-4 hours. Very likely, the gas formation of the dark colour is the dust created as a result of destruction of the moraine material or/and vulcanic slag from the glacier slope. However, it can not be excluded gas-thermal action of the volcano as well.

Concerning the catastrophic event on Devdoraki glacier in August 20, 2014 it can be noted that the natural disaster was provoked by heavy rain and the lake created below the glacier tongue accompanied by the fall of a large mass of ice from the glacier tongue. The consequences of this tragedy were heavy. Casualties totaled to 2 persons, newly constructed hydropower station was destroyed. The existing infrustructure was damaged: Mtskheta-Stepantsminda-Larsi road was ruined and traffic was closed for a long time, the gas pipeline was broken, the customs was damaged.

It should be noted that throughout the history of the survey of Devdorak glacier such big scale and strong catastrophe was not observed *twice a year*.

In fig. 9 Devdoraki glacier for February 10, 2015 is shown. Empty locations of big depth in Devdoraki Glacier's body are observed. The places (ponds), where the ice, fused are presented with red arrows, apparently as a result of thermal exposure.

***Figure 9.** Devdoraki glacier. February 10, 2015. The places (ponds), where the ice fused are presented with red arrows.*

So the place of the collapse of Devdoraki glacier (fig 6) may have the same explanation.

These issues are the subject of further researches.

5. Conclusion

The problem of the transportation safety for Mtskheta-Stepantsminda-Larsi road is overviewed. At present the main problem for the safety of transportation route are glacial and hydrological disasters.

The study area is mainly concentrated on the territory occupied by the glaciers from Kazbegi glaciation that are located in Georgian territory and adjacent terrain, where the Mtskheta- Stepantsminda-Larsi road passes.

The natural disasters that had significant impact on Mtskheta- Stepantsminda-Larsi road are glacier falls from the Kazbegi glaciation, accompanied with the river bed blocking and natural dam formation, with a consequent break and passing of catastrophic mudflows. The biggest catastrophes from Georgian territory took place due to Devdoraki glacier fall in 1776, 1778, 1785, 1808, 1817, 1832 and 2014 (twice).

The detailed research of the Kazbegi glaciers: Devdoraki, Mna, Gergeti, Abano etc. are made to study, analyze and forecast these processes.

In the last century the terrestrial observations on glaciers were carried out. Due to the difficulties of organization and conducting of terrestrial observations the outputs resulted in different serries of various temporal duration characteristics of glaciers. The data received contain an uncertainities. From another hand carring out of such field works are expensive and data gathered have no sufficient spatial and temporal resolution.

With the launch of the Earth's satellites it was determined that satellite remote sensing is the best technology allowing to receive data with needed regularity in terms of both time and space resolution. Some uncertainities remain in the data as the observational tool is too far away from the Earth's surface. So, the necessity for the strong quality assessment/quality control (QA/QC) remains. Sometimes the satellite data often are hard to interpret and determine the needed glacier characteristics. A lot of studies show that the best method for investigation of glaciers are application of satellite remote sensing data combined with terrestrial observations and expert knowledge of separate glaciers. Actually, this is the important steps for the QA/QC.

It would be noted that whenever the glacial catastrophic events early warning system is to be realized the special operational space borne system i.e. specifically satellite images are to be applied.

Based on the detailed comparison of the output data received using the satellite remote sensing provided with similar data from the glaciers catalogue it can be concluded that during the last 50 years all the characteristics of the main glaciers from East Georgia change:

- The minimum elevation and equilibrium line altitude are increasing;
- The length and area as well as the ablation area are decreasing.

Researches and corresponding analysis show that the impact of modern climate change, i.e., glaciers retreat due to modern warming is well expressed for the considered glaciers from Kazbegi glaciation.

Consequences of the glaciers subduction are the main reasons of the glacial mudflows genesis. Most of the mountainious glaciers are characterized by pulsation. Snow-ice glacier mass is fed through the gravitational movement. Pulsation of mountain glaciers is characterized by a different velocity of spreading ahead and retreat of glaciers periodically. Obviously, the mudflows that took place in the past (from the glaciers observations starting from the end of the 19th century) were the result of the glaciers pulsation. At that time periods characteristics of the Georgian glaciers were steadily increasing.

During 2014 two strong catastrophes were observed and from the second half of the last century the characteristics of the Devdoraki glacier are steadily diminishing.

It should be noted that throughout the history of the survey of Devdoraki glacier such big scale and strong catastrophe was not observed twice a year.

Analysis show that the creation of the May 17 and August 20, 2014 catastrophic mudflows is not connected with the pulsation of the glacier. In May 17, 2014 from Kazbegi slope at 4, 700 m elevation was cut ice block, with the volume of 1 million cubic meters. This mass fell down along 80-85 degrees inclined slope below 2, 500 m. Then it jumped over the Devdoraki glacier tongue and stroke the left side of the narrow gorge. The red trace of the volcanic slag is well reflected in the area's of the strike of the gorge sides and debris flow trace from the source to the confluence.

The torrential semi-viscous passed through narrow Devdoraki-Amali gorge blocked r. Terek riverbed and artificial reservoir was established. It was accompanied by the creation of the white colour gas in the foot of the areal of ice block fall. The next day it was changed by the black color fume. Despite the wind the smoke (maybe dust) did not change its shape and covered the surface for 3-4 hours. Very likely, the gas formation of the dark colour is the dust created as a result of destruction of the moraine material or/and vulcanic slag from the glacier slope. However, it can not be excluded gas-thermal action of the volcano as well.

This last conclusion is promoted by the latest data of Devdoraki glacier from the February 10, 2015 representing the places (ponds) fused in Devdoraki glacier that may have the same explanation.

The consequences of the May 17 and August 20, 2014 catastrophic mudflows was heavy: human deaths were reported, existing infrustructure was damaged, Mtskheta-Stepantsminda-Larsi road was ruined, traffic was closed for a long time, the gas pipeline exploded, newly constructed hydropower station was destroyed etc.

6. Acknowledgment

The research is carried out in frames of the Shota Rustaveli National Foundation of Georgia that supported research Grant Project FR/586/9-110/13.

The authors are grateful to the specialists of Hydrometeorological Department of National Environmental Agency of Georgia for materials (mainly Figures: 7-9) related to the catastrophic events having place in Devdoraki glacier and surrounding area in May 17 and August 20, 2014.

References

[1] Abikh G.V. (1877). Neskolko slov o sostoianii v nastoiachee vremia Devdorakskogo lednika. – Izvestia Kavkazskogo otdela IRGO, t. V, 57-64 (in Russian).

[2] Zaporojchenko E.V., Chernomorets S.S. (2004). Istoria i izuchennost Kazbegskikh zavalov. // Vestnik Kavkazskogo gornogo obchestva, Piatigorsk, № 5, 33-54 (in Russian).

[3] V. Tsomaia (1983), Opasnye gidrometeorologicheskie javlenija na Kavkaze. L., 158-184 (in Russian)

[4] A. N. Bojinski, S.S. Chernomorets 2007). Modelirovanie Kazbekskikh lednikovikh zavalov XVIII-XIX vekov. Materiali gliatsiologicheskikh issledovanii, vip. 103, 91-95.

[5] Tavasiev R. A., Galushkin I.V. (2014). Kamenno-ledovi obval s gori Kazbek 17 Maia 2014 goda. Vestnik Vladikavkazskogo nauchnogo tsentra, tom 14 №2, 43-45.

[6] Chris R., Stokes V., Stephen D. Gurney, Maria Shangedanova, Victor Popovnin (2006). Late-20th century changes glacier extent in the Caucasus Mountains, Russia/Georgia. Journal of Glaciology, Vol. 52 No. 176, 99-109.

[7] Magnus Sylvén, Rasmus Reinvang, Žanete Andersone-Lilley (2008).Climate Change in Southern Caucasus: Impacts on nature, people and society, WWF Norway, WWF Caucasus Programme,10-31.

[8] [Von Ina Keggenhoff, Tatjana Keller, Mariam Elizbarashvili, Ramin Gobejishvili und Lorenz King (2011). Naturkatastrophen durch Klimawandel im Kaukasus? Hochwasser und Hanginstabilitaten in Georgien immer haufiger. Spiegel der Forschung, N2, 16-23.

[9] Shengelia L.; Kordzakhia G.; Tvauri G.; Davitashvili T.; Begalishvili N. (2012). Possibilities of the use of remote sensing technologies for the estimation of modern climate change impact on the Caucasus glaciers. Monthly Scientific-Reviewed Magazine of Georgian National Academy of Sciences „Metsniereba da Technologiebi" (Science and Technologies"), №4-6, 25-30, (in Georgian).

[10] Petri Pellikka, Gareth Rees W. (2010). Remote Sensing of Glaciers. Taylor & Francis Group, London, UK, 340.

[11] Bolch, T., and Kamp, U. (2005). Glacier Mapping in High Mountains Using DEMs, Landsat and ASTER Data, Grazer Schriften der Geographie und Raumforschung, In: Proceedings 8th International Symposium on High Moutain Remote Sensing Cartography, vol 41, 20-27 March 2005. La Paz, Bolivia, 13-24.

[12] Raup, B.H.; A. Racoviteanu; S.J.S. Khalsa; C. Helm; R. Armstrong; Y. Arnaud (2007). "The GLIMS Geospatial Glacier Database: a New Tool for Studying Glacier Change". Global and Planetary Change 56, 101-110.

[13] Kordzakhia G., Shengelia L., Tvauri G., Tsomaia V., Dzadzamia M., (2014). Satellite Remote Sensing Outputs of the Certain Glaciers in theTerritory of East Georgia. Egyptian Journal of Remote Sensing and Space Sciences. www.elsevier.com

[14] Tsomaia, V.S., Drobishev, O.A., (1977). Glaciers Catalog of the USSR, the North Caucasus, Leningrad,Gidrometeoizdat, Vol. 8. (In Russian), 71. G. Eason, B. Noble, and I. N. Sneddon, "On certain integrals of Lipschitz-Hankel type involving products of Bessel functions," Phil. Trans. Roy. Soc. London, vol. A247, pp. 529–551, April 1955. (references)

Assessment of Technical and Economic Feasibility of Activated Charcoal Removal of Organic Matter from Different Streams of Grey Water Through Study of Adsorption Isotherms

Ababu T. Tiruneh[1], Amos O. Fadiran[2], William N. Ndlela[1], Jonna Heikkilä[3]

[1]Department of Environmental Health Science, University of Swaziland, Mbabane, Swaziland

[2]Department of Chemistry, University of Swaziland, Kwaluseni, Swaziland

[3]Turku University of Applied Sciences, Turku, Finland

Email address:

atiruneh@uniswa.sz (A. T. Tiruneh), fadiran@unisw.sz (A. O. Fadiran), wndlela@uniswa.sz (W. N. Ndlela), jonna.heikkila@turkuamk.fi (J. Heikkilä)

Abstract: The feasibility of household level treatment of grey water with activated charcoal was performed using laboratory batch adsorption testing on locally available charcoal media. The study results indicated that the potential for removal of organic matter was significantly high for the high pH cloth wash water compared to the low pH kitchen wastewater which also contained non-adsorbed organics. The addition of ash considerably improved the removal and projected life length of adsorption media for kitchen wastewater treatment. The adsorption isotherms obtained were all modeled adequately using the Freundlich isotherm while the isotherm shapes display different types of adsorption for the different streams of grey water because of the heterogeneous nature of the adsorbates in grey water. The replacement life length of activated charcoal for single drum household level treatment ranged between 7 and 15 months. For family daily flow rates up to 400 lit/day, the replacement costs of a single drum charcoal per cubic meter of grey water treated were calculated to be below the current tariff levels for acquiring water in cities in Swaziland. A considerable part of the grey water pollutant can be removed through pretreatment by sorption alone such as by filtration through sand or other cheap media before adsorption. For complete household level treatment of grey water, a three-step treatment consisting of sand pre-filtration, activated charcoal adsorption and sand post-filtration are recommended.

Keywords: Grey Water Treatment, Activated Charcoal, Water Recycling, Wastewater Treatment, Adsorption Isotherm, Pollutant Removal

1. Introduction

Grey water recycling has been practiced worldwide since the 1970s and is being increasingly considered as feasible option for providing readily available water for reuse at household level while at the same time providing environmental protection and water conservation [1]. The pollutant load, such as the organic matter content of mixed grey water streams, may be comparatively low compared to the wastewater from toilets. However, where the grey water streams are dominated by the less diluted kitchen and wash water, the organic matter content can be quite high. In addition, the biodegradability of the different grey water streams is variable. For example, Edwin et al [2] reported COD/BOD_5 ratio of 8.2 for laundry wash water. Such ratio indicates poor biodegradability of the waste stream. By contrast grey water from the kitchen was reported to have COD/BOD_5 ratio of only 1.2. These kinds of differences in biodegradability of grey water can pause a challenge to the task of treating grey water by biological means. For this reason, other treatment alternatives such as activated charcoal adsorption are worthy of consideration for grey water treatment.

Carbon media as an adsorbent has been in use for quite long time. Historical records show that carbon has been used for medical purposes as far back as 1550 BC. Further in 200 BC, carbon in the form of wood char was used to remove disagreeable odour. In 1785, chemists learnt that carbon could remove unwanted contaminants from water. Carbon in activated form was first used as filter medium in the late 1800s [3].

Experiments involving the use of activated charcoal for wastewater treatment indicated significant percentage removal of organic matter and of other pollutants from wastewater. For example, Nansubuga et al [4] reported that, for a wastewater pretreated with activated sludge and subjected further to charcoal filtration, COD removal percentages up to 60% were achieved. The mechanism of adsorption of organic compounds from grey water onto activated charcoal may in general be due to one or a combination of the three forms of adsorption. These are namely: Physical adsorption driven by the weak London Forces; electro-static interactions; and adsorption resulting from chemical interactions between the adsorbent and the adsorbate. However, the predominantly non-polar nature of organic compounds in grey water may favor the physical type of adsorption due to the lesser polar attraction of the solvent water. Even then, because of the general heterogeneity of the compounds present in grey water, the adsorption mode is expected to be mixed consisting of all the three mechanisms [5].

The adsorption of pollutants on activated carbon is modeled through a number of isotherm models. The Langmuir adsorption model is suited for homogenous, mono-layer adsorption [6]. In Langmuir adsorption isotherm models, it is assumed that adsorption is only on local sites whereby the interaction between adjoining adsorbed molecules is limited and the adsorption is represented by a single layer [7]. The adsorption of compounds from grey water containing heterogeneous mix of compounds on activated charcoal may better be represented by the more versatile, though empirical, Freundlich adsorption model which is suited for a mixture of heterogeneous adsorption with multi-layer adsorption capabilities.

The effect of pH on adsorption depends partly on the level of hydrophobicity of the adsorbate. Hydrophobic organic compounds that are often present in grey water streams are not expected to be influenced by variation in pH. However, for pollutants in grey water that may display some level of polarity, pH can have a significant influence on the extent of adsorption [8]. The effect of pH can be considerable particularly for grey water streams containing significant concentrations of organic acids. At low pH the competing influence of the presence of excess H^+ ions in the solute can make adsorption of organic acids difficult [9]. On the other hand, the increase in surface acidity of activated carbon as a result of the presence of functional groups such as the carboxylic acid groups on the charcoal media has been implicated in the reduced capacity of activated carbons in attracting and removing hydrophobic compounds [10].

The kinetics of adsorption influences the extent to which equilibrium is reached and whether the equilibrium models can be applicable. The inter-particle diffusion may be the limiting rate especially when fast advective flows occur in adsorption columns. In such situations, the time available for adsorption may not be enough for equilibrium to form as it is constrained by the limited rate of inter particle diffusion. Some studies asserted that even a slow advection velocity as low as 1 $cm.hr^{-1}$ may not be enough to attain equilibrium in terms of adsorption [11]. For example, for the modeled flow in this research, the velocity of flow in a single drum activated charcoal treatment is estimated to be around 0.8 $cm.hr^{-1}$. This is apparently low advection velocity but not so low as to ensure complete equilibrium [11]. However, this may relate to the theoretical 100% equilibrium and for most hydrophobic compounds which constitute the grey water components, equilibrium may be reached faster. Because of this, equilibrium models are applied even at much faster velocity of transport of solutes [12]. The assertion that equilibrium is slow to attain serves to illustrate the theoretical limitation of equilibrium adsorption models in terms of the slow kinetics but may not be of great practical significance.

The economic feasibility of household level recycling of grey water through activated charcoal adsorption should consider treatment costs which include both running and capital costs [13]. This cost system depends on the complexity of the grey water system and the extent to which such system is affordable by households concerned. For household level grey water treatment in communities such as Mnyamatsini in Swaziland where the research project was undertaken and in which the majority of the water supply is in the form of yard taps, the installation of pumps may not be considered an appropriate choice. Doing away with pumps also reduces the running cost. The cost system is, therefore, kept simple and consists of cost of components such as the charcoal holding drum, pipe accessories and the replacement cost of the activated charcoal.

The possibility of regeneration of activated charcoal is limited by the energy intensive nature of the recovery. Because of the release of excess energy in the form of heat during the adsorption process, adsorption is considered as exothermic process. By contrast, regeneration, the reverse process, is endothermic and requires external heat energy input for it to take place [14]. The technology of regeneration can be complex with elevated heating as the only technically viable regeneration alternative at household level. However such energy intensive procedure is uneconomical. Regeneration is, therefore, not an economically feasible alternative especially for household level treatment of grey water.

2. Materials and Methods

2.1. Experimental Design

The experimental design for the research study was based

on laboratory batch testing of powdered activated charcoal adsorption of organic matter and development of adsorption isotherm for each batch of grey water tested. The technical and cost feasibility of the treatment were also evaluated based on the resulting isotherms developed. Separate streams of grey water from cloth washing and the kitchen were taken for the tests. The effect of pH on activated charcoal adsorption was evaluated by treating the low pH kitchen grey water was ash.

For the batch adsorption study, 200 mL sample volumes of the grey water to be treated were placed in each of a series of beakers. Measured weights of activated charcoal ranging between 1 g and 100 g were then added to the beakers and the mixtures were stirred for sufficient period of time. The sample mixtures were then filtered using filter paper and were taken for the COD testing.

Organic matter removal was selected for evaluation of the adsorption efficiency of activated charcoal through Chemical Oxygen Demand testing. Chemical Oxygen Demand (COD) was determined using potassium dichromate oxidation. For the COD test, the closed reflux titrimetric method was followed [15]. For each sample tested for COD before and after treatment with activated charcoal, 2.5 mL of the sample was pipetted into the digestion tube, followed by addition of 2.5 mL of potassium dichromate reagent (0.1 M) and 3.5 mL of sulfuric acid reagent. The digestion tubes were capped, placed into block digester and heated for 2 hours at 150°C. The digested samples were then cooled and quantitatively transferred to conical flasks after which few drops of ferroin indicator were added. The samples were then titrated with ferrous ammonium sulphate titrant (0.005 M). The COD was then determined using the formula:

$$COD\ as\ mg\frac{O_2}{L} = \frac{(A-B)*M*D*8000}{mL\ of\ sample}$$

Where:

A = mL FAS used for blank

B = mL of FAS used for sample

M = molarity of ferrous ammonium sulphate titrant (0.005 M)

D = dilution factor for the sample,i.e., the ratio of volume of sample to total volume of dilution from which the 2.5 mL sample was taken for digestion.

8000 = Milli-equivalent weight of oxygen X 1000 mL/L

Because of the lack of homogeneity of grey water samples and the small sample size used for digestion, a minimum of three repetitions were made for each of the COD test determined. For the determination of pH of raw grey water samples as well as treated ones, a pH meter was used that was periodically calibrated with standard buffers of pH = 4, pH = 7 and pH = 10.

Adsorption isotherms

For modeling the adsorption isotherms, the Freundlich model has been found fitting in most of the experiments. The Freundlich adsorption model takes the form:

$$\frac{X}{M} = KC^{\frac{1}{n}} \tag{1}$$

Where (X/M) is the pollutant adsorbed per unit mass of charcoal, C is the pollutant in equilibrium with the adsorbent (COD in mg/L) and K and n are constants evaluated from the log-log plot of (X/M) Vs. C.

2.2. Mass Transfer Model of Adsorption

The mass transfer model of adsorption [16] is used to model the adsorption onto the activated charcoal using the equation:

$$\frac{dC}{dh} = \frac{k_a}{Q}(C - C*) \tag{2}$$

The height of the adsorption zone h_z is calculated using the equation:

$$h_z = \frac{Q}{K_a}\int_{C_B}^{C_E}\frac{dC}{(C-C*)} \tag{3}$$

Where Q is the wastewater flow rate, K_a is the overall mass transfer coefficient which includes the resistances offered by film and pore diffusion, C is the COD concentration at particular height $h < h_z$ and C_B and C_E are the solute COD concentrations in equilibrium with the adsorbent at breakthrough and exhaustion respectively.

For a given height of adsorption column which is assumed to be H = 0.6 m and for a single drum household level charcoal treatment with drum diameter of 0.572 m and flow rate of 50 lit/day, the mass transfer coefficient K_a is computed from the following equation [17]:

$$K_a = \frac{G}{H}\int_{C_B}^{C_E}\frac{dC}{(C-C*)} \tag{4}$$

Where G is the mass flow rate of the wastewater in Kg/min- m^2, H is the depth of the adsorption column (m) and the integral expression is as defined above.

It will be noted in the subsequent presentation and discussion of the research results that the mass transfer rate calculated is low due to the low flow rate adopted and this low rate of mass transfer results in low value of the adsorption zone, h_z. The low value of adsorption height h_z is useful for complete utilization of the available depth of the media.

The fractional capacity of the adsorption zone in the phase between breakthrough and exhaustion is calculated using the equation:

$$F = \int_0^1\left(1-\frac{C}{C_0}\right)\Delta\left(\frac{V-V_B}{V_E-V_B}\right) \tag{5}$$

Where F is the fraction of adsorption zone available for adsorption, C_0 and C are the COD concentrations before and after adsorption respectively and V, V_B and V_E are the volume of wastewater treated at current fractional capacity, at break through and exhaustion respectively.

For calculation of the ultimate capacity of the activated charcoal, the percentage saturation is calculated from the equation;

$$\%\ saturartion = \frac{H+(F-1)h_z}{H} * 100 \tag{6}$$

Where H is the depth of the activated charcoal, F is the fractional capacity and h_z is the depth of the adsorption zone. This equation has been used to calculate the length of run of the activated charcoal design as will be shown later.

3. Results and Discussion

3.1. Characteristics of the Wastewater Samples Tested

Different streams of grey water samples were taken for the study and their characteristics in terms of pH and COD were evaluated and are presented in Table 1. The grey water samples were taken from two different households and separate samples were taken from each house representing wastewater samples from the kitchen and wash water from cloth washing. The wastewater sample from one of the kitchens was further modified with the addition of ash with a view to increasing the pH of the sample. According to the results shown in Table 1, grey water samples from kitchen and cloth washing all have higher organic strength as are evidenced by the high COD values of all the samples. Comparison of grey water samples from kitchen and cloth washing also shows that grey water samples from the kitchen have relatively higher organic strength while their pH is low. The low pH is expected due to the formation of organic acids during the initial storage time of the more biodegradable kitchen water samples. By contrast, grey water samples from cloth washing have relative higher pH ranging between 8 and 9.5. As will be shown later, the adsorption of organics by activated charcoal shows a bias in favour of better adsorption at higher pH. This fact motivated the modification of one of the grey water samples from kitchen with readily available ash.

Another important observation revealed through the study is the ease with which significant portion of the organic matter in both the kitchen and cloth washing samples were removed by filtration prior to adsorption by activated charcoal. All the samples were filtered through #100 filter paper and the CODs of the raw water and filtered water samples were analyzed.

Table 1 again shows the percentage removal of organics by filtration. Organic matter from cloth washing, measured as COD, was easily be removed 60-70% by sorption alone while for kitchen grey water the percentage COD removal ranged between 36 and 70%.

Table 1. *Characteristics of the grey water samples tested for charcoal adsorption.*

Waste water type	pH range	Initial COD	Filtered COD	Percentage removal by sorption
	(pH units)	mg/L	mg/L	(%)
Cloth wash water-1	8.00-9.25	1800	714	60%
Cloth wash water-2	8.31-9.5	2630	770	70%
Kitchen wastewater-1	3-68-5.00	2320	1470	36%
Kitchen wastewater-2	4.00-4.38	2740	830	70%
Kitchen wastewater + ash	8.03-8.52	2000	867	56%

Since there is no significant chemistry involved in removal of organic matter by filtration, the removal can only conceivably considered to be sorption. This finding is useful for activated charcoal adsorption as it is recommended to provide filtration as pretreatment prior to charcoal adsorption since the activated charcoal adsorption is best reserved for adsorption of soluble organics

3.2. Adsorption Characteristics Among Different Wastewater Streams

For each stream of grey water tested, the organic matter adsorbed per unit weight of charcoal adsorbent (X/M) was plotted against the grey water solute COD (C) that is in equilibrium with the adsorbent. Figures, 1, 2 and 3 show the different nature of the adsorption isotherms obtained from the test results. As is evident from Figure 1, grey water from cloth washing was amenable to complete adsorption as the isotherm curves tend to pass through the origin as the mass of the adsorbent increases. By contrast for grey water from the kitchen (Figure 2), the adsorption curves do not pass through the origin. It can be seen from Figure 2 that the curves intersect the COD line at around 500 mg/L which represents the non-adsorbable part of COD. Modification of the samples through the addition of ash to a certain extent decreased the non-adsorbed COD from 500 to about 400 mg/L.

The adsorption curves also vary between the Type I ($n<1$) and Type III ($n>1$) curves and show mixed nature of adsorption of the grey water sample tested. This mixed nature of adsorption may be expected considering the heterogeneous nature of constituents of grey water. The convex downward adsorption isotherm of wash water shown in Figure 1 with $(1/n) <1$ in the adsorption isotherm is typical of cases where the marginal adsorption energy decreases with increasing surface concentration [18]. By contrast the rest of the curves in Figures 1 and 2 have $(1/n)>1$ with the Freundlich isotherm model. This isotherm is typical of cases where the marginal adsorption energy increases with increasing surface concentration [19]. It may represent a result of strong intermolecular forces in the adsorbent layer and penetration of the solute further in the adsorbent media. Part of the difference in the nature off adsorption shown in Figure 1 and Figure 2 between $(1/n) < 1$ and $(1/n)>1$ may be explained by the solute COD concentrations whereby the higher solute organic compound concentrations influence intermolecular forces resulting in increase in the marginal energy of adsorption at higher surface concentration of adsorbed compounds. This cooperative adsorption may become dominant in situations where the activated charcoal surface, possessing increased surface acidity and preventing adsorption of non- polar organics is covered and facilitates further adsorption through cooperative adsorption mechanism [20].

However, the adsorption for cloth wash water-1 shown in Figure 1 with (1/n =0.612) <1 has its (1/n) sufficiently lower than the n=1 linear adsorption range. Thus the adsorption is influenced by decreasing surface energy of adsorption which happens at higher solute COD values typical of grey water streams. In short the adsorption is different from the linear adsorption isotherm that usually happens at low solute COD concentrations such as in natural non-point source pollutions from runoff [21].

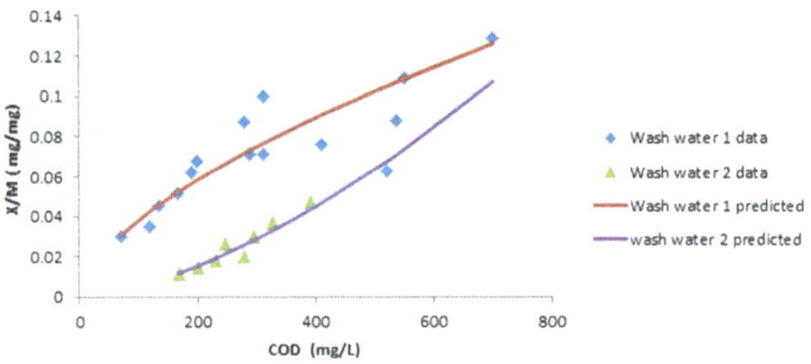

Figure 1. Mixed nature of the adsorption isotherm of grey wash water from cloth washing.

The comparison of the adsorption efficiency between grey water samples from cloth washing and the kitchen as shown in Figure 3 indicates that generally grey water from cloth washing show greater efficacy of adsorption compared with grey water samples from the kitchen. The efficiency of adsorption of the kitchen grey water samples was increased by about 30% when these samples were modified with the addition of ash to increase the pH environment of the adsorption. This step was motivated by the apparent difference in extent of adsorption at high pH environment of the cloth water samples. The decrease in adsorption of organics at low pH may be explained possibly because of the competing influence of the presence of excess H^+ at low pH competing for adsorption with the organic acids present in kitchen wastewater [9].

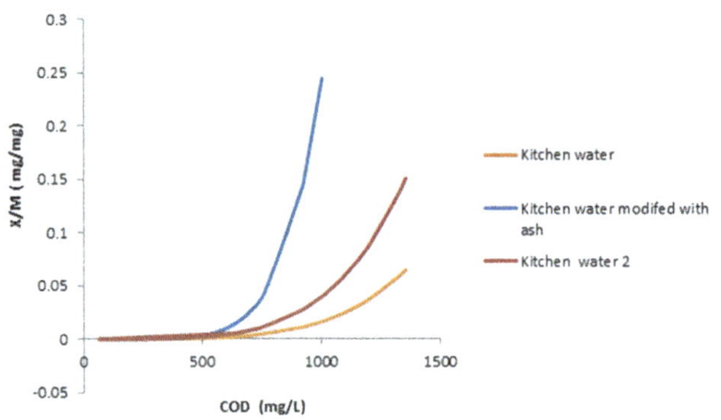

Figure 2. Comparison of isotherms of grey water from kitchen including the effect of adding ash.

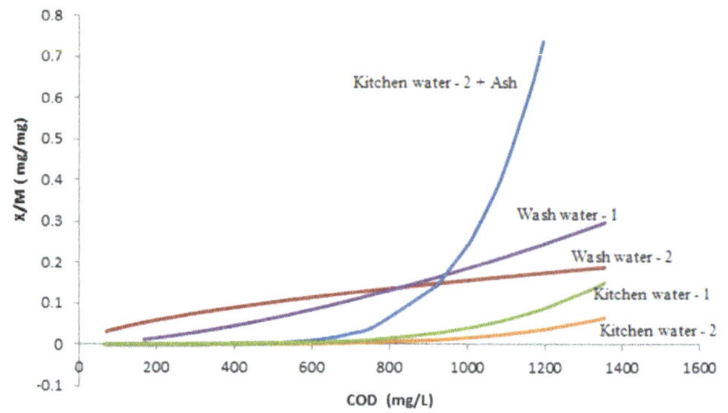

Figure 3. Comparison of isotherms of different streams of grey water.

Apart from the mixed nature of the adsorption, the test results shown in Figure 2 illustrate the difficulty of adsorption of some streams of kitchen water with activated charcoal. This indicates that not all streams of grey water are adsorbed to a favorable degree by activated charcoal.

3.3. Development of Adsorption Isotherms

The adsorption isotherms for all the grey water samples tested were found to be adequately modeled through the Freundlich adsorption model. The model constants K and n were determined by regression of the logarithm of (X/M) against logarithm of COD values in equilibrium with the adsorbent (C value). Figures 4 and 5 show the regression plot of the Freundlich isotherm for typical cloth washing and kitchen grey water samples.

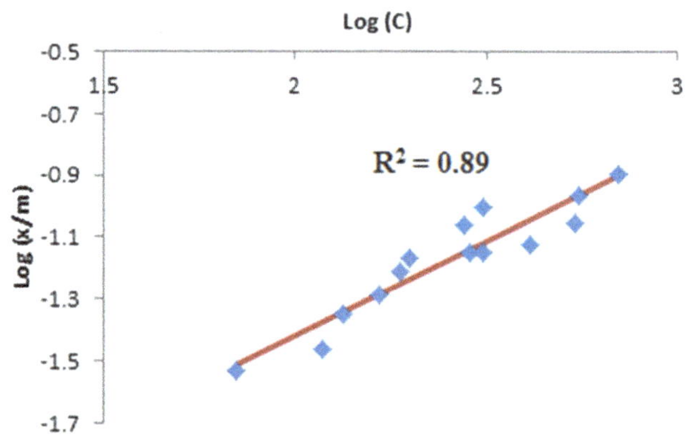

Figure 4. Freundlich isotherm of activated charcoal adsorption of cloth wash water -1.

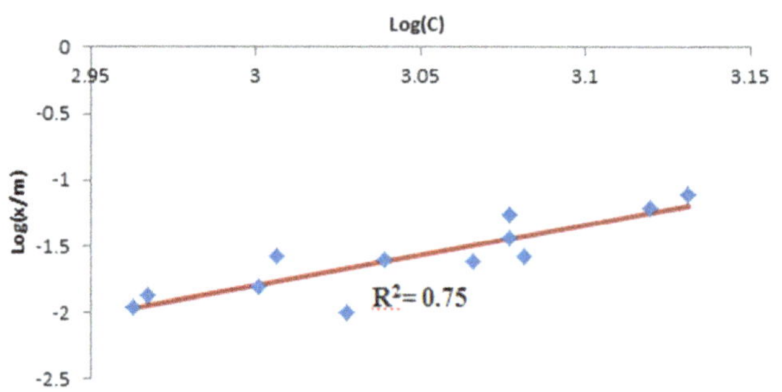

Figure 5. Freundlich isotherm of activated charcoal adsorption kitchen wastewater-2.

3.4. Operation Lines and Fractional Capacities

Assuming that saturation adsorption zone moves uniformly through the adsorption column the operation lines and equilibrium lines for the different grey water streams are plotted and shown in Figures 6 and 7. It can be seen from Figure 6 that the lines are close for the cloth wash water whereas Figure 7 shows that a marked variation in the operation lines for the different grey water streams. Modification of the second grey water sample from kitchen with ash is seen to improve the operation lines considerably. This improvement shows also in the marked increase in the length of operation of the adsorption column as calculated from the isotherm models.

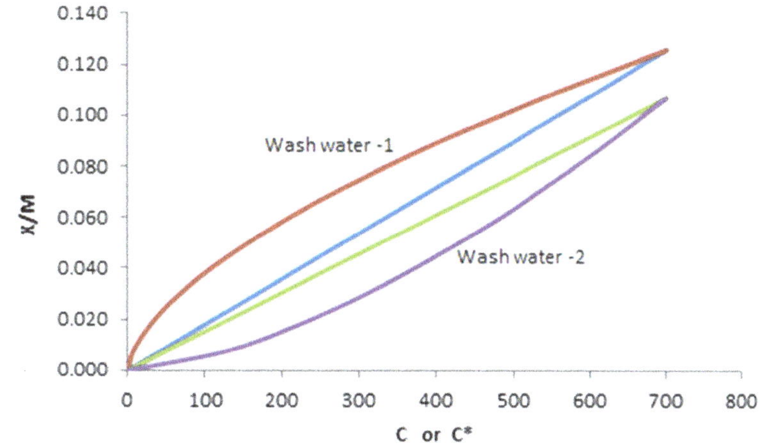

Figure 6. *Operation lines of grey water from cloth washing.*

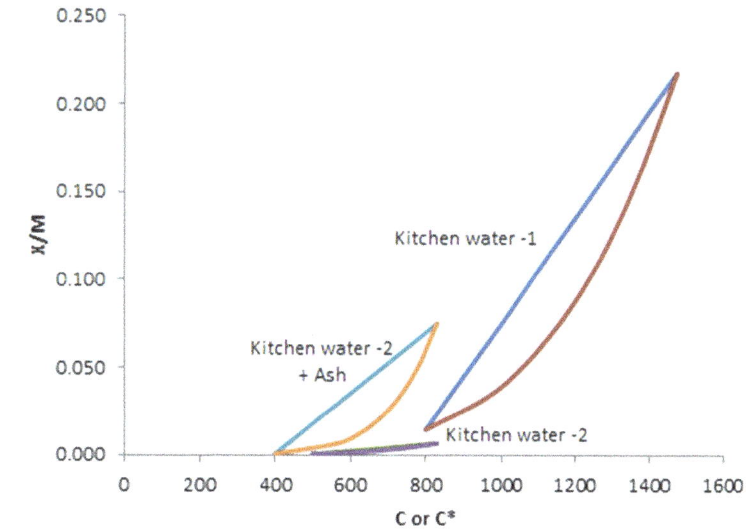

Figure 7. *Operation line of grey water from kitchen including ash modification.*

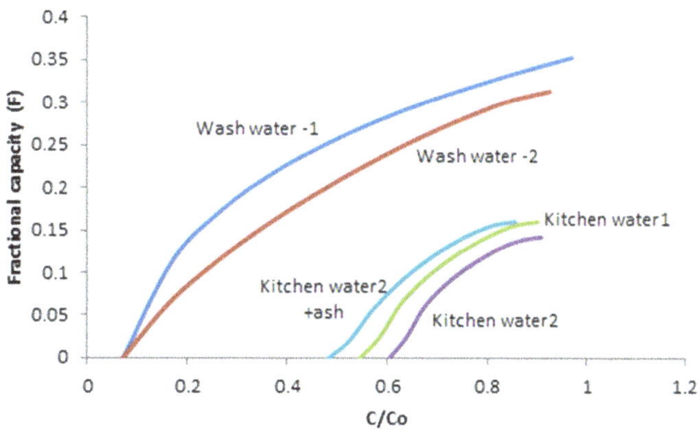

Figure 8. *Fractional capacity at breakthrough for the different streams of grey water.*

The fractional capacities of the different stream of grey water are shown in Figure 8. The differences in the available fraction of adsorption column within the adsorption zone between the different streams of grey water are apparent from Figure 8. The calculation of percentage saturation of the adsorption zone for a 60 cm depth of charcoal in a single drum when breakthrough occurs was worked out and is shown in Table 2. In all cases, the percentage saturation is very high and it can be safely stated that most the charcoal will be utilized for adsorption at breakthrough and this fact can be taken into consideration for the subsequent computation of the length of life of the drum of charcoal and the running cost per cubic meter of wastewater treated as shown below.

Table 2. *Percentage saturation of the adsorption zone at column breakthrough.*

Type of grey water	drum diameter	Depth of charcoal	Area	Flow rate	Flow rate (Q)	Mass transfer, K_a	Fractional capacity	Depth of adsorption zone	(%) saturation
	(m)	(m)	(m^2)	(lit/day)	(lit/min-m^2)	Kg/(min-m^3)		(m)	
Wash water-1	0.572	0.6	0.257	50	0.009	2.145	0.353	0.03962	95.73
Wash water-2	0.572	0.6	0.257	50	0.009	2.044	0.313	0.03962	95.46
Kitchen water-1	0.572	0.6	0.257	50	0.009	0.914	0.16	0.03962	94.45
Kitchen water-2	0.572	0.6	0.257	50	0.009	0.989	0.142	0.03962	94.33
Kitchen water 2 + ash	0.572	0.6	0.257	50	0.009	0.513	0.16	0.03962	94.46

3.5. Economic Feasibility of Using Charcoal Adsorption for Grey Water Treatment

The hypothetical replacement cost for a single drum of charcoal with depth 60 cm and diameter of 57.2 cm for treating a typical daily grey wastewater of 50 lit/day from a household was worked out for the different streams of grey water using the adsorption modeling information presented earlier. The result is summarized in Table 3. It can be seen from Table 3 that, barring the second kitchen water sample, most of the grey water samples studied show a longer run length for a single drum treatment ranging between 7.2 and 15.1 months. The single drum charcoal life length for the difficult to treat stream of kitchen wastewater was also extended from 1 month to 8.1 months which is a very significant improvement. The replacement cost in the feasible cases is below the current baseline tariff provided by the Swaziland Water Services Corporation of about 3 USD/ m^3 of water. This comparison, however, is tentative as there are several other running costs involved in treating grey water.

The variation with daily flow rate of the length of life of activated charcoal and its replacement cost per cubic meter of water treated are given in Figure 9 and Figure 10 respectively. Comparing the costs with the current tariff of acquiring one cubic meter of water in the city which is about 3 USD, it can be seen from Figure 10 that the single drum charcoal can run economically up to a daily flow rate of 400 lit/day. The corresponding economical length of life as seen from Figure 9 can be as low as 0.5 months. This means that replacing the charcoal even every two seeks is economically justifiable compared with the per cubic meter cost of acquiring water from the city water supply.

Post treatment of the grey water stream from the kitchen following activated charcoal adsorption is necessary because of the significant proportion of non-adsorbed organics from the kitchen wastewater streams. It is, therefore, generally recommended that a post-adsorption sand filter drums be provided to remove the remaining organics further by anaerobic/aerobic biological means. The suitable recommended pilot combination of treatment for household grey water streams would, therefore consist of drums of coarse sand filter acting as pretreatment by sorption, followed by activated charcoal adsorption and finally drums of fine sand filter for removal of the residual soluble organics by a combination of /anaerobic treatment.

The study results presented above give a useful insight into the technical and economic feasibility of treating grey water using activated charcoal as well as forming a design basis for further pilot scale study of treatment with the mentioned combination of sand filters and activated charcoal adsorption.

The spent charcoal cannot be regenerated economically and may either be disposed off or possibly be burnt to generate energy for the household. However, the health impact of burning spent charcoal which contains potentially harmful substances adsorbed from grey water need to be taken into consideration before choosing such alternative.

Table 3. *Cost of Length of run of charcoal with replacement cost for different types of grey water.*

Type of grey water	Initial COD	Effluent COD	COD removal	(X/M)sat	Mass of charcoal	COD adsorbed	Waste water flow rate	Run time	Replacement cost
	mg/L	mg/L	(%)		(Kg)	(Kg)	(lit/day)	(Month)	(USD per m^3)
Wash water-1	700	50	93%	0.126	74	8.901	50	8.5	2.3
Wash water-2	700	50	93%	0.107	74	7.553	50	7.2	2.7
Kitchen water-1	1400	500	64%	0.217	74	15.194	50	15.1	1.3
Kitchen water-2	830	500	40%	0.007	74	0.474	50	1.0	20.6
Kitchen water -2 + Ash	830	400	52%	0.075	74	5.246	50	8.1	2.4

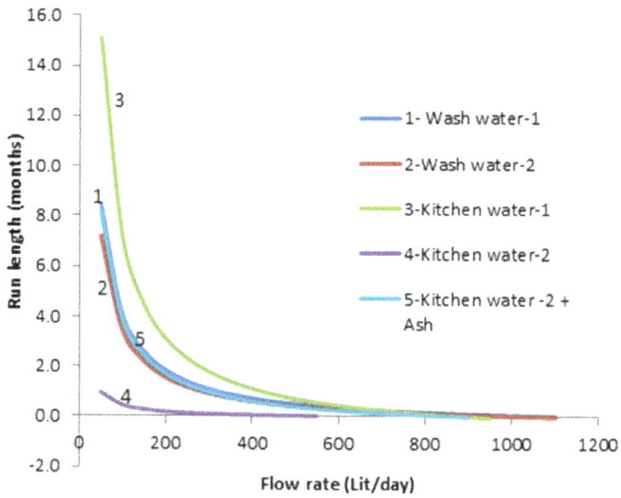

Figure 9. Length of run of a single drum of charcoal at different daily flow rates.

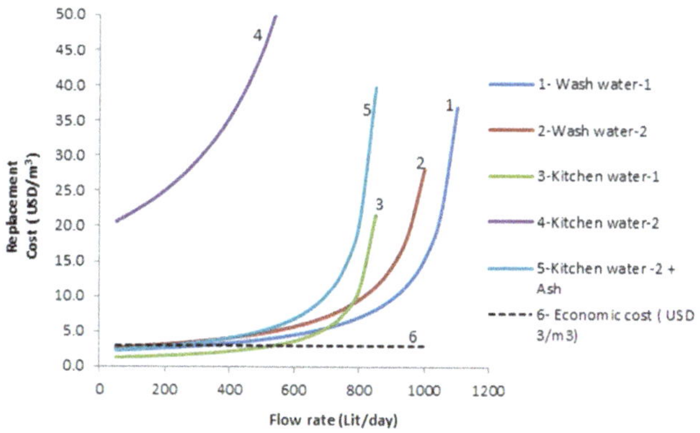

Figure 10. Cost comparison of replacement cost of charcoal at different daily flow rates.

4. Conclusion

Adsorption of organics from grey water streams using activated charcoal at household level provides a feasible treatment alternative. The research study indicated that activated charcoal can remove over 95% of the organic matter from cloth wash water while the efficiency of removal of organics from kitchen wastewater ranges between 30 and 70%. The efficiency of removal of organic matter using activated charcoal adsorption is seen to increase with increase pH as grey water with higher pH are removed more easily than the low pH kitchen wastewater. In addition modification of kitchen wastewater pH through the addition of ash improves the performance and considerably extends the life length of charcoal columns to economical ranges.

The adsorption isotherms display a mixed isotherm ranging between Type I and Type III adsorption which is typical as grey water represents a heterogeneous mixture of different organic compounds. The adsorption isotherm of cloth wash water passes through the origin were as the ones for kitchen wastewater intersect the COD axis at considerable COD values which represents the non-adsorbed part of the organic matter from kitchen waste streams.

It has also been seen that a considerable portion of the organics from the different grey water stream scan be removed by the physical process of sorption alone. Any type of media that presents a filtration mechanism including readily available sand can be employed for this sorption process. Such pretreatment is useful prior to activated charcoal adsorption as the adsorption is best reserved for the soluble part of the organic matter.

For mixed grey water streams that may be containing a combination of wash water and kitchen water, it is recommended to have a three step treatment consisting of a pretreatment coarse sand filter drum followed by drums of activated charcoal adsorption and finally sand filter drums for anaerobic/aerobic treatment of the residual organics following adsorption by activated charcoal.

Calculation of the length of life of a single drum of activated charcoal for household level treatment of a 50 lit/day grey water stream for the different types of grey water shows that the single drum can serve for a period ranging from 7 to 15 months depending on the type of grey water. Modification of the difficult to treat kitchen wastewater streams with ash also significantly improves the life length of a single drum from a non-feasible one month to almost eight months. In addition the economic life length of the charcoal can be as low as half a month while at the same time keeping the per cubic meter cost charcoal replacement below the tariff need to acquire the same volume of water from the city supply. In addition, a daily wastewater flow of up to 400 lit/day can be economically treated by a single drum activated charcoal treatment.

Comparison of the replacement cost of the charcoal at the current market cost indicates that the per-cubic replacement cost is below the current tariff levels of a unit cubic meter of water in cities in Swaziland. In general, the study indicated that activated charcoal represents a potentially feasible treatment alternative for the household level treatment of grey water.

Acknowledgment

This research study was supported by the government of Finland as part of the Mbabane Dry Sanitation and Waste Management Project (MDSWMP) research activity. The researchers would like to thank the financial assistance provided for carrying out the research work.

References

[1] Pidou MA, Memmon FA, Stephenson T, Jefferson B, Jeffrey P (2007). Grey water recycling: A review of treatment options and applications. Engineering Sustainability, 160: 119-131.

[2] Edwin GA, Gopalsamy P, and Muthu N (2014) Characterization of domestic gray water from point source to determine the potential for urban residential reuse: a short review. Appl Water Sci. 4: 39–49.

[3] USACE: Us Army Corps of Engineers (2001) Adsorption design guide. Design Guide No. 1110-1-2.

[4] Nansubuga I, Meerburg F, Banadda, N, Rabaey K, Verstrete W (2015). A two stage decentralized system combining high rate activated sludge with alternating charcoal filters for treating small community sewage to reusable standards in agriculture. African journal of Biotechnology Vol (14) 7, pp. 593-603.

[5] Thomas CV, Walter J, Weber JR (1983) Sorption of hydrophobic compounds by sediments, soils and suspended solids - theory and background. Water Res. Vol. 17, No. 10, pp. 1433-1441.

[6] Nethaji S, Sivasamy A, Mandal, AB (2013) Adsorption isotherms, kinetics and mechanism for the adsorption of cationic and anionic dyes onto carbonaceous particles prepared from Juglans regia shell biomass. Int. J. Environ. Sci. Technol. 10:231–242.

[7] Site AD (2001) Factors Affecting Sorption of Organic Compounds in Natural Sorbent Water Systems and Sorption Coefficients for Selected Pollutants: A Review. Phys. Chem. Ref. Data, Vol. 30, No. 1

[8] Harris CI and Warren GF (1964) Adsorption and desorption of herbicides by soil. Weeds 12: 120-126.

[9] El-Khaiary MI, Hameed BH (2008) Malachite green adsorption by rattan sawdust: Isotherm, kinetic and mechanism modeling. J Hazard Mater 159:574–579.

[10] Li QL, Snoeyink VL, Mariaas BJ, Campos C (2003) Elucidating competitive adsorption mechanisms of atrazine and NOM using model compounds. Water Res. 37(4): 773-784.

[11] Brusseau ML, Larsen T and Christensen TH (1991) Rate-limited sorption and non-equilibrium transport of organic chemicals in low organic carbon aquifer materials. Water Resour. Res. 27, 1137.

[12] Picer N, Picer M, Strohal P (1977) The interaction of DDT with suspended particles in sea water. Water, Air, Soil Pollut. 8: 429-440.

[13] Hernandeza OR (2004). To treat or not to treat? Applying chemical engineering tools and a life cycle approach to assessing the level of sustainability of a clean-up technology. Green Chem., 6: 395-400.

[14] Flagan J, Richard C. and Seinfeld John H (1988) Fundamentals of air pollution engineering. Prentice-Hall Inc., Englewood Cliffs, New Jersey ISBN 0-13-332537-7, pp. 479-520.

[15] APHA (1999) Standard Methods for the examination of Water and Wastewater: Chemical Oxygen Demand, 5220, # (102). American Public Health Association, American Water Works Association, Water Environment Federation.

[16] Weber WJ. Jr (1972) Physicochemical processes for water quality control, Wiley-Interscience, New York. 640 pp.

[17] Klobucar JM, Pilat MJ (1992). Continuous flow thermal decomposition of VOC's from activated carbon. Environmental progress 11(1): 11-17.

[18] Farrell J, Reinhard M (1994) Desorption of halogenated organics from model solids, sediments, and soil under unsaturated conditions. Environ. Sci. Technol 28 (1): 53–62.

[19] Haque R. and Coshow WR (1971) Adsorption of isocil and bromacil from aqueous solution onto some mineral surfaces. Environmental Science and Technology 197; 5(2): 139-141.

[20] Pignatello JJ (2000) The measurement and interpretation of sorption and desorption rates for organic compounds in soil media. Adv. Agron. 69: 1–73.

[21] Rao PSC, Davidson JM (1979) Adsorption and movement of selected pesticides at high concentrations in soils. Water Res., 13: 375-380.

Glaciers Dynamics Over the Last One Century in the Kodori River Basin, Caucasus Mountains, Georgia, Abkhazeti

Levan G. Tielidze[1], Lela Gadrani[1], Mariam Tsitsagi[1], Nino Chikhradze[1,2]

[1]Vakhushti Bagrationi Institute of Geography, Ivane Javakhishvili Tbilisi State University, Tbilisi, Georgia
[2]Ilia State University, Tbilisi, Georgia

Email address:
levan.tielidze@tsu.ge (L. G. Tielidze), levan_tielidze@yahoo.com (L. G. Tielidze)

Abstract: This paper considers the last one century's dynamics of the glaciers in the Kodori River basin, which is located on the southern slope of the Greater Caucasus in Georgia. The latest statistical information is also given about the glaciers located in the individual river basins; Their morphological types, exposition and the dynamics are considered according to the individual years. In our research, we used the Catalogue of the glaciers of Georgia compiled by K. Podozerskiy in 1911. We also used the military topographic maps with the scale of 1:25 000 and 1:50 000 drawn up in 1960, where there are mapped in detail the glaciers and the ends of their ice tongues on the southern slope of Greater Caucasus of those times. We also used the remote sensing method, which is the best modern way for the study of the dynamics of glaciers and this method is remarkable by a quick obtaining of results. The spectral images of the Landsat L5/TM and ASTER L1/A are the necessary materials for our study. As a result of the use of this method, we have identified the dynamics of the glaciers in the Kodori River basin.

Keywords: Glaciers Dynamics, Glaciers of Georgia Caucaus Mountains, Remote Sensing, Climate Change in Georgia, Glaciers of Abkhazeti

1. Introduction

It is known that the glaciers are the most important resource of the world's fresh water, that's why they always play an important role in the evolution of natural conditions and therefore, in the evolution of the biosphere. Melting of glaciers due to modern climate change is observed in all mountainous regions of the world.

Mountain glaciers play an important role in the regional hydrological cycle by modulating the storage and release of freshwater so, as they retreat, the availability of runoff for irrigation and hydro-power generation is altered [1].

In Georgia research of the glaciers on the southern slope of the Greater Caucasus has a long time history. The first scientific reports are given by a great Georgian scientist Vakhushti Bagrationi in the beginning of the 18th century as yet [2].

The compilation of the first reliable map of the Greater Caucasus at a scale of 1:420,000 and depicting the largest glaciers was completed by 1862. Topographic surveys of the Great Caucasus at a scale of 1:42,000 were accomplished 30 years later. Having analyzed these maps, K.I. Podozerskiy in 1911 published the first catalogue of the Great Caucasus glaciers, compiling data on the lengths and areas of all of the glaciers [3].

Observation on the glaciers of the Western, Central and Eastern Caucasus is continuous since 30s of the 20th century to the present day. The scientists from the Vakhushi Bagrationi Institute of Geography conduct the permanent monitoring of the glaciers of Svaneti, Abkhazeti (by means of modern space images) Racha and Khevi Caucasus.

In Georgia contemporary glaciers are mainly concentrated in the Enguri, Rioni, Kodori and Tergi River basins. Main centers of the glaciation are related to the higher elevated Greater Caucasus and Kazbegi massif. Also, there are the individual centers in the branch ranges of the Greater Caucasus. Nowadays there are 637 glaciers in Georgia with the total area of 355.8 km^2 [2, 4].

2. Study Area

Kodori River basin is located on the southern slope of the Western Caucasus (Fig.1). Kodori gorge is divided into several sections from the headwaters to the confluence due to the morphological, lithological and tectonic features.

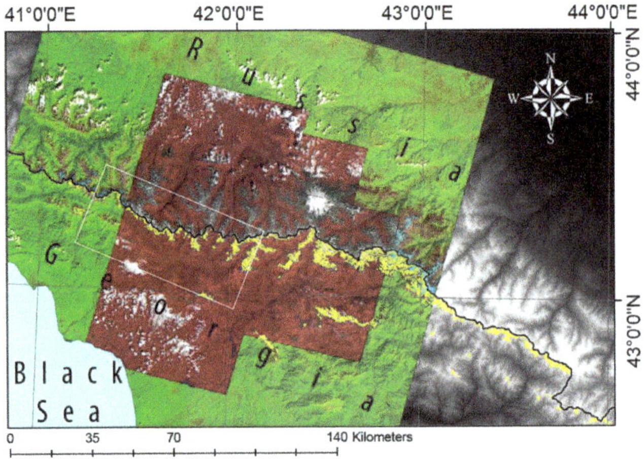

Fig. 1. Location of the Kodori River Basin, Landsat and ASTER Imagery

Gorges of the Kodori River tributaries (Ptishi, Khetskvara, Klich, Ghvandra and Sakeni) is located on the southern slope of the Greater Caucasus are of meridional and submeridional directions. Morphologically they are trough valleys and are deeply sat in the relief and are slightly modified by the erosion processes. Glacial deposits-moraines are spread on the bases and slopes of the valleys. Erosion forms are poorly represented (debris cones). Numerous snow avalanches are involved in the modeling of the valleys [5].

Study of the glaciers of the Kodori River basin began in the last century as yet. Famous researchers of the Greater Caucasus gave us information about glaciers: G. Radde (1873), N. Bush (1905), K. Podozerskiy (1902, 1911), P. Ivankov (1959), D. Tabidze (1965, 1966, and 1975) and R. Gobejishvili (1989).

Southern slope of the Western Caucasus from Marukhi to Dalari Passes presents the main center of glaciation in the Kodori River basin. High peaks such as Marukhi, Ertsakho, Sopruju, Dombai Ulgeni, Khakeli and Ghvandra are located in this section. Their height is over 3800-4000 m. Isolated glaciation enters are presented in the Kodori, Chkhalta (Abkhazeti), Khutia and Klich ranges. Especially should be noted the western and south-western slopes of the Kharikhra range, which is a main glaciation center in the Sakeni River basin. Peaks higher than 3700 m are located in this range, they are: Kharikhra, Maguashikhra and Okrilatavi.

3. Data and Methods

In this research we used the catalogue and maps of glaciers compiled by K. Podozerskiy in 1911. Although in the mentioned maps in some cases there are mistakes made in visualization of areas and number of glaciers, but this is the most reliable and adequate source to get the information about the glaciers conditions of those times.

Also we used the military topographic maps of 1:25,000 and 1:50,000 scales created in 1960, where the glaciers, the ends of their ice tongues and the firn line of the southern slope of the Greater Caucasus of those times are depicted in detail. These maps are in the Pulkovo-1942 projection. We used standard transformation parameters to re-project the map in Universal Transverse Mercator (UTM), zone 38-North on the WGS84 ellipsoid, to facilitate comparison with modern datasets.

We used the Remote Sensing research method as well. Remote sensing is an deft tool for increasing the number of monitored glaciers on a global scale, especially in remote or politically sensitive areas. It has the added advantage of allowing a single user to determine changes in glacier surface characteristics over very large areas (thousands of km^2), and is extremely rapid when compared to field investigation [6, 7, 8, 9]. Landsat L5/TM (Thematic Mapper), with 30 m horizontal resolution available since 1982 and 1999 respectively, and Advanced Spaceborne Thermal Emission and Reflection Radiometer (ASTER) imagery with 15 m resolution available since 2000 [10]. Together with historical topographical maps, this imagery enabled assessments at time steps of 50-55 years during which the magnitude of glacier change significantly exceeded measurement errors.

Therefore, except the map of 1911 by Podozerskiy and topographical maps of 1960, the spectral image Landsat L5 TM (Thematic Mapper) and ASTER L1/A taken in August of 2014 is a necessary material for investigation. The mentioned case corresponds to the clear weather conditions, when the impact of snow cover on glaciers and the absence of clouds facilitate to perfect visibility of objects on the Earth's surface. Then we used the Composite Bands function with the help of ArcGis program and put the images in the sequence of 7-4-2 (for Landsat) and 3N-2-1 (for ASTER). As a result of the application of the mentioned method we identified the variability in the area of the glaciers of the Kodori River Basin in 1911-1960-2014, as well as the altitudinal variability in the glaciers length and ice tongues according to the mentioned years.

4. Results and Discussion

According to the data of K. Podozerskiy, at the end of the 19th century there were 118 glaciers in the Kodori River basin with the area of 73.2 km^2 [11]. according to the data of D. Tabidze (according to the 1946 aero images) – there were 141 glaciers with the area of 60.0 km^2 [12]. And by the data of R. Gobejishvili (based on the topographic maps of 1960) – there were160 glaciers with the area of 64.5 km^2 [13]. By the data of 2014 there are 145 glaciers in the basin with a total area of 40.1 km^2 [2]. The glaciers are unevenly distributed in the Kodori River basin, not only according to the orographical units, but also according to the individual tributary river basins.

Here dominate small glaciers by a number. The glaciers are distributed according to the tributary river basins as follows (Fig. 2).

The glaciers of all morphological types are distributed in the Kodori River basin except of the compound-valley type. Morphological and morphometric characteristics of the relief and climatic conditions are favorable for the existence of multiple cirque type glaciers. Cirque type glaciers occupy the first place according to the number and area. 66.2% of the total number of glaciers of the Kodori River basin is the share of them, as well as 48.1% of the total area of the glaciers of the Kodori River basin is a share of the cirque type glaciers (Fig. 3).

According to the area, the second place occupy the glaciers of cirque-valley type and according to the number, the second place occupy the cirque-hanging glaciers.

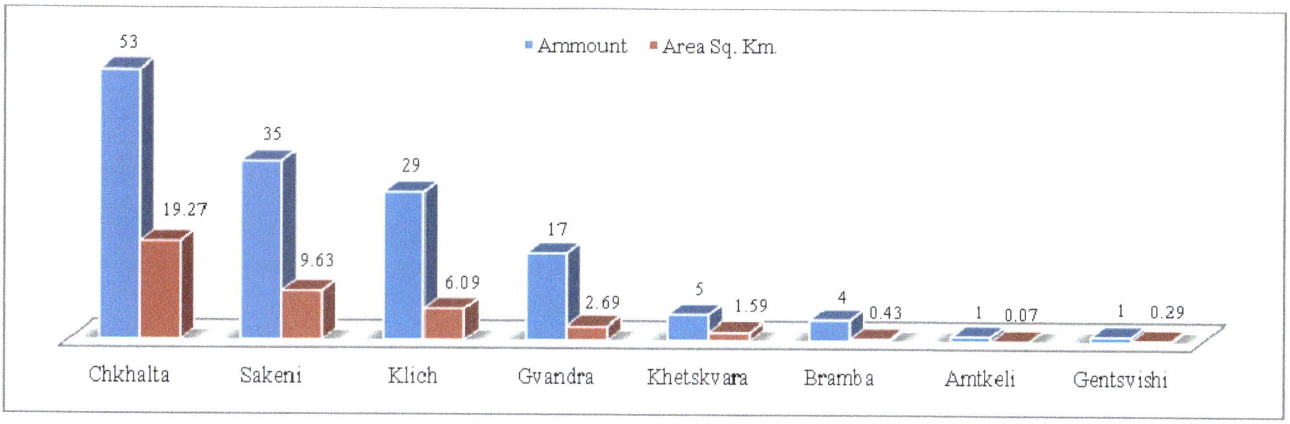

Fig. 2. Distribution of the Glaciers of the Kodori River Basin according to the Tributary River Basin

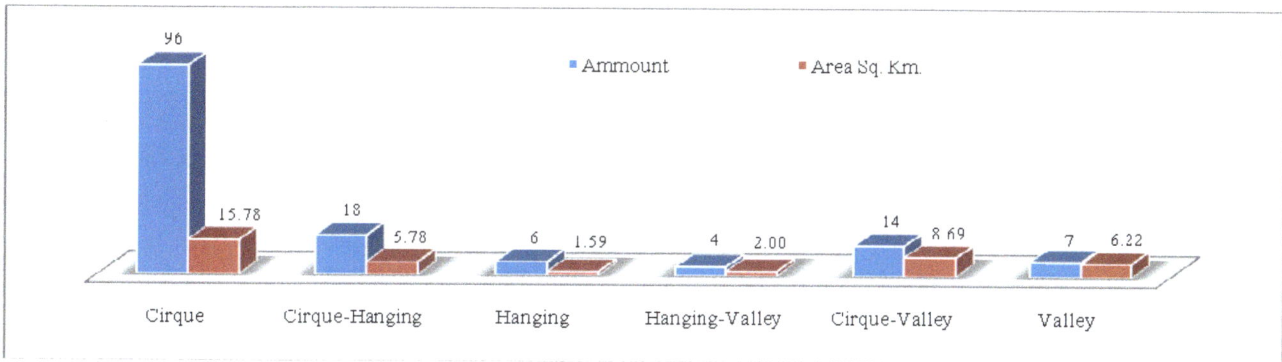

Fig. 3. Distribution of the Glaciers in the Kodori River Basin according to the Morphological Types

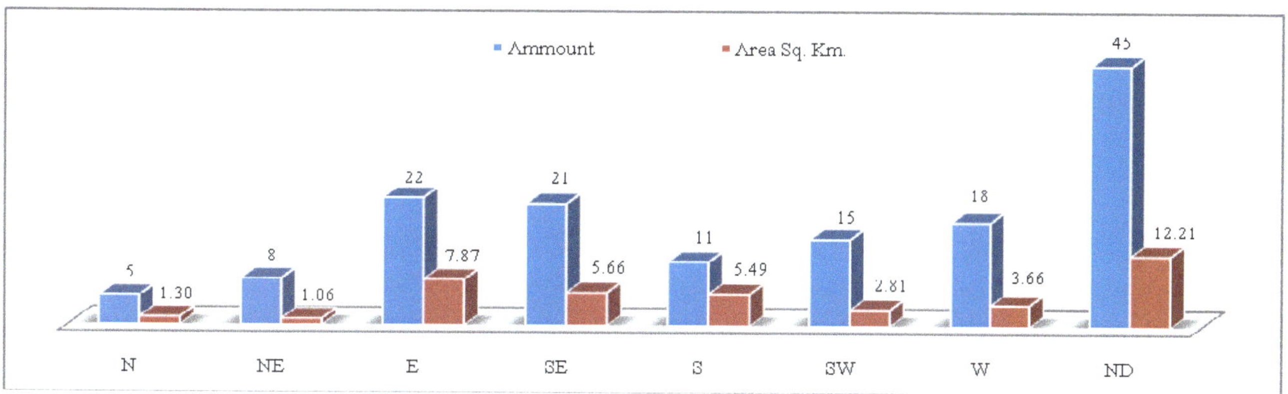

Fig. 4. Distribution of the Glaciers in the Kodori River Basin according to the Exposition

There are the glaciers of all expositions in the Kodori River basin (Fig. 4). By number exceed the glaciers of north-western exposition – 31.0%, then comes the glaciers of the eastern exposition – 15.2% and south-eastern expositions – 14.5%. According to the occupied area, there is a same sequence: north-western exposition – 30.5%, eastern exposition – 19.6% and south-eastern – 14.1%.

Kodori River basin is located on the southern slope of the Greater Caucasus and the exposition should be mainly of southern direction, but the conducted study shows that the glaciers of northern exposition dominate, both in number and area. Predominant distribution of the glaciers of northern exposition can be explained by the sublatitudinal or latitudinal direction of branch ranges.

As it was already mentioned, the glaciers of the Kodori River basin are not equally distributed, not only according to the orographical units, but also according to the individual river basins; therefore, to have a clearer idea on the contemporary glaciers of the Kodori River and its tributary river basins and their variability in time and space, it is necessary to consider these basins separately.

Chkhalta River basin. 36.6% of the amount and 48.1% of the area of the glaciers of the Kodori River basin are the shares of the Chkhalta River Basin. Similarly as in the entire Kodori basin, here also, observed reduction the total amount and area of glaciers (Fig. 5).

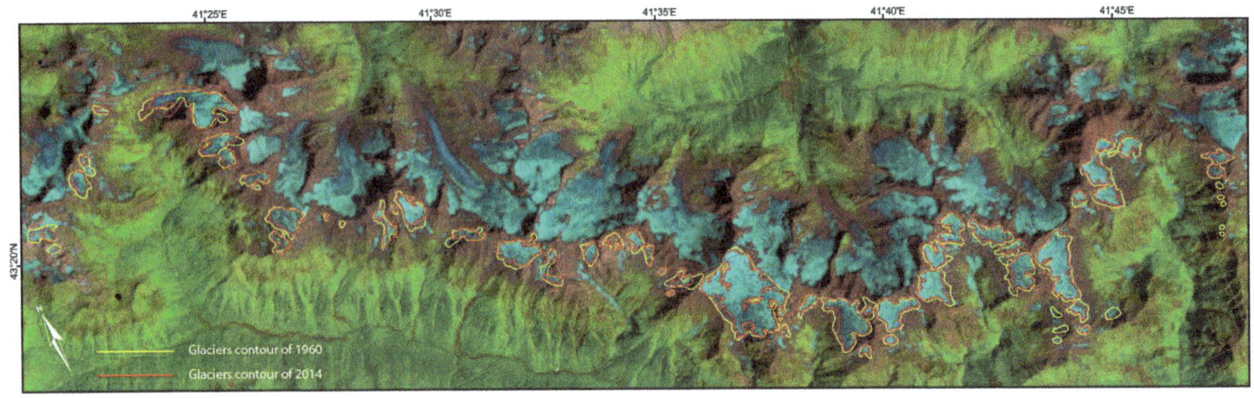

Fig. 5. *Glaciers Retreat in the Chkhalta River Basin in 1960-2014 (Landsat L5/TM)*

The glaciers are distributed unequally by the morphological types and exposition. Two centers of the glaciation can be distinguished in the Chkhalta River basin.

A separate description of the Marukhi River basin has its preconditions: 1. it is a completely separate center in terms of the contemporary glaciation. 2. According to the old glaciations this basin was independent as well.

By Podozerskiy (1911) 12 glaciers were in the Marukhi River basin, and their area was 6.4 km^2. By the data of 1960 year 10 glaciers with the area of 4.8 km^2 were presented in this basin. 9 glaciers with the total area of 3.9 km^2 are by 2014 data.

Marukhi glacier is the largest one in this basin, which belongs to the type of valley glacier. Its area is 1.7 km^2, and the length is ~4.1 km. Marukhi has a clearly defined ice tongue and ~300 m high icefall. The ice tongue is about ~1.5 km long, which is weakly fractured and characterized by a slight inclination. The central part of the tongue is clean, only its last section is covered with loose material. The surface of the ice tongue is rugged by melting waters due to its weak fracturing and slight inclination. Water-raised grooves are quite large (~2-5 cm wide) and they can be found only on the pure glacier surface. In front of the ice tongue at a distance of ~800-900 m long, in the valley (the glacier trough) there are moraine hillocks of different sizes, which allows us to rebuild the location of the glacier during the Fernau stage and discuss its dynamics for the last ~200 years. It is true that Marukhi glacier has no lateral stade moraines, but the trace of the stade glacier is distinguished on the both slopes of the valley by a sharp line (according to the moraine material spread on the slope, vegetation cover, slope color tone and other morphological parameters). The last stade and microstade moraines are well-expressed in the valley. The last Fernau stade moraine covers the top of the ledge and is presented by ~10-15 m thick loose material. Thickness of the microstade moraines is ~1.3 m. There is a ~150-200 m long straight surface between the microstade moraines, which is built by the fluvioglacial deposits. As a whole, there are three microstade moraines in the valley with two rows per each.

In August of the years of 1975-1982 the phototheodolite survey of the Marukhi ice tongue was conducted [2]. The plan of 1:5000 scale was compiled. The mentioned map clearly explaines the morphometric and morpographic peculiarities of the ice tongue and the area in front of it.

The rest of the glaciers of Marukhi River Basin is the glaciers of small cirque and cirque-hanging types. They are mainly located in the headwaters of the rivers of Buloni and Chvakhra. These basins fed the large Marukhi glacier during the Wurm glaciation together with the Marukhi River basin.

The left section of the Chkhalta River gorge from the Mount Ertsakho to Mount Ptishi covers the southern slope of the Great Caucasus. The peaks create good conditions for existence of glaciers. By 1960 data there were 32 glaciers with the area of 15.6 km^2. By 2014 data there are 25 glaciers with the total area of 8.1 km^2. The comparison between the mentioned data well depicts the degradation of glaciers in recent decades.

In the given section the biggest glacier is the Southern Sopruju (Fig. 6). The glacier has an extensive firn field, and the ice tongue is characterized by small parameters and it is hanging over the ledge. The glacier is slightly fractured, let alone the last section of the ice tongue. Due to morphological conditions of the area the glacier suffers the lack of moraine material (both the bed moraine and surface moraine material). The glacier exarates the bedrocks. The Sopruju glacier is the largest in size in the whole Kodori River basin; Its area is 3.7 km^2.

Atsiash-Atsgara River headwater is distinguished by the independent center of glaciation, which morphologically represents an old cirque. By 2014 data here are the eight glaciers. All eight glaciers are of cirque-hanging type. Their total area is 3.8 km^2

Ptishi River is a left tributary of the Chkhalta River. The morphology and morphometric features of the relief of its headwaters create better conditions for the existence of glaciers. By the data of 1960 there were 12 glaciers with the area of 4.0 km^2, and by the data of 2014 there are 7 glaciers with the total area of 1.8 km^2.

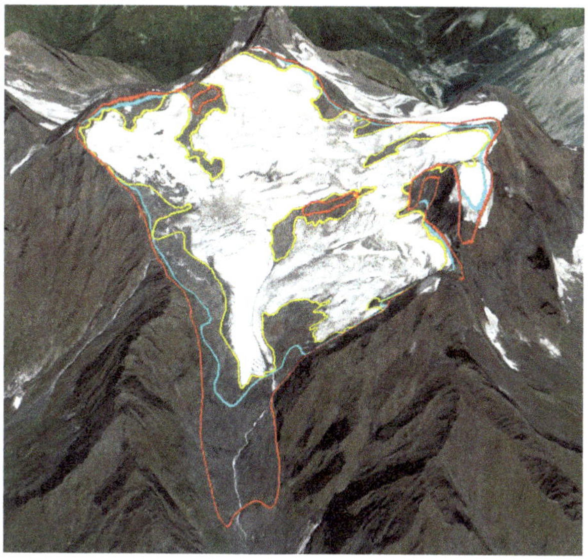

Fig. 6. Sopruju Glacier Retreat in 1911-1960-2014 (Googla Earth Image)

The glaciers in the Khetskvara River basin are developed on the western slope of the Khutia range. By K. Podozerskiy (1911) there were 6 glaciers in this basin with the area of 2.8 km^2. By 1960 data the number of the glaciers and their area remained unchanged. By 2014 data there are 5 glaciers with the total area of 1.6 km^2. Four glaciers are of cirque type morphologically, and one – is of hanging type.

Due to the fact that the glaciers are distributed in the western slope of the Khutia range, their exposition is northwestern and western.

According to the topographical maps of 1960 there were two small cirque type glaciers of southern exposition in the Gentsvishi River basin. There is no information on them in the work of K. Podozerskiy. By 2014 data there is a small cirque type glacier of south-eastern exposition.

Klich River basin is the only gorge in the Kodori River basin, which maintains the shape of the trough valley along the whole distance. Contemporary glaciers descend to a very low elevation. The glaciers are mainly represented in the basins of the rivers of Achapara, Klich and Nakhari.

According to K. Podozerskiy (1911) there were 25 glaciers with the area of 16.6 km^2 in the Klich River basin at the end of the 19th century. By the data of 1960 the number of the glaciers was 28 with the area of 9.5 km^2. We should indicate that the distribution of the glaciers in the topographical map of 1960 and of K. Podozerskiy do not coincide with each other. The glaciers given in the Catalogue of K. Podozerskiy were not reflected in the data of 1960. It is related to the small glaciers or the snow spots, which have disappeared due to the intense retreat of the glaciers or they still are the snow spots (in the Achapara gorge and near the Klukhori Pass). By the data of 2014 the number of glaciers is 29 with the total area of 6.1 km^2. It follows that after 1960 the total area of the glaciers has reduced by 35.6% and the number has increased by the 1 glacier.

There are mainly small cirque type glaciers in the Klich River basin. The glaciers located on the Khutia range are of eastern exposition; as for the glaciers located on the Klich range, their exposition is of western or north-western direction.

20.0% of the total number and 15.2% of the total area of the glaciers of the Kodori River basin is a share of the glaciers of the Klich River basin. It is behind of the basins of the rivers of Chkhalta and Sakeni by the thickness of the ice sheet. In the Klich River basin the largest is the Klich glacier. Its length is ~3.1 km and the area – 0.8 km^2. The glacier starts from the Mount Klich and spreads to the northeastern direction. Its ice tongue ends at a height of 2450 m in the shape of a forehead; and in front of it there are well-expressed annual moraine hillocks, the height of which does not exceed ~0.5-1.0 m. These forms are created due to every year movement of the glaciers and fix the size of the glacier's retreating in time. The number of the arcwise hillocks is 15 and they end at the top of the ledge, but the annual moraines below are washed away and the bedrocks are outcropped. The moraines are located in parallel to each other and sometimes they cover each other [13].

There are three cirque type glaciers to the left of the Klich glacier. The glaciers exarate the bedrocks. Their bottom is quite inclined and the glaciers are not able to develop the moraines. The glaciers end nearly at a similar elevation.

Cirque type glaciers have retreated by 400-500 meters after the Fernau stage, on which indicate the well-developed lateral moraine hillocks in front of the glaciers and the lateral moraines which are extremely washed away. Among these three glaciers the right-hand glacier has very weakly developed moraines. This right-hand glacier was a Klich glacier's branch range during the Fernau stage. Its bottom is very inclined and does not allow the formation of moraines.

All early researchers mention the two glaciers to the right of the Klukhori Pass. Today there are powerful snow covers instead of them. Morphostructural character of the relief indicates that in the past (XIX century) here, for sure, should be the small glaciers of a cirque type, which are now melted and there are thick snow covers instead of them.

In the Ghvandra River basin there were 24 glaciers with the area of 7.2 km^2 by K. Podozerskiy; by the data of D. Tabidze – 20 glaciers with the area of 6.0 km^2, and by the data of 1960 – 22 glaciers with the area of 6.2 km^2. By the data of 2014 there are 17 glaciers in this basin with a total area of 2.7 km^2.

The Ghvandra River gorge has a shape of typical trough above the site of Mindora. Nevertheless, the Late Holocene stade moraines are weakly remained due to active erosion and gravitation processes [14].

Sakeni River basin is located between the ranges of the Greater Caucasus and Kodori. Its area is ~233 km^2. Glaciers are mainly located in the southern slope of the higher elevated Greater Caucasus and the western slope of the Kharikhra range. There are small glaciers in the northeastern slope of the Ghvaghva range. Sakeni River gorge is a second by the number and area of the glaciers among the Kodori

River tributaries. 24.1% of the number and 24.0% of the area of all glaciers of the Kodori basin is a share of the Sakeni River basin.

By K. Podozerskiy (1911) there were 17 glaciers in the Sakeni River gorge with the area of 15.5 km^2; by the data of 1960 there were 36 glaciers with the area of 19.1 km^2, and by the data of 2014 there are 35 glaciers with the total area of 9.6 km^2.

As for morphological types, here can be found all types of glaciers except of valley complex type.

Glaciers are mostly of commonly north exposition (N, NE, NW). This is due to the abundance of glaciers on the western slopes of the Kharikhra range.

Sakeni glacier, which is located in the southern slope of the Greater Caucasus, is the largest by area in the Sakeni River basin. It is a valley type glacier and has the southeastern exposition. In 1911 the glacier area was 2.8 km^2; In 1960 – 2.5 km^2 and by the data of 2014 its area is 2.0 km^2 (Fig. 7).

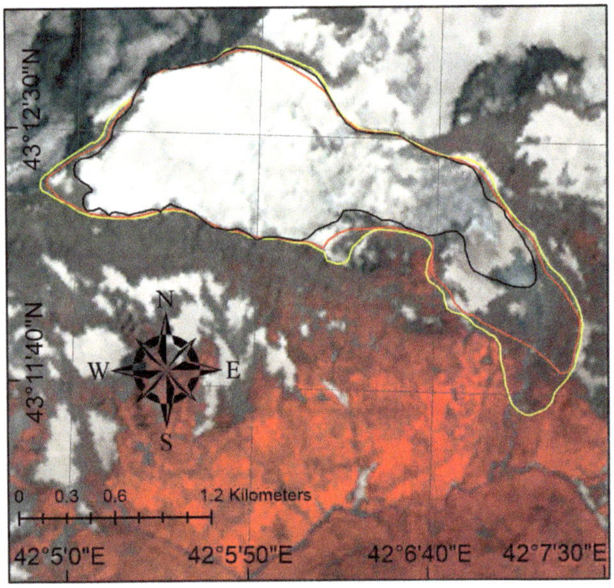

Fig. 7. The Retreat of Sakeni Glacier in 1911-1960-2014

Brakhma River basin is located on the north slope of the central section of the Kodori range. The range has a horizontal direction and is not high hypsometrically (Mount Khojali - 3314 m), which causes a lack of glaciers. There are 4 glaciers in this basin with the area of 0.4 km^2. Three glaciers are of cirque type, and one – of cirque-valley. Glaciers are distinguished with their small sizes. Their expositions are of northern and northwestern direction. K. Podozerskiy did not mention the glaciers in this basin. But by the data of 1960 there were 4 glaciers with the total area of 1.2 km^2 in this basin.

Amtkeli River basin. The Amtkeli River originates in the southwestern slope of the Chkhalta range. An old vast cirque of the northwestern direction is well represented in its headwaters. Today, there is one small cirque type glacier of the northwestern exposition. In 1960 its area was 0.08 km^2, and by the 2014 data it is 0.07 km^2.

5. Conclusions

The research revealed that increasing in number of glaciers in the years of 1911-1960 was caused by the degradation of glacial cover during their division (Fig 8). If we compare the data of the years of 1946-1960, we will see that in this period both the increase in area and number took place, which is very suspicious to us, because the increase in glacial cover is not recorded in other basins of the Great Caucasus in the mentioned years.

Data of 1946 is taken from the materials of D. Tabidze, who had conducted a great work to research the glaciers in the study region [12]. He studied the Kodori River basin glaciers in detail and gave us a new catalogue, which is mainly based on the 1:50,000 scale topographic maps compiled in 1952. It should be also specified that the glaciers are very poorly depicted on the mentioned maps. Maps are drawn based on the aeroimages of 1946. This may be caused the fact that D. Tabidze mapped the reduced glacial cover.

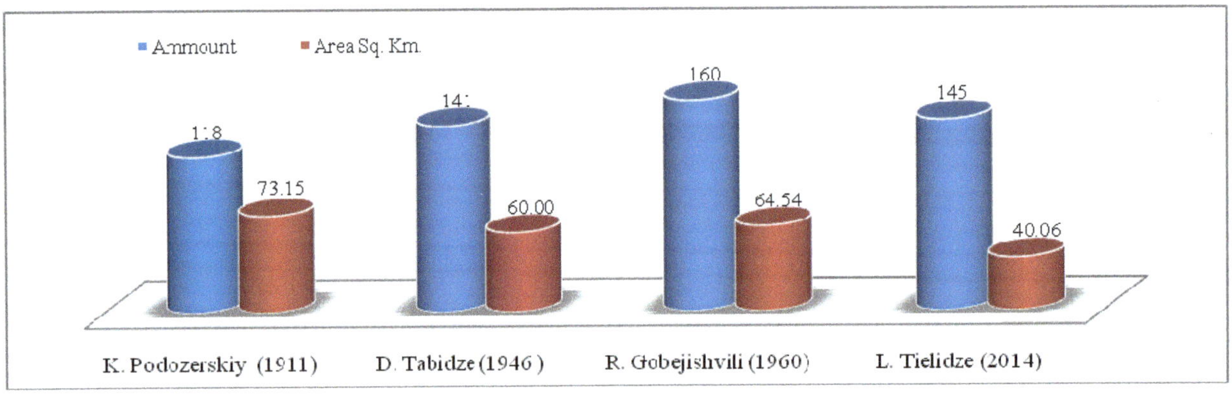

Fig. 8. Dynamics of the glaciers in the Kodori River basin in 1911-1946-1960-2014

Data of 1911-1960 well reflects the situation which was typical for evolution of glaciation in the Greater Caucasus till 1960-1980. In the years of 1960-1980 the area of the glaciers was reduced and their number was increased at the same time. We can't say the same today, because as recent studies have shown, along with the reduction in area of the glaciers the decrease in number of them takes place during the last half a century. The same situation is in the Kodori River basin, when the area of the glaciers has reduced by 37.9% in 1960-2014 (24.5 km^2), while the number of glaciers has decreased by 15 (9.4%) (Fig. 8). Overall in 1911-2014 the area of the glaciers has decreased by 45.2% (33.1 km^2) in the Kodori River Basin.

Acknowledgements

We are grateful to the Shota Rustaveli Georgian National Science Foundation for the financing our research.

References

[1] Kaltenborn B.P., Nellemann C., Vistnes I.I. (Eds). (2010). High mountain glaciers and climate change - Challenges to human livelihoods and adaptation. United Nations Environment Programme, GRID-Arendal, http://www.grida.no

[2] Tielidze L. G. (2014). Glaciers of Georgia, Monography, Publ. "Color" 254 p. Tbilisi. (in Georgian).

[3] Kotlyakov V.M., Dyakova A.M., Koryakin V.S., Kravtsova V.I., Osipova G.B., Varnakova G.M., Vinogradov V.N., Vinogradov O.N. and Zverkova N.M. (2010). Glaciers of the former Soviet Union. In: Williams, R.S. Jnr and Ferrigno, J.G., (eds.) Satellite image atlas of glaciers of the world - Glaciers of Asia: U.S. Geological Survey Professional Paper 1386–F-4 (doi:1386-F-4)

[4] Tielidze, L.G., Gadrani, L. and Kumladze, R. (2015) A One Century Record of Changes at Nenskra and Nakra River Basins Glaciers, Causasus Mountains, Georgia. *Natural Science*, 7, 151-157. http://dx.doi.org/10.4236/ns.2015.73017

[5] Gobejishvili R.G. (2011). Relief of Georgia, Monography, Publ. „Universali" 258 p. (in Georgian).

[6] Bayr K.J., D.K. Hall and W.M. Kovalick (1994). Observations on glaciers in the eastern Austrian Alps using satellite data. Int. J. Remote Sensing, 15(9), 1733-1752.

[7] Paul F. (2000). Evaluation of different methods for glacier mapping using Landsat TM. In Proceedings of EARSeL SIG Workshop, Land Ice and Snow, June 16–17, 2000, Dresden, Germany. Paris, European Association of Remote-Sensing Laboratories Special Interest Group Land Ice and Snow, 239-245.

[8] Paul F. (2002). Changes in glacier area in Tyrol, Austria, between 1969 and 1992 derived from Landsat TM and Austrian glacier inventory data. Int. J. Remote Sensing, 23(4), 787–799.

[9] Paul F., C. Huggel, A. Kaab, T. Kellenberger and Maisch M. (2002). Comparison of TM-derived glacier areas with higher resolution data sets. In Proceedings of EARSeL LISSIG Workshop, Observing our Cryosphere from Space, March 11–13, 2002, Bern, Switzerland. Paris, European Association of Remote-Sensing Laboratories Special Interest Group Land Ice and Snow, 15–21.

[10] Shahgedanova M., Nosenko G., Kutuzov S., Rototaeva O. and Khromova T. (2014). Deglaciation of the Caucasus Mountains, Russia/Georgia, in the 21st century observed with ASTER satellite imagery and aerial photography The Cryosphere, 8, 2367–2379, doi:10.5194/tc-8-2367-2014

[11] Podozerskiy K.I. (1911). Glaciers of the v Caucasus Mountain Range. "Zap. KORGO", vol. 14, issue 1, (in Russian).

[12] Tabidze D. (1965). Geomorphology of Kodori River basin. Thesis for a Phd's degree, 120 p. Tbilisi. (in Russian).

[13] Gobejishvili R.G. (1995). Present day glaciers of Georgia and evolution of glaciation in the mountains of Eurasia in late Pleistocene and Holocene. Thesis for a Doctor's degree, 320 p. Tbilisi. (in Georgian).

[14] Gobejishvili, R., N. Lomidze, and L. Tielidze (2011), Late Pleistocene (Wurmian) glaciations of the Caucasus, in Quaternary Glaciations: Extent and Chronology, edited by J. Ehlers, P. L. Gibbard, and P. D. Hughes, pp. 141–147, Elsevier, Amsterdam. DOI: 10.1016/B978-0-444-53447-7.00012-X

Importance of Design Factor in Improvement of Fishway Efficiency

Prashant Mandal[1, 2], Zhiying Tu[1], Xi Yuan[1], Yong Gao[3], Yingping Huang[1, 2, *], Hui Peng[1, 2]

[1]Engineering Research Center of Eco-Environment in Three Gorges Reservoir Region, Ministry of Education, China Three Gorges University, Yichang, China
[2]Collaborative Innovation Center for Geo-Hazards and Eco-Environment in Three Gorges Area, Hubei Province, Yichang, China
[3]Hubei Key Laboratory of Three Gorges Project for Conservation of Fishes, Chinese sturgeon Research Institute, China Three Gorges Corporation, Yichang, China

Email address:
prashantmandal12@yahoo.com (P. Mandal), chem_ctgu@126.com (Yingping Huang), hpeng1976@163.com (Hui Peng)

Abstract: Properly designed fishway plays an important role to mitigate the migrating problem of fish and help them meet their life cycle's basic requirements, especially the spawning activities with enhancing the local ecosystem as well. However, the fishways are still constricted at the small section of the dam or weir giving less priority and the designs of fishways are further limited by engineering, hydraulic and economic constraints. Thus, this paper presents an overview of the fishway design history and their consequences following different available literatures till date. Furthermore this paper also suggests on considering some important parameters during design such as turbulence, roughness coefficient, minimum head difference, slope which have a key role in improving the working efficiency of fishway. In addition, this review provides following recommendations: i) Need to improve downstream design of fishway and associated experimental methodology, ii) Consideration of river temperature with dissolved oxygen and their effect on fish behavior during design work, iii) Importance of further research work on coarse species along with other economically important fish, and iv) Updating the fishway design by considering fluctuating water level condition in the river. Hence this paper can contribute in the enhancement and restoration of fisheries resources from the perspective of fish passage design problems and their solution. Furthermore it may help the new researchers and designers to upgrade the existing design concept for the better result in fishway efficiency in coming future.

Keywords: Attraction Efficiency, Dam, Discharge, Fishway, Passage Efficiency, Velocity

1. Introduction

Barriers across river such as dam and weir usually have adverse impact on natural fish population and may contribute to the extinction of many aquatic species. Dams are threatening many aquatic species in Europe and North America, as well as in other continents due to less awareness and concerns on biology, behavior, fishery and population dynamics of the fish species. In recent time, the major concern is that the fisheries and the associated livelihoods are in risk due to an effect of construction of dams, which generally have effect on the normal pattern of water temperature, flow regime, water chemistry, nutrient transport, fish movement, and community structure in a river system. This in turn will have an impact on spawning. Obstacles generally change the hydraulic and morphological properties of the river and it can create many threats to aquatic life such as (i) Dams and weirs significantly reduce flow velocity and the variability of the current, (ii) Water temperature rises in the reservoir due to reduced flow velocity and the longer retention time of the water in the impoundment, (iii) Energy flow is interrupted by increased sedimentation of organic matter causing the metabolic processes in rivers, (iv) Spawning migrations are blocked by impassable obstruction, e.g., dam, and the fish may spawn in parts of the river where conditions are less suitable, and (v) Channel reaches below dams that fall dry due to the diversion of maximum water by bypass power stations constitute a further problem in downstream for aquatic organisms [1].

A fishway is any type of riverine channel segment created to passively facilitate fish migration across an elevated barrier without any direct human or mechanical intervention [2]. Fish passage facilities are an effective means of mitigating the impact of barriers, e.g., dams, weirs, and road crossings, on migrating fish. Fish migration takes place mostly in three directions: upstream, downstream and laterally. The general principle of upstream fish passage facilities, which is also called fish passes or fishways or sometimes even fish ladders, is to attract fish and help them move up from the obstruction by opening a waterway.

Fishways have a long history, with the earliest one recorded more than 300 years ago in France [3-5] when the southern province of France, Bearn, made it mandatory that weirs and dams construction must take into consideration of a passage for fish for their smooth movement. In 1883, Scotland has built the world's first pool fishway on a tributary of the River Teith explained by Nanjing Hydraulic Research Institute in 1982, however, their attempt to pass fish was not successful because the fishway was different from the fish's habits [6].

China has a vast system of reservoirs about 86000 and the fisheries of these reservoirs are intensively exploited and maintained by stocking from hatcheries so that little need has been realized to construct fish passes [5, 7]. The first fish passes are only 40 years old [7] and around 60 to 80 fish passes have been built [8]. The main target species are potamodromous species, mainly four species of carp and in catadromous species, mostly Japanese eel. Most of the used fish passes were pool-type.

The Yangtang fishway exists on the Mishui River, which passes 45 species and more than 580000 fish per year. This fish pass effectiveness was monitored quite well i.e., 5000 hours of observation annually. The effect of the Yangtang fish pass was found to be significant as the statistics of fish harvest showed that the annual fish output in the upstream part of the Mishui River increased to 3.5 times compare with that in the years before the fishway building. Similarly this fish pass has been specifically designed to pass very small fish, with very low turbulence in pools and low drops about 0.05 m between pools. The attraction flow of 16 m^3s^{-1} and the collection gallery above the turbines are considered to play an essential role in the effectiveness of the fish pass facility. Yangtang fish pass is one of the few examples of a well-designed fish pass, adapted to native species and well monitored in developing countries [9]. Since the fishway construction in different countries including China has a long history but the real status and construction scenario is looking different compare to its total number and the effectiveness (Table 1) [10].

Table 1. *(Fishway Research Status) [10].*

Country	Year	Location	Fishway	Remarks
France	1662	Bern province	Fishway Construction	The earliest fishway was constructed
The United States, Canada	Early 1960's		Fishway Construction	100 numbers or more
Japan	Early 1960's		Fishway Construction	35 numbers
Japan	Late 20th century		Fishway Construction	More than 1400 numbers
North America	Late 20th century		Fishway Construction	400 numbers
England and Wales.	Late 20th century		Fishway Construction	About 380 numbers
United States	Late 20th century		North Branch dam Fishway	Climbing the height of 60m
United States	Late 20th century		The Pardon Fishway	Total length of 4.8 km
China having following three stages:				
	(1958-1980)		Fishway Construction	Initial stage
	(1980-2000)		Fishway Construction	Stagnant period
	(2000-present)		Fishway Construction	Secondary development period
China	1958	Zhejiang's Fuchun river	Fishway Construction	First fishway when designing Seven Mile Ridge hydropower
China	1960	Heilongjiang's Xingkai Lake	Xingkailiu and Liyukang fishways	Two numbers
China	1966	Jiangsu Province	Doulongwa Fishway	
China	1980's		Anchui's Yuxi flood gate Fishway, Jiangsu Liuheay fishway and Tuanjie river fishway including Hunan's Yangtang fishway and other.	Fish passage structures were more than 40 numbers
China	(1980-2000)		Fishway Construction	Stagnant period, no construction
China	2003	HuiChun city, Jilin province	Fishway Construction	When the LaoLongKou water conservancy was built
China	2004	Zhejiang province	Fishway Construction	Fishway with the length of 500 m and the width of 2 m in each side of the Cao'e River brake
China	2005	Han xinglong water conservancy hub		Fishway with the length of 461.6m and the width of 3 m and the depth of 2m built on its right bank beach

Since 1930, physical and numerical hydraulic studies and improved biological assessments have provided a more comprehensive scientific knowledge and a more reliable basis for designing and operating fish passage systems. Current development in biology, particularly with enhanced understanding of fish behavior, advanced hydraulic concepts, more nature-like solutions, improved interdisciplinary or ecohydraulic approaches, and increasing focus on fish passage efficiency in field assessments, indicate that fish passage systems are becoming more effective and more adaptable to small and large bodied fish, as well as freshwater and multiple species [11-13]. However, the earliest fish bypass structures were generally poorly designed for local hydraulic conditions and fish species. Thus, it is required to focus more on the importance of design parameters that will enhance the effectiveness and efficiency of fishway in coming future. This paper contributes to the enhancement and restoration of fisheries resources from the perspective of fish passage design problems and their solution. It further highlights the importance of upgrading the existing design concept for the better result in enhancing fishway efficiency.

2. Overview of Fishway and Its Design Parameters

In general, most of the papers have discussed the fishway design used in upstream migration compared to the downstream. However, the design of downstream passage also needs to take into consideration about the swimming ability and behavior of the target species and the physical and hydraulic conditions at the water intake as in case of upstream fish passes [14]. Fishways success depends on interaction of fish behavior and swimming ability with hydraulic characteristics (Table 2) [1].

Table 2. *(Different Fishway Structures and Effectiveness) [1].*

Type	Dimensions and discharge	Range of application	Advantages and disadvantages	Effectiveness
Slot passes	Pool dimensions: lb > 1.90 m; b > 1.20 m; h > 0.5 m; Slot width: s > 0.17 m. Discharge can be from Q = 140 l/s up to several cubic metres per second.	Generally used for small and medium heads including variable impounding heads. Can be used for small streams and large rivers. The minimum tailwater depth must be h > 0.5 m.	Relatively high discharges can be sent through, thus good attraction currents can form. More reliable than conventional pool passes because of the lower risk of clogging of the slots.	They are currently the best type of technical fish pass, being suitable for all species of fish and are passable for invertebrates if a continuous bottom substrate is built in.
Pool passes	Pool dimensions depend on the river zone; lb > 1.4 m; b > 1.0 m; h > 0.6 m. Submerged orifices: b_S/h_S > 25. 25 cm. Discharge Q = 80 to 500 l/s.	Generally used for small and medium heads, at melioration dams and at hydroelectric power stations.	Only relatively low discharges allowed; there is great risk of clogging with debris.	Suitable for all species of fish if the dimensions of the pools and orifices are chosen as a function of the fish size that can be expected to ccur. There might not be ufficient attraction current at low discharges.
Denil passes	Channels: b = 0.6 to 0.9 m; h > 0.5 m; < 1:5; Q > 250 l/s. Channel lengths can be 6 to 8 metres; resting pools are required for heights > 1.5 to 2 m.	Generally suitable for small heads, particularly for retrofitting of old milldams when there is not much space.	Relatively high discharges; should not be used for variable headwater levels; not sensitive to varying tailwater levels; need little space; cheap; good formation of ttraction current.	According to present knowledge, less suitable for weak swimmers or small fish. Selective. Benthic fauna cannot pass.

Q: Discharge or flow.
h: Water level difference.
b: Pool width.
lb: Pool length.
b_S: Submerge orifice width.
h_S: Submerge orifice height.

The performance of fishway varies greatly with their type, design and operating regime, and with the species concerned, and is often the product of experience in dealing with these variables [15]. Construction of instream barriers and the subsequent impacts on migration is one of the major threats to freshwater fish diversity worldwide [16, 17]. Barriers restrict access to spawning grounds and preferred habitats, thus preventing dispersal and recolonisation [18-20]. Providing passage for both small and large fish requires specific engineering solutions. The appropriateness of any solution is largely dependent on what direction fish are attempting to move. Biologically oriented fishway research has focused mostly on anadromous fish species (e.g. salmonids) [21, 22]. Therefore, considerably limited information on coarse species has been available. Recent studies have proven that these species can travel considerable distances for reproduction, refuge and feeding purposes [22-24]. Study of swimming energetics and kinematics of juvenile S. chongi at a single temperature provide data useful for the design of fishways [25]. Likewise, Chinese sturgeons conserve energy by swimming efficiently and have high fatigue recovery capability which will be important during fishway designing [26]. Therefore, there is an increasing need to conduct many studies to accommodate movements and behavior of different fish within fishway and to assess the effect of potential key-variables that should be considered for the successful development of future designs.

A number of studies have addressed the flow circulation patterns, the jet characteristics and the turbulence generated by the energy dissipation in pools for different configurations, and their relevance for the development of suitable hydraulic criteria for passage of salmonid species [27-29]. However, passage studies focused on coarse species are limited. Therefore, extensive studies still need to be carried out on coarse species [22, 30] for the real effectiveness of fishway. The modification of internal flow characteristics in the pools by placement of submerged structures to examine the extent to which turbulence, particularly the horizontal Reynolds shear stress component can be reduced, and thus facilitate and shortening the passage time of small individuals, should be considered. To improve the effectiveness of fish passage for larger species, it can be modified by incorporating orifices into the traditional pool and weir design, reported in the studies of swimming behavior for sturgeon [31-33]. Likewise other important design considerations are appropriate flow speed, passage time [34], maximum speed for sustained swimming [35], hydraulic conditions [36], and swimming performance [37]. Similarly protocols for testing swimming performance shall be well-established and the results are of interest to fish physiologists as they provide design criteria for construction of fishways [38-40]. The relevance of swimming behavior and performance to fishway issues is a motivated work in this field [32].

3. Correlation Between Effective Fishway and the Design Parameters

Some important parameters need to be consider during effective fishway designing are discussed below.

3.1. Suitable Position for a Fish Pass

In general, in the natural river when there is no any obstructions such as weir, dam then fish can easily move upstream without giving much attention to the flow direction and other factors. However, when there exists any barrier across the river, fish find the difficulties to confirm the proper waterway to move upstream. Similarly the possible dimensions of any fishways are usually severely limited by engineering, hydraulic and economic constraints, particularly in larger rivers. Thus, the position of fishway at the dam plays a significant role. Fish usually migrate upstream following the path in or along the main current possibly the bank of the river where the water current is maximum. With a position near the bank, the fish can be more easily linked to the bottom or bank substrate. Usually at hydroelectric power stations, fish pass is placed on the same side of river as the powerhouse [1].

A fish pass, which extends far into the tail water below the dam, considerably limits the possibility that fish find the entrance. Thus, a design fault that has been responsible for the failure of many fish passes. Where dams or weirs are placed diagonally across the river and overflow along their entire crest, upstream migrating fish usually concentrate at the upstream, narrow angle between weir and bank. Therefore, the fish pass should clearly be sited in this area. Facilitating movement of fish back to mainstream habitats can be achieved through the construction of specialist floodplain fishway or potentially by manipulating the river hydrograph [41]. Improving lateral connectivity not only benefits native species, but also facilitates the movement of non-native species utilizing floodplain habitats [42, 43].

In case of a wide river it may be necessary to provide not only several entrances but also more than one fish pass because a single fish pass cannot be expected to attract certain species from the opposite bank. Migrating fish may arrive either at the bank where the powerhouse is located or at the opposite bank where the spillway is discharging and it is therefore advisable to design two separate fish passes, each with one or more entrances [44]. Hence, the good design that comprises of experience and knowledge in fish behavior in Natural River also plays an important role.

3.2. Fish Pass Entrance and Attraction Flow

For a fish pass to be considered efficient, the entrance must be designed so that fish find it with a minimum of delay as "No fish in = No fish out" [45]. It is explained that the perception of the current by aquatic organisms plays a decisive role in their orientation in rivers. Fish that migrate upstream as adults are usually swim against the main current. However, they do not necessarily migrate within the maximum flow but, depending on their swimming abilities, they may swim along its edge. If migration is blocked by an obstruction, the fish seek onward passage by trying to escape laterally at one of the dam's sides. In such case they continue to react with positive rheotaxis and, in perceiving the current coming out of a fishway, are guided into the fish pass. The properties of the tailrace below a dam where water velocity and degree of turbulence influence the attracting current that forms at the entrance to the fish pass. The velocity at which the attracting current exits the fish pass should be within the range of 0.8 to 2.0 m s^{-1}. Particularly where the tailwater level fluctuates, a special bypass can be used to channel additional flow directly from the headwater to the entrance of the pass in order to boost the intensity of the attracting current. The bypass can be in the form of a pressure pipe, but it is usually better to have an open channel. Except for special cases flow velocity should not exceed 2m s^{-1}. In many cases this is directly below the weir or dam, at the foot of the barrage or at the turbine outlets. A critical problem is how to construct the fish pass entrance so that fish can swim into the fishway even at low water levels. Entry into the fish pass can be made, even for bottom-living fish species, by linking the fish pass to the natural river bottom. This can be done with a ramp with a maximum slope of 1:2. Some existing fish passes have their entrances oriented towards the weir and thus at an angle of 180° relative to the river current. In such cases the entrance is unsuitable and it cannot establish an attracting current to enable the fish to find the entrance to the fishway [1].

Overall passage effectiveness are determined by properly utilized the method of attracting fish to locate systems for upstream movements. No fish passage system, will work successfully if fish do not find, or are not attracted to the entrance regardless of type, configuration or internal passage efficiency. Proper flow management at a hydroelectric project is an important methodology that can be used to attract fish and flow releases near a fishway entrance which help the migrants to locate it effectively. Approaches to attracting fish to passage systems need to consider species, hydraulics and site-specific conditions. Usually site-specific biological and hydraulic studies are needed to develop concepts into effective fish passage [46]. New Technology like tracking fish through telemetry [22, 47] has helped in quantification of fish guidance, attraction and fish passage efficiencies mainly for upstream fish passage systems [46]. Fishway with attraction current is explained in Figure 1 [1].

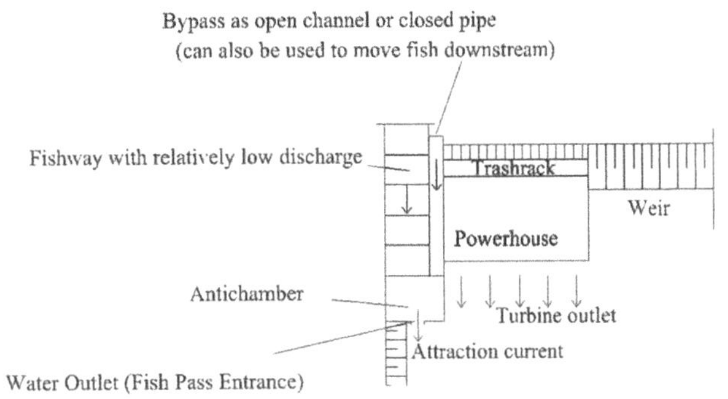

Figure 1. Additional discharge is sent through a bypass into an antechamber downstream of the first pool of the fish pass to increase the attraction current at the fish entrance [1].

3.3. Fish Passage Exit and Exit Conditions

When designing of fish passage, its water inlet (exit into the headwater) must be located far enough from the weir or turbine intake to prevent the fish coming out of the passage from swept into the turbine from the water current. At least 5m distance should be kept between the fish passage and the turbine intake or trash rack. If the water velocity of the headwater is greater than 0.5 m s^{-1}, the exit area of the fish pass has to be prolonged into the headwater by a partition wall. [1].

3.4. Discharge and Velocity Conditions in the Fish Pass

The discharge required to ensure optimum hydraulic conditions for fish within the pass is generally less than that needed to form an attracting current. However, during dry period when the availability of water is less, subsequently the total discharge available should be used in the fish passage to allow unhindered passage of migrants. This method is used for the dams that are not used for hydropower generation. If more water is available to supply the fishway than is needed for the hydraulically sound functioning of the existing or planned fish pass, alternative designs should be carried out. In general, current velocity in fish pass should not exceed 2.0 m s^{-1} at any narrow point such as in orifices or slots and this limit to velocity should be assured by the appropriate design of the pass. The pass should incorporate structures that form sufficient resting zones to allow weak swimming fish to rest during their upstream migration [1].

Water flow velocity is an important factor governing the upstream movements of migrating fish [48, 49], and the provision of suitable flow condition at the downstream of a dam has been identified as an effective method of improving passage [50]. For fish to be attracted towards a fishway, it is generally held that flows originating from a point near the fishway entrance should be high relative to those released elsewhere along the dam face [50-52].

3.5. Dimensions, Slopes, Resting Pools

Proper dimensions of fishway include information on such features as slope, width, length and water depth as well as the dimensions of orifices and resting pools. These parameters depend mainly on the particular type of fish pass including the available discharge. The body length of the biggest fish species that occurs or could be expected to occur in accordance with the concept of the potential natural fish fauna is an important consideration in determining the dimensions of fish passes. However maximum sizes, such as that of the sturgeon that can grow to 6.0 m in length, are not used. The average body length of the largest fish species expected in the river as well as the permissible difference in water level must be considered in defining the dimensions of a fish pass. For more technical constructions the maximum permissible slope ranges from 1:5 to 1:10, depending on the construction principle chosen, while close-to-nature constructions should show maximum slopes less than 1:15 corresponding to the natural form of rapids. The fishway has a bed slope of 10% is typical [5, 53]. In any case, the requirements of the weakest species, or of the weakest life stages, must be considered when defining the dimensions of a pass. Resting zones or resting pools should be provided in fishways at intervals of such lengths as defined by the difference in level of not more than 2.0 m between pools [1]. The turbulence of the flow through the fish pass should be as low as possible so that all aquatic organisms can migrate through the pass independently of their swimming ability. The volumetric energy dissipation in each pool of a fish pass should not exceed 150 to 200 Watt per cubic meter, of pool volume [1]. Turbulence is frequently identified as the major factor limiting the passage of small-bodied species and can be manipulated in a fishway pool [54], either by: (i) improving dissipation of energy in the pool or (ii) reducing the amount of energy entering the pool [28, 55, 56].

3.6. Effectiveness and Efficiency

The concept of effectiveness and efficiency may be used to clarify the degree of mitigation provided by a fish pass. Effectiveness is a qualitative concept, which checks that the fish pass is capable of letting all target species through within the range of environmental conditions observed during the migration period. Effectiveness may be measured through inspections and checks like visual inspection, trapping, video checks [57].

The efficiency of a fish pass is a more quantitative description of its performance. It may be defined as the proportion of stock present at the dam which then enters and successfully moves through the fish pass in what is considered an acceptable length of time. The methods giving an insight into the efficiency of a pass are more complicated than those for effectiveness. Marking and telemetry are valuable techniques to assess the overall efficiency of fish passes and the cumulative effect of various dams along a migration path [44].

Effective passage systems for both upstream and downstream movements have three basic characteristics such as (i) fish passes are easy to locate by the fish community, (ii) fish passes have hydraulic conditions that match biological needs and that species seek rather than avoid, and (iii) fish passes provide suitable and efficient transport [46]. However, attraction and passage efficiencies are associated mainly with upstream fish passage systems. Attraction efficiency refers to the probability that fish approaching the barrier will locate the fishway entrance, while passage efficiency refers to the probability they will move through the fishway and pass upstream [46].

Effectiveness and efficiency may be limited by how easily fish can locate or be guided to upstream or downstream fish passage systems. Fish passage system choice basically depends on species type and fish habitat management objectives including site conditions, fish attraction and guidance, water level difference, hydraulic characteristics, operational constraints, construction materials, maintenance, and economics. Fish attraction and guidance aspects, biological requirements, and hydraulics are the most critical aspects for effectiveness and efficiency.

Efficiency of fish movement at an area of difficult passage involves knowing how many fish of a particular species attempt to pass upstream relative to the number of fish that successfully pass through [36, 58].

Attraction efficiency was defined as the proportion of fish tagged and released during the study that were subsequently located within less than approximately 3m from a fishway entrance [59] or at the base of a barrier to fish movement and near enough to a fishway entrance for fish to detect fishway attraction flow [60].

Passage efficiency was calculated by dividing the number of fish of a particular species that exited a fishway by the number that was detected at the fishway entrance [59, 60]. Passage efficiency may relate to type of fish passage, its slope and hydraulic head, while attraction efficiency may relate to biological factors such as migratory characteristics and temperature tolerance of the fish species present [2]. The topology of the flow also influences the fish passage efficiency. [61- 63]

Three sequential components of fish passage relevant to both up and downstream migrations are attraction, passage itself, and post passage effects [64], which are very important during design. Fishway entrance, i.e. attraction is a two-step process, consisting of guidance to the fishway entrance and actual entry into the fishway [65]. Indeed, this distinction may be important, as some studies have reported fish approaching the entrance but failing to actually enter [66, 67]. Since guidance, attraction, or passage can independently limit fishway efficiency, evaluating these different components of passage is necessary to understand mechanisms of passage failure, and identify potential mitigation measures.

It is highlighted the importance of assessing physiological consequences of passage that may affect survival or reproduction and not just efficiency at the barrier [68, 69]. The studies of physiology may also be important in determining how passage by unhealthy fish, when they have low energy reserves or high stress levels, may differ from the passage by healthy fish [36].

4. Observation and Suggestions

It is observed that very few papers have discussed about the downstream fishway designing compared to the fishway in upstream migration. Likewise, all management efforts are on promoting upstream migration than downstream movement, however, fish passage requirements within river channels are bidirectional and equally important in all directions [70]. Similarly, large-bodied fish moving downstream can be delayed or even abandon migration if an appropriate migration pathway is not found [71]. It is thus suggested to have further research to be carried out on downstream fishway access at or near the dam/weir to meet the requirement of entire complete life cycle activities of species.

As every type of species has its important role in ecosystem but the fishway design research has found to be carried out mostly with economically significant species like salmonids while the species considered to be of low commercial value have been generally ignored may be due to limited information available of such species. Hence, it is suggested to have depth study and research on different type of fish regardless of its economic value for the effective design of fishway.

Most of the researchers have only emphasized the importance of fish passage efficiency, i.e., number of fish passes through the fishway but limited research tends to carried out on fish mortality rate after passing through the fishway as this factor can also guide the real design effectiveness of the fishway. Hence passing efficiency and consequences of fish mortality rate shall be considered precisely.

Generally in pool designing, difference in water head in pool of about 0.2 m have followed which impose a maximum current velocity of 2.0 m s^{-1} for instance at orifices and cross walls. It is recommended that the water level difference between pools in a fishway shall be kept below 0.2 m or that may differ according to river flow pattern and velocity required inside the pool, so that it can comfortably pass the fish having weak swimming performance as well [1] In addition, the designer shall also concentrate on different turbulence factor and vortices created mainly at the slot opening as these play an important role in finding the required dissipation of energy in the pool.

While swimming performance is impaired under the highly unnatural conditions where behavior is restricted to rectilinear swimming [72] however many research on fish swimming capability are generally carried out in swim chambers model and designers are using those data for the fishway designing. Hence, it may not always give the good result while comparing with the fish swimming in the natural condition. Thus, it is suggested to use large experimental channels, which are more compatible with natural environment for better output.

Similarly during fishway model test, roughness coefficient of bottom surface and surrounding areas of fishway shall be selected according to the roughness of the material generally used in the real field scenario to achieve the better performance of fishway. The type and base slope of the fishways are also needed to be selected according to the flow pattern, upstream and downstream water level differences including swimming behavior of the locally available native fish rather than simply imitating the design from other similar type of projects.

During fishway research, it seems the fish are usually acclimated in acclimation tank for certain time and released it to the fishway chamber directly for the experiment purpose. However, in the natural river, fish have its own natural movement and attraction behavior toward the fishway inlet guided by the attraction velocity. The attraction efficiency may not only be compatible during the laboratory test carried out in fishway model. This paper suggests to conduct experiments in large experimental channels taking detailed information of fish behavior found in Natural River. Hence, this is also one of the factors needed to be considered during model test.

Fish passage efficiency is influenced by discharge [24] when the water level in the river is comparatively lower or excessively higher than the swimming capabilities of the fish in general. Consequently the effectiveness of fishway may decrease during such events. Therefore, this paper suggests to modify the design in such a manner that it can always maintain the proper required flow at the inlet of the fishway regardless of flow depth variation in the river.

As reported in [2]: "variation in fish attraction is determined by biological characteristics of the fish", which is contradictory with the expression "fish attraction efficiency is developed through attraction velocity of water", as discussed in [1]. Thus, it is suggested to have further research on it to mitigate this contradiction for the improvement of fishway design in future.

Due to global warming effect, normal temperature is also gradually rising every year, which may also have an effect on the natural river environment and different aquatic life living there. At higher temperature, it seems fish swimming tends to become continuous with a constant increase in muscle activity over time [73]. This will result in an increase in maximum swimming speed, but a decrease in endurance, which will affect the ability of fish to ascend fish passes. Thus, the behavior of migratory fish can be influenced by temperature as well [74]. Consequently, it is suggested to consider also the temperature effect as one of the key factors in designing of the fishway.

Hydraulic factors like discharge, velocity and flow patterns of the river are found to be an integral part of the fishway. In addition, it is also suggested to consider some other environmental parameters during design such as dissolved oxygen content, water quality, noise and light effect, which are indirectly affecting the fish movement and behavior in Natural River.

The importance of fish behavior on encountering conditions associated with fish passes is also significant when facilities fail to sufficiently attract fish to their entrance and in some cases may actually repel those from the inlet [75]. Thus, it is suggested to improve fish pass design with quantifying the behavioral response to factors, such as prevailing discharge and temperature regimes that are likely to impact the degree to which structures impede fish movements.

Furthermore fishway designs depend greatly on the interplay between hydraulics and biomechanics, yet very little data are available on the responses to specific hydraulic settings for these species. Therefore, there are still many factors that may need to be considered for the effective and efficient fishway design in future studies.

5. Conclusion

Fishways are very much important for the restoration of free passage for fish and other aquatic species in rivers from the obstacle like dam, weir. The fishways are seemed to be one of the key elements for the ecological improvement of running water. Their efficient functioning is also very much required for the restoration of free passage in rivers. However, studies of existing devices have shown that many of them do not function in an efficient manner. Thus, group of experts e.g. engineers, biologists and administrators are required to show their common interest to work together for the efficient design criteria. Hence, for the effective fishway design, it needs to consider the fish biology, life stage, behavior, space, swimming ability, and hydraulic conditions, including velocity and turbulence patterns according to the different species and their living environment condition. With the improved design, the fish can easily locate the fish pass or be guided to them and their life cycle basic requirements are easily met with timely migration. Therefore all the design parameters discussed above if taken into consideration can help to make the fishway more functioning with the improvement of its effectiveness and efficiency. Instead of increasing the number of experiments and research works on the fishway, it is recommended to carry out less but more precise work on quality research on fishway designing which is well - suited with fish behavior and natural river characteristics.

Acknowledgments

This work was supported by grants from the National Major Science and Technology Program for Water Pollution Control and Management (2012ZX07104-003-04), the National Nature Science Foundation of China (No. 51309140) and Fish Resources Protection of Three Gorges Project in Hubei Province Key Laboratory Open Projects Fund (0704102). We also like to thank all the researchers and authors for their great effort and valuable information in making the people more aware on designing the more efficient fishway.

References

[1] DVWK (2002) Fish Passes - Design, dimensions and monitoring.

[2] Bunt C. M., Castro-Santos T. Haro A (2012) Performance of fish passage structures at upstream barriers to migration, River Res. Applic 28: 457–478.

[3] Larinier M., Travade F., Porcher J. P. (2002) Fishways: biological basis, design criteria and monitoring, Bull. Fr. Pêche Piscic., 364 (Suppl.), p. 208.

[4] Kamula R. (2002) Flow Over Weirs with Application to Fish Passage Facilities, Oulu University Press, Finland.

[5] Clay C. H. (1995) Design of Fishways and Other Fish Facilities, (second ed.) Lewis Publishers, Boca Raton, FL, U. S. A.

[6] Chen Kai - Qi, Tao Jie, Chang Zhong-Nong, Cao Xiao-Hong, Ge Huai-Feng (2014) Difficulties and prospects of fishways in China: An overview of the construction status and operation practice since 2000, Ecological Engineering 70: 82–91.

[7] Wang Y. (1990) Design and Application of Fish Passage and Protection Facilities in China. In: Proceedings of the International Symposium on Fishways, Gifu, Japan.

[8] Nakamura S. (1993) A review of fish passage facilities in East Asia. Fish Passage Policy and Technology. In: Proceedings of the Symposium. Portland, Oregon, USA.

[9] Zhili G., Qinhao L., Keming A. (1990) Layout and Performance of Yangtang Fishway, Proceedings of the International Symposium on Fishways, Gifu, Japan.

[10] Xiao BL, Hui CD, Zhu G, Yong C. (2014) Research progress of fish protection in water resources and hydropower engineering. International Conference on Water Resource and Environmental Protection (WREP 2014).

[11] White L. J., Harris J. H., Keller R. J. (2010) Movement of three non-salmonid fish species through a low-gradient vertical-slot fishway. River Res. & Appl. 27: 499-510.

[12] Katopodis C., Williams J. G. (2011) The development of fish passage research in a historical context, J. ecol. eng., Published online DOI: 10.1016.

[13] Silva A. T, Santos J. M., Ferreira M. T., Pinheiro A. N., Katopodis C. (2011) Effects of water velocity and turbulence on the behavior of Iberian barbel (Luciobarbus bocagei, Steindachner 1864) in an experimental pool-type fishway. River Research and Applications 27: 360-373.

[14] Larinier M., Marmulla G. (2003) Fish Passes: Types, Principles and Geographical Distribution an Overview, Proceedings of the Second International Symposium on the Management of Large Rivers for Fisheries, Volume II.

[15] Katopodis C. (1992) Introduction to Fishway Design, Working Document. Winnipeg, Freshwater Institute: Winnipeg.

[16] Lucas M. C. and Baras E. (2000) Methods for studying spatial behaviour of freshwater fishes in the natural environment. Fish and Fisheries 1: 283–316.

[17] Lucas M. C., Baras E., Thom T. J., Duncan A. and Slavik O. (2001) Migration of Freshwater Fishes. Blackwell Science Ltd, London, UK. 192.

[18] Gehrke P. C., Brown P., Schiller C. B., Moffatt D. B. and Bruce A. M. (1995) River regulation and fish communities in the Murray-Darling River system, Australia. Regulated Rivers: Research & Management 11: 363–375.

[19] Pelicice F. M. and Agostinho A. A. (2008) Fish passage facilities as ecological traps in large neotropical rivers. Conservation Biology 22: 180–188.

[20] Welcomme R. L. (1995) Relationships between fisheries and the integrity of river systems. Regulated Rivers: Research and Management 11: 121–136.

[21] Laine A, Jokivirta T, Katopodis C. (2002) Atlantic salmon, Salmo salar L., and sea trout, Salmo trutta L., passage in a regulated northern river - fishway efficiency, fish entrance and environmental factors. Fisheries Management and Ecology 9: 65–77.

[22] Katopodis C. (2005) Developing a toolkit for fish passage, ecological flow management and fish habitat works. Journal of Hydraulic Research 43: 451–467.

[23] Lucas MC, Frear PA. (1997) Effects of a flow-gauging weir on the migratory behaviour of adult barbel, a riverine cyprinid. Journal of Fish Biology 50: 382–396.

[24] Ovidio M, Philippart JC. (2002) The impact of small physical obstacles on upstream movements of six species of fish - Synthesis of a 5-year telemetry study in the River Meuse basin. Hydrobiologia 483: 55–69.

[25] Tu Z. , Yuan X., Han J, Shi X., Huang Y., Johnson D. (2011) Aerobic swimming performance of juvenile Schizothorax chongi (Pisces, Cyprinidae) in the Yalong River, southwestern China, Hydrobiologia 675: 119–127.

[26] Cai L. et al. (2014) Integrating water flow, locomotor performance and respiration of Chinese Sturgeon during multiple fatigue-recovery cycles, PLOS ONE 9(4): e94345.

[27] Ead SA, Katopodis C, Sikora GJ, Rajaratnam N. (2004) Flow regimes and structure in pool and fishways. Journal of Environmental Engineering and Science 3: 379–390.

[28] Liu M., Rajaratnam N. and Zhu D. Z. (2006) Mean Flow and Turbulence Structure in Vertical Slot Fishways. Journal of Hydraulic Engineering 132: 765–777.

[29] Barton AF, Keller RJ, Katopodis C. (2008) A free surface model of a vertical slot fishway to numerically predict velocity and turbulence distributions. In Advances in fisheries bioengineering, Amaral SV, Mathur D, Taft EP (eds). American Fisheries Society Symposium 61, Bestheda, Maryland 39–52.

[30] Stuart IG, Mallen-Cooper M. (1999) An assessment of the effectiveness of a vertical-slot fishway at a tidal barrier on a large tropical/subtropical river. Regulated Rivers: Resource & Management 15: 575–590.

[31] Adams, S R, Adams G, L, Parsons G R. (2003) Critical swimming speed and behavior of juvenile shovel nose sturgeon and pallid sturgeon. Transactions of the American Fisheries Society, 132(2): 392-397.

[32] Cheong T S, Kavvas M L, Anderson E K. (2006) Evaluation of adult white sturgeon swimming capabilities and applications to fishway design. Environmental Biology of Fishes, 77(2): 197-208.

[33] He X, Lu S, Liao M, *et al*. (2013) Effects of age and size on critical swimming speed of juvenile Chinese sturgeon Acipenser sinensisat seasonal temperatures. Journal of Fish Biology, 82: 1047-1056.

[34] Larinier M. (2008) Fish passage experience at small-scale hydro-electric power plants in France. Hydrobiologia, 609(1): 97-108.

[35] Castro-Santos T. (2005) Optimal swim speeds for traversing velocity barriers: an analysis of volitional high-speed swimming behavior of migratory fishes. The Journal of Experimental Biology, 208(3): 421-432.

[36] Roscoe D. W., Hinch S. G. (2010), Effectiveness monitoring of fish passage facilities: historical trends, geographic patterns and future direction. Fish and Fisheries 11: 12–33.

[37] Hatry C, Thiem J D, Binder T R, *et al*. (2014) Comparative Physiology and Relative Swimming Performance of Three Redhorse (*Moxostoma spp.*) Species: Associations with Fishway Passage Success. Physiological and Biochemical Zoology, 87(1): 148-159.

[38] Jain K E, Hamilton J C, Farrell A P. (1997) Use of a ramp velocity test to measure critical swimming speed in rainbow trout (*Onchorhynchus mykiss*). Comparative Biochemistry and Physiology Part A, 117(4): 441-444.

[39] Kerr S R, Nelson J A. (2000) Aerobic and anaerobic swimming performance of individual Atlantic cod. Journal of Experimental Biology, 203(2): 347-357.

[40] Yagci O. (2010) Hydraulic aspects of pool-weir fishways as ecologically friendly water structure. Ecological Engineering, 36(1): 36-46.

[41] Jones M. J. and Stuart I. G. (2008) Regulated floodplains a trap for unwary fish. Fisheries Management and Ecology 15: 71–79.

[42] Jones M. J. and Stuart I. G. (2009) Lateral movement of common carp (Cyprinus carpio L.) in a large lowland river and floodplain. Ecology of Freshwater Fish 18: 72–82.

[43] Stuart I. G. and Jones M. J. (2006) Movement of common carp, Cyprinus carpio, in a regulated lowland Australian river: implications for management. Fisheries Management and Ecology 13: 213–219.

[44] Larinier M. (2001) Environmental Issues, Dams and Fish migration, Published in dams, fish and fisheries opportunities, challenges and conflict resolution, FAO fisheries technical paper.

[45] Bates, K. (1992). Fishway design guidelines for Pacific salmon. Report of Washington Department of Fish and Wildlife.

[46] Katopodis Ecohydraulics Ltd. (2013) Fish passage considerations for developing small hydroelectric sites and improving existing water control structures in Ontario.

[47] Colavecchia M., Katopodis C., Goosney R., Scruton D. A., McKinley R. S. (1998) Measurement of burst swimming performance in wild Atlantic salmon (Salmo salar L.) Using Digital Telemetry. Regul. Rivers-Res. Manage. 14: 41–51.

[48] Banks, J. W. (1969) A review of the literature on the upstream migration of adult salmonids. Journal of Fish Biology 1: 85–136.

[49] Alabaster, J. S. (1970) River flow and upstream movements and catch of migratory salmonids. Journal of Fish Biology 2: 1–13.

[50] Andrew, F. J., and G. H. Geen. (1960) Sockeye and pink salmon production in relation to proposed dams in the Fraser River system. International Pacific Salmon Fisheries Commission Bulletin 11.

[51] Leman, B., and G. J. Paulik. (1966) Spill-pattern manipulation to guide migrant salmon upstream. Transactions of the American Fisheries Society 95: 397–407.

[52] Larinier, M. (1998) Upstream and downstream fish passage experience in France. Pages 127–145 in M. Jungwirth, S.

[53] Wu S, Rajaratnam N, Katopodis C. (1999) Structure of flow in vertical slot fishway. Journal of Hydraulic Engineering 125(4): 351–360.

[54] Lee Baumgartner, Brenton Zampatti, Matthew Jones, Ivor Stuart and Martin Mallen-Cooper. (2014) Ecological Management & Restoration vol 15 no s1.

[55] Mallen-Cooper M. (2005) Monitoring fish in Innovations of the Vertical-Slot Fishway Design. Fishway Consulting Services, Canberra 66 pp.

[56] Tarrade L., Texier A., David L. and Larinier M. (2008) Topologies and measurements of turbulent flow in vertical slot fishways. Hydrobiologia 609: 177–188.

[57] Travade F., Larinier M., Boyer-Bernard S., Dartiguelongue J. (1998) Performance of four fish pass installations recently built in France. In: Fish migration and fish bypasses (eds M. Jungwirth, S. Schmitz nd S. Weiss). Fishing News Books, Oxford, UK: Blackwell Science Ltd. Publisher.

[58] Castro-Santos T., Haro A (2010), Fish guidance and passage at barriers. Fish Locomotion: An Eco-ethological Perspective, Science Publishers.

[59] Bunt C. M., Katopodis C., McKinley R. S. (1999) Attraction and passage efficiency of white suckers and small mouth bass by two Denil fishways. North American Journal of Fisheries Management 19: 793–803.

[60] Aarestrup K., Lucas M. C., Hansen J. A. (2003) Efficiency of a nature-like bypass channel for sea trout (Salmo trutta) ascending a small Danish stream by PIT telemetry. Ecology of Freshwater Fish 12: 160–168.

[61] Calluaud D., Pineau G., Texier A., David L. (2014). Modification of vertical slot fishway flow with a supplementary cylinder. Journal of Hydraulic Research; Vol. 52 (5), pp 614-629.

[62] Calluaud D., Cornu V, Baran P., David L. (2014). Relationship between fish behaviour, turbulence and unsteady flow in experimental vertical slot fishways. 10th International Symposium on Ecohydraulics, Trondheim, Norway.

[63] Tarrade L., Pineau G., Calluaud D., Texier A., David L., Larinier M. (2011). Detailed experimental study of hydrodynamic turbulent flows generated in vertical slot fishways. Journal of Environmental Fluid Mechanics. Vol. 11, pp 1-21.

[64] Odeh, M. (1999) Innovations in fish passage technology. American Fisheries Society, Bethesda, Maryland.

[65] Castro-Santos, T., Cotel, A. and Webb, P. W. (2009) Fishway evaluations for better bioengineering: an integrative approach. In: Challenges for Diadromous Fishes in a Dynamic Global Environment. American Fisheries Society Symposium, Bethesda, MD.

[66] Moser, M. L., Matter, A. L., Stuehrenberg, L. C. and Bjornn, T. C. (2002) Use of an extensive radio receiver network to document Pacific lamprey (Lampetra tridentata) entrance efficiency at fishways in the Lower Columbia River, USA. Hydrobiologia 483: 45–53.

[67] Naughton, G. P., Caudill, C. C., Peery, C. A., Clabough, T. S., Jepson, M. A., Bjornn, T. C. and Stuehrenberg, L. C. (2007) Experimental evaluation of fishway modifications on the passage behaviour of adult Chinook salmon and steelhead at Lower Granite Dam, Snake River, USA. River Research and Applications 23: 99–111.

[68] Ferguson, J. W., Sandford, B. P., Reagan, R. E., Gilbreath, L. G., Meyer, E. B., Ledgerwood, R. D. and Adams, N. S. (2007) Bypass system modification at Bonneville Dam on the Columbia River improved the survival of juvenile salmon. Transactions of the American Fisheries Society 136: 1487–1510.

[69] Schreck, C. B., Stahl, T. P., Davis, L. E., Roby, D. D. and Clemens, B. J. (2006) Mortality estimates of juvenile spring-summer Chinook salmon in the Lower Columbia River and Estuary, 1992–1998: Evidence for delayed mortality? Transactions of the American Fisheries Society 135: 457–475.

[70] Calles O. and Greenberg L. (2009) Connectivity is a two-way street—the need for a holistic approach to fish passage problems in regulated rivers. River Research and Applications 25: 1268–1286.

[71] O'Connor J. P., O'Mahony D. J. and O'Mahony J. M. (2005) Movements of Macquaria ambigua, in the Murray River, south-eastern Australia. Journal of Fish Biology 66: 392–403.

[72] Haro A, Castro-Santos T, Noreika J, Odeh M. (2004) Swimming performance of upstream migrant fishes in open-channel flow: a new approach to predicting passage through velocity barriers. Canadian Journal of Fisheries and Aquatic Sciences 61: 1591–1601.

[73] Booth RK, McKinley RS, Okland F, Sisak MM. (1997) In situ measurements of swimming performance of wild Atlantic salmon (Salmo salar) using radio transmitted electromyogram signals. Aquatic Living Resources 10: 213–219.

[74] Linton E, Jonsson B, Noakes D. (2007) Effects of water temperature on the swimming and climbing behaviour of glass eels, Anguilla spp. Environmental Biology of Fishes 78: 189.

[75] Kemp PS, Gessel MH, Williams JG. (2008) Response of downstream migrant juvenile Pacific salmonids to accelerating flow and overhead cover. Hydrobiologia 609: 205–217.

Evaluation of Sewerage System Sustainability Technically Around Condominiums Areas: A Case Study in Debre Berhan, Ethiopia

Fraol Abebe Wudineh*, Nigatu Chala Kuke

Department of Civil Engineering, University of Debre Berhan, Debre Berhan, Ethiopia

Email address:
fraolabebe21@gmail.com (F. A. Wudineh), nigatuchala200@gmail.com (N. C. Kuke)

Abstract: Sewage discharges from centralized water-borne collection systems can pollute surface water, and discharge from sewers; septic tanks and pit toilets pollute living environment and groundwater which directly affects the health of poor people at downstream [13]. Poor sanitation facilities and low sewerage system continues to be a critical problem in rural and urban areas of Ethiopia. This study aims to evaluate the sewerage system in terms of technical aspect by identifying the cause and extent of failures on the schemes then forwarding the mitigation measures for the problem explored. This study is conducted in Debre Berhan town from February, 2014-January, 2015. In order to gather information's for the study, the researchers used pre-prepared questionnaire (for the condominium lives), soil sampling and the other was data collection checklists (variable and context information checklists), semi open ended interview and physical instruments such as GPS, Axes soil sample bags. From the study, some important conclusions are made. The technical performance evaluation indicators show that sewerage system especially around condominium area is poorly implemented, inadequate in quantity, or poorly constructed or both. Of the total users around 31.54% have no any awareness about sewerage system operation, sanitation and hygiene. Of constructed floor drainage ditches only 24.8% have good appearance and functional, in the rest of scheme, the structure is destructed and non-functional. Only 36.47% of septic tanks, gives good function for users properly as required, in the rest of the schemes, there is many defects and problems observed. The technical problems noticed in the sample sites are includes O&M, construction, and design respectively put in the order of causing considerable failure.

Keywords: Sustainable Sewerage, Water, Health, Hygiene, O&M, Capacity Building, Grey Water Reuse

1. Introduction

Sanitation facilities with safe sewage disposal methods not only reduce the numerous diseases but are very often the first step towards to other development activities (MDGs).

Poor sanitation facilities and low sewerage system continues to be a critical problem in rural and urban areas of Ethiopia. Even if from a very low base, access to improved water and sanitation is raising rapidly the coverage of sanitation facilities is only 39% throughout the country up to 2010 compared with 48% for benchmark [1]. Lack of sufficient wastewater management in a community can face several risks on public health and the environment.

In addition to constructing new sanitary schemes, attention should be given to operation and maintenance of the existing schemes, in order to achieve the envisaged objectives. Substantial achievements will not be achieved unless the factors contributing to poor implementation of the old schemes are fully explored. Previous experiences should also be taken into account in the new schemes to avoid repetition of problems. Evaluation of impact is a longer-term activity. It is the process by which results of development projects are measured against their targets or objectives to see whether they have had the intended impacts or any undesired negative effects [20].

Debre Birhan is one of the town, which faced to problem of sewerage system management, especially around low cost houses found in this town. According 2014, D/Berhan town health office report , more than 82% of top ten diseases registered are directly connected to unsafe waste management around residential buildings[2].

This study aims to evaluate the sewerage system in terms of technical aspect by identifying the cause and extent of failures on the schemes then forwarding the mitigation measures for the problem explored.

2. Materials and Methods of the Study

2.1. Study Area and Period

This study is conducted in Debre Berhan town from February, 2014 - January, 2015. Debre berhan is one of fast growing towns found in Ethiopia located in north shewa zone, Amhara regional state, about 130kms north east of Addis Ababa on paved highway to Desse [10]. Debre Berhan town is determined by highland climatic zone with an avg. elevation of 2,840 meters and receives an annual rainfall of 920mm while the temperature varies from 2.4 co in November to 23.3 co in June[11].

2.2. Methods

The paper commences with background information on the sewerage infrastructure status found in the condominium compound and the causes of damage to the infrastructure. In the second section the steps that were followed to determine functionality level for infrastructure are given. This is followed by a short discussion of the results obtained. Conclusions are given in the final section of the paper.

Data collection tools: In order to gather information's for the study, the researchers used pre-prepared questionnaire (for the condominium lives), soil sampling and the other was data collection checklists (variable and context information checklists), semi open ended interview and physical instruments such as GPS, Axes soil sample bags.

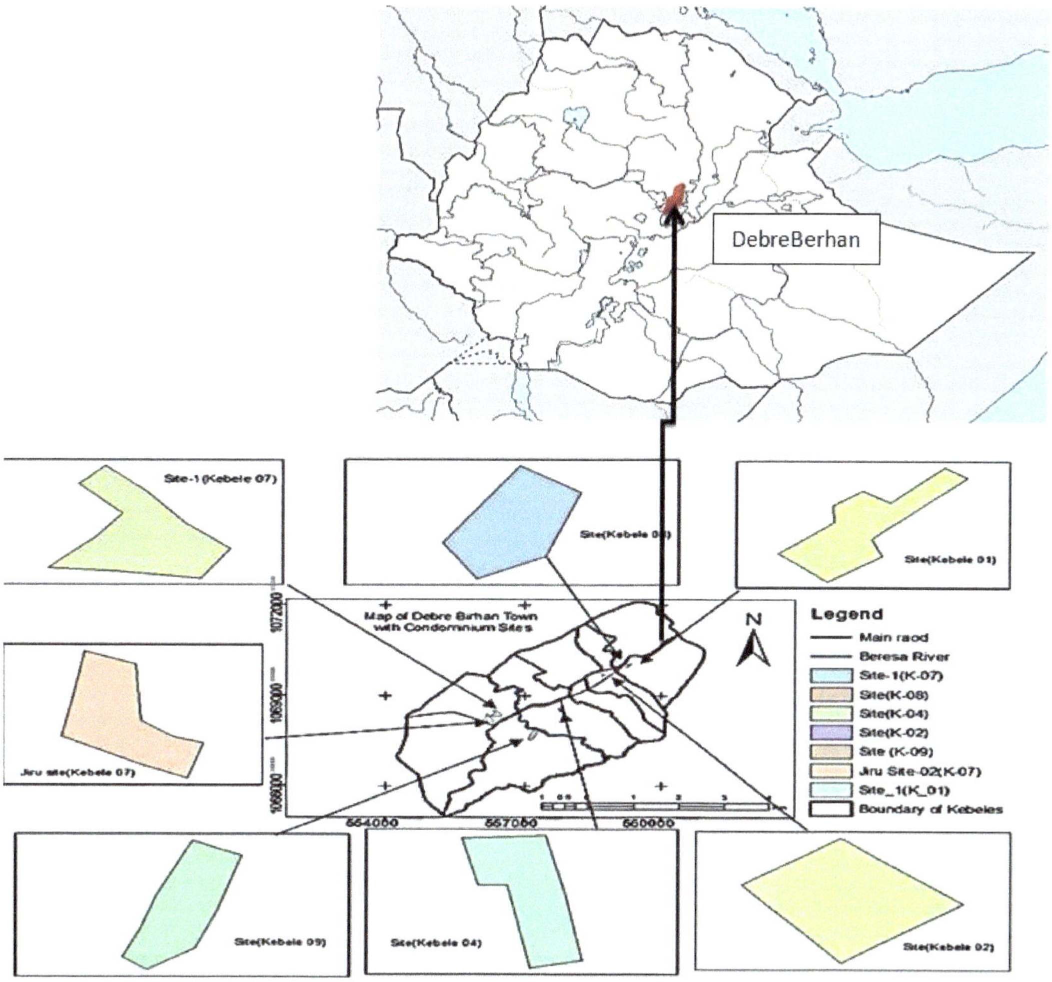

Figure 1. *Map of study area.*

2.3. Sampling Methods and Data Collection

Based on information about the problem on sewerage system, the study was carried out on all condominium sites found in the town which faced most visualized problems on sanitary system of town. The field data collection consisting of: interviews and data gathering from different stakeholders, field observation (on different infrastructures, sewerage system and other related schemes) and soil sample taken from each sites and tested at laboratory.

2.4. Methods of Data Analysis

Both quantitative and qualitative methods of data analysis and interpretation were employed. Data obtained by questionnaires were analyzed using descriptive statics through percentage, mean and grand mean. In addition to this qualitative data gathering through observations, interviews were qualitatively analyzed. The data collected on the functionality and utilization of the sewerage system at local level of the seven study areas were analyzed for determination of the existing situation in the study areas. These findings on situations of the sewerage system at local level is analyzed and hence further suggestions and recommendations that will help to improve the sewerage system in the study areas was done.

3. Results and Discussions

3.1. Household Characteristics

Table 1. Household Characteristics (N=76).

Variables	Frequency (percent)
Sex of the respondents(N=76)	
Male	43(56.58)
Female	33(43.42)
Age group of the respondents (N=76)	
<=35	63(82.89)
>35	13(17.11)
Education status (N=76)	
Literate	76(100)
Health and hygiene awareness level	
Un aware	31.54%
Aware	68.46%

N=Sample population

As indicated in Table 1, (56.58%) of the respondents are male whereas (43.42%) are females are participated on this study. Table 1 shows that (82.89%) of respondents age are 35 and above whereas the left (17.11%) are below 35 years. The result also shows that all respondents had literate. Also table shows that even if the literacy level of the family in study is good the awareness level on health and hygiene is different, that (68.46%) of respondents have good awareness on health and hygiene whereas (31.54%) of respondents have no sufficient awareness on health and hygiene which is the consequence of safe waste water disposal.

3.2. Overview of the Sewerage Systems in the Study Areas

Many of national regional and town's housing development projects have set ambitious goals of fulfilling residential houses and advanced waste disposal in past decade. In the study area the condominium sewerage systems are implemented by government and hand over to users, with cost recovery agreement. Inadequate delivery of condominium sewerage services, due to institutional constraints, user's way of services and fiscal difficulties is inhibiting the access of communities to basic sewerage system infrastructure services. Existing facilities are faced to different technical and institutional problems observed. According to discussions held with community and questionary result indicates that, in buildings the numbers of non-functional and non-operative fixtures are huge. Due to this the environment around the condominium sites are to some extent polluted, facing the living community to the health problem.

In Debre berhan town, the level of services ranges from internal sanitary system and external sewerage systems to septic tank with some subsurface disposal method. The septic tanks with subsurface soak away pits system serve few consumers and sites through condominium sewerage systems.

3.3. Internal Sanitary System Physical Status, 2014

Estimated 58 condominium buildings exist in the study town. These buildings internal sanitary systems are visited categorizing into seven sites. For the wastewater collection from each room upvc pipe is installed in all buildings. The system is single fixture pipe system. The diameter of each specific pipes are fixed based the flow quantity they collect according to Ethiopian Building Code Standards version 9, 1995 [8]. The diameter of waste water collector for internal sanitary system, in general, ranges from 50mm to 110mm.

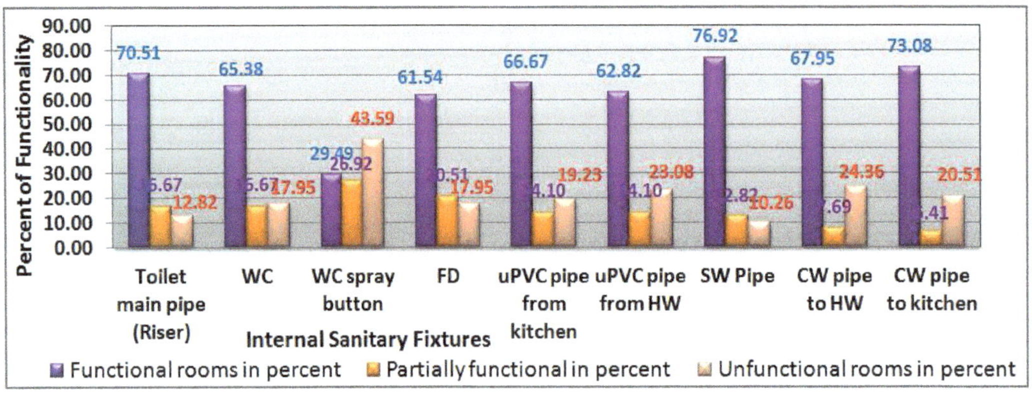

Figure 2. *Functionality status of fixtures in internal sanitary system.*

The physical status of fixtures installed in study buildings varies on different buildings as observed on figure 2 above. As can be seen on figure 2 among 58 toilet riser (truck pipe) only 41(70.51%) are functional; 9 (16.67%) are partially functional and the rest 8 (12.82%) are not functional as required. In the same manner the functionality status of all fixtures are indicated on figure 2 above in percent.

3.4. External Sewerage System Physical Status, 2014

There are 7(seven) condominium sites exist in Debre berhan town. The external sewerage systems of the condominiums vary with size, topography of the sites, and typologies of the condominiums. External sewerage system of buildings composes storm line, underground buried sewers, and appurtenance structures such as man holes, septic tanks, launching aprons, filter Medias. For the wastewater collection from each room upvc pipe and concrete pipe is installed out of the buildings. The diameter of each specific pipes are fixed based the flow quantity they collect according to Ethiopian Building Code standard version 1995 Number nine [8]. The diameter of waste water collector for external sanitary system, in general, ranges from 110mm to 200mm. The graph below shows that the physical status of schemes in percent.

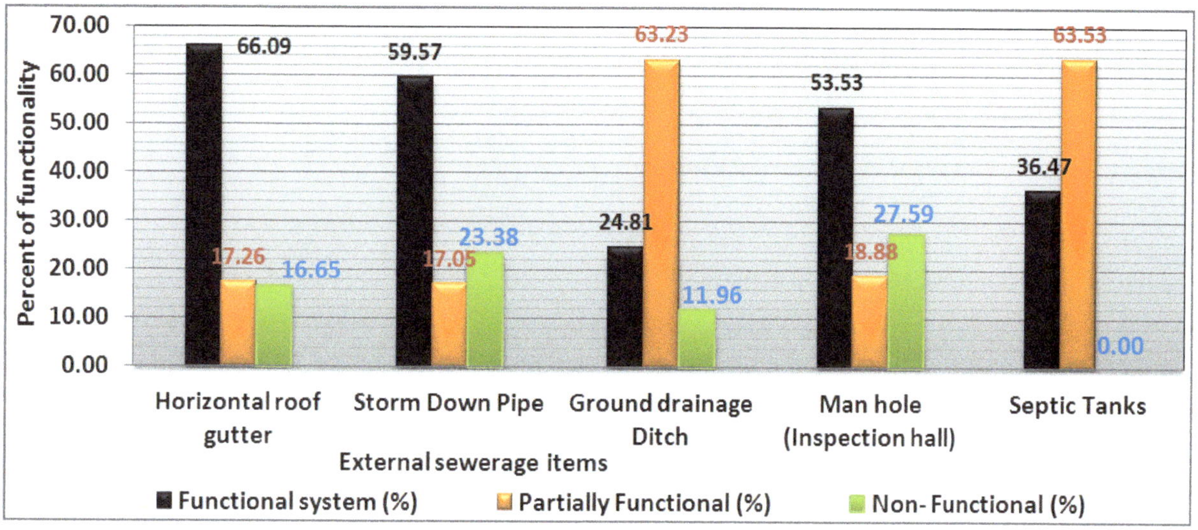

Figure 3. *External Building sewerage scheme functionality status.*

As indicated on graph above, around 74% of drainage ditch constructed around building could not give required function. This is the main cause observed for poor waste flow collection to dispose area. Around the area of buildings the numbers of surcharged sewers are high.

4. Common and Major Problems on Sewerage Systems of Sample Sites

Technical Problems: The significant technical problems noticed in the sample sites include operation and maintenance (O&M), construction, and design respectively put in the order of causing considerable failure.

Institutional problems: Poor implementation of water and sanitation policies, lack of capacity to keep the system running after project completion and Lack of village level organization, like hygiene user association and water committee the main institutional problems investigated in the study area.

Social and Financial problems: Lack of awareness between users about benefits of improved sewerage system, absence of satisfaction between users with the service they get from sewerage facilities and less capability of users to generate efficient revenue for maintenance and operation (M&O) purpose are the social and financial problems observed.

Environmental problem: Potential environmental problems included drainage problems around living apartments; pollution of living compound due to poor solid and liquid wastes and the sitting of latrines close to apartments; pounded wastewater around living areas as breeding grounds for mosquitoes and water borne and water related diseases.

5. Major Causes for Sewerage System Failures of Sample Sites

Figure 4. *Leakage on main track (soil pipe).*

Leakage of pipe at fittings: During construction/installation sewer lines and fittings on the sewer line are not appropriately installed. This implies that the two line which connect each other is not fitted correctly even by using any other fixture liquids such as solvent cement 500mg and fibers.

Blockage during construction: Poor solid waste management is seen in the room and outside the room. This cause blockage of sewer line and floor drains which consequently cause surcharge of the flow on the floor.

Construction problem: miss of necessary construction item in building.

Figure 5. *Missed shower tray from construction.*

Design problem:

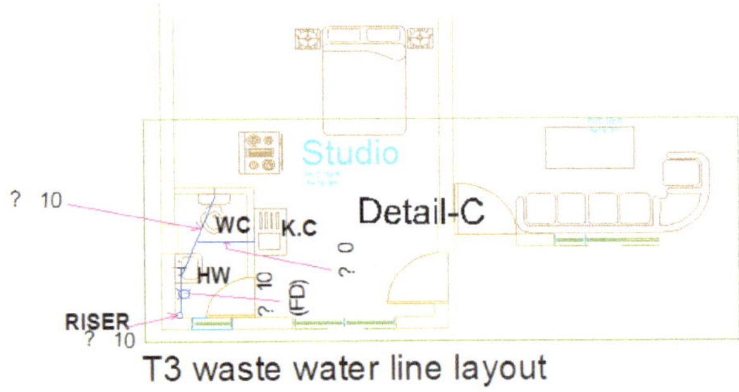

Figure 6. *Deign problem observed on sanitary line layout.*

The alignments of line and fixture sit are loosely planed, during design. The line of sanitary sewer takes large space in the room *(for eg, in T3 typology, studio room, the line of sanitary sewer takes large space in the room).*

Loosely tighten of gutter and down pipe: so that it releases rain water to wall.

Leakage of Septic tanks:

Figure 7. *Leakage and over flow of sewage from septic tanks.*

There is no periodical suck way implemented as per design, there is lateral inflow through the soil. Due to this, the flow from septic tank is simply leak to external road, side ditch without any secondary treatment which is the cause of health problem to communities. Eg Kebele 2 (site 3) and Kebele 7(melikt academy area).

Poor solid waste management

Figure 8. *Community collects solid wastes from manholes manually which blocks sewers.*

Solid wastes deposited on sewer line, in drainage ditches, manholes and septic tanks. Due this there is surcharge of wastes and over flow to the ground surface. In addition composite of solid waste around sewer, makes very difficult for sewer supervision.

Insufficient physical protection of floor drainage ditch, manholes and septic tanks:

Drainage ditches, manholes and septic tanks are totally buried under solid wastes (i.e., Stones, animal bones, and different garbage's) easily enter to the box and block the out let pipe frequently and fill manholes and block the flow line. Due this there is surcharge of wastes and over flow to the ground surface. In addition composite of solid waste around sewer, makes very difficult for sewer supervision.

Figure 9. *Overflow of sewage from manholes due to solid wastes filled in holes.*

6. Proposed Short and Long Term Strategically Remedies

- The initial task has to be make awareness between the communities; how to use sanitary fixtures, way of dispose solid waste.

- A regular checking of the sewerage system, manhole cleaning and application of dilution are essential in order to ensure safety of community health.
- Giving maintenances for fixtures which leak flow and replacing fittings which show any rupturing and break.
- For those the sanitary alignment seen sever problem regarding space and material quality realignment is preferable based on the willingness and cooperation of households, who the line coincide their room.
- Community organizations responsible for the O&M of sewerage system should be legally constituted and registered. The tariff collection should be regular, and tariff levels should be set according to the level of service and cover expenses for regular and periodic maintenance.
- Controlling infiltration using grouting or sealing of soils surrounding the sewer pipe, pipe relining, and sewer replacement, all of them are costly [14].
- Providing alternative sewage disposal methods;

Sand Filter Media and Soak away pits (based on soil percolation and topography of the area)[7].

According to laboratory result, soil type of Kebele 1,2,4 and 9(site1,site3,site4 and site7) show us sand and gravel which have greater percolation capacity than other types. So we can use such type disposal method in these a sites.

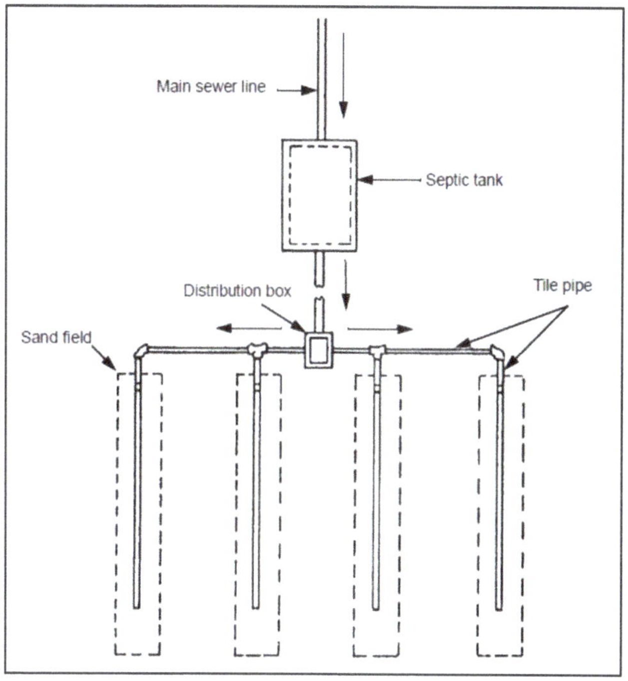

Figure 10. *Sand Filter Media.*

Vacuum Systems: 50 to 150 mm (2 to 6 inch) plastic pipe is buried deep in a vacuum, the source is connected to the vacuum through entry valve which opens automatically when a sufficient volume of sewage has accumulated and recloses when time interval reach to be drawn into the main. Then the vacuum is maintained by pumps at a central collecting tank [18].

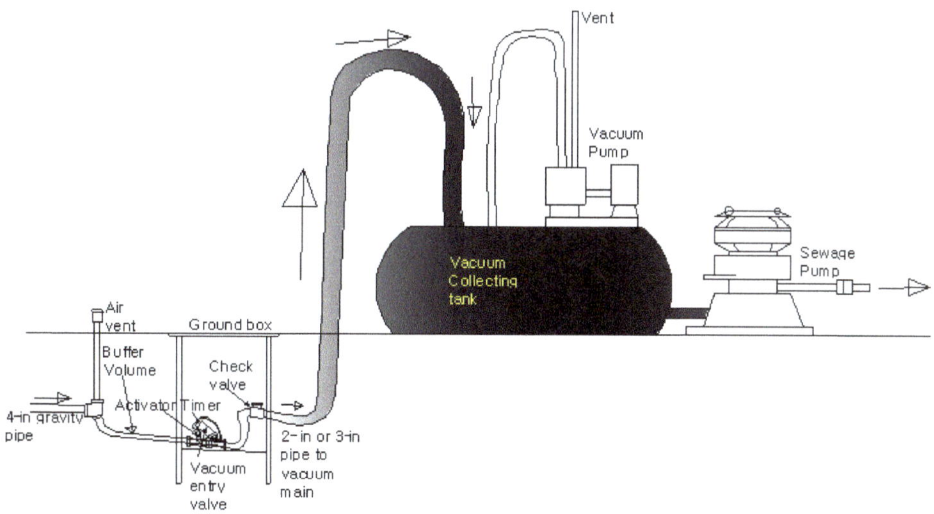

Figure 11. *Vacuum system.*

7. Conclusions and Recommendations

In the study area Projects were formulated through supply-driven approach and without consultation with representatives of communities. These factors have resulted in: absence in sense of ownership between communities and non-effective O&M on the schemes.

To increase the level of ownership of future project, appropriate measures should be incorporated in the project design and implementation, such as capacity-building program for sector agencies and communities, and clear delineation of responsibilities among central, local agencies and communities.

From this study we noticed that O&M is by far the weakest aspect at most of the sample projects. Poor administrative and technical supports, lack of operating fund are cited as the most frequent cause of failure.

Quality of construction is a crucially important parameter to be monitored in sewerage systems [6]. Though it looks that construction quality monitoring is neglected in the study areas. Construction materials checks and control (i.e. physical and chemical) before constructing the system, as well as on a regular basis should be considered in the future.

The other problems we have seen in the sample schemes are problems associated to the utilization and protection of the system and construction problem as a whole. Although technical sustainability depends largely on design and construction, it also depends on economic, social, political, cultural, financial, technological, and management aspects. Addressing these aspects calls for a well-designed, long-term, sewerage system monitoring and evaluation program complemented by all stakeholders.

Institutional weakness is the most significant problem in the study sites. Stated objectives are no set for the utilities. There is no NGOs take part on O&M. Therefore we would like to forward that to undertake O&M responsibility, integration of efforts under sanitation hygiene team composed of sectors like town water supply & sewerage bureau, town health centers, NGOs and communities is required.

References

[1] Tessema, Yitbarek. 2013. Ethiopia - Ethiopia Water Supply and Sanitation Project: P076735 - Implementation Status Results Report: Sequence 19. Washington, D.C.: World Bank Group.

[2] Debre berhan town health Office (2014) Report of top Ten Diseases, Debre berhan, Ethiopia.

[3] 2012 Population and Housing Census of Ethiopia: Results for Amhara Region, Vol. 1, part 1, Tables 2.1, 2.7, 2.10, 2.13, 2.17, Annex II.2 [2012, April].

[4] A.P. Moser and Steven Folkman (2008) Buried Pipe Design (Third edition).

[5] Cairncross S, Carruthers I, Curtis D, Feachem R, Bradley D, Baldwin G (1980) Evaluation for village water supply planning, John Wiley and Sons, New York.

[6] David Bufler and John W. Davis (2004), Urban Drainage (Second edition).

[7] Departments of the Army (August, 2001) Plumbing, Pipe Fitting, and Sewerage, Washington DC.

[8] EBCS-9 (1995) Ethiopian Building Code Standards Number Nine, Sectoral manual, Addis Ababa.

[9] Ethiopian News Agency [2009, May]"42-km Debrebirhan-Ankober road being maintained".

[10] http://www.en"Local History in Ethiopia" The Nordic Africa Institute website [2007, Dec.].

[11] http://www.enWikipedia.org/wiki/Debre_Berhan#local_economy.

[12] Jerrya. Nathanson(2009). Basic Environmental Technology Water supply, Waste Management, and Pollution control (Fifth edition), New Delhi _11001.

[13] Knut Werner Alsen, Petter D. Jansoon (2004) Ecological Sanitation (for mankind and nature), Norway.

[14] Mark. J. Hammer (2005) Water and Waste Water Technology (Fourth edition), New Delhi_11001, 2005.

[15] Metcalf & Eddy, Inc. (1960). Wastewater Engineering-Treatment and Reuse (2nd edition). McGraw - Hill. ISBN-13:978-0070418783.

[16] Pradnya Thakur and D.B. Panse (2012) Success Stories of Sustainable Sanitation Initiatives in India by ESF.

[17] S.K. Garg (March, 2008) Sewage Disposal and Air Pollution Engineering (21st edition).

[18] Terence J. McGHEE (1960) Water Supply and Sewerage (6th edition), McGRAW-HILL INTERNATIONAL EDITIONS.

[19] Unpublished document, March 2006 Adapted from Sanitary Fixtures Specifications, Avalon Pharma factory, Riyadh, Saudi Arabia.

[20] Selamawit Aseffa (2007) Evaluation of Technical sustainability of Rural water supply projects in the Amhara Regional state, case study in south Wollo, Addis Abeba, Ethiopia.

28

An Approach to Removing Large Quantities of Atmospheric Greenhouse Gases

John B. Cook[1], Svetlana R. Amirova[2], Edwin A. Roehl Jr.[1], Paul A. Comet[3], Tamara V. Tulaykova[4]

[1]Advanced Data Mining International, Greenville, S.C, USA

[2]Brody School of Medicine, East Carolina University, Greenville, North Carolina, USA

[3]Comet Environmental Consulting, Houston, Texas, USA

[4]Wave Research Center at A. M. Prokhorov's General Physics Institute, Moscow, Russia

Email address:

john.cook@advdmi.com (J. B. Cook), tulaikova@gmail.com (T. V. Tulaykova), ed.roehl@advdmi.com (Edwin A. R. Jr.), pcomet@prodigy.net (P. A. Comet), amirova.svetlana@yahoo.com (S. R. Amirova)

Abstract: A method for removing atmospheric carbon dioxide (CO_2) and water vapor is proposed. The method sprays clouds with alkaline compounds to significantly increase the solubility of CO_2 in the cloud water, providing for much higher than normal levels of CO_2 to be absorbed by rain droplets. The CO_2 is transported to the ground for sequestration in surface and/or ground water, and available for carbon fixation by plants and organisms. Presented calculations estimate that 38 gigatonnes of atmospheric CO_2 could be removed per year by applying the process over 0.08% to 2.4 % of the Earth's surface. Laboratory experiments that grew multiple edible plant species irrigated with the modified rainwater indicated yield benefits. A concept for removing atmospheric methane (CH_4) is also presented. Powerful lasers would ionize the CH_4 to form CO_2 that could then be removed by the alkaline-enhanced rainfall method.

Keywords: Climate Change, Greenhouse Gas, Carbon Dioxide, Methane, Water Vapor, Removal, Alkali, pH Adjustment, Precipitation, Cloud Seeding, Laser

1. Introduction

Current global emissions of greenhouse gases are exceeding the Earth's assimilative capacity, causing them to accumulate in the atmosphere [1], [2]. Water vapor comprises 95% of all greenhouse gases [3], however, carbon dioxide (CO_2) and methane (CH_4) are of greater concern. Higher greenhouse gas concentrations are widely believed to contribute to climate change, extreme weather events, rising sea levels, and other marked changes, such as the acidification of surface waters and soils.

Over the past century the carbon dioxide (CO_2) concentration rose from 280 to 400 parts per million (ppm), the methane (CH_4) concentration rose from 0.9 to 1.8 ppm, and both continue to rise rapidly [4], [5]. The oceans' natural alkalinity allows them to store a large amount of CO_2, including more than a quarter of that produced anthropogenically. This has increased ocean acidity ~26% from preindustrial levels, which corresponds to a decrease in pH from an average of 8.2 to 8.1, or less. A further decrease of ~0.3 units by 2100 is likely unless CO_2 emissions are reduced rapidly. Higher ocean acidity negatively impacts marine ecosystems and the economic benefits they provide [6], [7]. Similarly, rains in industrial regions since the mid-20th century have caused soil deterioration due to higher acidity, with current pH ≈ 3 – 4 [8], reducing alkalinity in the environment.

The development of methods that can efficiently remove greenhouse gases already in the atmosphere would be beneficial. Precipitation enhancement methods, which *de facto* remove water vapor, have been developed primarily for agriculture, water supply, and reducing air pollution [9], [10]. Typically, aircraft, rockets, or other projectiles are used to disperse hygroscopic or glacial particles to alter cloud microstructure, inducing precipitation [11], [12]. Seeding along a narrow flight path can be sufficient to restructure an entire cloud. Acoustic wave generators have been investigated as an alternative means to inducing precipitation [13], [14].

A new method proposed here expands the purpose of precipitation enhancement to include climate modification through the removal of large amounts of under-cloud CO_2. The solubility of CO_2 in water is highly sensitive to pH [15], [16], so the first step is to disperse an alkaline compound such as potassium hydroxide (KOH) inside clouds over selected areas to increase the pH of the clouds' constituent microscopic water droplets to 10 – 11. The alkaline particles should be consistent in size and composition, and dispersed uniformly inside clouds to produce droplets formed by nucleating water vapor which are uniform in size and pH.

The method's second step is to induce rain, allowing the high pH rain to absorb substantially more CO_2 than a normal pH rain or a two-dimensional (2-D) patch of surface water whose specific surface area is less than rain's three-dimensional (3-D) droplets. The second step would be unnecessary if alkalis having strong hygroscopic properties are used. Once on the ground, the dissolved CO_2 is sequestered in surface and/or ground water, and available for carbon fixation by plants and organisms.

Below are calculations that estimate the efficiency of the proposed method for atmospheric CO_2 removal, and results from laboratory experiments that grew different crop species irrigated with the modified rainwater. Also described is a concept that uses laser-induced oxidation to convert atmospheric CH_4 into CO_2 that can then be removed by the alkaline-enhanced rainfall method.

2. Atmospheric CO_2 Removal

The typical pH of normal rainwater ranges from 5.6 to a neutral 7.0, at which the concentrations of dissociated ions are relatively small. The insoluble and dissociated portions of the weak acid that remains after attaining the equilibrium of saturation for water by CO_2 are described by (1), where H_2CO_3 is carbonic acid, H^+ is hydrogen ion to determine pH, and HCO_3^- and CO_3^{2-} are bicarbonate and carbonate ions.

$$CO_2 + H_2O = H_2CO_3$$

$$H_2CO_3 = HCO_3^- + H^+ \quad (1)$$

$$HCO_3^- = CO_3^{2-} + H^+$$

The equilibrium concentrations of H_2CO_3, HCO_3^-, and CO_3^{2-} at a neutral pH = 7.0 are approximately 0.71, 3.3, and 0.001 milligrams per liter (mg/l), respectively [17]. CO_2 solubility increases as pH increases due to an increasing H^+ concentration, with HCO_3^- and CO_3^{2-} concentrations increasing by one and two orders of magnitude, respectively, for each unit change in pH. The fractions of carbon in the molar masses of H_2CO_3, HCO_3^-, and CO_3^{2-} are 0.1935, 0.1967, and 0.2, respectively, so all fractions are equal to or approximately equal to 0.2.

The mass of carbon $M_C(pH)$ in the dissolved CO_2 in a rainfall volume equivalent $U_w = h_wA$ is estimated from (2), where h_w and A are the volume's height and cross-sectional area. For brevity, the H_2CO_3, HCO_3^-, and CO_3^{2} concentrations are referred to as C_1, $C_2(pH)$, and $C_3(pH)$.

$$M_C(pH) \approx U_w\{0.2\ [C_1 + C_2(pH) + C_3(pH)]\} \quad (2)$$

The concentration of atmospheric carbon C_{a-} in an air volume with a height h and a unit (1 m^2) cross-section can be estimated from (3), which is derived from (2). The vertical mixing ratio of CO_2 is estimated to be constant for altitudes of interest. The fraction of carbon in the molar mass of CO_2 is 12/44 = 0.27. For a cloud at an altitude of 1 km, the cloud water pH should be increased from a typical 5.6 to 10.3 for complete CO_2 removal (C_{a-} = 420 ppm).

$$C_{a-} \approx (h_w/h)[0.2(C_1 + C_2(pH) + C_3(pH)]/0.27 \quad (3)$$

The deployment of KOH in a liquid water aerosol is used to illustrate how to estimate the amount of an alkali needed to raise the pH of cloud water to an effective level. The dissociation reaction of KOH in water represented by (4) indicates that the molar concentrations of KOH and OH^- are equal.

An alkali facilitates the dissociation of ionic species in water according to the constant $K_w = 10^{-14}$ at 20°C [17]. According to (4) and (5), the KOH mass concentration n_{KOH} needed to raise the pH of the cloud water from 5.6 to 10 is 10^{-4} mol/l, or 56×10^{-4} g/l when converted to a molar mass. For an example 1 km^3 cloud having a liquid water concentration of 1 g/m^3, the mass of KOH to be dispersed would only be 5.6 kg. Note that little of the KOH would be spent within the cloud over a short period prior to rainfall inducement due to the small resident concentration of CO_2 molecules (~400 ppm).

$$KOH = K^+ + OH^- \quad (4)$$

$$H^+ + OH^- = H_2O,\ K_w = [H^+]\ [OH^-] \quad (5)$$

The volume of water precipitated by a cloud U_I is estimated by (6), where I is the rainfall rate, A_s is the 2-D receiving area on the Earth' surface, and T is the duration of the rainfall event. The number of spherical droplets N having a radius r_d in U_I and the droplets' total surface area A_d are estimated by (7) and (8) respectively.

$$U_I = I\ A_sT \quad (6)$$

$$N = 3U_I\ /\ (4\pi r_d^3) \quad (7)$$

$$A_d = N \times 4\pi r_d^2 \quad (8)$$

If given a moderate rainfall characterized by $A_s = 1\ km^2 = 10^6\ m^2$, I = 6 mm/h, T = 1 hour, and r_d = 0.5 mm, then U_I = 6,000 m^3, N = 10^{13}, and $A_d = 3.6 \times 10^7\ m^2 = 36 \times A_s$. Alternatively, the Marshall-Palmer approximation accurately estimates droplet radii spectra for drizzle, rain, or storm events, and can be applied to perform more detailed droplet surface area estimates [18], [19]. The parameters and measurement methods used to characterize precipitation inside clouds and atmosphere volumes are described in [20]–[22].

Rapid CO_2 saturation of droplets is made possible by their small size and slow speeds. The terminal speed of a droplet V_g is estimated by (9) by equating the forces of gravity and air resistance, where ρ_a and ρ_w are the densities of air and water respectively, and g is the acceleration due to gravity. A droplet having $r_d = 0.5$ mm falling from an altitude h = 1 km has a V_g = 2.9 m/sec (ρ_a = 1.2 kg/m^3 at 12°C). The travel time t_h = 350 seconds is estimated by (10).

$$V_g = (2r_d g\rho_w / \rho_a)^{1/2} \quad (9)$$

$$t_h = h / V_g \quad (10)$$

The time t_{aw} needed for a droplet to become saturated with CO_2 depends on the ratio of the droplet's aerated volume to its surface area, and is estimated by (11), where k_{aw} is a measured constant, called the gas transfer velocity, for air and water [23]–[26]. For $k_{aw} \approx 0.5\times10^{-4}$ m/sec and r_d = 0.5 mm, t_{aw} = 6 seconds, indicating that the much longer travel time $t_h \approx 350$ seconds is sufficient for nearly complete removal of under-cloud CO_2 molecules.

$$t_{aw} = r_d / 3k_{aw} \quad (11)$$

2.1. Impact on Plants

a

b

Fig. 1. First crop plants after 19 days (a) and 31 days (b) days. The blue pot (left) was irrigated daily with distilled water. The brown pot (right) was irrigated with an alkaline solution.

A potential benefit to plant growth is illustrated by results from indoor bench-scale experiments [27] in which two simulated crops were irrigated with and without alkaline-enhanced water for a period of one month (Figs. 1 and 2). The blue pots were irrigated daily with distilled water, and the brown pots were irrigated using the same regime with a stable KOH-distilled water solution having a pH=12. For the first crop both pots were identically planted with beetroots (10 grains), carrots (20 grains), and parsley (2 grams). For the second crop the pots were planted with dill (20 grains) and cucumber (5 grains). As shown, both crops responded favorably to the alkaline-enhanceme irrigation.

Fig. 2. Second crop plants after 19 days.

2.2. Global Application

Table 1. Estimated Outcomes of Global Application.

pH	5.6	10	10.5	11
M_C^1 (kg/m^2)	1.6×10^{-4}	0.849	4.05	26.5
M^A_C (kg)	8.2×10^{10}	4.3×10^{14}	2.1×10^{15}	1.4×10^{16}
M^A_{CO2} (kg)	3×10^{11}	1.6×10^{15}	7.7×10^{15}	5×10^{16}
A_p (%)	-	2.4	0.5	0.08
Q_{KOH} (kg)	-	6.9×10^{10}	4.8×10^{10}	2.3×10^{10}

pH = pH of induced rainfall, M_C^1 = mass of atmospheric carbon removed by 1 meter of rainfall per m^2 of the Earth's surface, M^A_C = mass of global atmospheric carbon removed, M^A_{CO2} = mass of global CO_2 removed, A_p = percentage of the Earth's surface area needed to completely remove annual global CO_2 emissions (E_A), and Q_{KOH} = mass of KOH needed to remove E_A each year.

Table 1 provides statistics that estimate outcomes of applying the method globally. The Earth's surface area A_E = 5.1×10^8 km^2, the average global annual rainfall h_A = 1 m [28], and the annual global CO_2 emissions $E_A \approx 3.8\times10^{13}$ kg [29]. Using (2), row 2 gives the masses of atmospheric carbon M_C^1 removed by 1 meter of annual rainfall per m^2 of the Earth's surface at the different pH levels shown in row 1. Row 3 gives the global masses of carbon removed M^A_C per (12). Row 4 gives the global masses of CO_2 removed M^A_{C02} per (13), 0.27 being the carbon ratio of CO_2. Using (14), row 5 gives the percentages A_p of A_E needed to completely remove annual emissions E_A, indicating the need is 0.08% – 2.4%. Row 6 gives the mass Q_{KOH} of KOH needed to remove E_A each year per (15), indicating that the amount needed is a fraction of the 19×10^{10} kg of nitrogen, phosphorous, and potassium (N/P/K) fertilizers used globally in 2014 [30].

$$M^{A}_{C} = M_{C}^{1}\ A_{E} \tag{12}$$

$$M^{A}_{C02} = M^{A}_{C}\ /\ 0.27 \tag{13}$$

$$A_{p} = 100 \times E_{A}/M^{A}_{C02} \tag{14}$$

$$Q_{KOH} = n_{KOH}\ h_{A}\ A_{E}\ A_{p} \tag{15}$$

3. Atmospheric CH_4 Removal

A concept is also proposed that would use lasers to enhance the conversion of atmospheric CH_4 into CO_2 and water, a process that occurs naturally through the photo-dissociation of CH_4 molecules by sunlight [31]. Once freed, the carbon ions bond preferentially with oxygen [32]. This process can be stimulated by using a powerful laser with an optimized wavelength. Subsequent alkaline-enhanced rainfall would remove the CO_2.

Measurements of atmospheric CH_4 concentrations can be performed at distances of 8 – 11 km using pulsed integrated-path differential absorption Lidar in the infrared spectrum at wavelengths $\lambda = 1{,}650$ nanometers (nm) [33], [34]; however, the photon energies are too low for ionization. For efficient ionization, the laser wavelength should coincide with a peak absorption frequency at the molecule's C-H bond dissociation energy of $E \approx 4.3$ eV electronvolts (eV) or $\lambda = 288$ nm due to the Planck–Einstein relation $\lambda E = hc = 1240$ eV–nm, where h is the Planck constant and c is the speed of light [35]. However, at UV irradiation other gases such as oxygen and nitrogen also could be ionized. Alternatively, multiple-photon excitation achieves photo-dissociation using a laser to bombard gas molecules with multiple photons, increasing their vibrational energy until their chemical bonds are broken [36]. CH_4 has a small absorption peak near $\lambda = 865$ nm (1.43 eV) [37], suggesting that three or four photons could be sufficient for dissociation to occur.

4. Conclusions

A proposed approach could be effective at removing significant amounts of the main greenhouse gases of concern, i.e., water vapor, CO_2, and CH_4. Estimates show that a method for alkaline-enhanced rainfall could remove more than the ~38 gigatonnes of anthropogenic CO_2 emitted each year if applied over a small portion of Earth's surface. The next step would be to field test the method's efficacy. An adjunct method for pre-treating atmospheric CH_4 by laser is also proposed. It would convert the CH_4 into CO_2 for subsequent removal by alkaline-enhanced rainfall.

Acknowledgements

The authors thank Dr. Yakov Ponurovskii and Dr. Imant Vinorgadov for sharing information about monitoring greenhouse gases with tunable diode laser spectroscopy and the photo-dissociation of methane with lasers.

References

[1] IPCC, Contribution of Working Group III to the Fifth Assessment Report of the Intergovernmental Panel on Climate Change. Summary for Policymakers, Cambridge University Press, New York, 2014.

[2] Kleidon, A., Malhi, Y. and Cox, P. M., "Maximum entropy production in environmental and ecological systems introduction," Philos. Trans. R. Soc. B 365, 1297-1302. 2010.

[3] Freidenreich, S. M. and Ramaswamy, V., "Solar radiation absorption by carbon dioxide, overlap with water, and a parameterization for general circulation models," J. of Geophys. Res. 98, 7255-7264. 1993.

[4] Latest reading, Mauna Loa Observatory. 2016. https://scripps.ucsd.edu/programs/keelingcurve/. [Accessed Apr. 4, 2016].

[5] Schmidt, G., "Methane: a scientific journey from obscurity to climate super-stardom," NASA Goddard Space Center, 2004. http://www.giss.nasa.gov/research/features/200409_methane/. [Accessed Apr. 4, 2016].

[6] Andersson, A. J. and Gledhill, D., "Ocean acidification and coral reefs: effects on breakdown, dissolution, and net ecosystem calcification," Annu Rev Mar Sci 5, 1.1–1.28, 2013.

[7] Secretariat of the Convention on Biological Diversity, Roberts, J. M. and Williamson, P. – ed., An Updated Synthesis of the Impacts of Ocean Acidification on Marine Biodiversity, Technical Series No. 75, Montreal, Canada, 2014.

[8] Izrael, Y. A. et al., Acid rain (Hydrometeoisdat, Russia, 1989).

[9] Langmuir, I., "Improved methods of conditioning surfaces for adsorption" J. Am. Chem. Soc. 59. 1762-1763. 1937.

[10] Dennis, A. S., Weather modification by cloud seeding, Academic Press, New York, 1980.

[11] Shmeter, S. M. and Berynlev, G. P., "Efficiency of cloud and precipitation modification with hygroscopic aerosols," Meteorology and Hydrology Rus., 2. 43-60. 2005.

[12] Tulaikova, T., Mihtchenko, A. and Amirova, S., "Micro physical model for glaciogenic particles in clouds for precipitation enhancement," Am. J. Env. Protection, 5(3-1). 10-14. 2016.

[13] Tulaikova, T. et al., Acoustic rains, Physmathbook, Moscow, 2010.

[14] Claus R. O. and Tulaikova, T. V., "New methods for helicopter for free flight inside clouds and precipitation enhancement," Am. J. Env. Protection, 5. 1-9. 2016.

[15] Yunge, H., Chemical compounds and radio-activity in the atmosphere, Clarendon, Oxford, 1965.

[16] Rasool, S. I. (ed.), Chemistry of the lower atmosphere, Plenum, New York, 1973.

[17] Sillen, L. G. (ed.), Stability constants of metal-ion complexes, Chemical Society, London, 1964.

[18] Borovikov, A. M., Physics of clouds, Hydromet-Press, Leningrad, Russia, 1961.

[19] Kobayashi, S. T. et al, "Backscattering enhancement on spheroid-shaped hydrometeors: considerations in water and ice particles of uniform size and Marshall-Palmer distributed rains," Radio Science, 42. Apr.2007.

[20] Bruintjes, R. T., "A review of cloud seeding experiments to enhance precipitation and some new prospects," BAMS, 90. 805-820. 1999.

[21] Daly, C. et al, "Observation bias in daily precipitation measurements at United States cooperative network stations," BAMS 88. 899-912. 2007.

[22] Taylor, J. W. et al, "Aerosol measurements during COPE: composition, size and sources of CCN and IN at the interface between marine and terrestrial influences," Atmos. Chem. Phys. Discuss., doi: 10.5194/acp-2016-84, in review. 2016.

[23] Hoover, T. E., "CO_2 exchange at the air-sea interface," J. Geoph.Res 74. 456-464. 1969.

[24] Liss, P. S., "Processes of gas exchange analysis an air-water interface," Deep-Sea Res. 20. 221-238. 1973.

[25] Broecher, H. C., "The influence of wind on CO_2 exchange in a wind-water tunnel including the effect of minelayers," J. Mar. Res. 36. 595-610. 1978.

[26] Wanninkhof, R., "Chemical enhancement of CO_2 exchange in natural water," Limnol. Oceanogr., 41. 689-687. 1996.

[27] Tulaikova, T., and Amirova, S., "The method for effective CO_2 purification in the atmosphere," Global J. Sc. Frontier Res., 15-H (1). 1-9. 2015.

[28] Rees, G. and Rees, W., Physical principles of remote sensing, Cambridge University Press, 2013, 125.

[29] Olivier, J., Janssens-Maenhout, G., Muntean, M. and Peters, J., Trends in global CO_2 emissions: 2015 report, PBL Netherlands Environmental Assessment Agency, 2015.

[30] Food and Agriculture Organization of the United Nations, "World fertilizer trends and outlook to 2018," Rome, Italy, 2015.

[31] Sheu, J., Mokheimer, E. and Ghoniem, A., "A review of solar methane reforming systems," Int. J. Hydrogen Energy, 40. 12929-12955. 2015.

[32] Clark, J., Calculations in AS/A Level Chemistry, Pearson Education, London, 2000.

[33] Dawsey, M. et al, "Optical parametric technology for methane measurements," Proc. 2015 SPIE, Lidar Remote Sensing for Environmental Monitoring XV, San Diego, USA.

[34] Riris, H. et al, "Airborne measurements of atmospheric methane column abundance using a pulsed integrated-path differential absorption Lidar," Applied Optics, 51 (34). 8296-8305. 2012.

[35] Schmidt, W., Optical spectroscopy in chemistry and life sciences, Wiley-VCH, Germany, 2005.

[36] Cantrell, C. D. (ed.), Multiple-photon excitation and dissociation of polyatomic molecules, Springer, New York, 1986.

[37] Brien, J. and Cao, H., "Absorption spectra and absorption coefficients for methane in the 750–940 nm region obtained by intracavity laser spectroscopy," J. Quant. Spec. Rad. Tran., 75 (3). 323-350. Oct. 2002.

Determination of Some Heavy Metals in Raw Petroleum Wastewater Samples Before and After Passing on *Australis Phragmites* Plant

Mohamed Ezeldin[1, 2], Sulieman A. G. Nasir[1], Ali M. Masaad[1], Nawal M. Suleman[3]

[1]Department of Chemistry, Faculty of Science and Technology, Omdurman Islamic University, Khartoum, Sudan
[2]Department of Quality Control, Almazaq Factory, South of Khartoum, Sudan
[3]Department of Chemistry, Zilfi Facultyof Education, Majmmah University, Zilfi, Saudi Arabia

Email address:
wadalmsna3.com@gmail.com (M. Ezeldin)

Abstract: The main objective of this project was determining of lead, cadmium, nickel and chromium in raw petroleum wastewater (RPWW) samples. The samples were taken before and after passage the RPWW on Australis Phragmites Plant (APP). Atomic absorption spectrophotometer was used to determine the heavy metals in RPWW samples before and after passing, The concentration of lead, cadmium, nickel and chromium before passing on APP was found to be 0.25, 0.9, 0.52 & 0.012 ppm respectively, but after passing on APP was fond to be 0.123, 0.75,0.45 & 0.01 ppm, respectively. The obtained results after passage were found within the range assigned by American Society for Testing and Materials ASTM, except the concentration of nickel.

Keywords: Heavy Metals, Petroleum Wastewater, Australis Phragmites Plant

1. Introduction

1.1. Petroleum Wastewater

Refineries can generate a significant amount of wastewater that has been in contact with hydrocarbons. Wastewater can also include water rejected from boiler feed water pretreatment processes (or generated during regenerations) [1]. Wastewater can also refer to cooling tower blowdown stream, or even once-through cooling water that leaves the refinery. Once-through cooling water typically does not receive any treatment before discharge [2]. Contaminated wastewater is typically sent to either a wastewater treatment plant that is located at the facility, or it can be pretreated and sent to the local publicly owned treatment works or third-party treatment facility for further treatment. Water that has not been in direct contact with hydrocarbons or which has only minimal contamination can be a source for reuse, a purge stream is removed from the water purification systems in order to prevent the buildup of contaminants. This purge stream is sent to wastewater treatment and is replaced by fresh makeup water [3].

1.2. Effect of Heavy Metals

Heavy metals are individual metals and metal compounds that can impact human health. Eight common heavy metals are discussed in this brief: arsenic, barium, cadmium, chromium, lead, mercury, selenium, and silver. These are all naturally occurring substances which are often present in the environment at low levels. In larger amounts, they can be dangerous. Generally, humans are exposed to these metals by ingestion (drinking or eating) or inhalation (breathing). Working in or living near industrial sites which utilize these metals and their compounds increases environmental risks, as does living near sites where these metals have been improperly disposed. Subsistence lifestyles can also impose higher risks of exposure and health impacts because of hunting and gathering activities [4].

1.2.1. Lead

As a result of human activities, such as fossil fuel burning, mining, and manufacturing, lead and lead compounds can be found in all parts of our environment. This includes air, soil, and water. Lead is used in many different ways. It is used to produce batteries, ammunition, metal products like solder and pipes, and X-ray shielding devices. Lead is a highly toxic metal and, as a result of related health concerns (see below), its use in several products like gasoline, paints, and pipe solder, has been drastically reduced in recent years. Today, the most common sources of lead exposure in the United States are Lead-based paint and possibly water pipes in older homes, contaminated soil, household dust, drinking water, lead crystal, lead in certain cosmetics and toys, and lead-glazed pottery. Health effects EPA has determined that lead is a probable human carcinogen. Lead can affect every organ and system in the body. Long-term exposure of adults can result in decreased performance in some tests that measure functions of the nervous system; weakness in fingers, wrists, or ankles; small increases in blood pressure; and anemia [4, 5].

1.2.2. Cadmium

Cadmium is a very toxic metal. All soils and rocks, including coal and mineral fertilizers, contain some cadmium. Cadmium has many uses, including batteries, pigments, metal coatings, and plastics. It is used extensively in electroplating [5].

1.2.3. Nickel

Nickel is a very abundant, naturally occurring metal and is used for a variety of industrial purposes. Nickel compounds, such as barium-nickel alloys are used for spark-plug electrodes and in vacuum tubes as a drying and oxygen-removing agent; nickel sulfide is used in fluorescent lamps; nickel sulfate is used in diagnostic medicine; nickel nitrate and chlorate give fire works a green color. Barium compounds are also used in drilling muds, paint, bricks, ceramics, glass, and rubber [4] [5].

1.2.4. Chromium

Chromium is found in rocks, animals, plants, and soil and can be a liquid, solid, or gas. Chromium compounds bind to soil and are not likely to migrate to ground water but, they are very persistent in sediments in water. Chromium is used in metal alloys such as stainless steel; protective coatings on metal (electroplating); magnetic tapes; and pigments for paints, cement, paper, rubber, composition floor covering and other materials. Its soluble forms are used in wood preservatives [6].

2. Materials and Methods

2.1. Materials

All chemicals used were of analytical reagent grade (AR) and of highest purity degree available. They included hydrochloric acid, stock lead (II) solution, stock cadmium (II) solution, stock nickel (II) solution, stock chromium(III) solution, deionized water.

2.2. Methods

Preparation of samples

In clean 100 mL volumetric flasks about 5mL of petroleum raw wastewater samples before and after passing on APP was diluted to the mark by using deionized water. Then the solutions were homogenized well.

Preparation of standard solutions

Standard solutions of lead (II), cadmium (II) and nickel (II) were prepared by concentrations 0.01, 0.1. 0.2. 1 ppm, but standard chromium solutions were prepared by concentrations 0.001, 0.002, 0.01, 0.1 ppm from the stock solutions [7].

Blank of instrument

Atomic absorption spectrophotometer was adjusted to zero point by using deionized water. Absorption of the standard lead (II), cadmium (II), nickel (II) and chromium (III) solutions were measured and recorded automatically by the instrument. The absorption of RPWW before and after passage was measured and recorded. The concentrations of heavy metals in samples were determined directly [7] by using AAS shown in Fig1.

Fig. 1. *Atomic Absorption spectroscopy (AAS).*

3. Results and Discussion

In this project the samples of raw petroleum wastewater before and after passage on *Australis Phragmites Plant* were taken from Alfola petroleum area, west of Sudan.

The experimental work was conducted at Central Petroleum Laboratories (CPL) ministry of Petroleum, marked heavy metals (lead, cadmium, nickel and chromium) was determined by using Atomic Absorption spectroscopic technique. The obtained results were shown in Table 1.

Table 1. *The concentration of marked heavy metals.*

Heavy metal	Concentration (ppm)(before)	Concentration (ppm) (after)
Lead	0.25	0.123
Cadmium	0.9	0.75
Nickel	0.52	0.45
Chromium	0.012	0.01

The obtained results represented that: APP showed high ability to sorption the marked heavy metals because the concentration of metals before passing RPWW on the APP higher than concentration after passing that attributed to the empty insides of APP.

The concentration of lead, cadmium, nickel and chromium in RPWW after passage on APP was reduced, the reducing process was very important because, lead can affect every organ and system in the body [7], long-term exposure of adults can result in decreased performance in some tests that measure functions of the nervous system; weakness in fingers, wrists, or ankles [8]; small increases in blood pressure; and anemia. Cadmium and cadmium compounds are known human carcinogens. Nickel is not known to cause cancer, short term exposure can cause vomiting, abdominal cramps, diarrhea, difficulties in breathing, increased or decreased blood pressure, numbness around the face, and muscle weakness, large amounts of nickel intake can cause, high blood pressure, changes in heart rhythm or paralysis and possibly death. Chromium compounds are toxins and known human carcinogens, whereas Chromium is an essential nutrient. Breathing high levels can cause irritation to the lining of the nose; nose ulcers; runny nose; and breathing problems, such as asthma, cough, shortness of breath, or wheezing, Skin contact can cause skin ulcers. Allergic reactions consisting of severe redness and swelling of the skin have been noted [8], Long term exposure can cause damage to liver, kidney circulatory and nerve tissues, as well as skin irritation.

ASTM has been assigned right permissible range to heavy metals found in petroleum wastewater, the permissible limits was shown in Table 2 [9].

Table 2. *Permissible range of marked heavy metals assigned by ASTM.*

Metals	concentration(ppm)
Lead	< 0.2
Cadmium	<0.8
Nickel	<0.4
Chromium	<0.01

The obtained results before passage on APP out of rang assigned by ASTM, but after passage was found within the range, except the concentration of nickel was found to be 0.45 but the permissible range <0.4.

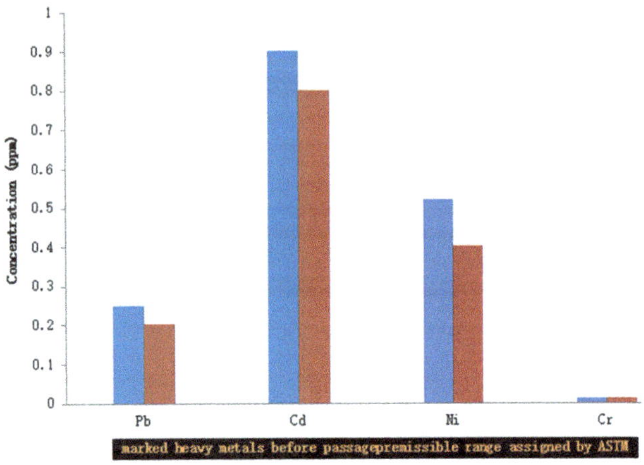

Fig. 2. *Comparison between obtained results before passage and a permissible range.*

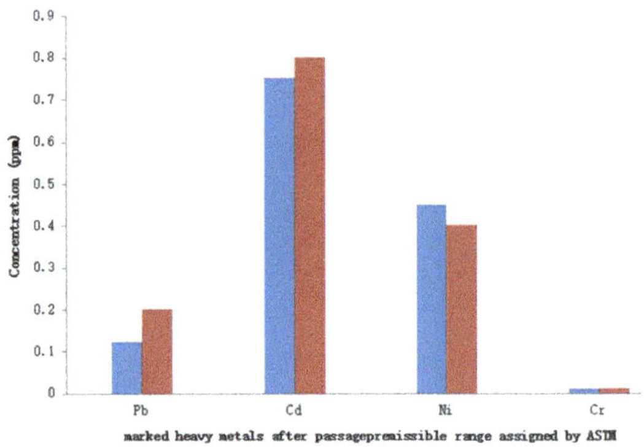

Fig. 3. *Comparison between obtained results after passage and a permissible range.*

Results of marked heavy metals after passing the RPWW to the APP represented that, the APP showed high ability to sorption because it reduced the marked heavy metals within assigned range except the nickel concentration out of range. The comparison between obtained results before and after passage and permissible range was shown in Fig 2 and Fig 3, respectively.

4. Conclusion

Based on the previously discussed analyses, the following conclusions can be summarized as.

- The raw petroleum wastewater samples were taken from Alfola petroleum, field west of Sudan.
- Lead, cadmium, nickel and chromium were determined in raw petroleum wastewater before and after passing on *Australis Phragmites Plant,* by using atomic absorption spectroscopic technique.
- *Australis Phragmites Plant* used to reduce the toxicity of raw petroleum wastewater.
- The concentration of marked heavy metals before passage was found out of range assigned by ASTM, on the other hand, the concentrations after passage within the right permissible limits except the concentration of nickel out of range.

Acknowledgement

Above all, praise is to *GOD* who has sustained me throughout this work. We would like to express my special thanks to prof. Hatel H. Alkamali, Dean of faculty of science and technology, Omdurman Islamic university. Special thanks also expressed to prof. Mohamed hamaad, Mr. Omar Abdalhameed Goda, Mr. Alsiddig Osama, Mr. Ahmed Mohamed Osman, and all staff at water and environment laboratories in Khartoum.

References

[1] American Petroleum Institute. (March 1996). A Guide to Leak Detection for Above-Ground Storage Tanks (first addition), USA: Publication 334.

[2] American Petroleum Institute. (June 2007). Welded Tanks for Oil Storage (11th ed), USA: Downstream Segment, API Standard 650.

[3] Wang & Howard. (2004)Handbook of Industrial and Hazardous Wastes Treatment (4th ed), USA.

[4] Jha, S, K., Krishnamoorthy, T, M., Pandit, G, G., Nambi, K, S, V. (1999). History of accumulation of mercury and nickel in Thane Creek, Mumbai, using 210Pb dating technique. Science of the total environment, 236(1): 91-99.

[5] Lokhande, R, S., Kelkar, N. (1999).Studies on heavy metals in water of Vasai Creek, Maharashtra. Indian J. En-viron. Protect, 19(9): 664-668.

[6] Czepiel, P., Crill, P., Harriss, R. (1995). Nitrous oxide emissions from domestic wastewater treatment. Environmental Science and Technology, 29(9): 2352-2356.

[7] Doorn, M, R, J., Strait, R., Barnard, W., Eklund, B. (1997). Estimate of Global Greenhouse Gas Emissions from Industrial and Domestic Wastewater Treatment, Final Report. Prepared for United States Environmental Protection Agency, Research Triangle Park, NC, USA.

[8] Doorn, M, R, J., Liles, D. (1999). Global Methane, Quantification of Methane Emissions and Discussion of Nitrous Oxide, and Ammonia Emissions from Septic Tanks. Latrines, and Stagnant Open Sewers in the World. Prepared for U.S. EPA, Research Triangle Park, NC, USA.

[9] Annual book of ASTM standards. (2005). American Society for Testing and Materials. West Conshohocken: Salvter. J. Rand.

PERMISSIONS

All chapters in this book were first published in AJEP by Science Publishing Group; hereby published with permission under the Creative Commons Attribution License or equivalent. Every chapter published in this book has been scrutinized by our experts. Their significance has been extensively debated. The topics covered herein carry significant findings which will fuel the growth of the discipline. They may even be implemented as practical applications or may be referred to as a beginning point for another development.

The contributors of this book come from diverse backgrounds, making this book a truly international effort. This book will bring forth new frontiers with its revolutionizing research information and detailed analysis of the nascent developments around the world.

We would like to thank all the contributing authors for lending their expertise to make the book truly unique. They have played a crucial role in the development of this book. Without their invaluable contributions this book wouldn't have been possible. They have made vital efforts to compile up to date information on the varied aspects of this subject to make this book a valuable addition to the collection of many professionals and students.

This book was conceptualized with the vision of imparting up-to-date information and advanced data in this field. To ensure the same, a matchless editorial board was set up. Every individual on the board went through rigorous rounds of assessment to prove their worth. After which they invested a large part of their time researching and compiling the most relevant data for our readers.

The editorial board has been involved in producing this book since its inception. They have spent rigorous hours researching and exploring the diverse topics which have resulted in the successful publishing of this book. They have passed on their knowledge of decades through this book. To expedite this challenging task, the publisher supported the team at every step. A small team of assistant editors was also appointed to further simplify the editing procedure and attain best results for the readers.

Apart from the editorial board, the designing team has also invested a significant amount of their time in understanding the subject and creating the most relevant covers. They scrutinized every image to scout for the most suitable representation of the subject and create an appropriate cover for the book.

The publishing team has been an ardent support to the editorial, designing and production team. Their endless efforts to recruit the best for this project, has resulted in the accomplishment of this book. They are a veteran in the field of academics and their pool of knowledge is as vast as their experience in printing. Their expertise and guidance has proved useful at every step. Their uncompromising quality standards have made this book an exceptional effort. Their encouragement from time to time has been an inspiration for everyone.

The publisher and the editorial board hope that this book will prove to be a valuable piece of knowledge for researchers, students, practitioners and scholars across the globe.

LIST OF CONTRIBUTORS

Lee Fergusson
Principal Consultant, Prana World Consulting, Gold Coast, Australia

Yawar Abbas, Muhammad Idrees and Syed Umairullah Jamil
Department of Earth and Environmental Sciences, Bahria University, Islamabad, Pakistan

Fiza Sarwar
Department of Earth and Environmental Sciences, Bahria University, Islamabad, Pakistan
Department of Environmental Sciences, Quaid-i-Azam University, Islamabad, Pakistan

Attarad Ali
Department of Biotechnology, Quaid-i-Azam University, Islamabad, Pakistan

Ishtiaque Hussain
Department of Environmental Sciences, Quaid-i-Azam University, Islamabad, Pakistan

Md. Redwanur Rahman
Ecology and Biodiversity Research Laboratary, Institute of Environmental Science, University of Rajshahi, Rajshahi, Bangladesh

George Kordzakhia, Lali Shavliashvili, Gulchina Kuchava and Nugzar Buachidze
The Institute of Hydrometeorology, Georgian Technical University, Tbilisi, Georgia

Justin R. Nayagam and Mani Varghese K. I.
Department of Botany, Union Christian College, Aluva-2, Kerala, India

Tibebu Yelemfrhat Simegn
Ethiopian Wildlife conservation Authority, Addis Ababa, Ethiopia

Teshome Soromessa
Center of Environmental Science, Addis Ababa University, Addis Ababa, Ethiopia

Chakoro Tamire
Environemntal Management Team, Ethiopian Railways Corporation/ERC, Addis Ababa, Ethiopia

Mekuria Argaw
Center for Environmental Science, Addis Ababa University, Addis Ababa, Ethiopia

Dancan Otieno Onyango and James Wambua Kaluli
Department of Soil, Water and Environmental Engineering, Jomo Kenyatta University of Agriculture and Technology, Nairobi, Kenya

Chritopher L. Kanali
Department of Biomechanical and Processing Engineering, Jomo Kenyatta University of Agriculture and Technology, Nairobi, Kenya

Mahendra Kumar Trivedi, Alice Branton, Dahryn Trivedi and Gopal Nayak
Trivedi Global Inc., Henderson, NV, USA

Ragini Singh and Snehasis Jana
Trivedi Science Research Laboratory Pvt. Ltd., Bhopal, Madhya Pradesh, India

Hamere Yohannes
Department of Natural Resource Management, College of Agriculture and Natural Resource Science, Debre Berhan University, Debre Berhan, Ethiopia

Teshome Soromessa and Mekuria Argaw
Center for Environmental Science, College of Natural Science, Addis Ababa University, Addis Ababa, Ethiopia

Abineh Tilahun
Department of Geography and Environmental Studies, Adigrat University, Adigrat, Ethiopia

Bogale Teferie
Department of Geography and Environmental Studies, Dilla University, Adigrat, Ethiopia

Maureen Nkemdilim Chukwu
Department of Biology, School of Science and Technology, National Open University of Nigeria, Lagos, Nigeria

Tauqir Nasir, Shafaqat Ali, Muhammad Rizwan, Fakhir Hannan and Rehan Ahmad
Department of Environmental Sciences and Engineering, Government College University, Faisalabad, Pakistan

Mujahid Farid
Department of Environmental Sciences, University of Gujrat, Hafiz Hayat Campus, Gujrat, Pakistan

Muhammad Shahbaz
Punjab Institute of Nuclear Medicine (PINUM) Hospital, Faisalabad, Pakistan

Fikre Belay
Department of Geography and Environmental Studies, College of Social Science and Humanities, Adigrat University, Adigrat, Ethiopia

Filomena Loffredo, Mariagabriella Pugliese, Maria Quarto and Vincenzo Roca
Dipartimento di Fisica, Università di Napoli Federico II, Naples, Italy
Istituto Nazionale di Fisica Nucleare (INFN), Sez. Napoli, Naples, Italy

Valerio Pisacane
Euro.Soft srl, Naples, Italy

Renato Aurigemma
Consorzio Società Aerospaziale Mediterranea (SAM), Naples, Italy

Mohammad Shafiqul Alam, Rexona Khanom and Mohammad Arifur Rahman
Department of Chemistry, University of Dhaka, Dhaka, Bangladesh

Simon Sambou, Idrissa Guiro and Bienvenu Sambou
Institute of Environmental Sciences, Faculty of Sciences and Technology, Cheikh Anta Diop University of Dakar. Dakar Fann, Senegal

Anne Mette Lykke
Department of Biological sciences, Aarhus University, Vejlsovej, Silkeborg, Denmark

Cheikh Mbow
Institute of Environmental Sciences, Faculty of Sciences and Technology, Cheikh Anta Diop University of Dakar. Dakar Fann, Senegal
Laboratory of Teaching and Research in Geomatics, Polytechnic High School, Cheikh Anta Diop University of Dakar, Dakar-Fann, Sénégal
World Agroforestry Centre (ICRAF), Nairobi, Kenya

Hyacinthe Sambou
Institute of Environmental Sciences, Faculty of Sciences and Technology, Cheikh Anta Diop University of Dakar. Dakar Fann, Senegal
Laboratory of Teaching and Research in Geomatics, Polytechnic High School, Cheikh Anta Diop University of Dakar, Dakar-Fann, Sénégal

Nino Tkhilava
Department of Environmental Policy and International Relations, Ministry of Environment and Natural Resources Protection, Tbilisi, Georgia

Eno Anietie Moses, Bassey Anie Etuk and Essien Daniel Udosen
Department of Chemistry, University of Uyo, Uyo, Nigeria

Laurice Codou Faye
Department of Civil Engineering, College of Engineering, Kwame Nkrumah University of Science and Technology, Kumasi, Ghana

Hyacinthe Sambou and Bienvenu Sambou
Institute of Environmental Sciences, Cheikh Anta Diop University of Dakar, Dakar, Senegal

Boateng Kyereh
Department of Silviculture and Forest Management, Faculty of Renewable Natural Resources, Kwame Nkrumah University of Science & Technology, Kumasi, Ghana

Nino Gokhelashvili
Department of Environmental Policy and International Relations, Ministry of Environment and Natural Resources Protection, Tbilisi, Georgia

Mengistu Asmamaw
Department of Biology, Debre Berhan University, Debre Berhan, Ethiopia

Argaw Ambellu and Seid Tiku
Department of Environmental Health Sciences and Technology, Jimma University, Jimma, Ethiopia

G. Kordzakhia and L. Shengelia
Georgian Technical University, the Institute of Hydrometeorology, Tbilisi, Georgia

G. Tvauri
I. Javakhishvili Tbilisi State University, Nodia Institute of Geophysics, Tbilisi, Georgia

M. Dzadzamia
National Environmental Agency of the Ministry of Environment and Natural Resources Protection of Georgia, Tbilisi, Georgia

Ababu T. Tiruneh and William N. Ndlela
Department of Environmental Health Science, University of Swaziland, Mbabane, Swaziland

Amos O. Fadiran
Department of Chemistry, University of Swaziland, Kwaluseni, Swaziland

Jonna Heikkilä
Turku University of Applied Sciences, Turku, Finland

Levan G. Tielidze, Lela Gadrani and Mariam Tsitsagi
Vakhushti Bagrationi Institute of Geography, Ivane Javakhishvili Tbilisi State University, Tbilisi, Georgia

Nino Chikhradze
Vakhushti Bagrationi Institute of Geography, Ivane Javakhishvili Tbilisi State University, Tbilisi, Georgia
Ilia State University, Tbilisi, Georgia

Zhiying Tu and Xi Yuan
Engineering Research Center of Eco-Environment in Three Gorges Reservoir Region, Ministry of Education, China Three Gorges University, Yichang, China

Yingping Huang, Hui Peng and Prashant Mandal
Engineering Research Center of Eco-Environment in Three Gorges Reservoir Region, Ministry of Education, China Three Gorges University, Yichang, China
Collaborative Innovation Center for Geo-Hazards and Eco-Environment in Three Gorges Area, Hubei Province, Yichang, China

Yong Gao
Hubei Key Laboratory of Three Gorges Project for Conservation of Fishes, Chinese sturgeon Research Institute, China Three Gorges Corporation, Yichang, China

Fraol Abebe Wudineh and Nigatu Chala Kuke
Department of Civil Engineering, University of Debre Berhan, Debre Berhan, Ethiopia

John B. Cook and Edwin A. Roehl Jr.
Advanced Data Mining International, Greenville, S.C, USA

Svetlana R. Amirova
Brody School of Medicine, East Carolina University, Greenville, North Carolina, USA

Paul A. Comet
Comet Environmental Consulting, Houston, Texas, USA

Tamara V. Tulaykova
Wave Research Center at A. M. Prokhorov's General Physics Institute, Moscow, Russia

Sulieman A. G. Nasir and Ali M. Masaad
Department of Chemistry, Faculty of Science and Technology, Omdurman Islamic University, Khartoum, Sudan

Mohamed Ezeldin
Department of Chemistry, Faculty of Science and Technology, Omdurman Islamic University, Khartoum, Sudan
Department of Quality Control, Almazaq Factory, South of Khartoum, Sudan

Nawal M. Suleman
Department of Chemistry, Zilfi Facultyof Education, Majmmah University, Zilfi, Saudi Arabia

Index